1007785775

Advances in Mathematics Education

Series editors
Gabriele Kaiser, University of Hamburg, Hamburg, Germany
Bharath Sriraman, The University of Montana, Missoula, MT, USA

International editorial board
Ubiratan D'Ambrosio (São Paulo, Brazil)
Jinfa Cai (Newark, NJ, USA)
Helen Forgasz (Melbourne, Victoria, Australia)
Jeremy Kilpatrick (Athens, GA, USA)
Christine Knipping (Bremen, Germany)
Oh Nam Kwon (Seoul, Korea)

More information about this series at http://www.springer.com/series/8392

Uwe Gellert • Joaquim Giménez Rodríguez
Corinne Hahn • Sonia Kafoussi
Editors

Educational Paths to Mathematics

A C.I.E.A.E.M. Sourcebook

Editors
Uwe Gellert
Fachbereich Erziehungswissenschaft
 und Psychologie
Freie Universität Berlin
Berlin, Germany

Corinne Hahn
Management de l'Information
 et des Opérations
ESCP Europe
Paris, France

Joaquim Giménez Rodríguez
Facultat de Formació del Professorat
Departament de Didàctica de les Ciències
 Experimentals i la Matemàtica
Universitat de Barcelona
Barcelona, Spain

Sonia Kafoussi
Department of Sciences of Pre-School
 Education and of Educational Design
University of the Aegean
Rhodes, Greece

ISSN 1869-4918   ISSN 1869-4926 (electronic)
Advances in Mathematics Education
ISBN 978-3-319-15409-1   ISBN 978-3-319-15410-7 (eBook)
DOI 10.1007/978-3-319-15410-7

Library of Congress Control Number: 2015939438

Springer Cham Heidelberg New York Dordrecht London
© Springer International Publishing Switzerland 2015
This work is subject to copyright. All rights are reserved by the Publisher, whether the whole or part of the material is concerned, specifically the rights of translation, reprinting, reuse of illustrations, recitation, broadcasting, reproduction on microfilms or in any other physical way, and transmission or information storage and retrieval, electronic adaptation, computer software, or by similar or dissimilar methodology now known or hereafter developed.
The use of general descriptive names, registered names, trademarks, service marks, etc. in this publication does not imply, even in the absence of a specific statement, that such names are exempt from the relevant protective laws and regulations and therefore free for general use.
The publisher, the authors and the editors are safe to assume that the advice and information in this book are believed to be true and accurate at the date of publication. Neither the publisher nor the authors or the editors give a warranty, express or implied, with respect to the material contained herein or for any errors or omissions that may have been made.

Printed on acid-free paper

Springer International Publishing AG Switzerland is part of Springer Science+Business Media (www.springer.com)

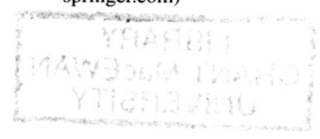

Acknowledgements

The volume collects work which originates from presentations and discussions at the 63rd and the 64th conference of CIEAEM in the summers of 2011 in Barcelona and 2012 in Rhodes. These two conferences were organised around the themes of *facilitating access and participation* and of *mathematics education and democracy*. In both years the focus was on educational practices of mathematics inside and outside the classroom. The conferences witnessed many presentations breaking new grounds and opening up new vistas for mathematics education practice. The presentations initiated collective discussions and, in turn, these discussions deepened our insight and understanding of the many ways in which children, youth and adults may find their paths to mathematics. We want to thank all participants of the 63rd and the 64th conference of CIEAEM for their perhaps invisible but nevertheless important contribution of ideas, from which the volume benefits.

As CIEAEM meetings are truly international conferences, non-native speakers of English author many chapters of this volume. Particular thanks go to Maja Schermeyer (Germany), who was very effective in improving the English of many chapters and to Catherine Whybrow (UK), who provided extensive native-English-speaker support. Birgit Abel (Germany) was highly efficient in formatting the papers.

Finally, all papers have been reviewed at least twice. We thank all those who, by reviewing chapter drafts, have significantly contributed to this volume: Gilles Aldon (France), Luciana Bazzini (Italy), Nina Bohlmann (Germany), Petros Chaviaris (Greece), Julie Cwikla (USA), Javier Díez-Palomar (Spain), Uwe Gellert (Germany), Joaquim Giménez Rodríguez (Spain), Sonia Kafoussi (Greece), Nielce Meneguelo Lobo da Costa (Brasil), Luís Menezes (Portugal), Zoi Nikiforidou (UK), Hélia Oliveira (Portugal), Maria Elisabette Brisola Brito Prado (Brasil), Cristina Sabena (Italy) and Hauke Straehler-Pohl (Germany).

Contents

Educational Paths to Mathematics: Which Paths Forward to What Mathematics?... 1
Uwe Gellert and Corinne Hahn

Part I Cultural Tensions in the Field of Mathematics Education

Re-interpreting Students' Interest in Mathematics: Youth Culture and Subjectivity 15
Paola Valero

Connecting Place and Community to Mathematics Instruction in Rural Schools 33
Robert Klein

Establishing Mathematics Classroom Culture: Concealing and Revealing the Rules of the Game... 67
Christine Knipping, David Reid, and Hauke Straehler-Pohl

Part II Working with Adults

Learning Mathematics In and Out of School: A Workplace Education Perspective.. 99
Gail E. FitzSimons

Mathematical Modelling and Bank Loan Systems: An Experience with Adults Returning to School... 117
Vera Helena Giusti de Souza, Rosana Nogueira de Lima,
Tânia Maria Mendonça Campos, and Leonardo Gerardini

Working with Adults: A Commentary 131
Javier Díez-Palomar

Part III Working with Pre-schoolers

'Number in Cultures' as a Playful Outdoor Activity: Making Space for Critical Mathematics Education in the Early Years 143
Anna Chronaki, Georgia Moutzouri, and Kostas Magos

Fairness Through Mathematical Problem Solving in Preschool Education 161
Zoi Nikiforidou and Jenny Pange

How Do Fair Sharing Tasks Facilitate Young Children's Access to Fractional Concepts? 173
Julie Cwikla and Jennifer Vonk

Working with Pre-schoolers: A Dual Commentary 191
Michaela Kaslová and Sixto Romero

Part IV Taking Spaces and Modalities into Account

Digital Mathematical Performances: Creating a Liminal Space for Participation 201
Susan Gerofsky

Participation in Mathematics Problem-Solving Through Gestures and Narration 213
Luciana Bazzini and Cristina Sabena

Considering the Classroom Space: Towards a Multimodal Analysis of the Pedagogical Discourse 225
Eleni Gana, Charoula Stathopoulou, and Petros Chaviaris

Commentary: Semiotic Game, Semiotic Resources, Liminal Space—A Revolutionary Moment in Mathematics Education! 237
Peter Appelbaum

Part V Criticising Public Discourse

Numbers on the Front Page: Mathematics in the News 247
Dimitris Chassapis and Eleni Giannakopoulou

On the Role of Inconceivable Magnitude Estimation Problems to Improve Critical Thinking 263
Lluís Albarracín and Núria Gorgorió

Criticizing Public Discourse and Mathematics Education: A Commentary 279
Charoula Stathopoulou

Part VI Organising Dialogue and Enquiry

Facilitating Deliberate Dialogue in Mathematics Classroom 289
Ana Serradó, Yuly Vanegas, and Joaquim Giménez Rodríguez

Inquiry-Based Mathematics Teaching: The Case of Célia 305
Luís Menezes, Hélia Oliveira, and Ana Paula Canavarro

**Using Drama Techniques for Facilitating Democratic
Access to Mathematical Ideas for All Learners** 323
Panayota Kotarinou and Charoula Stathopoulou

Organising Dialogue and Enquiry: A Commentary 341
Lambrecht Spijkerboer and Leonor Santos

Part VII Providing Information Technology

**Educational Laptop Computers Integrated
into Mathematics Classrooms** 351
Maria Elisabette Brisola Brito Prado and Nielce Meneguelo Lobo da Costa

**Technology and Education: Frameworks to Think Mathematics
Education in the Twenty-First Century** 365
Gilles Aldon

**Technology in the Teaching and Learning of Mathematics
in the Twenty-First Century: What Aspects Must Be Considered?—
A Commentary** 383
Fernando Hitt

Part VIII Transcending Boundaries

**Family Math: Doing Mathematics to Increase the Democratic
Participation in the Learning Process** 393
Javier Díez-Palomar

Service-Learning as Teacher Education 409
Peter Appelbaum

**The Learning and Teaching of Mathematics as an Emergent
Property Through Interacting Systems and Interchanging Roles:
A Commentary** 425
Fragkiskos Kalavasis and Corneille Kazadi

Appendices 431

Index 471

Author Index 481

Contributors

Lluís Albarracín Facultat de Ciències de l'Educació, Universitat Autònoma de Barcelona, Barcelona, Spain

Gilles Aldon Ecole Normale Supérieure de Lyon, Lyon, France

Peter Appelbaum Department of Curriculum, Cultures and Child/Youth Studies, School of Education, Arcadia University, Philadelphia, PA, USA

Luciana Bazzini Dipartimento di Filosofia e Scienze dell'Educazione, Università di Torino, Torino, Italy

Tânia Maria Mendonça Campos Post-Graduate Program in Mathematics Education, Anhanguera University, São Paulo, Brazil

Ana Paula Canavarro Universidade de Évora, Évora, Portugal

Dimitris Chassapis School of Educational Sciences, University of Athens, Athens, Greece

Petros Chaviaris University of the Aegean, Mytilene, Greece

Anna Chronaki Department of Early Childhood Education, University of Thessaly, Volos, Greece

Julie Cwikla Director of Creativity & Innovation in STEM, Office of the Vice President for Research, University of Southern Mississippi, Ocean Springs, MS, USA

Javier Díez-Palomar Department of Mathematics and Science Education, University of Barcelona, Barcelona, Spain

Gail E. FitzSimons Melbourne Graduate School of Education, University of Melbourne, Parkville, VIC, Australia

Eleni Gana Department of Special Education, University of Thessaly, Volos, Greece

Uwe Gellert Fachbereich Erziehungswissenschaft und Psychologie, Freie Universität Berlin, Berlin, Germany

Leonardo Gerardini Colégio Dante Alighieri, São Paulo, Brazil

Susan Gerofsky Department of Curriculum and Pedagogy, University of British Columbia, Vancouver, BC, Canada

Eleni Giannakopoulou Educational Studies Unit, Hellenic Open University, Athens, Greece

Joaquim Giménez Rodríguez Facultat de Formació del Professorat, Departament de Didàctica de les Ciències Experimentals i la Matemàtica, Universitat de Barcelona, Barcelona, Spain

Vera Helena Giusti de Souza Post-Graduate Program in Mathematics Education, Anhanguera University, São Paulo, Brazil

Núria Gorgorió Facultat de Ciències de l'Educació, Universitat Autònoma de Barcelona, Barcelona, Spain

Corinne Hahn Management de l'Information et des Opérations, ESCP Europe, Paris, France

Fernando Hitt Département de Mathématiques, Université du Québec à Montréal, Montréal, QC, Canada

Fragkiskos Kalavasis Faculty of Humanities, Department of Sciences for Preschool Education and Educational Design, University of the Aegean, Rhodes, Greece

Michaela Kaslová Faculty of Education, Charles University, Prague, Czech Republic

Corneille Kazadi Université du Québec à Trois-Rivières, Trois-Rivières, Canada

Robert Klein Ohio University, Athens, OH, USA

Christine Knipping Universität Bremen, Bremen, Germany

Panayota Kotarinou Department of Special Education, University of Thessaly, Volos, Greece

Nielce Meneguelo Lobo da Costa Post-graduation Program of Mathematics Education, Anhanguera University of São Paulo, UNIAN, São Paulo, Brazil

Kostas Magos Department of Early Childhood Education, University of Thessaly, Volos, Greece

Luís Menezes Escola Superior de Educação de Viseu, Viseu, Portugal

Georgia Moutzouri Department of Early Childhood Education, University of Thessaly, Volos, Greece

Zoi Nikiforidou Department of Early Childhood, Liverpool Hope University, Liverpool, UK

Rosana Nogueira de Lima Post-Graduate Program in Mathematics Education, Anhanguera University, São Paulo, Brazil

Hélia Oliveira Instituto de Educação da, Universidade de Lisboa, Lisbon, Portugal

Jenny Pange Department of Preschool Education, University of Ioannina, Ioannina, Greece

Maria Elisabette Brisola Brito Prado Post-graduation Program of Mathematics Education, Anhanguera University of São Paulo, UNIAN, São Paulo, Brazil

Applied Information Technology Center of Education, State University of Campinas, UNICAMP, Campinas, Brazil

David Reid Universität Bremen, Bremen, Germany

Sixto Romero Escuela Técnica Superior de Ingeniería, Universidad de Huelva, Huelva, Spain

Cristina Sabena Dipartimento di Filosofia e Scienze dell'Educazione, Università di Torino, Torino, Italy

Leonor Santos Instituto de Educação, Universidade de Lisboa, Lisbon, Portugal

Ana Serradó Department of Science, Mathematics and Technology, La Salle-Buen Consejo, Cádiz, Spain

Lambrecht Spijkerboer APS-International, Utrecht, The Netherlands

Charoula Stathopoulou Department of Special Education, University of Thessaly, Volos, Greece

Hauke Straehler-Pohl Freie Universität Berlin, Berlin, Germany

Paola Valero Aalborg University, Aalborg, Denmark

Yuly Vanegas Departament de Didàctica de la Matemàtica i de les Ciències Experimentals, Universitat Autònoma de Barcelona, Barcelona, Spain

Jennifer Vonk Associate Professor, Department of Psychology, Oakland University, Rochester, MI, USA

Educational Paths to Mathematics: Which Paths Forward to What Mathematics?

Uwe Gellert and Corinne Hahn

Abstract Most people involved in mathematics education agree that it is complex, multi-layered, dynamic, multi- and interdisciplinary. To study and to improve mathematics education on the various levels of its curricula and its practices has been a goal of the Commission Internationale pour l'Etude et l'Amélioration de l'Enseignement des Mathématiques (CIEAEM, the International Commission for Study and Improvement of Teaching Mathematics) since it was created and established in the 1950s. CIEAEM continues to investigate the actual conditions and the possibilities for the development of mathematics education. This introductory chapter provides the rational for the book by looking at historical developments in school mathematics. The structure of the sourcebook is explained at the end of the introduction.

Mathematics education is a multi-facetted endeavour that has been regarded from many theoretical points of view. Some believe that it has to do with the transmission and the acquisition of mathematical knowledge, while others emphasise the learners' mathematical constructions and the teachers' role in providing appropriate mathematical environments. Mathematics education is concerned with the formation of the learners' identities, but also with the institutional (re-)production of a mathematically-educated workforce. It is about the transposition, or recontextualisation, of academically produced mathematics into a mathematics curriculum, and it is also about how people activate and integrate mathematical skills and knowledge in everyday contexts. Mathematics education has been regarded as a process, a product, a discourse, a practice, an activity system, a material reality, a research domain, a field of academic research and an area of study. Most people involved in mathematics education agree that it is complex, multi-layered, dynamic, multi- and interdisciplinary. Some argue that mathematics education is the key to the

---

U. Gellert (✉)
Fachbereich Erziehungswissenschaft und Psychologie,
Freie Universität, Berlin, Germany
e-mail: ugellert@zedat.fu-berlin.de

C. Hahn
Management de l'Information et des Opérations, ESCP Europe, Paris, France
e-mail: hahn@escpeurope.eu

development of individual, national and global prosperity. But, is this all true, and if so, in which sense?

Looking Back: Developments in School Mathematics

Mathematics education is political, as can be seen in disputes over the question: "What is school mathematics?" 300 years ago, this question would have been quite difficult to pose, let alone to answer. The mathematics that was taught varied heavily across countries and the different institutions of learning according to their education and formation purposes. But then, a process of international modernisation started, and, according to Gispert and Schubring (2011), two countries were leading this process: France and Germany.

The Construction of School Mathematics

In France, an important and influential development was realised in the early eighteenth century: The idea of a generalised science education, including mathematics, was propagated. Until then, mathematics as such was taught mostly in military schools and only to a very small extent at the university within philosophical studies. The publication and distribution of textbooks in mathematics promoted the development of the discipline. The authors, who have been teachers in different types of schools, advocated for it, emphasising the moral and cultural value of mathematics.

The development reached a new level when schooling shifted from being a privilege for a social elite to being a part of an overall system of education, organized by state authorities, with formalized teacher education programmes and compulsory for all children. From this moment on, various stakeholders were, and still are, involved in contesting the very nature of school mathematics. As Gispert (2011) claims, these stakeholders can be grouped as follows: experts (mathematicians), field professionals (high school mathematics teachers), political and economic actors. She further argues that these three spheres of actors are associated with three registers of modernity: a mathematical, an educational, and a socio-economic register.

During the nineteenth century, the status and importance of mathematics in secondary education in France remained marginal within the "classical canon" of a humanistic secondary education which, at that time, formed the country's elite. In contrast to the elite's education, the school curriculum for the children of the middle classes (called the "upper primary") included a strong emphasis in science education, in which "practical" mathematics played a significant role. A very important reform took place in 1902. It unified secondary education. It implemented, in parallel with the prestigious traditional pathway focused on teaching classical humanities, a "modern" sector based on language and science education. This sector combined two visions of mathematics, cultural and practical, and incorporated new contents: among others, breaking with Euclidean geometry, and introducing the quasi-experimental study of functions in connection with physics. The reform was mainly driven by mathematicians; high school teachers seized it a little later.

This reform challenged the theoretical and disinterested view of the formation of the elite conveyed by the classical secondary education. After World War I, the reform was accused of being inspired too much by the German model, that had been developed during the nineteenth century in Prussia, and betraying the spirit of the French classical humanities. These developments in Prussia are an interesting historical case in which the controversies about the nature of school mathematics came to a certain end.

Jahnke (1986) discusses the origins of the dispute over what school mathematics should be in early nineteenth-century Prussia—a quite decisive moment in the development of mathematics curricula. He shows how, during a short period after a military defeat against the French revolutionary troops in 1806, the debate about the constitution and form of school mathematics affected a radical change in the perception of what the difference between school mathematics and academic mathematics should be. During the dispute it became clear that it would no longer be possible to simply define school mathematics as academic mathematics on a lower level and a lower degree of difficulty. There is more to do than merely to simplify academic mathematical knowledge in order to build up a school mathematics curriculum. Jahnke reconstructs the historical process of the development of what he calls "the 'deep structure' of school mathematics" (p. 86), a structure that "has remained essentially the same since the early nineteenth century" (ibid.). He shows how mathematicians were extremely active in promoting the view that school mathematics should be uniform in all schools and free from any practical concerns. However, their insistence was not successful. State administration, particularly school inspectors and headmasters of prestigious secondary (Gymnasium) schools in Berlin, argued that it is not "appropriate to exclude 'common arithmetic' from the curriculum" (p. 91) and that "it is necessary to find ways and means of linking 'common arithmetic' and higher mathematics, everyday knowledge and scientific knowledge" (ibid.). Ultimately, the structure of the school mathematics curriculum was based on the concept of 'mathematical operation' thus founding higher school mathematics on elementary arithmetic: School algebra was constructed as the study of the formal properties of the arithmetical operations; infinitesimal calculus was constructed as formal school algebraic theory. The extension of the number concept, the 'principle of permanence', defined the macro-structure of the school mathematics curriculum from the early nineteenth century to the present.

Towards the Problem-Centred Curriculum

Much later, in the mid- to late-twentieth century, new initiatives of curriculum development in mathematics have brought about an orientation towards 'problem-solving' and 'mathematical modelling'. Pólya's conception of mathematics as an essential problem solving activity has often been quoted as the main root for the first of the two orientations (e.g., 1962). It was his idea to convert the ontological concept of mathematics as problem solving into an epistemological programme. The student should learn mathematics in a way that is analogous to the way mathematicians work. Less related to 'pure mathematics' than the first, mathematical modelling is often taken as an idealisation of the developmental activities within applied mathematics. In contrast to the field of applied mathematics, a mathematical modelling approach to school mathematics only rarely aims at the development of new mathematical algorithms and technology that can be used to solve real problems or to engage mathematically in real situations. Instead, as English and Sriraman (2010) adhere to, mathematical modelling can be conceptualised as an advance on existing classroom word problem solving. Arguably, the "problems" in mathematical modelling differ from the "problems" in mathematics as problem solving.

Both orientations only tacitly challenge the macro-structure of the school mathematics curriculum. They are not meant to re-evaluate the importance of mathematical operations and the principle of permanence. The recontextualisation of problem solving and mathematical modelling within the institutional frame of schooling brings about a transformation of ontological statements into didactic principles and pedagogic strategies. By this mechanism, problem solving and mathematical modelling, as didactic principles and pedagogic strategies, appear as official curricular paths to mathematics (cf. Jablonka and Gellert 2012). It is part of the self-concept of the mathematics research community to regard the resulting modifications of curriculum material, classroom activities, attainment descriptions, etc. as topics for empirical research and as impulses for design activities.

Fit to and Fit for the Data-Driven Society

A seemingly different kind of curricular renovation occurred during the last decades in numerous countries with the introduction (or expansion) of statistics, of chance and probability in the primary and secondary mathematics curriculum. This mathematical area has been integrated into official curriculum descriptions and attainment standards, thus actually bearing the potential to shift or diversify the macro-structure of the school mathematics curriculum. Why did the shift occur, or: in which way is it a shift of the macro-structure? Note that the introduction of stochastics in mathematics education fits well to the three dimensions of "modernity" defined by Gispert above: mathematical "modernity" as it takes into account recent developments in academic mathematics, pedagogical "modernity" through the use of "real problems" to motivate students and build interdisciplinary links, and the socio-political importance of statistics and data analysis in "modern" societies. Indeed, it can be argued that the focus of the traditional school mathematics curriculum on mathematical operations (during the primary grades) and school algebra (during the secondary grades) is in a critical relation to the concept of the Western democracy or, more precisely: of technocracy. In a technocracy, political decisions based on calculations require a populace that is used to trust the legitimacy of calculations—and not necessarily a populace that is able to critically evaluate the mechanism by which political decisions are made legitimate. A curriculum focussed on mathematical operations and school algebra may perfectly contribute to customise and familiarise the student population with the imposed legitimacy of calculations. In the second half of the twentieth century, partly due to the advancement of computer technology, more and more of the calculations, that inform political decisions and by which political decisions are communicated, became of probabilistic and statistic character. Introducing statistics, chance and probability in the school mathematics curriculum might then be regarded less a challenge to its macrostructure than an attempt to repair the familiarisation with the mathematical operations and representations mentioned in legitimacy claims in political decisions. However, some scepticism seems appropriate here about the way stochastic is introduced in the school mathematics curriculum and the claim that it can offer new paths to mathematics. As Fabre (2010) holds, knowledge is multi-dimensional. These dimensions can be classified as historical, systematic and operational. School curricula mostly emphasise the last one. Consequently, statistics is often reduced to a set of techniques that students need to master, at worst to mathematics-in-contexts that are supposed to motivate students. School statistics, chance and probability seems to be a recontextualisation of stochastics that does not take into account its complex epistemology, in particular the tension between a data analysis approach and a modelling approach. Apparently, there is a constant threat that the new and the different is systematically recontextualised and, thus, subordinated to the traditional foci of the mathematics curriculum. In any case, research in mathematics education is concerned with scrutinising the impact of the curricular change on mathematics classroom practice and beyond.

ICT Challenging the Mathematics Curriculum

The technological development of the last decades is a factor that might alter the forms of mathematics education practices both on the curricular and the non-curricular level. Research in mathematics education has extensively focussed on the curricular potential of ICT, perhaps best illustrated by the attempts to render school geometry more dynamically. Although an initial period in which ICT had been promoted like the silver bullet for a mathematics education for the twenty-first century has faded away, ICT still seems to have the potential of rendering some mathematics classroom traditions obsolete. This is another area for research and design activities—and this volume is further exploring the topic. From a political perspective, we might ask whether the turn to ICT is related to new mechanisms by which political decisions are generated, communicated and made legitimate—and the respective ethics involved—or if the broad availability of ICT is related in some way or another to a form of democracy that considers technocracy a risk. Anyway, even if schools are simply taking up technological standards, the potential of ICT for new educational paths to mathematics seems to be beyond doubt.

Looking Forward: Why and Where?

A profound criticism has arisen recently. What are we actually doing when always looking for new educational paths to mathematics? Are we uncritically bound to the ideology that we have to continuously reform mathematics education, because the whole enterprise of mathematics education is not running as we wish it would? But, could it ever? Can the reasons for mathematics education not being a fully developed success story be suppressed by an improved mathematics education? Or are we to face eternal frustration—education cannot compensate for society (Bernstein 1970)—but cannot stop producing new ideas, new strategies, new theories and paradigms, because … because of what? Because mathematics educators and others have indeed been successful in constituting mathematics education as an ethical system in which mathematical knowledge is "good" (Lundin 2012), "mathematics for all" even a Lacanian "supreme good" (Pais 2012) and, at the same time, establishing the mathematics educators as the key producers of knowledge in mathematics education? This is, of course, a reflection from a cynical point of view. From this perspective, mathematics education appears as an ingenious self-reproducing machinery. However, the metaphor of the machinery is essentially modern and suppresses all personal motives, uninterested commitment and illogical behaviour of those who like to improve mathematics education. There are many mathematics educators who do not wish to make a better world by means of mathematics, many who do not believe more fair and just mathematics education would cause a more fair and just world, many who do not see our economical and ecological problems resolved once the students achieve a better understanding of mathematics. In lieu thereof, many mathematics educators do not stop producing new ideas because they simply seek to make the learning of mathematics, under the conditions of institutionalised schooling, more meaningful to students. As the students of the twenty-first century seem to be different from those of e.g. the nineteenth century, and because the students are exactly the ones who decide about the meaningfulness of mathematics education activities, there indeed is a never-ending necessity to study and improve mathematics education. Stopping this endeavour can hardly be an alternative.

In a similar spirit, to study and to improve mathematics education on the various levels of its curricula and its practices has been a goal of the *Commission Internationale pour l'Etude et l'Amélioration de l'Enseignement des Mathématiques*

(CIEAEM, the International Commission for the Study and Improvement of Mathematics Education) since it was created and established in the 1950s. CIEAEM continues to investigate the actual conditions and the possibilities for the development of mathematics education. The commission regularly organises conferences characterised by exchange and discussion of research work and of experiences based on the craft knowledge of teaching at all levels. It fosters the dialogue between researchers and educators in all domains of practice. Whereas the founding members of CIEAEM—mathematicians, mathematics educators, psychologists and philosophers—had focused on interrelating an academic mathematical perspective with a modern pedagogy based on psychological models of cognitive development. They missed recognising the social dimension of mathematics education, this missing dimension has since been largely integrated into the work of the commission—as you will immediately notice when reading this sourcebook.

The Structure of the Sourcebook

The volume begins with three chapters that exemplarily illustrate the obstacles that any attempt to promote mathematics education might face. These obstacles are no minor ones. They are written in the social fabric of mathematics education. The first chapter raises a fundamental concern: Why should the students be interested in learning mathematics at all? Paola Valero argues that there is "a cultural gap between the forms of subjectivity promoted by mathematics as areas of schooling, and the forms of subjectivity experienced by students in their everyday life". In order to counter the referred students' lack of interest in mathematics, it seems necessary not to restrict the remedial activities to the pedagogic and the didactic but to consider the students' identity constructions as closely connected to the cultural politics of schooling. The second chapter investigates how the cultural politics of mathematics education play out in rural areas. The study, reported by Robert Klein, is based on the expectation that these places "would be engaged in meaningful efforts to connect mathematics instruction to local places and communities". Although the study found a variety of support for making local connections between mathematics education and locally relevant issues, it concludes that the support remained mainly on the level of rhetoric. Instead, mathematics education interacts in a rather alienating way on the students' developing identities: mathematics "inhabits nowhere rather than round here". The third chapter exposes a kind of cultural micro-politics of schooling. Christine Knipping, David Reid and Hauke Straehler-Pohl adjust their analytical lenses on the micro-dynamics of the mathematics classroom. By drawing on mathematics instruction practices in 'officially selective' and 'officially inclusive' school systems, they trace how the basic political principle of meritocracy translates into disparity producing interactional mechanisms in the classroom. A classroom culture is generated in which the conflicting nature of the distribution of access to mathematical knowledge, and thus to mathematical identities, is naturalised.

The three chapters constitute the first part of the volume and serve as a horizon for the following seventeen chapters. These seventeen educational paths to mathematics try—not to attack but—to understand, to redefine and to engage with the reported, and other, obstacles. What they essentially do is look for resources.

We grouped the seventeen paths to mathematics in seven parts of two to three chapters each. All parts end with a commentary. The intention behind this second part of the volume is to offer new ideas of educational paths to mathematics. The chapters differ from standard research articles in mathematics education that aim at the development of methodologies or theories. Although many of the chapters are indeed outcomes of systematically controlled research projects (and reference to research papers in this respect is given in the chapters), the focus here is not on the details of data construction and analysis etc., but on new mathematical activities and conceptions enriching the repertoire of educational paths to mathematics.

The volume is remarkably international. Teachers and researchers from 14 countries authored 20 chapters and 7 commentaries. The reader is invited to reflect on the particular effect of presenting avenues to mathematics contrived in diverse national settings in which the praxis of mathematics education might look different compared to what happens in the reader's place.

'Working with adults' is the heading of the second part. Gail FitzSimons reflects on her experience with pharmaceutical operators engaged in vocational education courses on 'Calculations' and 'Basic Computer Skills'. She shows that, if the workers' knowledge and experience, their artefacts and practices, their voices and stories are taken into account and incorporated in the course material, then a path to the up to now unthinkable is offered. In contrast, Vera Helena Giusti de Souza, Rosana Nogueira de Lima, Tânia Maria Mendonça Campos and Leonardo Gerardini face a situation of young adults returning to school in order to catch up on their school-leaving certificate. As these learners do not dispose of a shared work experience, the authors design a mathematical modelling activity with bank loan systems that might be important for the young adults in their near future. In his commentary on both chapters, Javier Díez-Palomar emphasises the importance of using, and further developing, strategies to bridge the gap between academic knowledge and common sense based on experience in order to generate more democratic mathematical activities.

'Working with pre-schoolers' is the focus of the third part. Anna Chronaki, Georgia Moutzouri and Kostas Magos open this part with an outdoor activity for Roma and non-Roma children. They designed 'Number in Cultures' as a counter event in which the correspondence amongst number words and symbols was explored in three languages: Greek, Romany, and Arabic. Their study concludes that such a counter event can open a space for marginalised children, mathematical knowledge and silenced identities. The following two chapters are about the issue of fairness and fair sharing in children aged 3–6. Zoi Nikiforidou and Jenny Pange discuss how logico-mathematical activities in pre-schoolers' classrooms may contribute to the children's developing understanding of fairness as an intersection of their cognitive, social and moral development. Julie Cwikla and Jennifer Vonk investigate if, and how, fair sharing tasks facilitate pre-schoolers' access to fractional concepts. They find evidence that children can comprehend or acquire fractional concepts before whole numbers are consolidated. In their commentary on the three chapters, Michaela Kaslová and Sixto Romero Sánchez expose the historical background to recent work with pre-schoolers. Particular attention is paid to the influence of Comenius' principles and to the developments in many European countries at the turn from the nineteenth to the twentieth century.

'Taking spaces and modalities into account' is the heading of the fourth part in which different 'spaces' and their relation to 'participation' are conceptualised. Susan Gerofsky commences with examples of digital mathematical performances, which may pave ways to complement, or antagonise, the "disembodied, antiperformative traditions in school mathematics pedagogy". An expansion of 'liminal spaces', i.e. "play in the spaces of paradox and contradiction, ambiguity and transition", may lead to deeper levels of mathematical understanding. Liminal performative spaces may offer ample opportunities for students' participation. Luciana Bazzini and Cristina Sabena use a teaching experiment to illustrate how space for interaction is often filled with gestures and narration. They expose the multimodal nature of teaching and learning of mathematics. Awareness of the multimodality of "the 'semiotic game' between the teacher and students" proves to be important for the teacher in order to understand, and facilitate, the students' participation in classroom interaction. Eleni Gana, Charoula Stathopoulou and Petros Chaviaris focus on the space as the material space, in which a classroom teacher and her/his students are spatially related. The authors argue from a social semiotic perspective that the teachers' use of classroom space is involved in enabling "students' social experience in the specific teaching and learning environment". Classroom space is considered as one of the teacher's resources for the delineation of semantically coherent learning activities. In his spirited commentary to the three chapters, Peter Appelbaum distinguishes the part as a revolutionary moment in mathematics education. The attentiveness towards performance, towards gestures and narration, and towards spaces allow for new ways "of collaboration, experimentation, aesthetic participation, and playful creation of new worlds to be in".

'Criticising public discourse' is the core of the mathematical activities depicted in Part V. Lluís Albarracín and Núria Gorgorió analyse a teaching experiment in which the school mathematical topic of inconceivable magnitude estimation is related to a critical understanding of media reports about political events. In the teaching experiment, the students disclose the political bias of numbers devised by political parties and other stakeholders, and published by the media. Their critical competence is fostered through their mathematical investigations. Dimitris Chassapis and Eleni Giannakopoulou draw on the role of public media in the legitimation of the recent austerity policies in Greece. They show how mathematical concepts are used in the media to convey policies and political views. From a critical mathematics education perspective, mathematics being used as a discursive instrument within the 'apparatus of truth' can be regarded as the linchpin of the school mathematics curriculum. In her commentary on both chapters, Charoula Stathopoulou puts emphasis on the educational potential of public discourse as a focal point for a critical mathematics curriculum by which the relationship of mathematics with issues of social justice, with manipulation of the public sphere, and with critical citizenship can be made explicit.

'Organising dialogue and enquiry' is the aim of the chapters in Part VI. Ana Serradó, Yuly Vanegas and Joaquin Giménez analyse an example of blended learning in which the students exchange strategies to solve open-ended tasks in on-line forums. They show how the distance produced by the internet can be both beneficial and obstructive to processes in which the students negotiate meaning. The role of the teacher in blended learning settings is particularly highlighted. Luís Menezes, Ana Paula Canavarro and Hélia Oliveira follow a teacher and her fourth-graders through collective mathematical discussions and syntheses of emerging mathematical ideas. They document the teacher's intentions and actions in order to understand educational practice in inquiry-based mathematics classrooms better. In a teaching experiment with eleventh-graders titled "Is our world Euclidean?", Panayota Kotarinou and Charoula Stathopoulou engage the students in discussions about axiomatic definition in Euclidean and non-Euclidean geometries. Their pedagogic technique is 'drama in education'. They show how this technique offers a viable way to foster students' active participation in mathematical enquiry and critical thinking. In their commentary on the three chapters, Lambrecht Spijkerboer and Leonor Santos distinguish between surface and deep approaches for learning and ask whether drama, open-ended tasks and collective mathematical discussions might contribute to deep learning, to dealing with differences in the classroom and to the formation of democratic citizens.

'Providing information technology' is the focus of the seventh part. In the first chapter of this part, Maria Elisabete Brisola Brito Pardo and Nielce Meneguelo Lobo da Costa analyse the challenge to teachers of the introduction of laptop computers in the mathematics classroom. They document that, although the teachers see the potential that technology offers for exploration and articulation with other areas of knowledge, they still find it difficult to deviate from a traditional teaching pattern in which the teacher explains and the students practice. In the second chapter of the part, Gilles Aldon argues for the development of a sufficiently complex theoretical framework necessary to understand better the dynamics and the complexity of using computer technology as a standard in the mathematics classroom. From his perspective, the new standard that technology offers to the teaching and learning of mathematics is crucially related to the ways in which teachers and students document their mathematical activities. Such a new documentary system is related to the processes of memorisation, of organisation of ideas, of creativity and of communication and has thus the potential to re-orientate the dynamics of knowledge construction. In his commentary on both chapters, Fernando Hitt points to the well-documented low impact that the development of ICT has until now on classroom practice. He stresses that empirical research is urgently needed for a systematic and substantial integration of technology into classroom practice.

'Transcending boundaries' completes the educational paths to mathematics. As in the other parts, the two chapters invite the reader to think about the teaching and learning of mathematics beyond the usual patterns of transmission and acquisition of knowledge in school. Javier Díez-Palomar opens the part with a call for family involvement in order to increase democratic participation. On one hand, "parental involvement is recognised as a crucial outside-school aspect in children's mathematics achievement". On the other hand, a broader involvement of the public in the mathematics education enterprise seems to go hand in hand with more participative and democratic practices. In the final chapter, Peter Appelbaum puts forward the idea of "service-learning as teacher education". In the form of 'intergenerational math circles' of secondary students, mathematics teachers and future teachers, the participants experienced radically different forms of teaching and learning. Their experience encourages reflection on "dominant educational structures grounded in competitive individualism". For the future teachers, a redefinition of 'teaching' and 'learning' relates to the possibility of active invention of pedagogical practices. "rather than being a passive implementer of prepackaged curriculum." In their commentary on both chapters, Fragiskos Kalavasis and Corneille Kazadi present a model of the complex structures in education, exemplified by considering parental involvement and service learning. They call for a "new epistemology [...], which valorizes the particular and the involvement of all partners in mathematics education."

The end of the volume provides information about the topics of past conferences of CIEAEM.

References

Bernstein, B. (1970). Education cannot compensate for society. *New Society, 15*(387), 344–347.
English, L., & Sriraman, B. (2010). Problem solving for the 21st century. In B. Sriraman & L. English (Eds.), *Theories of mathematics education: Seeking new frontiers* (pp. 263–290). Berlin: Springer.
Fabre, M. (2010). Problématisation des savoirs. In A. Van Zanten (Ed.), *Dictionnaire pédagogique* (pp. 539–541). Paris: Presses Universitaires de France.
Gispert, H. (2011). Enseignement, mathématiques et modernité au XXème siècle: réformes, acteurs et rhétoriques. *Bulletin de l'APMEP, 494*, 286–296.
Gispert, H., & Schubring, G. (2011). Societal, structural, and conceptual changes in mathematics teaching: Reform processes in France and Germany over the twentieth century and the international dynamics. *Science in Context, 24*(1), 73–106.
Jablonka, E., & Gellert, U. (2012). Potential, pitfalls, and discriminations: Curriculum conceptions revisited. In O. Skovsmose & B. Greer (Eds.), *Opening the cage: Critique and politics of mathematics education* (pp. 287–308). Rotterdam: Sense.
Jahnke, H. N. (1986). Origins of school mathematics in early nineteenth-century Germany. *Journal of Curriculum Studies, 18*(1), 85–94.
Lundin, S. (2012). Hating school, loving mathematics: On the ideological function of critique and reform in mathematics education. *Educational Studies in Mathematics, 80*(1), 73–85.
Pais, A. (2012). A critical approach to equity. In O. Skovsmose & B. Greer (Eds.), *Opening the cage: Critique and politics of mathematics education* (pp. 49–92). Rotterdam: Sense.
Pólya, G. (1962). *Mathematical discovery: On understanding, learning and teaching problem solving*. New York: Wiley.

Part I
Cultural Tensions in the Field of Mathematics Education

Re-interpreting Students' Interest in Mathematics:
Youth Culture and Subjectivity 15
Paola Valero
Aalborg University, Denmark

Connecting Place and Community to Mathematics Instruction
in Rural Schools 33
Robert Klein
Ohio University, Athens, USA

Establishing Mathematics Classroom Culture: Concealing
and Revealing the Rules of the Game 67
Christine Knipping and David Reid
Universität Bremen, Germany
Hauke Straehler-Pohl
Freie Universität Berlin, Germany

Re-interpreting Students' Interest in Mathematics: Youth Culture and Subjectivity

Paola Valero

Abstract This chapter argues that in Western, developed societies, young people's decreasing engagement in mathematics has to do fundamentally with a cultural gap between the forms of subjectivity promoted by mathematics as areas of schooling, and the forms of subjectivity experienced by students in their everyday life. While the organization of mathematics in schooling is deeply rooted in Modernity and requires students to embody its core values in order to be successful, current culture offers a myriad of projects of becoming that compete effectively with school forms of subjectivity. An understanding of the youth's lack of interest in terms of the cultural gap places mathematics education as a field of practice in the realm of the cultural politics of our time. Such displacement may offer alternative ways of thinking and acting with respect to youth's engagement with mathematics.

Introduction

Since the second industrial revolution at the end of the nineteenth century, mathematics and science education have become central components of national educational systems in the Western developed world (Aronowitz and De Fazio 1997). The need for more qualified labor in all fields of science, technology, engineering and mathematics (STEM) has been a constant concern for national governments, which have the responsibility to provide education for their population and, thereby, secure the necessary human resources for economic development (Ashton and Sung 1997). Professional organizations of scientists state for more than a century that high qualifications in mathematics and science are central for the development of a society. This notion was transformed according to the dominant political agenda at different moments during the last century. In 1899, on the introductory pages of the first issue of the journal L'Enseignement Mathématique, the official communication organ of the International Commission on Mathematical Instruction (ICMI; part of IMU, the

---

P. Valero (✉)
Aalborg University, Aalborg, Denmark
e-mail: paola@learning.aau.dk

International Mathematical Union), Laisant and Fehr (1899) envisioned the important mission of mathematics and science for the generations and the world to come:

> L'avenir de la civilisation dépend en grande partie de la direction d'esprit que recevront les jeunes générations en matière scientifique; et dans cette éducation scientifique l'élément mathématique occupe une place prépondérante. Soit au point de vue de la science pure, soit à celui des applications, le xxe siècle, qui va s'ouvrir, manifestera des exigences auxquelles personne ne doit ni ne peut se dérober. (p. 5)

In the times of the Cold War, a similar argument emerged, however the justification was related to, on the one hand, keeping the supremacy of the Capitalist West in front of the growing menace of the expansion of the Communist Soviet Union (Kilpatrick 1997). And on the other hand, the argument also had to do with economic and technological development, and the increase in numerical-based decision-making:

> Tout d'abord, d'une manière générale, la société actuelle exige de plus en plus de tous les citoyens la connaissance de notions élémentaires de mathématiques et l'appréciation de l'importance du point de vue numérique. Les dirigeants des grandes organisations sont appelés de nos jours à prendre de plus en plus souvent des décisions dans lesquelles les jugements quantitatifs jouent un rôle essentiel. (Fehr et al. 1961, p. 11)

Nowadays, professional associations all around the globe argue that the low numbers of people in STEM fields can severely damage the competitiveness of developed nations in international, globalized markets (e.g., National Academies 2007). These voices resonate with political agendas that clearly connect these fields and social and economic welfare:

> Two weeks ago Infomedia presented the following message: "The decline for the natural science programs continues." Denmark needs more young people who want to be engineers, biotechnologists and teachers of science. We need that many choose a career in the scientific and technologic areas. A great part of Denmark's welfare and competitiveness in technology build on these areas. Therefore public as well as private companies are completely dependent, now and in the future, of the sufficient availability of qualified workforce with education in science and technology. (Haarder 2009, p. 8)

For this reason, national governments have invested significant resources in the development of school reforms and the improvement of teacher qualifications in general, and particularly in mathematics (Wood et al. 2008) and science (Keeves and Darmawan 2009). However, it has been documented that since the 1990s the recruitment of students at North American and European universities in STEM fields decreased relatively in relation to the growing amount of the population entering tertiary education. Therefore, it is expected that a shortage of people with STEM qualifications at high levels will soon occur. Even if the serious concerns of the end of the 1980s for the situation in the 1990s were not completely proven right (Pearson 2008), shortages are likely to happen, and this is a risk that highly technological societies cannot afford to take. In this historical time of globalization and knowledge-based economy, a relative decrease in the numbers of people majoring in STEM or a definitely not growing number of qualified people in these areas is worrying, whether they result in a real shortage or not. This potential problem has been treated by professional associations, national governments and international organizations as a serious threat to the competitiveness, economic stability, and supremacy of Western industrialized nations (European Commission 2004; Fox 2001) in the face of the rapid growth of the so-called "rising economies" such as China, India and Brazil.

Critical readings of the sustained discourse on the centrality of STEM fields during the twentieth and now the twenty-first century argue that there is no solid evidence to actually expect a shortage of STEM graduates (Smith and Gorard 2011). Actually, critics argue that STEM recruitment has been relatively stable since the 1970s when these fields boomed in developed countries. A closer look at the numbers shows that the decrease is present when the actual numbers are compared to the prognoses of governments, professional organizations, economic organizations and industry for the amount of qualified population that would be *desirable* to have in order to meet certain expectations of knowledge-intensive production and technological development (Pearson 2008).

In other words, whether the problem of decrease in STEM areas is a "real" or only an "expected" problem, there seems to be an agreement between both critics and proponents of the problem on the fact that, in the context of informational, capitalist societies and economies, where manufacturing is no longer the broad basis for economic activity because basic production has been transferred to developing and emerging countries, the demands on knowledge production justify the concern for a relative decrease or a constant number of people entering these fields. That more people have to be attracted to pursue careers in STEM fields is the overall conclusion to which there seems to be a general adherence.

Mathematics education research and practice have certainly taken this concern up. Recommendations as well as evidence-based proposals for increasing young people's interest in mathematics, from small to large scale programs (e.g., PRIMAS 2013), have been provided. But this "problem" of the lack of interest for the STEM fields—and within that the interest in learning more mathematics—is not new, as argued above. Why is it persistent? In this paper I will take a critical stance towards the dominant discourses on this matter and offer an alternative reading of the problem. I will argue that an understanding of this "problem" requires the displacement of our thinking from the field of didactics to the field of the cultural politics of schooling and education of which mathematics—and the other fields in the STEM cluster—are a part. The strategy for unfolding my argument is to look at existing trends in mathematics education research and examine their assumptions about the learners. Then I will bring the learners to the fore by thinking about how youth culture meets the culture of school mathematics. I propose the thesis of the cultural gap between the forms of subjectivity that these two cultures invite young people to develop as a plausible way of grasping the problem. I conclude by drawing attention to forms of researching mathematics education practices that may advance our ways of thinking about the cultural clashes and tensions in the projects of subjectivity formation in the educational cultures of school mathematics.

What Existing Research Has to Offer

Research in mathematics and science education developed in the fields of scholarly research with the intention of understanding teaching and learning in these areas in all levels of schooling, but mainly with the central aim of providing evidence-based solutions to the problems of practicing teachers (Bishop et al. 1996, 2003; Fraser and Tobin 1998; Grouws 1992; Roth and Tobin 2009; Tobin 2006). In mathematics education research, the overall assumption is that an improved understanding of teaching and learning, and a research-based effort to support teachers' practice lead to the improvement of students' understanding, performance, interest and engagement in learning this school subject. In this way, research in mathematics education helps securing an adequate number of young people who would like to pursue further studies in STEM areas, and who will constitute a highly qualified work force in these fields. Besides diagnosing the systematic lack of success of many students and explaining why this is the case, mathematics and science education research have also devised methods for cognitive development, suggested and implemented curricular adjustments, developed and tested better teacher qualification programs, and even initiated broader different popularization campaigns for making science and mathematics part of the everyday life of people.

Mathematics education research has helped identifying causes for the "leakage" in the educational pipeline. Drawing on the image of a pipeline whose purpose is to channel as many of the intakes—children entering education—as an output—adults educated in STEM fields—, this type of research has emphasized the deficiencies in the processes of teaching and learning of mathematics and science, leading to systematic low performance and disengagement with the subjects. In existing research in mathematics and science education, three types of dominant explanations have been offered to the failures in the "pipeline". Firstly, research points to the cognitive challenge that acquiring mathematical and scientific knowledge represents for students. It is difficult for many to reach the deep understanding needed to attain the expected achievement levels that the school curriculum demands (e.g., Sierpinska 1994). Secondly, teachers' effectiveness for generating adequate learning is also highlighted. In primary schooling, teachers' lack of subject-matter knowledge accounts for poor teaching. In high schools, teachers' lack of pedagogical and didactical training is associated with a highly de-contextualized and formalized teaching form that fails to develop students' learning. The qualification of teachers and what teachers should know in order to improve their practice have been central topics of research associated with students' engagement in the subjects (e.g., Cobb et al. 2003; Ostermeier et al. 2010). Thirdly, the public views of science and mathematics (e.g., Moreau et al. 2010), the support of the family (e.g., Winbourne 2009) as well as of the authorities to both students and teachers (e.g., Goos et al. 2004), and even deficiencies in national curricula (e.g., Morgan 2009) are also highlighted as contributing factors to the many lacks and failures in the effectiveness of mathematics and science in schooling children and youth in a way that will lead them to successful STEM university studies.

More recently, a fourth plausible explanation has been set forward. Building on existing international comparative data of both quantitative (such as TIMSS and PISA) and qualitative (such as the ROSE study in Sjøberg and Schreiner 2010) studies, students' *interest* in mathematics and science in affluent Western countries has been identified as an important factor impacting on further study choice (e.g., OECD 2006; Troelsen 2005). Together with other notions such as motivation and affection, research has attributed the problem to a lack of emotional and personal attachment and engagement with school mathematics and science. Research on the affective dimensions of students' engagement is not new (e.g., Carr 1996). The limitation of these types of studies on people's engagement with mathematics and science is that, more often than not, the constructs, borrowed from psychology (Krapp 1999) and recontextualized into mathematics and science education research, posit the problem of interest, motivation or affection at the level of individual traits and interpersonal communication. The assumption is made that there is an intrinsic mechanism of personality that triggers engagement and thereby learning. Therefore, at the end, if teachers or students themselves are not capable of "turning on that mechanism", nothing can be done. Similar examples and critical research on these shortcomings can be found when engagement in learning in the sciences is related to gender or social class (e.g., Boaler 1998).

That people's engagement with learning mathematics is more than an individual matter, but a matter of the constitutive relationship between the person and the social, cultural and political context which s/he is a part of is a proposition explored by research which adopts a socio-cultural and political perspective to study mathematics and science learning and education (Alrø et al. 2008). From this point of view, interest and engagement are relative to social practice. This viewpoint is evidenced by the results of international comparative studies such as TIMSS and ROSE. The studies show that there seems to be a connection between students' expression of like and desire to engage in the subjects, and the type of society where they are in. Students from developing countries, despite the fact of not performing highly, express their like for the subjects and their desire of pursuing the fields in further studies. For many of them, the possibility of pursuing a university career in these fields represents a means for social mobility (e.g., García et al. 2010; Skovsmose et al. 2008). In contrast, students from Western developed countries in general score higher in the tests and have a tendency to express that they are aware of the importance of the subjects but definitely do not want to engage in further studies in the field, except in the case of life-sciences where many students see the possibility to make a contribution to the betterment of human life (Sjøberg and Schreiner 2010). In the case of East Asian students, the ones who systematically score highest in the international comparisons, they express the most negative feelings towards the subjects when asked about their like and interest in science and mathematics. They are good at the subjects, they will study them, but they do not like them (Leung 2006). As part of the Confucian culture, study, dedication and practice are values that in families and schools guide the engagement with schooling. The sense of "duty" that is part of this culture contrasts sharply with the culture of individual choice that has evolved in Western developed countries. In the light of this discussion, it becomes clear that grounding the analysis of mathematics education practices in the broader realm of education, culture and society is necessary.

This general overview of the way mathematics and science education research have dominantly addressed the problem of youths' interest in STEM fields reveals the shortcomings of mathematics and science education research, as well as their disciplinary boundaries. With a tradition of defining objects of study inside the didactic triad of teacher—student—content (mathematics or science) and together with the predominance of theories of learning and teaching to address the relationship in the didactic triad, the fields of mathematics and science education have tended to internalism (Skovsmose and Valero 2001; Valero 2010). Together with this, the restricted focus on the "utility" of research to engineer learning that has dominated educational research in the last decades (e.g., Biesta 2005; Popkewitz 2009) as well as educational research in science and mathematics has a strong impact on the direction of research towards solving the problems of practice. The result of this tendency is that fundamental problems such as the one under examination here—as well as many others such as the differential access of different types of students to success and high achievement in mathematics and science—are treated as problems of subject-matter pedagogy that need to be solved by subject-matter pedagogy. This focus has been productive in providing important understandings about the micro-pedagogical and interactional aspects of mathematics teaching and learning. However, it has prevented mathematics education from being placed within the larger realm of economic and cultural politics (Pais and Valero 2012; Valero et al. 2012).

Let me clarify my claim so far. I am not claiming that pedagogical practices are not an important contributor to the problem of young people's apparent lack of interest in mathematics and science, particularly when dedicating professional lives to these subjects is at stake. I do not want to diminish either the many contributions from decades of research to providing a better understanding of the micro-pedagogical aspects involved in the teaching and learning of mathematics. My claim is rather that it is necessary to go beyond the limits imposed by the very same development of the scientific endeavor in order to build disciplinary fields that can have a grasp of the full functioning of mathematics and science as school subjects in society. Such an attempt demands generating a strong rooting of the fields in sociological, historical, cultural and political studies.

Beyond Disciplinary Boundaries in Mathematics Education Research

When placing mathematics and science in the realm of culture, politics and society, it is important to mention existing research that offers a stepping-stone for this project. There are different types of research both inside the field and in other fields that need to be brought together.

Inside mathematics and science education research, there has been a "social turn" in the types of theories adopted to see learning during the previous two decades (Lerman 2006). A "political turn" has also been identified (Gutierrez 2013; Valero 2004b). A growing number of studies start showing a move from individualistic, cognitive theories of learning towards socio-cultural theories of human thinking (e.g., Radford 2008; Sfard 2008). There is also a growing adoption of sociological and discursive approaches that explore the intersections between school mathematical and scientific practices and larger processes of socialization, enculturation and subjectification (e.g., Black et al. 2009; Skovsmose 1994). In the sociopolitical trend, the adoption of post-modern and post-structural ideas to view learning, education, society and the role of knowledge in society has opened new possibilities of thinking the connection between educational practices, mathematics and science as areas of knowledge also in schooling and society (e.g., Appelbaum 1995; Walshaw 2004).

Latest conceptualizations of identity and identification offer interpretations of the connection between the individual and the social, which seem to be productive for understanding how learners relate to and experience school mathematics (e.g., Sfard and Prusak 2006; Stentoft and Valero 2009). However, within the prolific use of the notion of identity, few existing studies have pointed to how processes of identity formation and subjectification in current forms of social organization can be related to students' disengagement in pursuing science at higher levels. In science education, Schreiner (2006), Schreiner and Sjøberg (2007) interpreted the results of the ROSE study for Norwegian students in relation to the advance of "post-modern" trends in Norwegian society. While Norwegian students express their understanding of the relevance of science and indicate they like science, they reject the possibility of pursuing science studies in further education. Youth culture in countries such as Norway comprises many of the characteristics of late modern societies, such as self-directedness, individualization of choice and a break with family patterns of occupation. These characteristics collide with the images that science as a school subject offers as a possible project of identity formation for youths in their future. In mathematics education, few studies of students' choice of mathematics in the transition into high-school or senior secondary school (+16 years old) point out students' relationships and identification (or lack of it) with their teachers, the public images of mathematics and the gendering that is socially and culturally associated to the field (e.g., Mendick 2005). Instead of using theoretical tools that allow a reading of identity construction in science and mathematics in relation to the socio-cultural-historical context of schooling, these studies restrict themselves to an analysis within the boundaries of pedagogy suggesting that if the school subjects do not succeed in attracting people, new forms of more open and progressive pedagogy should be adopted. The studies point towards an interesting direction but remain within the notion that identifications can be modified with appropriate pedagogical change in science and mathematics classrooms. In this sense, they do not take the radical step of placing identity construction and subjectification in school mathematics and science in the broad realm of cultural dynamics.

Outside mathematics and science education, other areas of study deploying analytical strategies offer some possibilities of seeing identification in different terms (Stentoft and Valero 2009, 2010). In general educational research, critical post-structural as well as cultural-historical studies provide tools for entering the discussion of the constitution of subjectivities in the discursive spaces of education (e.g., Biesta and Egéa-Kuehne 2001; Popkewitz 2008). Youth culture studies from a postmodern perspective offer a way of interpreting the experience of young people in late-modern societies (Blackman 2005; Kahane and Rapoport 1997). Different studies in the area of technology, science and society, carrying out post-structural feminist analysis, are also sources of inspiration for studying the impact of the mathematical and scientific rationality in the construction of modern subjectivities (e.g., Bauchspies 2009; Harding 1998).

Instead of explaining the decrease of engagement with the subjects as a problem that is caused and needs to be solved inside the realm of good pedagogy, I consider the problem as a central cultural, social and political one. There is a gap between the forms of knowledge and associated ways of being that school mathematics as a subject in the curriculum puts in operation for young people in schooling, and the forms of knowing and being that young people experience outside schools in current practices. In Western developed societies, the fact that few young people show interest and choose to study STEM related fields in higher education is not (exclusively and mainly) a matter of deficient learning and interest or pedagogy and public images of science. Economic, demographic and in particular cognitive and instructional reasons associated with the pedagogy of the subjects are unsatisfactory explanations for this phenomenon. Rather, the phenomenon is fundamentally a cultural, social and political one that evidences the contradictions and ruptures of the cultural project of the modernity, in face of young people's experiences of knowing and becoming in current highly developed Western societies. More concretely, the phenomenon evidences a cultural gap between the forms of subjectivity promoted by mathematics and science as subjects of study in the educational system, and the forms of subjectivity experienced by students in other social and cultural spaces outside schooling.

Exploring the thesis of the cultural gap as a plausible and productive way of addressing young people's lack of interest in pursuing further studies in STEM areas demands paying attention to a series of interconnected reflections. I will explore some of these reflections, which challenge dominant assumptions constructed in the field of mathematics education research.

School Mathematics as a Social, Cultural, Political and Historical Battlefield

Mathematics teaching and learning became part of massive educational systems in the turn from the nineteenth to the twentieth century. As documented in studies about some of the first organizations of mathematics education—namely the journal

L'Enseignement Mathématique (Coray et al. 2003) and ICMI (Menghini et al. 2008)—reflections about the teaching and learning of mathematics emerged out of the interest of mathematicians and their concern for the passing on of mathematical knowledge. Even though different social sciences and humanistic fields had contributed to the endeavor through the twentieth century, the core of the interest in the field revolved around the idea of the specificity of mathematics in teaching and learning practices (Pais and Valero 2012).

This historical constitution built strongly on the assumption that the mathematical content of schooling, being the core element of instructional relationships and of students' thinking processes, needs to be strongly related to the discipline of mathematics. Examples of such an assumption are Chevallard's (1985) notion of the didactic transposition. Such theory views school mathematics as a transposed and recontextualized form of mathematical knowledge to be adjusted to the purposes of schooling. Ernest's (1991, p. 85) model for the relationship between objective and subjective knowledge of mathematics supposes that the objective knowledge of mathematics is being represented in order to enter a personal reformulation in the learners' process of mathematical enculturation, which will lead to new knowledge in a private realm of subjective knowledge of mathematics. In many other theories of mathematical learning the referent to the disciplinary body of knowledge of mathematics—Bishop's Mathematics with capital M (Bishop 1988)—seems to be a common element and assumption. From such a perspective, the task of mathematics education research is to find understandings and solutions to the practical problems of teachers to transmit, re-construct or enculturate children into the contents and ways of thinking of mathematicians.

It is possible, however, to adopt a different basic assumption: What difference would it make in the approach and the investigation of mathematics teaching and learning if the strong relationship with the discipline of mathematics was not seen as a fundamental condition for the study of mathematics education but as one among the many constitutive elements of school mathematics? Let us play with this idea. In a new attempt to reconceptualize ethnomathematics, Knijnik and collaborators (Knijnik 2008; Knijnik and Wanderer 2010) formulated a different thesis about ethnomathematics. When "mathematical practices" are researched in different social and cultural contexts and are seen as immersed and built in intricate relationship within people's activity, it becomes evident that the artifacts and resources of knowledge deployed are bound to the characteristics of the social practices. Using the second Wittgenstein, Knijnik argues that each "mathematical" practice in a particular context can be seen as a "language game" with its rules and functions. Language games are distinct but keep a family resemblance. The family resemblance allows the observer to identify distinct language games as being "mathematical". For the case of school mathematics such a view implies that school mathematics, as constituted in the realm of schooling, is a particular construction that has a family resemblance with the practices of academic mathematics—and that is why a mathematical view can identify it as "mathematics". However, school mathematics is governed by a different series of rules, uses, and technologies than those of the language game that we call "mathematics".

Such a view supposes that there is a much weaker connection between school mathematics and academic mathematics than what the majority of research in the field seems to assume. Yet, if one assumes that there is no strong connection between the two "language games", then what else does constitute the language game of "school mathematics"? I have argued elsewhere that seeing school mathematics as a network of social practices (Valero 2010) allows understanding that school mathematics is neither only a field of knowledge that is part of the school curriculum, nor a simplified and adjusted version of (old and fundamental) mathematics. Rather it is a sustained, collective activity that acquires meaning and is shaped by multiple participants, institutions and interests. School mathematics, therefore, is a field of practice bound to schooling as an important social institution, being also formed by politicians and policy makers, textbook writers, economic interest groups, international agencies, etc. Consequently, school mathematics is a place for the construction of culture and of politics, where students sometimes meet contradictory agendas on what to think and know and how to "become". School mathematics education is a cultural battlefield, and not a clean space for the transmission of mathematical culture.

One very strong implication of this alternative assumption is that the field of mathematics education becomes open to be studied as a field of interest for the construction of culture, in a historical context where the mathematical rationality has been highly valued and promoted. Out of this, the question arises: Why students do not seem to want to engage more with further mathematics, although they face strong voices—of teachers, parents, experts, labor organizations—arguing for the importance of mathematics for their lives and for society?

School Mathematics as a Field of Modern Subjectivity

Schools are not only institutions for the learning of concepts and ideas, skills and qualifications. Schools are also spaces where children are made *governable* (Foucault 1982). That is, children are inscribed in the dominant systems of reason that support social practices, institutions and discourses (Popkewitz 2009). In schools, children are formed as subjects: they form views of themselves as they relate to the norms and values of schooling as well as to objectified forms of knowing present in schooling.

Mathematics education is a field of practice that, within schooling, inserts children into particular forms of knowing and being. As an example, Walkerdine (1988) shows how school mathematics education and its practices of teaching and learning of arithmetic inscribe in the child the norms of what it means to be rational and reasonable. Radford (2008, p. 229) argues that the mathematics classroom is an "ethicopolitical space of the continuous renewing of being and knowing". In mathematics classrooms learners not only meet with the culturally created objectifications of school mathematics, that is, with "fixed patterns of reflexive activity incrusted in the ever-changing world of social practice mediated by artifacts", as they are created and recreated in the networks of practices of mathematics education. Students also become subjects as they meet others and the objects of the school mathematical culture. The important issue here is, which kinds of processes of subjectification do school mathematics set in operation?

Bishop (1988) formulated the thesis that mathematics education is in charge of educating children in, with and through the values of mathematics, in creating a relationship between children and their mathematical culture. Bishop identified six core values in mathematics education: *Rationalism* is the trust on the logico-deductive functioning of the human mind to address humans' own existence and their relationship with the environment. *Objectism* is a world-view "dominated by images of material objects", which makes possible for human beings to get detached from the creations of their thinking and see them as independent entities with an existence of their own. *Control* is a value associated to the value of *progress*, both of which relate to the view that the rational, objective way of thinking of the human mind can organize and tame nature and even itself. A form of knowledge that allows operating and manipulating the objects of the world leads to the advancement of human kind. *Openness* refers to the view that mathematics is open to the examination of anybody since it relies on well-defined, rational mechanisms of proof. Contrary to assertions provided by authoritative ideologies and opinions, the mathematical truth can be deduced from transparent systems. Finally, *mystery* is the value that, opposing openness, places mathematical knowledge as a special creation of a highly specialized group of people, whose work is quite unknown to the layperson (Bishop 1988, pp. 60–81).

Seen from a different theoretical perspective, the values that mathematical enculturation "brings" to school children are inscribing in children particular ideas of what it means to be the rational, desired child who can competently align his/her thoughts and actions to the norms of a highly valued form of being in society. These values embed some of the central values of Modernity (Popkewitz 2004a, b). Being successful in school mathematics requires students to develop a strong mathematical identity and demands students to be conversant in these forms of knowledge and all its values. It is in this sense that school mathematics make students become a subject within the project of Modernity. Such type of subjects are desirable to sustain the progress and advancement of Western developed societies. Indeed, mathematics educators and mathematics education researchers have formulated their mission to contribute to society through the mathematical empowerment of younger generations. Such expressions of a mission are present not only in textbooks and research documents, but also in policy documents promoted by professional organizations and also governments.

What is the problem then if there is a school subject that more than any other can help fabricate the people who are functional in a desired form of society and who can offer to realize the promises of the enlightenment and of modernity? The problem is that the form of becoming that mathematics education offers and effects in classrooms is only *one* particular form of becoming subject—in competition with many other alternative projects of being. It is one way of becoming and being in the world that younger generations do not necessarily find appealing and to which they do not want to surrender anymore. In the cultural configuration of late modernity or even post-modernity, young people may want to invest themselves in forms of knowing and being—and their related practices—that resonate with their experience in the world. Better pedagogy could help to make school mathematics more appealing—as suggested by the results of many recent research projects. But that will only be a palliative to a more fundamental rejection of the types of subjects that school mathematics fabricates.

Youth in a Postmodern Cultural Field of Becoming

The "postmodern" has been widely discussed as a variety of recent cultural forms and epistemologies configuring since the 1960s (Lyotard 1984). In the area of youth studies, it is argued that the characteristics of current forms of social, cultural, political and economic organization place youth in a quite unique configuration in which to conform their projects of identity. Best and Kellner (2003, p. 76) argue that:

> Today's youth are privileged subjects of the postmodern adventure because they are the first generation to live intensely in the transformative realms of cyberspace and hyperreality where media culture, computers, genetic engineering, and other emerging technologies are dramatically transforming all aspects of life [...] It is a world where multimedia technologies are changing the very nature of work, education, and the textures of everyday life, but also where previous boundaries are imploding, global capital is restructuring and entering an era of crisis, war, and terrorism, while uncertainty, ambiguity, and pessimism become dominant moods.
>
> Consequently, the youth of the new millennium are the first generation to live the themes of postmodern theory. Entropy, chaos, indeterminacy, contingency, simulation, and hyperreality are not just concepts they might encounter in a seminar, but forces that constitute the very texture of their experience, as they deal with corporate downsizing and the disappearance of good jobs, economic recession, information and media overload, the demands of a high-tech computer society, crime and violence, identity crises, terrorism, war, and increasingly unpredictable future. For youth, the postmodern adventure is a wild and dangerous ride, a rapid rollercoaster of thrills and spills plunging into the unknown.

Such dramatic formulations highlight that not only the fast-changing entanglements of digital technologies in the lives of young people are associated with emerging forms of knowing and being—as discussed by scholars tracing new cognitive abilities of the "Millennials" (Oblinger 2003). They also point to the political and economic changes that critical scholars have long ago identified to be shaping youth's current cultural experiences as much as future life opportunities (Aronowitz and Giroux 1991). All these conditions shape the way young people become subjects in culture and society (Kahane and Rapoport 1997), and challenge what the historical institution of schooling has to offer to this new generation and its forms of life.

At this stage in my argument the following question emerges: What is the significance of these studies for mathematics education? In my previous research I have argued that mathematics education practices and research have built quite restricted views on the mathematics learner as a cognitive subject, a *schizomathematicslearner* (Valero 2004a) who looks like:

> [A]n outerspace visitor, with a big head, probably a little heart, and a tiny chunk of body. That being would be mainly alone and mostly talk about mathematics learning, and would see the world through his school mathematical experience. That would be a 'schizo-being' since she has a clearly divided self—one that has to do with mathematics and the other that has to do with other unrelated things. (pp. 40–41)

The discursive reduction of the learner to a cognitive subject in fact is consistent with particular views of mathematics education as the set of practices of the teaching and learning of mathematics. It seems problematic that the disciplinary understandings provided by mathematics education research limit the issue of who is the learner and which of his/her "dimensions" are of relevance for the discipline result in the construction of a very narrow way of thinking about children and youth in mathematics classrooms.

What could then offer to our understanding the unlocking of youth from the cage of cognition and placing youth in the field of cultural politics of the "postmodern adventure"—as Best and Kellner suggest in the citation above? It is my contention that the limitations to the way of reasoning about the intrinsic need of interest for learning as a requisite for success in school mathematics and later on in choosing a study and career into the STEM fields would simply be opened. The constitution of students' motives to engage in the forms of subjectivity offered by school mathematics takes place as an entanglement between themselves, their cognitive possibilities and a cultural field plagued with multiple possibilities to form their identity and subjectivity. Thus, students' engagement with or rejection of the forms of subjectivity offered by school mathematics are not the result of a trait of personality—as for example Krapp (1999) argued to be the case of interest as an explanatory factor of learning. In systematic studies of patterns of study choice of young Danish high school students, Ulriksen (2003) has found that processes of individualization, the loss of orientation in family structures and tradition, and the autonomous and complex organization of everyday life are trends that impact the way youngsters identify in society. Changes in identification patterns open new configurations for what young people attribute as key reasons for their choice or lack of choice of mathematics and science. Illeris and collaborators (Illeris et al. 2002) provide an illustration of such changes with the transformation of the question asked of youth about their considerations about education. The question of "What do you want to become when you grow up?" now has been replaced by the question "Who do you want to be when you grow up?" Furthermore, some studies also point to the significance of ICT in bringing new possibilities of communication, identification, thinking and, thereby, new forms of subjectivity (Pickering 2011).

Young students in their everyday life and also in other spaces within schools are learning to become a type of self whose forms of being and expression simply cannot resonate with the modern project that school mathematics seem to make available for them. The sharp contrast between these forms of being can be seen as a gap, a rupture between a series of values and worldviews that mathematics as school subject has historically constructed, and a series of different values rooted in the development of postmodern youth cultures in technology intensive, highly developed, rich Western societies. If the disengagement and rejection of young people with the study of STEM areas is not simply a matter of lack of "interest" but rather a manifestation of a rupture in forms of subjectivity made evident in the gap between school practices and life experiences, then pedagogical improvements and changes in forms of teaching the same school subjects are not enough as a "remedy" to bring closer these forms of subjectivity. In other words, the traditional disciplinary interest of fields of study such as mathematics education has little to offer to this situation. The question of what to do then remains open.

Conclusion

So far I have argued that the perceived problem of the lack of interest of the youth in STEM areas may be understood in terms of a gap between the forms of subjectivity made available by school (science and) mathematics and the forms of subjectivity young people meet in their life in current late modern societies. I have suggested that generating such an understanding requires moving the study of mathematics (and science) education from the realm of pedagogical and didactical studies to the realm of the cultural politics studies of schooling. While in the last decade the political concerns with the modest numbers of STEM graduates have directed attention and funding to an avalanche of developmental initiatives and research programs to find the magic solution for raising "interest" in youth for the STEM fields, little has been done in analyzing the "problem" from a perspective that does not reproduce the same logic which creates it. Few existing studies using notions of identity construction and identification (e.g., Schreiner and Sjøberg 2007) have shown that the postmodern identities of junior secondary school students can be related to the lack of interest of students in pursuing science at higher levels. In mathematics education, few studies of students' choice of mathematics in the transition into high-school or senior secondary school (+16 years old) point to students' relationships and identification (or lack of it) with their teachers, the public images of mathematics and the gendering that is socially and culturally associated to the field (e.g., Mendick 2005). In spite of taking theoretical tools that could allow a reading of identity construction in science and mathematics in relation to the cultural politics of schooling, these studies stay in an analysis within the boundaries of pedagogy suggesting that if the school subjects do not succeed in attracting people, new forms of more open and progressive pedagogy should be adopted. The studies point towards an interesting direction, but do not go beyond a pedagogical advice.

In this sense, they do not take the radical step of placing identity construction and subjectivity in the dynamics of the cultural politics of schooling. However, only taking this step may allow us to understand how school mathematics has become an important field connected to culture and society. It is also in this way that we may face the challenges of teaching and learning school mathematics in a changing society. How the educational practices in the mathematics and science school curriculum are important elements in the constitution of modern subjectivities is a question that remains largely unexplored, but that deserves attention from scholars and practitioners as well.

References

Alrø, H., Skovsmose, O., & Valero, P. (2008). Inter-viewing foregrounds: Students' motives for learning in a multicultural setting. In M. César & K. Kumpulainen (Eds.), *Social interactions in multicultural settings* (pp. 13–37). Rotterdam: Sense.
Appelbaum, P. (1995). *Popular culture, educational discourse, and mathematics*. New York: SUNY Press.
Aronowitz, S., & De Fazio, W. (1997). The new knowledge work. In A. H. Halsey, H. Lauder, P. Brown, & A. S. Wells (Eds.), *Education: Culture, economy, and society* (pp. 193–206). Oxford: Oxford University Press.
Aronowitz, S., & Giroux, H. (1991). *Postmodern education: Politics, culture and social criticism*. Minneapolis: University of Minnesota Press.
Ashton, D. N., & Sung, J. (1997). Education, skill formation and economic development: The Singaporean approach. In A. H. Halsey, H. Lauder, P. Brown, & A. S. Wells (Eds.), *Education: Culture, economy, and society* (pp. 207–218). Oxford: Oxford University Press.
Bauchspies, W. (2009). Potentials, actuals and residues: Entanglements of culture and subjectivity. *Subjectivity, 28*, 229–245.
Best, S., & Kellner, D. (2003). Contemporary youth and the postmodern adventure. *Review of Education, Pedagogy, and Cultural Studies, 25*(2), 75–93.
Biesta, G. (2005). Against learning: Reclaiming a language for education in an age of learning. *Nordisk Pædagogik, 25*(1), 54–55.
Biesta, G., & Egéa-Kuehne, D. (2001). *Derrida & education*. London: Routledge.
Bishop, A. J. (1988). *Mathematical enculturation: A cultural perspective on mathematics education*. Dordrecht: Kluwer.
Bishop, A. J., Clements, K., Keitel, C., Kilpatrick, J., & Laborde, C. (Eds.). (1996). *International handbook of mathematics education*. Dordrecht: Kluwer.
Bishop, A. J., Clements, M. A., Keitel, C., Kilpatrick, J., & Leung, F. K. S. (Eds.). (2003). *Second international handbook of mathematics education*. Dordrecht: Kluwer.
Black, L., Mendick, H., & Solomon, Y. (2009). *Mathematical relationships in education: Identities and participation*. New York: Routledge.
Blackman, S. (2005). Youth subcultural theory: A critical engagement with the concept, its origins and politics, from the Chicago school to postmodernism. *Journal of Youth Studies, 8*(1), 1–20.
Boaler, J. (1998). Nineties girls challenge eighties stereotypes: Updating gender perspectives. In C. Keitel (Ed.), *Social justice and mathematics education: Gender, class, ethnicity and the politics of schooling* (pp. 278–293). Berlin: Freie Universität Berlin.
Carr, M. (1996). *Motivation in mathematics*. Cresskill: Hampton.
Chevallard, Y. (1985). *La transposition didactique*. Grenoble: La Pensée Sauvage.
Cobb, P., McClain, K., Silva Lamberg, T. D., & Dean, C. (2003). Situating teachers' instructional practices in the institutional setting of the school and district. *Educational Researcher, 32*(6), 13–25.
Coray, D., Furinghetti, F., Gispert, H., & Schubring, G. (Eds.). (2003). *One hundred years of L'Enseignement Mathématique: Moments of mathematics education in the twentieth century*. Geneva: L'Enseignement Mathématique.
Ernest, P. (1991). *The philosophy of mathematics education*. London: Falmer.

European Commission. (2004). *Europe needs more scientists*. Report by the high level group on increasing human resources for science and technology in Europe. Brussels: European Commission.
Fehr, H. F., Bunt, L. N. H., & OECE. (1961). *Mathématiques nouvelles*. Paris: OECE.
Foucault, M. (1982). The subject and power. *Critical Inquiry, 8*(4), 777–795.
Fox, M. A. (2001). *Pan-organizational summit on the U.S. science and engineering workforce: Meeting summary*. Washington: National Academy of Sciences.
Fraser, B. J., & Tobin, K. G. (1998). *International handbook of science education*. Dordrecht: Kluwer.
García, G., Valero, P., Camelo, F., Mancera, G., Romero, J., Peñaloza, G., & Samaca, M. (2010). *Escenarios de aprendizaje de las matemáticas: Un estudio desde la perspectiva de la educación matemática crítica*. Bogotá: Universidad Pedagógica Nacional de Colombia.
Goos, M., Kahne, J., & Westheimer, J. (2004). Learning mathematics in a classroom community of inquiry. A pedagogy of collective action and reflection: Preparing teachers for collective school leadership. *Journal for Research in Mathematics Education, 35*(4), 258–292.
Grouws, D. A. (Ed.). (1992). *Handbook of research on mathematics teaching and learning*. New York: Macmillan.
Gutierrez, R. (2013). The sociopolitical turn in mathematics education. *Journal for Research in Mathematics Education, 44*(1), 37–68.
Haarder, B. (2009). Naturfag er almen dannelse. In Undervisningsministeriet (Ed.), *Natur, teknik og sundhed. For alle og for de få. I bredden og i dybden* (pp. 8–13). Copenhagen: Undervisningsministeriet.
Harding, S. (1998). *Is science multicultural? Colonialisms, feminisms, and epistemologies*. Bloomington: Indiana University Press.
Illeris, K., Katznelson, N., Simonsen, B., & Ulriksen, L. (2002). *Ungdom, identitet og uddannelse*. Frederiksberg: Roskilde Universitetsforlag.
Kahane, R., & Rapoport, T. (1997). *The origins of postmodern youth: Informal youth movements in a comparative perspective*. New York: Walter de Gruyter.
Keeves, J. P., & Darmawan, I. G. N. (2009). Science teaching. In L. J. Saha & A. G. Dworkin (Eds.), *The new international handbook of research on teachers and teaching* (pp. 975–1000). New York: Springer.
Kilpatrick, J. (1997). *Five lessons from the New Math era*. Paper presented at the conference Reflecting on Sputnik: Linking the past, present, and future of educational reform, National Academy of Sciences of the USA, Washington, DC.
Knijnik, G. (2008). Landless peasants of Southern Brazil and mathematics education: A study of three different language games. In J. F. Matos, K. Yasukawa, & P. Valero (Eds.), *Proceedings of the fifth international mathematics education and society conference* (pp. 312–319). Lisbon: Centro de Investigaçao em Educaçao, Universidade de Lisboa.
Knijnik, G., & Wanderer, F. (2010). Mathematics education and differential inclusion: A study about two Brazilian time–space forms of life. *ZDM – The International Journal on Mathematics Education, 42*(3–4), 349–360.
Krapp, A. (1999). Interest, motivation and learning: An educational-psychological perspective. *European Journal of Psychology of Education, 14*(1), 23–40.
Laisant, C.-A., & Fehr, H. (1899). Préface. *L'Enseignement Mathématique, 1*(1), 1–5.
Lerman, S. (2006). Cultural psychology, anthropology and sociology: The developing 'strong' social turn. In J. Maasz & W. Schloeglmann (Eds.), *New mathematics education research and practice* (pp. 171–188). Rotterdam: Sense.
Leung, F. K. (2006). Mathematics education in East Asia and the West: Does culture matter? In F. K. Leung, K. D. Graf, & F. Lopez-Real (Eds.), *Mathematics education in different cultural traditions: A comparative study of East Asia and the West* (pp. 21–46). New York: Springer.
Lyotard, J.-F. (1984). *The postmodern condition: A report on knowledge*. Minneapolis: University of Minnesota Press.
Mendick, H. (2005). Mathematical stories: Why do more boys than girls choose to study mathematics at AS-level in England? *British Journal of Sociology of Education, 26*(2), 236–251.

Menghini, M., Furinghetti, F., Giacardi, L., & Arzarello, F. (Eds.). (2008). *The first century of the international commission of mathematical instruction (1908–2008): Reflecting and shaping the world of mathematics education.* Rome: Istituto della Enciclopedia Italiana.

Moreau, M.-P., Mendick, H., & Epstein, D. (2010). Constructions of mathematicians in popular culture and learners' narratives: A study of mathematical and non-mathematical subjectivities. *Cambridge Journal of Education, 40*(1), 25–38.

Morgan, C. (2009). Questioning the mathematics curriculum: A discursive approach. In L. Black, H. Mendick, & Y. Solomon (Eds.), *Mathematical relationships in education: Identities and participation* (pp. 97–106). New York: Routledge.

National Academies. (2007). *Rising above the gathering storm: Energizing and employing America for a brighter economic future.* Washington: National Academies Press.

Oblinger, D. (2003). Boomers, gen-xers, and millennials: Understanding the "new students" *EDUCAUSE Review, 38*(4), 36–47.

OECD. (2006). *Evolution of student interest in science and technology studies.* Policy report. Paris: OECD.

Ostermeier, C., Prenzel, M., & Duit, R. (2010). Improving science and mathematics instruction: The SINUS project as an example for reform as teacher professional development. *International Journal of Science Education, 32*(3), 303–327.

Pais, A., & Valero, P. (2012). Researching research: Mathematics education in the political. *Educational Studies in Mathematics, 80*(1), 9–24.

Pearson, W. J. (2008). *Who will do science? Revisited.* Paper presented at the annual meeting of the commission on professionals in science and engineering, Baltimore. www.cpst.org/2008meeting/presentations/pearson-chubin-davis.pdf. Accessed 24 Mar 2014.

Pickering, A. (2011). H-: Brains, selves and spirituality in the history of cybernetics. *Metanexus, 9*(3). Retrieved from metanexus website: http://www.metanexus.net/essay/h-brains-selves-and-spirituality-history-cybernetics

Popkewitz, T. S. (2004a). The alchemy of the mathematics curriculum: Inscriptions and the fabrication of the child. *American Educational Research Journal, 41*(1), 3–34.

Popkewitz, T. S. (2004b). School subjects, the politics of knowledge, and the projects of intellectuals in change. In P. Valero & R. Zevenbergen (Eds.), *Researching the socio-political dimensions of mathematics education: Issues of power in theory and methodology* (pp. 251–267). Boston: Kluwer.

Popkewitz, T. S. (2008). *Cosmopolitanism and the age of school reform: Science, education, and making society by making the child.* New York: Routledge.

Popkewitz, T. S. (2009). Curriculum study, curriculum history, and curriculum theory: The reason of reason. *Journal of Curriculum Studies, 41*(3), 301–319.

PRIMAS. (2013). *Promoting inquiry-based learning in mathematics and science across Europe.* www.primas-project.eu/en/index.do. Accessed 23 Apr 2013.

Radford, L. (2008). The ethics of being and knowing: Towards a cultural theory of learning. In L. Radford, G. Schubring, & F. Seeger (Eds.), *Semiotics in mathematics education: Epistemology, history, classroom, and culture* (pp. 215–234). Rotterdam: Sense.

Roth, W.-M., & Tobin, K. (2009). *The world of science education: Handbook of research in North America.* Rotterdam: Sense.

Schreiner, C. (2006). *Exploring a ROSE garden: Norwegian youth's orientations towards science – Seen as signs of late modern identities.* Unpublished Ph.D. thesis, University of Oslo.

Schreiner, C., & Sjøberg, S. (2007). Science education and young people's identity construction: Two mutually incompatible projects? In D. Corrigan, J. Dillon, & R. Gunstone (Eds.), *The re-emergence of values in science education* (pp. 231–248). Rotterdam: Sense.

Sfard, A. (2008). *Thinking as communicating.* Cambridge: Cambridge University Press.

Sfard, A., & Prusak, A. (2006). Telling identities: In search of an analytic tool for investigating learning as a culturally shaped activity. *Educational Researcher, 34*(4), 14–22.

Sierpinska, A. (1994). *Understanding in mathematics.* London: Falmer.

Sjøberg, S., & Schreiner, C. (2010). *The ROSE project: An overview and key findings.* Oslo: Oslo University.

Skovsmose, O. (1994). *Towards a philosophy of critical mathematics education.* Dordrecht: Kluwer.

Skovsmose, O., & Valero, P. (2001). Breaking political neutrality: The critical engagement of mathematics education with democracy. In B. Atweh, H. Forgasz, & B. Nebres (Eds.), *Sociocultural research on mathematics education: An international perspective* (pp. 37–55). Mahwah: Lawrence Erlbaum.

Skovsmose, O., Scandiuzzi, P. P., Valero, P., & Alrø, H. (2008). Learning mathematics in a borderland position: Students' foregrounds and intentionality in a Brazilian favela. *Journal of Urban Mathematics Education, 1*(1), 35–59.

Smith, E., & Gorard, S. (2011). Is there a shortage of scientists? A re-analysis of supply for the UK. *British Journal of Educational Studies, 59*(2), 159–177.

Stentoft, D., & Valero, P. (2009). Identities-in-action: Exploring the fragility of discourse and identity in learning mathematics. *Nordic Studies in Mathematics Education, 14*(3), 55–77.

Stentoft, D., & Valero, P. (2010). Fragile learning in mathematics classrooms: Exploring mathematics lessons within a pre-service course. In M. Walshaw (Ed.), *Unpacking pedagogies: New perspectives for mathematics* (pp. 87–107). Charlotte: IAP.

Tobin, K. G. (2006). *Teaching and learning science: A handbook.* Westport: Praeger.

Troelsen, R. (2005). Unges interesse for naturfag: Hvad ved vi, og hvad kan vi bruge det til? *MONA, 2,* 7–21.

Ulriksen, L. (2003). Børne- og ungdomskultur og naturfaglige uddannelser. In H. Busch, S. Horst, & R. Troelsen (Eds.), *Inspiration til fremtidens naturfaglige uddannelser. En antologi* (pp. 285–318). Copenhagen: Undervisningsministeriet.

Valero, P. (2004a). Postmodernism as an attitude of critique to dominant mathematics education research. In M. Walshaw (Ed.), *Mathematics education within the postmodern* (pp. 35–54). Greenwich: IAP.

Valero, P. (2004b). Socio-political perspectives on mathematics education. In P. Valero & R. Zevenbergen (Eds.), *Researching the socio-political dimensions of mathematics education: Issues of power in theory and methodology* (pp. 5–24). Boston: Kluwer.

Valero, P. (2010). Mathematics education as a network of social practices. In V. Durand-Guerrier, S. Soury-Lavergne, & F. Arzarello (Eds.), *Proceedings of the sixth congress of the European society for research in mathematics education* (pp. 54–80). Lyon: Institut National de Récherche Pédagogique.

Valero, P., García, G., Camelo, F., Mancera, G., & Romero, J. (2012). Mathematics education and the dignity of being. *Pythagoras, 33*(2), 171–179.

Walkerdine, V. (1988). *The mastery of reason: Cognitive development and the production of rationality.* London: Routledge.

Walshaw, M. (2004). *Mathematics education within the postmodern.* Greenwich: IAP.

Winbourne, P. (2009). Choice: Parents, teachers, children, and ability grouping in mathematics. In L. Black, H. Mendick, & Y. Solomon (Eds.), *Mathematical relationships in education: Identities and participation* (pp. 58–70). New York: Routledge.

Wood, T., Sullivan, P., Tirosh, D., Krainer, K., & Jaworski, B. (Eds.). (2008). *The international handbook of mathematics teacher education* (Vol. 4). Rotterdam: Sense.

Connecting Place and Community to Mathematics Instruction in Rural Schools

Robert Klein

Abstract This chapter uses a mixed-method approach to address practices of mathematics education in the context of various rural schools in the US. It reports on connections of mathematics instruction to place and community by developing issues of relevance, sustainability and social-class interaction. Special attention is paid to place-based education and the university-intending students, to rural insufficiency and rural affordance and to the egalitarian local/elite cosmopolitan continuum.

This chapter[1] addresses two questions: (a) How do rural schools connect mathematics education to local communities and places? and (b) What conditions enable and constrain their efforts? It draws from work by rural sociologists, mathematics educators, and rural educators in formal and informal educational settings. A mixed-methods approach began with a seven-site comparative case study model (Phase One) to cast as broad a net as possible to facilitate the identification of emergent patterns and themes, validated by triangulation techniques. Subsequently (Phase Two), a random sample of N=237 rural mathematics teachers in grades 6–12 (teaching ages ~12–18) completed a questionnaire probing the generality of the Phase One themes. Results suggest three primary themes that characterize how rural schools connect mathematics instruction to local communities and places and things that enable and constrain their efforts. These were coded *Relevance*, *Sustainability*, and *Social-class Interactions*. Analysis of the survey data confirmed the stability of these categories across the sample. The social, economic, political, and cultural contexts of rural schools and the communities they serve are sufficiently complex to suggest that efforts to connect mathematics instruction to place and community are acts of resistance and reinhabitation (Gruenewald 2003, 2006), even if the teachers, students, and community members doing the work do not view it explicitly as such. This chapter further responds to a need for increased focus on research in/for/by

---

[1] This work draws on the efforts of a talented team of colleagues including Aimee Howley, Craig Howley, Daniel Showalter, John Hitchcock, and Jerry Johnson. This paper makes use of Howley et al. (2010, 2011), and Klein et al. (2013) but focuses more acutely on mixed-methods results than do those papers.

R. Klein (✉)
Ohio University, Athens, OH, USA
e-mail: kleinr@ohio.edu

© Springer International Publishing Switzerland 2015
U. Gellert et al. (eds.), *Educational Paths to Mathematics*,
Advances in Mathematics Education, DOI 10.1007/978-3-319-15410-7_3

rural communities and those who dwell therein. So rare is the attention to rural contexts and mathematics education in prominent journals, that Edward Silver, while editor of one such prominent journal, scolded the research community for a "rural attention deficit disorder" (2003, p. 2). Harmon et al. (2003) reported the outcomes of a conference held to define a research agenda for studying rural mathematics and science education. They found that "Scholarship on rural education in the United States is relatively sparse (DeYoung 1987). Rural education issues rarely attract the attention of education professors at prestigious universities (DeYoung 1991)" (p. 52). They further found that, "Most of the usual solutions provided by educational policymakers fail to recognize the uniqueness of rural settings (Harmon 2003; Larsen 1993)" (p. 53). Rural schools are no less responsible to state accountability measures and must adhere to the same mandate for evidence-based strategies, yet with so little research on the contextual factors unique to rural areas, teachers and school districts may not find the same support that their colleagues in urban and suburban school districts enjoy.

Hence, the research team that undertook this work could similarly be characterized as engaging in the work of resistance and reinhabitation. There is an underlying activism here that seeks to support rural schools and communities by listening to their community members and focusing on opportunities to construct strategies that celebrate rural factors as assets rather than challenges.

Motivating Assumptions and Their Basis in Extant Literature

Despite significant social, cultural, geographic, and economic differences, rural communities around the globe face many of the same challenges: rural-to-urban migration, economic viability, sustainability of "rural lifeways" that value attachments to community and land, and community members seeing formal education as a mechanism for training their sons and daughters to leave the community, often never to return ("brain drain") (Carr and Kefalas 2009; Corbett 2007).

This study proceeded from five assumptions rooted in research from rural sociology, mathematics education, and critical pedagogy. First, *rural communities are important to the future of every region.* The world's food, energy, and material resources will continue to be focal points of policy discussion at regional, national, and global levels (International Fund for Agricultural Development 2010). Solutions to managing these resources effectively, including alternative energy development, likely will be found *in* rural places (Berry 1977). I contend that the development of local talent in math-intensive subjects supports solutions that respect and draw from the affordances of rural places *by those who know them best.* Rural communities in the United States (and elsewhere) possess cultural resources reflecting a "national character" rooted in rural ways of living and 'reckoning' (Berry 1977; Carr and Kefalas 2009).

Second, *"place" and "place-based education" (PBE) are conceived loosely so as to be open to different interpretations of it arising in the study.* Nevertheless, others have defined these terms, including David Sobel of the place-based "Community-Based School Environmental Education" (CO-SEED) program: "the process of using local community and environment as a starting point to teach concepts" (Sobel 2004). Yet the potential scope of impacts of PBE is wide, as Paul Theobald suggests:

> Allow teachers to mine the curricular and instructional potential of the local community... and you have taken a major step toward raising the consciousness of the next generation with respect to the full range of circumstances—political, economic, social—affecting one's home, family, neighbors, and neighborhood. (2009, p. 130)

Following Kunstler (1993), this study proceeded from the idea that "place" is a distinctly rural affordance and is *not* a feature of urban and suburban locales where "land" becomes "real estate." Though something like a place-based approach seems possible in urban and suburban locales, the authors sought exemplars of approaches that were inspired and motivated by (perhaps "rooted in") place rather than just conscious of it. Rural locales seemed the ideal place to inquire about place- and community-connections in mathematics instruction given the centrality of place, community, and kin/family to rural places.

Third, *mathematics is important to sustaining rural communities* as it underlies all sciences and therefore figures greatly to the success and sustainability of rural communities (Klein 2007). Nevertheless, connecting place and mathematics instruction is fraught with a number of implicit and explicit challenges (Klein 2008). Among these is the American rejection of an "integrated" curriculum (Math 1, 2, 3, etc.) so prevalent in other parts of the world in favor of "compartmentalized" approaches that treat subdisciplines as isolated (Algebra I ⇨ Geometry ⇨ Algebra II ⇨ Calculus). The kinds of problems that arise naturally in the world outside of the classroom usually demand more integrated approaches. Also, rooting good mathematics instruction in place demands that the educator is well prepared in both mathematics and mathematics instruction, and elementary educators (ages 5–11 typically) in the United States often have minimal preparation in those areas (Ma 1999). Moreover, teachers of advanced secondary mathematics courses generally celebrate the placelessness of mathematics, exhibiting Platonic or formalist leanings that result in enacted curricula largely devoid of, if not completely hostile to, connections to place (Hersh 1997). Hence the work of sustaining rural communities involves not only *resistance* to deficit discourses and cosmopolitan value structures that prize mobility over provincialism—discourses that have fostered associations of mathematics success with leaving rural communities, but also to *reinhabitation* of mathematics education discourse in rural areas (Gruenewald 2003). When mathematics instruction connects to place, it proclaims the value of both, together. Peter Berg and Raymond Dasmann have cast reinhabitation aptly as "learning to live-in-place in an area that has been disrupted and injured through past exploitation" (Berg and Dasmann 1990, quoted in Gruenewald 2003, p. 10).

Fourth, *place can motivate mathematics learning as important to modern rural life*. Attachments to place are a central feature of rural communities (Theobald 1997) and represent an affordance for instruction (Klein 2007, 2008). Bauman describes how, as a result of globalization, "localities are losing their meaning-generating and meaning-negotiating capacity and are increasingly dependent on sense-giving and interpreting actions which they do not control" (1998, p. 3). Issues of access and participation matter on geographic and community scales. Rural communities must be "spaces of resistance" (Castells 2004) instead of nostalgic fictionalizing. The research team for this study harbored biases not toward white picket fences and pastoral longing, but rather toward supporting the right of rural citizenry to define modern rural lifeways that can use high-level mathematics knowledge without having to adopt values and assumptions attached to that knowledge that may operate contrary to those lifeways. In Corbett's words,

> Place-based education is not nostalgic or constructed in terms of immersion in a 'traditional' way of life that sits outside in modernity and in which young people learn to stay. Rather, [it is] an immersion in the contemporary transformations of rural places. (2007, p. 269)

Mathematics should not be defined as a distinctly urban affordance just as attachments to rural living should not be defined as a barrier to success in school mathematics or its use outside of school.

Fifth, *social class tensions arise between rural schools and communities*, especially where schools become "travel agencies for those who can afford tickets" (Corbett 2007, p. 271). Corbett's generational study of a rural fishing community evidences perceptions of rural community members that education is about "learning to leave" (Corbett 2007). Carr and Kefalas document a class-driven, rural American "brain drain" arising from a school sorting of students into *stayers* (receiving few community resources) and *leavers* (receiving significant community resources). Schools are central to a social sorting: "individuals acquire and deploy their cultural assets to manage their position in the social order. One of the most important marketplaces for spending and earning this special sort of wealth is in the setting of a school" (2009, p. 33).

When rural communities are categorized by "durable agrarian," "resource extraction," or "suburbanizing rural," different patterns emerge for how social class is negotiated within and between those categories. Durable agrarian communities exhibit relative income equality, leading to a more egalitarian mindset. In contrast, resource-extraction communities and suburbanizing rural communities exhibit significant income disparity and attendant class distinctions in schooling and the community (Howley and Howley 2010). Schools and schooling have a role in maintaining or changing those patterns and, as such, the roles of schools and the values that drive curriculum (hidden or exposed) are of incredible importance to the future of rural communities of all kinds. When educational institutions promote, intentionally or not, the out-migration of talented youth, they become agents supporting resource extraction. It is also conceivable that the out-migration of rural youth may reflect the educational institution's underlying values coming not from within the local community's values, but rather from "elsewhere," including corporatized curriculum and testing, state mandates, and perhaps even popular culture, which is often tied to cosmopolitan values. The present study, therefore cannot ignore the assumed value systems that educational institutions inherit, and the ways that these values impact the construction of social class, (re)location, and the form of mathematics instruction engaged.

Contrary to the popular association of "rural" with "deficient," (Howley and Gunn 2003) mathematics test scores in rural schools, in the aggregate, show no significant difference from national averages and in some cases exceed the national average (Fan and Chen 1999) though regional variation is significant (Lee and McIntyre 2000). Such broad snapshots often overlook the complexities that play out in more regional or local contexts. Recently, Klein and Johnson (2010) compared ninth-grade state mathematics test scores between rural and non-rural schools and found that rural schools erased the influence of poverty on mathematics achievement for females and minorities.

Rural schools use the same textbooks and must adhere to the same state-level standards as urban and suburban schools. Teachers (in all locales) rely heavily on textbooks and attend to "meeting" the standards (often in checklist style) so the influence of these on the educational mission and the communicated values are often those of population centers generally, and more particularly, populous states that wield political influence nationally (such as Texas and California). The current U.S. environment of national accountability resulting from No Child Left Behind legislation makes the creation of a more "local" curriculum a radical one. Teachers who create place-based curriculum may risk significant investment of human and political capital (Klein 2008). Hence, the relative absence of research on place-connected approaches to mathematics instruction seems understandable though lamentable.

Methods

This study addressed two primary research questions: (a) How do rural American schools connect mathematics education to local communities and places? and (b) What conditions enable and constrain their efforts? The breadth of such questions suggested a mixed-methods design involving two overlapping phases: a cross-case comparative study and a broader survey to probe the generalizability of themes in the first phase. The Phase One data set was considerably large (>1,500 pages of transcribed interviews and focus groups) and involved a multiple-case study design with a two-stage analysis consisting of an inductive coding for emergent themes at each site followed by a cross-case analysis using multiple readers, a cross-site matrix, and detailed audit trails (Miles and Huberman 1994). For Phase Two, the research team used Phase One themes to design and construct a survey instrument that was administered to $N=237$ rural mathematics educators. Exploratory factor analysis (EFA) indicated four latent factors consistent with themes derived from Phase One themes. The two phases are outlined below, followed by analysis of the results and implications.

Data Collection

Phase One: Cross-Case Comparison

The Sample

A number of ongoing projects formed the bases for a network of nominators for research sites though this involved others besides mathematics educators. Work in place-based *science* education is more present in the literature and is often tied to environmental education more specifically. The COmmunity-based School Environmental EDucation (CO-SEED) project at Antioch University has engaged pre- and in-service teachers in place-based education development resulting in several schools active in engaging place in the service of science education. A group of Alaskan educators has worked with pre- and in-service teachers to engage the Yup'ik cultural resources actively in mathematics instruction. The Appalachian Collaborative Center for Learning and Instruction in Mathematics (ACCLAIM) worked with multiple cohorts of PhD students in mathematics education and developed a network of scholars interested in improving mathematics instruction in rural Appalachia. The present study was funded by a grant from the National Science Foundation to ACCLAIM.

Together, the network of 58 active place-based educators in science and mathematics education constituted a rich source of nominations for the present study. Scholars and educators from this network nominated sites they thought might evidence place-based approaches to mathematics instruction across a variety of grades and in a variety of locations across the United States. Sixty-one sites were proposed for investigation, from which the research team selected a smaller subset to reflect variety in geographic locations, and grade range. These sites were contacted by phone to determine their willingness to participate and to get a preliminary sense of the extensiveness of engagement with place-based pedagogies. This resulted in identification of the final seven sites studied here.

Capturing the variety represented in the seven sites is not possible in the space provided, so the research team has constructed a monograph that compiles individual site case studies that detail site-specific data and findings (Howley et al. 2010). A snapshot of each site is presented below to contextualize the data and the findings. State names are accurate but all other names have been changed consistent with ethical practices appropriate to this type of research.[2]

Underwood, Ohio

Underwood Local School District, located in rural Appalachian Ohio, has just one campus containing an elementary school and a high school and serving ~900 students in grades K–12 (typical ages: 5–18). Approximately 60 % qualify for

---

[2] Descriptions are modified from those given in Howley et al. (2010).

subsidized meals (as a result of relatively low family income) and the median household income for the district in 2000 was ~30,000 USD. Middle school students (typical ages: 11–14) have consistently performed close to the Ohio average and have met Ohio standardized proficiency requirements for many years. The teacher, Ms. Miller graduated from Underwood High School and now teaches sixth-grade mathematics at Underwood Elementary. Ms. Miller's students engage in community-inspired mathematics projects throughout the year, including Relay for Life (students collect and track money and graph lap speeds for the cancer charity run/walk), Pi Day, and a stained-glass project that focuses on geometry and measurement in creating artwork.

Gladbrook, Alabama

Gladbrook City School District is located in a rural county of southern Alabama. The city of Gladbrook has a population of ~2,000. As of 1999, 40.7 % of families with related children of 18 years or younger live in poverty, and 55.65 % of students in the county receive free or reduced-price lunches. Gladbrook City School District enrolls 162 students in the high school and 84 students in the middle school. Gladbrook High School received a bronze medal rating from *U.S. News and World Report*'s list of America's Best High Schools. The average ACT composite score for 2006 was 19.7 (nationally in 2006, 1.2 million students, or 38 % of high school graduates took the test for a composite score mean=21.1, sd=4.8).[3] Students participate in an aquaculture program, actively maintaining fish environments (monitoring pH, population size, health), tracking and fostering fish growth, and eventually selling the fish at a community fish fry that generates funds to sustain the program. A turf management program at the school similarly responds to local economic strengths (Fig. 1).

Hanover, Kentucky

Hanover High School enrolled 1,091 students in 2007–2008, with approximately half of the student population participating in subsidized meals. The school is situated on the edge of a city of more than 11,000 residents but is far from an urban area. On the Kentucky-mandated assessment, Hanover's mathematics scores were below both district and state averages in core content measures of 11th grade

---

[3] The ACT is a college readiness assessment administered in the United States to students in secondary school wishing to enter post-secondary education. It is used by many institutions of higher education in the United States, along with the similar SAT, as part of college admissions decisions and sometimes placement into mathematics and English courses. The ACT reports an overall composite score (1–36), with 36 being the highest score, as well as subject-specific composite scores (also 1–36) for English, Mathematics, Reading, Science Reading, and Writing.

Fig. 1 A Hanover student holds the ukulele he built

mathematics, and below the college[4] readiness (ACT-PLAN) scores of the district and state in mathematics. However, in 2007–2008, the school had a 91.5 % graduation rate and a 57.3 % college attendance rate, both significantly above district and state averages. Agriculture programs are popular at the school, and a collaboration between a mathematics teacher with musical interests and the agriculture "shop" teacher led to a foundation- and community-supported "lutherie" program where students built stringed instruments from raw lumber (not kits). The program is unique among American high schools and brings together university-intending students with agriculture program students—groups that would otherwise likely not take a class together in their junior and senior years.

Hamilton Collaborative, Nebraska

The Hamilton District Collaborative is now a consortium of four independent school districts led by a single superintendent, through a unique arrangement among the local boards. Nine schools are involved, and each district operates its own high school, with high school (grades 7–12) enrollment ranging from 45 to 197 students.

---

[4] The word "college" used throughout this chapter, refers to post-compulsory graduation. It may be used synonymously with "university" as is the practice in the United States.

The subsidized meal rate for all elementary schools combined was about 31 % in 2007–2008, and family incomes in the four communities vary from about 35,000 to 46,000 USD while poverty rates for families with children under 18 range 0–12 %. ACT composite scores in the two longest associated districts were ~22. The study conducted interviews at all high school locations. This site is the most complex "case" in the study—in many ways a counter-example, or perhaps representing place-based practices in decline. Most teachers in the study viewed place-based approaches as ill-suited to the distance-learning technologies being used to address concerns about itinerant teaching across the districts. Hence, the site offered few examples of specific place-based approaches but a rich set of reactions to them.

Twin Oaks, Vermont

A number of small town New England districts operate only a single K–6 school. In such circumstances, these very small town districts pay public funds to nearby private institutions to school their children. Twin Oaks, a K–8 school, accepts such tuition students from neighboring towns. Hence, unlike the other schools in this study, Twin Oaks (founded in the 1960s) is a non-profit private school. In 2007–2008, the school enrolled a total of 127 students, about 35 % of whom receive scholarships (school reported data). Median household income in the area was about 5 % below the national average (for 1999) of 42,000 USD, whereas just 8.2 % of families with children under the age of 18 existed on incomes below the official poverty line (as of 1999). Twin Oaks students engaged in "tree plot math," a 6-week project related to the local industry of timbering. They were assigned plots of land—often triangles, circles, or quadrilaterals—from which they gathered data about the trees in their plots, graphed the results, calculated the worth of their trees, visited the sawmills, and occasionally presented their findings to the school's board of trustees (e.g., to help the trustees decide whether or not to log the entire tract owned by the school).

Edgewater, Maine

Edgewater Community School is on an island more than 10 miles from the coast of Maine. Though not affluent, Edgewater is not a classically impoverished rural community—family incomes were just 5 % lower than the state average (U.S. Census Bureau 2000). The poverty rate for residents under the age of 18 was just 2.8 %. These figures may reflect the relative equality of income distribution among the island's year-round residents: 29 % of the families subsist on incomes of less than 35,000 USD; 58 % have incomes from 35,000 to 75,000 USD; and just 13 % of families have incomes of 75,000 USD or greater. The upper grades outperform the state on mathematics proficiency by a substantial margin. Clamming is a significant economic activity for the island though tourism is the primary one. While place-based approaches were seen across the grades, and largely due to staff affiliations

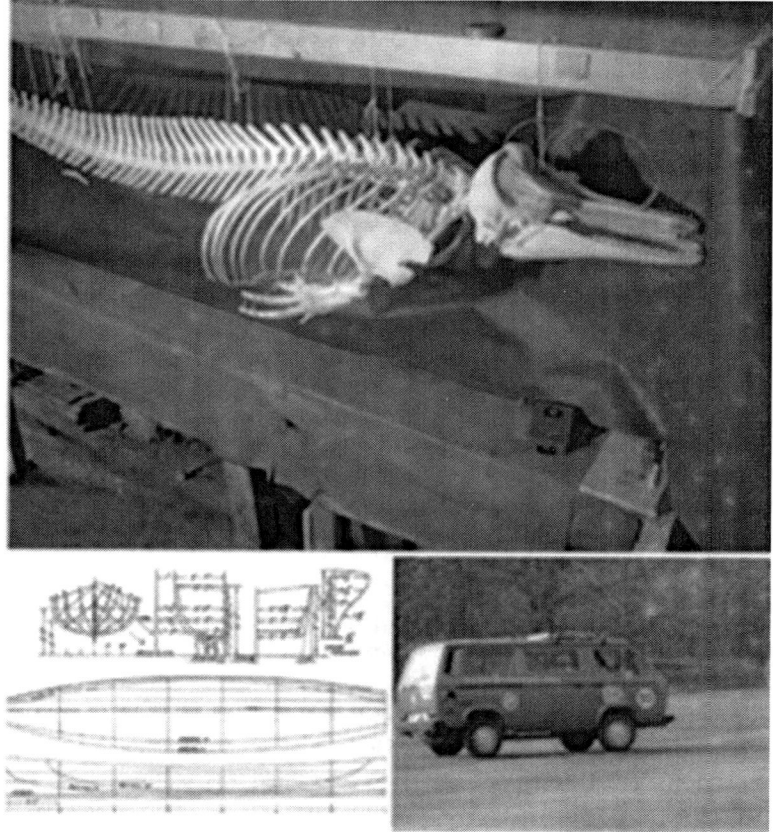

Fig. 2 Sea-life, pea-pod boats, and an electric car from Edgewater

with the aforementioned CO-SEED program, the upper grades implemented a more extensive place-based effort than the lower grades. These included the design and construction of pea-pod boats and an all-electric vehicle that was demonstrated in Washington, D.C. (Fig. 2).

Grover, Washington

Grover Junior Senior High School is located between two Pine Valley towns along the valley's primary highway (2002 populations were around 1,000 and 400 respectively). In 2007–2008 the school enrolled about 250 students in grades 7–12, with ~31 % of students eligible for subsidized meals (vs. 38 % statewide). Grover graduated 92.1 % (vs. state average of 72.4 %). Grover and the Pine Valley Elementary School share a campus and are the only schools in the district. Teachers come to know students in the district and their families well over the 6 years students spend at Grover and students do well on standardized state tests. Ms. Engels, the only middle-school mathematics teacher, invites community members to her classes to describe the mathematics they use in their daily work. She has received local foundation support to subsidize parental and community member visits and her "Careers in Math" presenters have included a local fiber artist, a bicycle shop owner, and a video game designer who is also an alumnus. A second initiative, "Math Communities," brings parents to school every 2–3 weeks to lead small groups of students through multi-step word problems. Parents are given educational "primers" on the problem in advance, and these primers include pedagogical hints, assessment details, and various examples of student work showing different correct and incorrect responses.

Protocols and Procedures

The research team developed semi-structured interview protocols derived from prominent *a priori* themes emerging from the research literature and from experience working with rural mathematics teachers, students, parents, and administrators. Interview protocols were developed for administrators, teachers, and other community members; and focus group protocols were generated for parents and students. Items were stated generally and akin to "Our study focuses on mathematics instruction that draws on or makes use of local places and communities. How does what you do with your students fit in with this way of thinking about instruction?" An observation template was also used across the sites to standardize formatting of field notes and observations of classes. The scope of data collection from 1-week site visits in 2007–2008 included 85 interviews, 27 observations, and 30 collections of fieldnotes. At the Twin Oaks Vermont site, the field researcher arrived to find the school shut down because of an outbreak of *pediculosis capitis* (head lice) that closed the school for 3 days of the week-long visit, limiting the data available at that site to just three interviews.

Interviews and focus groups were transcribed verbatim (~1,500 pages) and relevant artifacts including field notes and collected documents were scanned and organized. Transcripts were analyzed using first, an inductive generation of codes and the cataloging of emergent themes from each case, then comparing codes and themes across cases. Three members of the research team organized the data into 27 categories and identified emergent themes through discussion wherein a proposed theme was then tested against the data for possible support using a cross-site matrix that charted the prevalence of a theme across the different sites and data sources within that site. Eventually, three themes emerged as salient—*Relevance, Sustainability,* and *Social-class Interactions* (with attendant subthemes). These were used to guide item development in Phase Two.

Phase Two: The National Survey

The Sample and Questionnaire Development

Beginning in Fall 2010, the research team developed a questionnaire based on the three Phase One themes (Relevance, Sustainability, Social Class Interaction, discussed in the results section below). The team used a "cognitive lab" (or "think-aloud") protocol to observe two participants complete the instrument (Willis et al. 1991). This provided insight into how participants might read and think about each item, if the item was clear, and if participants had relevant thoughts or perspectives not captured by items in the survey. One change made as a result of the cognitive lab was that the research team added Item 1: "To what extent do you agree with the statement 'Math is everywhere.'?" This decision arose from Phase One findings that nearly all informants accepted the math educator's war cry "Math is everywhere" yet struggled to identify *which* math was *where* and being used *by whom*. In total, the survey totaled 41 items comprising 30, five-point Likert-type ("Strongly Agree" to "Strongly Disagree") items focusing on the three Phase One themes; 10 demographic items, and one open response items to capture miscellaneous comments.

After 30 participants piloted the revised questionnaire, the research team used National Center for Education Statistics (NCES) data sources to list all grade 6–12 public schools with Locale Codes of 41, 42, or 43 (Rural-Fringe, Rural-Distant, or Rural-Remote). The sample from this frame was 3,000 participant schools chosen randomly using SPSS statistical software package (IBM Corp. 2012). This set was further reduced by removing charter schools and schools with zero enrollments, leaving 2,923 schools. Respondents provided data via an online survey tool. Response rates were poor (4 %) initially so the research team used school websites to identify teacher names and emails corresponding to a sub-sample of the schools listed in the original sample and incentivized teachers' participation with a chance to win a prize valued at 100 USD. Complete responses rose to $n=237$ using this approach and data collection was subsequently halted once this phase ran its course. As is true of most surveys of this type, sample bias is a concern though demographic results suggest geographic representativeness of the data (Klein et al. 2013).

Results

In addressing the research questions of how rural American schools connect mathematics instruction to community and places and the conditions that sustain or constrain their efforts, analysis of the qualitative Phase One data identified three sets of dynamics explaining place- and community-based initiatives at all sites: (a) how *relevance* justified PBE initiatives, (b) which conditions *sustained* certain initiatives while jeopardizing others and (c) how initiatives embedded opposing perspectives on the ways each site responded to *social class interactions* (Howley et al. 2011).

Exploratory factor analysis (EFA) of Phase Two survey data suggested four factors (labeled A-D, Appendix A) should be extracted. Bartlett's test of sphericity was not significant and Kaiser-Meyer-Olkin measures did not change. A parallel analysis using polychoric factor analysis served as a further sensitivity check since the survey did not yield continuous items. The difference between the two approaches was trivial so EFA results are given in Appendix A with polychoric factor analyses reported in parentheses.

At first glance, Factor A confirms the theme of *Relevance to Students (their lives, their futures, their talents)*. Factor B comprises only three items, all of which relate to *Higher Level Mathematics and College*. Factors C and D both relate to *Sustainability* though Factor C has more to do with *Community Support and Resources*, whereas Factor D items probe the ease of connecting instruction to place and community, and the barriers and *teacher*-perceived challenges to sustainability. As such, the items in Factor D cohere rather well around a theme of *Barriers and Challenges to Sustainability*. That is:

Factor A: *Relevance to Students*
Factor B: *Higher Level Mathematics and College*
Factor C: *Sustainability: Community Support and Resources*
Factor D: *Sustainability: Barriers and Challenges*

Reliability of the factors was calculated as

Factor A: Alpha = .859 (increases to .863 if A10 dropped)
Factor B: Alpha = .702
Factor C: Alpha = .777 (increases to .790 if C7 dropped)
Factor D: Alpha = .543 (increases to .548 if D7 dropped).

Hence the first three factors have moderate to high reliability while Factor D has relatively low reliability. Correlation between the factors was relatively low, as indicated in Appendix B. The results confirm the relative stability of Phase One themes, each of which is further discussed below.

Theme-by-Theme Analysis

Relevance

Relevance was cited as a rationale for place-based mathematics many times at all sites in the Phase One study. Phase Two survey work confirmed the salience of the theme, with 99.2 % of the respondents indicating some agreement (or strong agreement) with the statement "Mathematics is everywhere" ($M = 1.11$, $s = .342$, $n = 246$). Participants also indicated high levels of agreement with the statement A6: "Students are more motivated to learn mathematics when they see it as being relevant to their daily lives" (Table 1). Phase One informants generally characterized this kind of connected mathematics as "concrete" (instead of abstract) and examples they gave

Table 1 Item response descriptives

	Item	Mean (SD)	% Agree (SA or A)	Response (N)
A1.	Connecting mathematics to students' everyday lives can help prepare them for study at 4-year colleges.	2.13 (.736)	79.8	243
A2.	Connecting mathematics to students' everyday lives can help prepare them for study at 2-year colleges.	2.06 (.680)	82.5	246
A3.	Using students' daily life experiences as part of instruction improves the learning of lower-level math topics (e.g., arithmetic, basic graphing, beginning algebra).	1.64 (.652)	92.7	247
A4.	Using students' daily life experiences as part of instruction improves the learning of higher-level math topics (e.g., trigonometry, advanced algebra, geometry).	1.81 (.714)	85.4	247
A5.	Connecting mathematics to students' everyday lives can help prepare them for coursework in a 4-year university or college.	2.01 (.874)	80.2	247
A6.	Students are more motivated to learn mathematics when they see it as being relevant to their daily lives.	1.68 (.748)	89.1	247
A7.	Connecting mathematics to the local community will improve instruction for all students.	2.18 (.749)	75.1	245
A8.	My remedial students like it when mathematics instruction is tied to their daily lives.	2.04 (.781)	76.7	246
A9.	My *advanced* students like it when mathematics instruction is tied to their daily lives.	1.98 (.756)	78.5	246
A10.	The mathematics I teach is directly applicable to my students' everyday lives.	3.34 (1.01)	67.2	247
A11.	Twenty-first century jobs will demand high-level mathematical skills of all students.	2.19 (.952)	73.6	246
B1.	All students can do higher-level mathematics.	3.34 (1.075)	25.6	246
B2.	Some students will never be ready for college-level mathematics.	3.68 (.900)	13.5	245
B3.	Schools should prepare *all* students to go to college.	3.24 (1.264)	32.1	246
C1.	My colleagues support efforts to connect mathematics instruction to the local community.	2.45 (.753)	56.9	246
C2.	My administration supports community-connected mathematics instruction.	2.46 (.791)	52.4	246

C3.	Parents of my students support community-connected mathematics instruction.	2.77 (.771)	31.7	246
C4.	My community offers opportunities for applying mathematics topics.	3.33 (.798)	15.9	246
C5.	Students who stay in the area will apply their math skills to address local needs.	2.94 (.834)	30.7	244
C6.	My students are accustomed to the sorts of projects that connect mathematics to everyday life.	2.92 (.951)	37.4	246
C7	Students with strong math skills are likely to settle in this area.	3.58 (.803)	6.9	246
D1.	It's easier to connect lower-level mathematics content to daily life than higher-level mathematics content.	2.22 (1.03)	66.7	246
D2.	There are barriers that make it difficult to connect mathematics to students' everyday lives.	2.27 (.927)	72.4	246
D3.	The mathematics of daily life matters more to the future of some of my students than it does to others.	2.21 (1.05)	74.1	247
D4.	My typical student will need only basic mathematics in his/her daily life.	3.05 (1.107)	37.4	246
D5.	Activities that connect math to the local community take up instructional time needed to address state standards.	2.51 (1.01)	58.9	246
D6.	Algebra I (or the equivalent) is where you start to notice which students will be ready for college level mathematics and which won't.	2.85 (1.069)	46.7	246
D7.	Some students receive better academic opportunities because of his/her family's local prominence.	2.56 (1.106)	56.7	245

for "relevant mathematics" almost always involved arithmetic and/or measurement and never anything beyond very basic algebra. More abstract mathematics was seen as less relevant to "real life." One high school mathematics teacher in the study commented:

> We don't really do anything in places ... I would say, maybe in the shop class is where you're going to find that. The closest we're going to come to that is just story problems, and we do an awful lot of application [in story problems].

Despite this, items A8 and A9 of the survey probed teachers' beliefs about remedial and advanced students' appreciation of place-connected approaches. Teachers had high levels of agreement that both groups "liked" such connections.

The Edgewater site was the only exception to this belief among mathematics teachers in Phase One—at Edgewater, higher-level mathematics courses used place in mathematics instruction routinely. Evident in the quotation above is another common feature of the data—without exception, vocational teachers (shop, agriculture) engaged place and community in instruction extensively. Still, the underlying assumption among those teachers was that their students were not likely to attend college and that place- and community-connected instruction was essential preparation for "staying" (cf. Carr and Kefalas 2009). Items C4, C5, and C7 directly addressed "stayers" and "leavers" relative to mathematics performance and use and, though read as loading onto a sustainability factor, are relevant here. Results for these items suggest neither firm agreement or disagreement regarding the opportunities for students to apply mathematics in local communities or for students who do well to leave, though item C7 did register the second-highest level of disagreement of any item, thus lending some credibility to claims from the literature of the effects of brain drain in rural areas and mathematics success as a particularly keen marker for leaving.

Data from all sites similarly suggested that *perceptions of relevance* motivated students to do mathematics. The middle school mathematics teacher at Grover commented:

> The number one thing would be that they see relevance in what they're learning and that it's sort of a motivational factor to stay engaged in math and to learn math. Of course ... maybe [it's] not the best instruction or whatever, but that's definitely secondary.

The "math as secondary" concern arose in the Hanover "lutherie" program as well where time pressures forced a choice between teaching the mathematics (logarithms, symmetry and measurement mathematics) that underlay instrument design and completing the instruments by the end of the year. The math was sacrificed. In the year the research team visited, the mathematics teacher (who partnered with the agriculture teacher) became the assistant principal for the school and had less of a role in the class, likely contributing to the de-emphasis of the mathematics.

The grandparent of a Gladbrook student involved in the aquaculture program thought that place-based mathematics lessons could motivate students to do math without their knowledge of doing math. He described this "covert motivation" in terms of:

> My grandson ... says he can't do math ... but he gave those fish one tenth of a milligram of ... pituitary hormone, but *he can't do math!* ... If you were showing him that he was doing math, he'd probably twitch. But if you don't tell him ... he just goes on and does it.

Phase Two data confirmed the generalizability of perceptions of relevance as motivational, with item A6, "Students are more motivated to learn mathematics when they see it as being relevant to their daily lives," receiving strong levels of support. Item A10 results suggest that most teachers (67.2 %) in the sample view the mathematics they teach as applicable to students' everyday lives. The Phase Two survey did not probe further for the reasons for this though Factor D barriers such as state standards coverage (item D5) or the amount of time it takes to create such lessons challenge item A10 findings.

Sustainability of Place-Based Approaches

In Phase One, three interrelated conditions were deemed most relevant to supporting sustainability of place-based approaches. These included: the presence of a "champion educator," the relative balance of obstacles and affordances, and participants' beliefs in the future of the local place and community. Phase Two findings found support for the theme of sustainability, divided into Factor C: Community Support and Resources (Alpha = .777), and Factor D: Barriers and Obstacles (Alpha = .543).

Some educators were willing to devote significant energy and time to sustaining the approaches. These *champion educators* were important to sustaining place-based approaches for many reasons. Place-based curriculum is inherently local and ill-fitting textbook teaching, so designing and implementing lessons involved significant effort. The Edgewater principal was far and away the strongest champion in the study. In fact, his support of place-and community-based education briefly cost him his job as principal when he disagreed with a powerful school board member who wanted more traditional approaches. Students and parents rallied and he was reinstated.

At Gladbrook, the founder of the aquaculture program returned to the school from retirement when he determined that the program was dying absent his stewardship. At Grover Middle School, Ms. Engels understood the balance that a champion must maintain between competing pressures:

> When you're in a small school, that's what happens: you teach every class and so you're scrambling, and then the state's asking you to teach all these things, and, then you're giving up [classroom] time for community members.

The *ease of implementation* is informed by these and other pressures that may deter would-be champions from pursuing place-based approaches. Still other factors contribute to the ease of implementation of place-based approaches including financial resources given the costs of field trips and consumable materials. Edgewater was the site most engaged in place-based approaches and this was facilitated by the nearly 23,000 USD spent per student per year by the school. In contrast, other schools, such as Hanover, spent under 8,000 USD per student. In the case of Hanover, the lutherie program resulted from several grants won by the two teachers and sustained by yearly community auctions of ukuleles built during the project.

Though money and administrative support were important factors, *time* was cited as the more critical resource determining ease of implementation. In the United States, a middle- or high-school teacher might have, on average, 1 h per day (during school hours) for planning, and this is often spent grading or tutoring students—offering little time to use for creative activity. But student time was also significant: students in Hanover's lutherie class met 2 h per day, 5 days per week for the class and, for university-intending students, this often meant choosing between the lutherie class and a university-preparatory class. Administrative support was a key factor in ease of implementation and it presented issues for teachers in the study, but also for the administrators (see Howley et al. (2010) and (2011) for further discussion).

In Phase Two, participants were divided about support from colleagues and administrators (C1–C2), and only 31.7 % of respondents saw parents as supportive of community-connected mathematics (C3).

Seeing value in local places precedes seeing them as valuable resources for instruction. A frequent theme in the data from all sites was how the informants perceived their local area and community, but also small towns and rural locales more broadly. One Hamilton Collaborative informant connected the declining populations of many small towns and rural areas to the inevitability of consolidated school districts, thus reorganizing the social landscape. His was a vision of a disappearing way of life.

Other communities, such as Grover, experienced an increase in population that brought about shifts in demographics and in values. Grover is located in a mountainous recreation area and was primarily a rural-recreation zone supported seasonally by cross-country skiers in the winter and hikers in the summer. The arrival of high-speed Internet in the valley brought with it wealthy telecommuters from suburbs of a nearby major city. The social and economic impacts of this in-migration were significant and played out in the schools. A school parent new to the valley saw peaceful cohabitation and described a "vibrant, open community," whereas others, including Ms. Engels, discussed the rising property values and cost of living and how it resulted in the rearrangement of families in the valley with wealthy families and expensive real estate to the north and less wealthy families to the south. Therefore many native residents and newcomers saw the future of the valley differently from each other, with newcomers talking about the new opportunities that were present in the valley and native residents worrying about the redistribution of resources and land.

Year-round Edgewater residents believed strongly in the future of their island and its continued way of life and, not surprisingly, offered a lot of support for place-based approaches that communicated value to both learning and rural lifeways simultaneously. They were most optimistic about the future of the island though they understood that the future of the island depended on the money from wealthy summer residents though that brought with it increasing property prices and different educational values.

Perspectives on the future of different communities interact strongly with social class interactions. In fact, one could characterize this in terms of "seeing one's place" which reads, in English, on a material level (perceptions of my physical locale, its organization and resources, and the people therein) but also on a social level (perceptions of my role in society, my economic or social limitations, and the interactions of people and their institutions).

Phase Two data were revealing. Only 15.9 % of respondents indicated some level of agreement that their community offered opportunities for applying mathematics topics (C4) and this was echoed in item C7 where only 6.9 % of respondents agreed that students with strong math skills would be likely to settle in their community.

Factor D, relating to barriers to sustaining place-connected approaches, registered two-thirds of participants agreeing that lower-level mathematics was easier to connect to daily life than higher-level mathematics (D1), and 74.1 % of respondents finding that the mathematics of daily life mattered more to the futures of some students than others (D3), despite low agreement (37.4 %) with the belief that their typical student would only need basic mathematics in his or her daily life. This factor had the lowest reliability (Alpha = .543) but this may have arisen from the broad range of grades surveyed. Future research will explore the distinction in these beliefs according to grades taught.

Social-Class Interactions

After finding 50+ pages of transcript data concerning perceptions of the disposition of resources, power, and institutional(ized) behavior, the research team concluded that the social and economic conditions described by the informants could be classified broadly as "social-class interactions" with equal emphasis on all three words. The web of relations constituting these interactions makes this the most complex dynamic and the team did not engage the data to specify distinct, cross-case "classes." Indeed, the organization (and reproduction) of "the social" emerged as a construct defined locally yet according to more global and cosmopolitan influences (cf. Bauman 1998; Theobald 2009). In the case of Grover and Edgewater, classes seemed to be defined according to "native" versus "newcomer" ("year-round resident" versus "summer resident" in Edgewater) though this was tied to measure of wealth. In the case of Hanover, there was a class distinction, not always tied to wealth, between "college-preparatory" and "agricultural/vocational" students. Hamilton Collaborative, like Grover, had income distinctions that mapped themselves geographically (wealthier residents to the south and less affluent residents to the north at Hamilton and the reverse at Grover).

Four overlapping subthemes structured the research team's Phase One findings within this theme: (A) Participation in place-based education as a barrier to the success of certain students wishing to enter university, (B) Rural insufficiency and rural affordances, (C) Egalitarian localism and (D) Elitist cosmopolitanism. This theme was not separate from the previous two, but infused throughout them. Indeed, the results of the exploratory factor analysis found only three items loading on this theme, with obvious social-class themes throughout many other items loading on the other factors. In both phases, "relevance" was defined differentially according to social class interactions at the various sites (relevance to whom, for what purposes). Similarly, "sustainability" involved political negotiation and the distribution of resources including time, money, and sanctioning of curriculum.

Subtheme A: Place-Based Education and the University-Intending Students

Parents and students at all high schools except Edgewater saw participation in place-based mathematics as an obstacle to or distraction from coursework necessary for admittance into a good university with a good placement into mathematics courses. Phase One Item C3, already mentioned, confirms this somewhat, with only 31.7 % of rural educators surveyed agreeing that parents support community-connected mathematics instruction. At the same time, the data from both phases evidenced perceptions that a place-based education was either appropriate to or important for the non-college bound. The concerns centered partially on the time that place-based activities "took away" from (traditional) classroom instruction, but also on the devaluation of some kinds of knowledge (including concrete and applied knowledge) and pedagogies (hands-on) in favor of other kinds of knowledge (abstract and pointed toward calculus) and pedagogies (individual learning, direct instruction).

The time commitment involved in place-based approaches was a common concern as reflected by a parent response: "a lot of people aren't going to be able to benefit from it because if they do they're not going to be able to fit in everything else that they need to have in order to go for that [university] degree." The assistant principal at Hanover described the stakes involved in the choice between participating in the lutherie class or not:

> One of our very, very bright students opted not to [take an Advanced Placement class] because he would have to have given up a lutherie class to take an Advanced Placement class in order to earn a Commonwealth Diploma [an advanced diploma]. And he said, "While I'm in high school this is something I want to do." And he gave up the opportunity for a Commonwealth Diploma just to participate in the lutherie program.

In Phase Two, 80.2 % of respondents agreed that "connecting mathematics to students' everyday lives can help prepare them for coursework in a four-year university." Hence survey respondents saw the choice above (place-connected approaches versus preparing for university) as a false choice even though most Phase One informants (educators, parents, and administrators alike) saw this as a very real tradeoff. For instance, an Edgewater school board member said,

> Some people were really worried that we were doing *integrated math type stuff*. That the kids were going to miss out and not do as well on tests or in getting into [universities] where they had to have—if they were going to do math. [emphasis added]

This could be read in a number of different ways according to pedagogy or mathematical content. In one reading, there is a negative association with "integrated" in which the perception that the *organization* of mathematical topics, independent of either the *pedagogy* or the mathematical (topical) *content*, defines the value of the mathematics preparation. Yet another reading centers the *content* as most valuable, with the "stuff" being the important feature of the sentiment. In this reading, the topics that a student sees are what determine the worth of the preparation for college. A (perhaps not) final reading would highlight the association of "integrated" with a *pedagogical* approach that handicaps a university-bound student who might see the same content as students in "traditional courses" but in ways that negatively impact his post-compulsory educational opportunities. Contextual clues in the surrounding dialogue did not offer insight into which of these (or other) intentions the speaker had. Given the second-hand nature of the comment, it perhaps excites less of an evidentiary note and more of a theoretical appreciation of the complicated nature of defining what is valuable and to/by whom.

The differentiated treatment toward the university-bound and the non-university-bound students, as well as the prejudicial sorting of students established a documented social-class structure (arguably one of many). Factor B items evidenced only one quarter of those surveyed agreeing that all students can do higher-level mathematics, and only a third suggesting that schools should prepare all students to go to college (not the same as "all students will go to college"). In strange contrast, only 13.5 % of teachers surveyed agreed that "some students will never be ready for college-level mathematics." This seeming contradiction is difficult to explain using the extant data and merits further study.

Subtheme B: Rural Insufficiency and Rural Affordance

This category bears obvious similarities to Factors C and D above but is worth noting here as a subtheme because it was not simply informants' perceptions of the future prospects of their community that mattered, but how they understood, lived, and evaluated "rural." In this way *"rural" is itself a class distinction* defined in contrast to and measured against "non-rural" urban and suburban or metropolitan. This distinction was seen in many cases in economic terms—between not having and having—and social terms—between progressive and backward. Though the team catalogued a number of reactions and perceptions in Phase One about rural/urban (people/places) and small-/big-town (people/places) generally, the focus here is in terms of the extent to which it informs education. The research team constrained Greeno's (actually Gibson's) concept of "affordance" as "whatever it is about the environment that contributes to the kind of interaction that occurs" (between agent and situation) to contributions to successful implementation of place-based approaches (Greeno 1994). "Insufficiency" here refers to a sense or perception of "not enough" or "lacking" relative to some minimum standard ("sufficient"). Moreover it reifies extant stereotypes of "rural" as synonymous to "deficient," "backwards," and "stupid."

A high school science teacher at Hamilton Collaborative described rural insufficiency in terms of "big" and "small": "when you live in a rural community, your sphere of what you see and know is small, until you can somehow reach out and maybe come back in. And then you can see that there's a lot more application that you can do if your vision was bigger." In this view ("until you can somehow reach out") education seeks the "big" picture and this involves leaving (Carr and Kefalas 2009; Corbett 2007). Her colleague, a high school mathematics teacher, further indicated, "I imagine there isn't the experiences of a variety of things available in your smaller community, as there is in your larger communities. You have more access to a larger ... things."

A middle school science teacher in Gladbrook associated "small" with "insufficient" and lamented that education prepared people to be successful "somewhere else":

> Gladbrook is really small as you've seen I'm sure. There's not a lot of industry or business here. So with Gladbrook being so small and the community being as it is, a lot of kids are not going to stay in the community. They're going to leave and go somewhere else because there's no jobs. So as applying it back to this community it's not very much but when they go out to other communities, that's when they're going to apply it. Which is sad.

Such perspectives inevitably hold little hope for place-based instruction in those areas given that educators are not likely to use an "insufficient" resource as the motivation for and site of instruction. A Hamilton Collaborative student saw a rural insufficiency directly in terms of mathematics learning:

> To get more advanced math, you would have to probably go somewhere other than [here] to get more, like, trigonometry or geometry, because sometimes people locally might not use that kind of math so you might have to go to a bigger community to find it.

For this student, his route through the mathematics curriculum was a route heading away from his home. Researchers at Underwood (Ohio) heard about local economic downturns and low prospects for jobs in the area motivating a more outward focus for applications. Opportunity was created by education and opportunities resided elsewhere. The lessons used by the elementary school instructor there were much more about a generic community "Race-for-a-Cure," "pi day" involvement than about involvement in activities central to Underwood.

Items C4, C5, and C7, in particular, evidenced more general equation of "rural" with "insufficient." It is perhaps the insufficiency of resources (to apply mathematics to, to draw from, to find employment with) that serves more than anything else to define class structures of any type. Sherman studied a small town that went from economically diverse with the presence of a logging industry, to economically equal (and poor) when that industry left. In the absence of economic distinctions to define class, the community engaged in a moral calculus to differentiate social classes within the community (Sherman 2009). The findings here add evidence to Sherman's work and to the importance of seeing latent community resources and an optimistic vision for one's community, but also for seeing how latent resources and their distribution (social, economic, or otherwise) are tied to the creation and maintenance of class structures.

In Grover, Edgewater, Twin Oaks, and Hanover, researchers found more instances of perceptions of how local places and communities were an affordance for good instruction. Twin Oaks had a large plot of land on which the school was situated and this allowed for a number of "tree-plot" investigations on site. Hanover students made two ukuleles during the year, one to keep and one to contribute to the end-of-year auction. The ukuleles sell out each year and fund another year of the program. Interest in the lutherie programs was sustained somewhat by the prevalence of boat-building and some lumbering in the area.

In contrast, at Grover, Ms. Engel equated "rural" with "small" but saw smallness as an affordance:

> I know a lot of people because I've lived here a long, long time and I know a lot of the community members, so that's one way, and I think, 'Oh yeah, that job', you know. For example, I bought a new bike a couple of years ago, and John Guy, who runs the bike shop, you know, he did all this math to fit my bike, and all this geometry and stuff, and I thought, 'What a great opportunity for him, and he's actually coming this week.'

Her connections to local business owners, fiber artists, and smoke-jumpers, arose from her connections to them as community members, and the smallness of the community afforded a diverse network from which she could recruit volunteers to present to her classes.

The data also evidenced some tension regarding what qualifies as an affordance. In Edgewater, during a student focus group, while one student praised a lesson tied to fishing and lobstering for how it allowed him to connect to older residents of the island, another student disagreed, seeing this in terms of his grade and his future:

> I kind of have the opposite from her. I am not from here. I moved here when I was, like, a baby, but ... it's still like, the family that I have that lives here ... is not ... like, fishing and all that. It's more of my less immediate family that does that, and so when we're focusing on fishing or nature or that stuff, it's totally irrelevant at home, and ... it's uninteresting and it has an effect on, like, my grades, because I'm less interested, so I ... and I feel like it's less important because it's not, like, what I want to pursue later on.

The importance of fishing to the island was an important instructional resource for one student but represented a form of cultural hegemony in which the school defines fishing as part of island identity even though it might not be part of the student's identity. Engaging context will inevitably confront questions of "whose context" and "whose values," emphasizing the need for sensitivity to contested contexts and values. It is not clear if students in non-place-based classes would cite more or less the irrelevance of mathematics to their lives though, anecdotally, this seems to be a common concern of students. The difference may be that the non-traditional (place-based) approach is unexpected or "different" from traditional approaches so the student's concern above may be anchored in a more widely accepted "norm."

Subthemes C and D: The Egalitarian Local/Elite Cosmopolitan Continuum

The use of "egalitarian localism" and "elite cosmopolitanism" calls on an ideological distinction that contrasts very different perspectives on schooling and life. The elitist side of the continuum "embeds the view that educational attainment, professional accomplishment, and wealth necessarily elevate their possessors above others" (Howley et al. 2011, p. 114). The idea of cosmopolitanism is relevant in that the ideology is not only pervasive (and urban-centric) but it is purported to apply to everyone as a modern world citizen. The egalitarian localist perspective instead embodies a "liberatory perspective on education, position[ing] critical thinking about cultural products (including ideas, knowledge, and artistic works) as central to the cultivation of informed citizens" (p. 114) but also a localist-vocationalist view that supports educating residents for roles in the local community with an eye toward the health of that community. The data can be read as evidence for a contest between egalitarian localism and elitist cosmopolitanism, but also as evidence that this contest shapes perspectives on the purpose of mathematics education, the roles of schools generally, and the importance of rural communities nationally and globally. Admittedly, this is not a clean categorization of a complex expression of values onto the proposed continuum. Indeed, it is a pragmatic attempt to sort and understand frequent references to social class tensions seen throughout the data, and understood within the other Phase One themes and across the Phase Two factors that emerged from the data.

Three Phase-One sites make the comparison of perspectives on this subtheme particularly evident. Hanover had a large number of respondents evidencing egalitarian localism perspectives. Edgewater data displayed the liberatory goals of egalitarianism but had elements of both localism and cosmopolitanism. Finally, data from Grover suggested the ways in which newcomers, armed with wealth and educational attainment, could replace the localist-vocationalist perspectives with elitist-cosmopolitanist views. Other sites such as Hamilton Collaborative evidenced strong perspectives on both sides depending on the informants. Twin Oaks had insufficient data for use here though its dual role as a private school enrolling public tuition students would make for an interesting investigation of this subtheme.

The Hanover "lutherie" program brought together students with different goals and backgrounds who ordinarily might not associate in their junior and senior years. Admittedly, this represents ~20 students from a much larger junior-senior class. The "social mixing" feature of the class arose in a focus group of students:

> Student 1: This class is really diverse and even outside of the classroom now, you know, different groups of students get together and talk and are friends because of this diverse class.
>
> Student 2: Yea, I never would have talked to "Belt buckle" here if it hadn't been for this class. I'll just be honest. So it helps like that.

"Belt Buckle" was an "ag student" who rode rodeo broncos and wore a large belt buckle. Student 2 identified as a "university prep student." The class, then, served as an equalizer where, according to the teacher, the ag students who might not otherwise be interested in music and math but knew how to use the tools well would mix with music and math students who might otherwise not know how to use a tape measure well. Despite the "lutherie" class being a site of egalitarian localism, the data still showed a number of instances of informants associating "rural" and "working class" with "redneck" and other derogatory terms.

This contrasts with Edgewater, where interviewees rarely made negative associations with "rural," but where economic and power relations told a different story. Island life in Edgewater is marked by year-round (often working class) residents and seasonal (generally wealthy summer recreationalist) residents. The wealthy summer residents have contributed financially to the Edgewater Community School, allowing the school to spend around 23,000 USD per student per year. Yet this wealth has also caused local property values to rise to the point where many year-round residents struggle to sustain households, thus threatening generational continuity. More elitist-cosmopolitan values came with the money and the principal admitted to having to attend to both localist and cosmopolitan values, as evident in his comment concerning a teacher (John) who was seen as too demanding by some year-round residents:

> You know, we started doing these programs, and eventually, that also angered people. They didn't like John. They felt that he was too demanding. They said, one person in the community said, um, "it doesn't have to be that good in Edgewater." You know.

"That good" indicates that it isn't simply about *what* or *how* things were taught, but to *what standards*. Factor B findings, and those of Item B1, in particular support this perspective broadly, with only one quarter of teachers surveyed agreeing that "all students can do higher-level mathematics." Relatedly, 74.1 % of teachers agreed that "the mathematics of daily life matters more to the future of some of [their] students than it does to others" (Item D3). Results from these items suggest a wider belief among rural educators that "it doesn't have to be that good" in many rural places.

The Edgewater principal further characterized some of the social class clashes in playful terms:

> It's hard to look at Edgewater and get a real picture of it. Because, when you're here now, it's a working class community. They're craftsmen, they're tradesmen … and then, in the summer, there are people from all over the world, and pretty wealthy and, in many cases, pretty influential and important people … I love it. I love it. I love the two, sort of, Edgewaters. I'm happy when the summer people leave and we can be our own little community. But, I'm happy when they arrive in the summer, because it's just, just a different place …
>
> You know, there's one song about the summer people. And … one of the lines is, "we can't live with them, and we can't live without them." … But, it's a very affectionate song. It's not disrespectful. I mean, you know, there are lines like, "Why are they always standing in the middle of the road?"

Island educators viewed the island as having important raw material for learning, but they also knew that students who left for further education and work were making a reasonable choice that they supported. They sought to balance respect for sustaining and valuing the local community with respect for students' right and need to make decisions about their future that may not involve living or working on the island. The Edgewater principal's comments illustrate this concern:

> ... that's also a big issue for our school because the state, like schools all over the country, are really being forced to sort of conform to a curriculum and to a view of the world that isn't about this place ... The focus, here, is preserving this community that has been in existence, you know, since before the, you know, establishment of the United States. And, these islands were the fishing outposts of Europe long before they were considered part of Maine. And, uh, you know, we're worried about whether the school can continue, and we're worried about whether the year-round community in Edgewater can continue. And, so, our focus is very different from what the state wants us to be focusing on. And, at the same time, we understand we have to prepare kids here to be able to go off to college or university ... And, in part, it is why we do things the way that we do, because this school can't be separate from the community.

Grover, in Washington State, sits in a river valley and the economic wealth of its residents maps itself onto the terrain from the upriver elite to the downriver poor, and students from across this span attend Grover Junior Senior High School. Prior to the arrival of wealthy telecommuters, the valley embodied a more egalitarian-localist perspective that has been supplanted by the elitist-cosmopolitan perspectives of the newcomers. This reinhabitation of the social and political space centered on the school. One newcomer parent formed an educational foundation that has raised money to support proposals by teachers for educational projects, including Ms. Engel's Math at Work program. But the foundation provides a mechanism of influence for a group of wealthy contributors and community members over educational decisions, programs, and initiatives. The parent-founder of the foundation evidenced the tension between egalitarian localism and elitist cosmopolitanism:

> There are kids here that have never left the valley; the parents have never left the valley. And there are kids that go to Europe every summer, and sort of everything in between ... There's a lot more in part because of the technological advances and the ease of ... working from here, in a remote way ... People who've lived here for generations [are] ... really open to new ideas and new things happening for their kids.

While well intended, the comment belies an ignorance of the shifting demographics in the valley or, perhaps, of their effects. Despite evidence to the contrary, informants seemed ready to characterize the valley as mostly made up of wealthy telecommuters. The shifting demographics brought a shift of values defined in terms of education and reproduced through influence and wealth. Though the research team focused on place-based mathematics in the middle school, interviews with the high school mathematics teacher indicated that there was little time for such place-connected mathematics in the high school curriculum. Citing the "high expectations" of parents and the need to do well on state tests and university entrance exams, the teacher said that he had dabbled with the idea but decided that it could not be a priority for him.

Perhaps the strongest elitist language came from the Hamilton Collaborative where administrators mandated place-based approaches that were losing traction and generating resentment from the teachers. One high school mathematics teacher cited global competitiveness in his defense against place-based approaches:

> You have to look at the competition that we have, global competition. The United States is falling so far behind in the mathematics, and it's not because Japan is out there doing community-based. They're in the classroom, and they're doing regular … Every foreign exchange student that I have is two or three years faster than our kids. And, so, for us to slow down or take more time to apply it, I'm not sure is going to keep us competing at the world level.

This subtheme highlighted the sharp sorting of students at the sites into the "university-bound elite" and the "rest." It is tied to national and global, rather than local interests and sustainability. Phase Two Item A11 saw nearly three-quarters of respondents agreeing that "Twenty-first century jobs will demand high-level mathematical skills of all students." Teachers' perceptions of the importance of global competitiveness and the role of mathematics to national economic and political prominence are translated into action by sorting students and selecting curriculum and pedagogy. In the case of the university-bound, results suggest that teachers felt that math should be abstract (universal?) and therefore connections to community and place would be concrete and local, hence taboo. This mirrors findings of Carr and Kefalas (2009) who paint a detailed picture of how one small town engaged in sorting students into achievers, stayers, and leavers by the community and through the schools. A high school teacher in Hamilton's (this study) most affluent high school expressed this sorting in even stronger terms:

> I teach upper math, so I don't know. I teach trig and calculus. So I doubt if we do any calculus in local places around here. I don't know how we have time to do that. I am not sure that the upper math classes—maybe a lower income average are going to need that—but not my upper math class … Because my kids that I teach now are interested 'in the math,' I don't have to make it flowery, I don't have to make it enjoyable, I don't have to make it fun, they just want to know, what's the math, the theory behind it.

The Hamilton Collaborative's commitment to prepare students for college and jobs elsewhere meant few resources dedicated to place-based approaches except in the lower grades and the business and vocational classes. The sentiment from upper-level mathematics teachers in the collaborative followed this elitist-cosmopolitan perspective without exception.

In fact, at no point in any of the sites did research team members witness place-based approaches involving mathematics beyond the level of basic geometry or basic trigonometry. Even at Gladbrook and Edgewater, where all students were exposed to some place-based approaches, the upper level mathematics courses did not engage these approaches. Several reasons are posited for this.

First, elitist-cosmopolitan ideologies trump egalitarian-localist ones in upper-level mathematics and thereby define what is important in mathematics instruction. It is hard to find logical justification for this given that a wide range of majors and careers require college algebra or statistics for application in their major field. Moreover, some jobs in rural areas demand university degrees and this level of mathematics. Either the requirements are being used as a filter or the reality of higher-level mathematics use in a wide range of careers and locales is underreported. Wong points out that the perception of "mathematics for all" is usually accompanied by "those with special talent" (Gates and Vistro-Yu 2003). Further, memorable experiences such as those that engage project-based or place-based learning often serve as important mental anchors for new knowledge (Conley 2010) regardless of ability, aspiration, or grade level.

Second, the Platonic view of abstract mathematics as "pure" and therefore more valuable is widely held by secondary and post-secondary instructors of mathematics, including teachers and those who prepare them (Hersh 1997). Anchoring instruction in place-connected activities could be devalued simply because the dominant philosophical paradigm in mathematics is a Platonic one (Ernest 1997). High-school teachers are prepared as discipline specialists, so there may also be less fluency with interdisciplinary pedagogical approaches (hence place-based approaches). Elementary and middle grades teachers are often licensed to teach in multiple areas, which may afford thinking in more interdisciplinary or cross-curricular terms, explaining the team's findings of no place-based approaches beyond basic algebra and geometry.

Third, it may be that *school mathematics* in particular, and at that level, is simply too difficult to connect to (or from) problems arising in (and from) place and community. In other words, the content itself may present the greatest obstacle to employing place-rooted approaches. If this is the case, then mathematics educators must commit to reexamining school mathematics, open to Abreu et al.'s admonition that, "Mathematics education curricula do not contain all the mathematics knowledge that exists. They contain the 'knowledge that should be taught' and they exclude other forms of mathematics" (de Abreu et al. 2002a, p. 12). Critical mathematics education perspectives may offer a starting point for structuring that inquiry and for reinhabiting the mathematics curriculum in schools.

Discussion

Data suggest that while place-based approaches are possible and might even improve mathematics instruction at some levels while aiding rural sustainability, a number of obstacles make the task daunting. Social class divisions and especially a conflict between egalitarian localist and elitist cosmopolitan perspectives were symptomatic of disagreement regarding what should be valued in mathematics education. All saw mathematics as important though many viewed abstract mathematics as important for the university-bound elite (and more valuable generally) and applied (or "integrated") mathematics as important for vocational-agriculture students (and less valuable generally). In some instances, class was defined according to abstract or concrete mathematics and this structured the set of opportunities available to the students. In other cases, socio-economic and political discourses shaped the purported values driving mathematics instruction and education more generally.

The research team began the study with the expectation that many sites around the United States would be engaged in meaningful efforts to connect mathematics instruction to local places and communities across the span of grades and mathematical topics. Moreover, given the team's perception that math teachers believed unquestioningly that "math is everywhere," later confirmed by 99.1 % agreement to the statement in Phase Two, it was therefore surprising that, even among the set of sites nominated by a trusted network of nominators knowledgeable about rural schooling, so few examples of well-established and meaningful place-based approaches existed. Among our sites, Edgewater represented the most advanced site regarding place-based education and it presented a number of salient issues that attend the use of place-based approaches. All of the sites demonstrated the importance of the community's belief(s) in the purpose of education generally and the purpose of mathematics education in particular. These beliefs seemed to be shaped primarily through global and national messages related to competitiveness and the value of professional jobs located primarily in urban centers. The data showed that communities were important arenas for negotiating those messages and shaping education. As such, this study should not be read as forcing a choice between (a) preparing students for and according to local values, or (b) global ones. There is no reason that education cannot attend to both. The data and the literature that preceded and informed this study show that, except in a few isolated cases, local values and perspectives are not attended to in mathematics instruction. The Phase Two survey results found rhetorical support for making local connections to instruction and the motivating power of such connections, but the importance of the Phase One case studies underscores the reality that rhetoric and action do not always coincide.

Guida de Abreu remarked that, "understanding of how particular social groups learn, use and transmit knowledge requires consideration of the link between knowledge and values" (de Abreu 1993, quoted in de Abreu et al. 2002b, p. 124). The present study demonstrates the importance of how social groups establish and sustain beliefs in the purpose of education generally and in mathematics in particular according to contests in values about the distribution and shape of educational resources and approaches. This study further points out that, despite some promise in the idea of place-based approaches to mathematics instruction, Skovsmose may be right that "Elitism might be a functional part of mathematics education" (Skovsmose 2005, p. 164) in its sorting mechanisms. The data presented here show that it functions variously at different levels of mathematics and different levels of connection to local versus global values. Moreover, mathematics sorts not just people, but places. In upper-level mathematics, in particular, mathematics inhabits *nowhere* rather than *'round here'* As Gruenewald (2003) suggests, reinhabitation may be in order. The results of this study suggest that doing so might promote more egalitarian, community-sustaining, and relevant education.

Appendices

Appendix A

Factor loadings A–D for exploratory factor analysis with polychoric factor analyses reported in parentheses

	Item	Factor 1	Factor 2	Factor 3	Factor 4
A1.	Connecting mathematics to students' everyday lives can help prepare them for study at 4-year colleges.	**.782 (.69)**	−.019	−.014	−.178
A2.	Connecting mathematics to students' everyday lives can help prepare them for study at 2-year colleges	**.769 (.82)**	−.072	−.074	−.112
A3.	Using students' daily life experiences as part of instruction improves the learning of lower-level mathematics.	**.698 (.84)**	−.020	−.037	.109
A4.	Using students' daily life experiences as part of instruction improves the learning of higher-level mathematics.	**.652 (.78)**	−.038	.021	−.103
A5.	Connecting mathematics to students' everyday lives can help prepare them for coursework in a 4-year university or college.	**.639 (.69)**	−.044	.096	−.096
A6.	Students are more motivated to learn mathematics when they see it as being relevant to their daily lives.	**.581 (.74)**	.007	−.028	.018
A7.	Connecting mathematics to the local community will improve instruction for all students.	**.558 (.63)**	.022	.159	.078
A8.	My remedial students like it when mathematics instruction is tied to their daily lives.	**.546 (.63)**	.089	.077	.259
A9.	My advanced students like it when mathematics instruction is tied to their daily lives.	**.517 (.59)**	.086	.022	.091
A10.	The mathematics I teach is directly applicable to my students' everyday lives.	**.366 (.37)**	.074	.189	−.164
A11.	Twenty-first century jobs will demand high level mathematical skills of all students.	**.321 (.35)**	.275	.068	.034
B1.	All students can do higher-level mathematics.	.057	**.725 (.76)**	−.007	−.093
B2.	Some students will never be ready for college-level mathematics.	−.014	**.665 (.75)**	−.134	−.167
B3.	Schools should prepare all students to go to college.	.001	**.621 (.68)**	.188	.135
C1.	My colleagues support efforts to connect mathematics instruction to the local community.	.075	−.091	**.710 (.77)**	.031

(continued)

	Item	Factor 1	Factor 2	Factor 3	Factor 4
C2.	My administration supports community-connected mathematics instruction.	.056	.048	**.683 (.722)**	.107
C3.	Parents of my students support community-connected mathematics instruction.	.137	.007	**.675 (.698)**	.094
C4.	My community offers opportunities for applying mathematics topics.	−.061	.066	**.539 (.701)**	−.121
C5.	Students who stay in the area will apply their math skills to address local needs.	.157	.031	**.418 (.54)**	−.088
C6.	My students are accustomed to the sorts of projects that connect mathematics to everyday life.	.199	−.074	**.389 (.498)**	−.141
C7.	Students with strong math skills are likely to settle in this area.	−.085	.154	**.258 (.34)**	−.066
D1.	It's easier to connect lower-level mathematics content to daily life than higher-level mathematics content.	.082	.023	.079	**.530 (.61)**
D2.	There are barriers that make it difficult to connect mathematics to students' everyday lives.	−.037	.033	−.147	**.463 (.59)**
D3.	The mathematics of daily life matters more to the future of some of my students than it does to others.	−.007	−.067	.024	**.416 (.55)**
D4.	My typical student will need only basic mathematics in his/her daily life.	−.169	.007	−.066	**.310 (.366)**
D5.	Activities that connect math to the local community take up instructional time needed to address standards.	.009	.008	−.099	**.305 (.32)**
D6.	Algebra I (or the equivalent) is where you start to notice which students will be ready for college.	−.003	−.103	.099	**.303 (.38)**
D7.	Some students receive better academic opportunities because of his/her family's local prominence.	.001	−.032	−.085	**.132 (.22)**

Note: Extraction method: Maximum likelihood. Rotation method: Oblimin with Kaiser normalization
a. Rotation converged in seven iterations

Appendix B

Factor correlation matrix

Factor	A	B	C	D
A	1.000	.120	.463	−.055
B	.120	1.000	.267	−.245
C	.463	.267	1.000	−.201
D	−.055	−.245	−.201	1.000

References

Bauman, Z. (1998). *Globalization: The human consequences.* New York: Columbia.
Berg, P., & Dasmann, R. (1990). Reinhabiting California. In C. Van Andruss & J. Plant (Eds.), *Home! A bioregional reader.* Philadelphia: New Society.
Berry, W. (1977). *The unsettling of America: Culture & agriculture.* San Francisco: Sierra Club.
Carr, P., & Kefalas, M. (2009). *Hollowing out the middle: The rural brain drain and what it means for America.* Boston: Beacon.
Castells, M. (2004). *The power of identity, the information age: Economy, society and culture* (2nd ed., Vol. II). Cambridge, MA: Blackwell.
Conley, D. (2010). *College and career ready.* San Francisco: Jossey-Bass.
Corbett, M. (2007). *Learning to leave: The irony of schooling in a coastal community.* Halifax: Fernwood.
de Abreu, G. (1993). *The relationship between home and school mathematics in a farming community in rural Brazil.* Unpublished PhD thesis, University of Cambridge, UK.
de Abreu, G., Bishop, A., & Presmeg, N. (2002a). Mathematics learners in transition. In G. de Abreu, A. Bishop, & N. Presmeg (Eds.), *Transitions between contexts of mathematical practices* (pp. 7–22). Dordrecht: Kluwer.
de Abreu, G., Cline, T., & Shamsi, T. (2002b). Exploring ways parents participate in their children's school mathematical learning: Cases studies in multiethnic primary schools. In G. de Abreu, A. Bishop, & N. Presmeg (Eds.), *Transitions between contexts of mathematical practices* (pp. 123–148). Dordrecht: Kluwer.
DeYoung, A. J. (1987). The status of American rural education research: An integrated review and commentary. *Review of Educational Research, 57*(2), 123–148.
DeYoung, A. J. (Ed.). (1991). *Rural education: Issues and practice.* New York: Garland.
Ernest, P. (1997). *Social constructivism as a philosophy of mathematics.* Albany: SUNY Press.
Fan, X., & Chen, M. (1999). Academic achievement of rural school students: A multi-year comparison with their peers in suburban and urban schools. *Journal of Research in Rural Education, 15*(1), 31–46.
Gates, P., & Vistro-Yu, C. (2003). Is mathematics for all? In A. Bishop, M. Clements, C. Keitel, J. Kilpatrick, & F. Leung (Eds.), *Second international handbook of mathematics education* (pp. 31–74). Dordrecht: Kluwer.
Greeno, J. (1994). Gibson's affordances. *Psychological Review, 101*(2), 336–342.
Gruenewald, D. (2003). The best of both worlds: A critical pedagogy of place. *Educational Researcher, 32*(4), 3–12.
Gruenewald, D. (2006). Resistance, reinhabitation, and regime change. *Journal of Research in Rural Education, 21*(9), 1–7.
Harmon, H. L. (2003). Rural education. In J. W. Guthrie (Ed.), *Encyclopedia of education* (2nd ed., pp. 2083–2090). New York: Macmillan.
Harmon, H., Henderson, S., & Royster, W. (2003). A research agenda for improving science and mathematics education in rural schools. *Journal of Research in Rural Education, 18*(1), 52–58.
Hersh, R. (1997). *What is mathematics, really?* New York: Oxford University Press.
Howley, C., & Gunn, E. (2003). Research about mathematics achievement in the rural circumstance. *Journal of Research in Rural Education, 18*(2), 86–95.
Howley, C., & Howley, A. (2010). Poverty and school achievement in rural communities: A social-class interpretation. In K. Schafft & A. Jackson (Eds.), *Rural education for the twenty-first century: Identity, place, and community in a globalizing world* (pp. 34–50). University Park: Pennsylvania State University.
Howley, A., Howley, C., Klein, R., Belcher, J., Tusay, M., Clonch, S., Miyafusa, S., Foley, G., Pendarvis, E., Perko, H., Howley, M., & Jimerson, L. (2010). *Community and place in mathematics education in selected rural schools.* Athens: Appalachian Collaborative Center for Learning, Instruction, and Assessment in Mathematics, Ohio University.

Howley, A., Showalter, D., Howley, M., Howley, C., Klein, R., & Johnson, J. (2011). Challenges for place-based mathematics pedagogy in rural schools and communities in the United States. *Children, Youth and Environments, 21*(1), 101–127.

International Fund for Agricultural Development (IFAD). (2010). *Rural poverty report 2011*. Quintily: IFAD.

Klein, R. (2007). Educating in place: Mathematics and technology. *Philosophical Studies in Education, 38*, 119–130.

Klein, R. (2008). Forks in the (back) road: Obstacles for place-based strategies. *Rural Mathematics Educator, 7*(1), 7–14.

Klein, R., & Johnson, J. (2010). On the use of locale in understanding the mathematics achievement gap. In P. Brosnan, D. Erchick, & L. Flevares (Eds.), *Proceedings of the 32nd annual meeting of the North American chapter of the international group for the psychology of mathematics education* (pp. 489–496). Columbus: The Ohio State University.

Klein, R., Hitchcock, J., & Johnson, J. (2013). *Rural perspectives on community and place connections in math instruction: A survey* (Working Paper No. 44). Appalachian Collaborative Center for Learning, Assessment, and Instruction in Mathematics (ACCLAIM). https://sites.google.com/site/acclaimruralmath/Home

Kunstler, J. (1993). *The geography of nowhere: The rise and decline of America's man-made landscape*. New York: Simon and Schuster.

Larsen, E. (1993). *A survey of the current status of rural education research (1986–1993)* (ERIC Document Reproduction Services No. ED 366 482).

Lee, J., & McIntyre, W. (2000). Interstate variation in the achievement of rural and nonrural students. *Journal of Research in Rural Education, 16*(3), 168–181.

Ma, L. (1999). *Knowing and teaching elementary mathematics: Teachers' understanding of fundamental mathematics in China and the United States*. New York: Routledge.

Miles, M. B., & Huberman, A. M. (1994). *Qualitative data analysis* (2nd ed.). London: Sage.

Sherman, J. (2009). *Those who work, those who don't: Poverty, morality, and family in rural America*. Minneapolis: University of Minnesota Press.

Silver, E. (2003). Attention deficit disorder? *Journal for Research in Mathematics Education, 34*(1), 2–3.

Skovsmose, O. (2005). *Travelling through education: Uncertainty, mathematics, responsibility*. Rotterdam: Sense.

Sobel, D. (2004). *Place-based education*. Great Barrington: Orion Society.

Theobald, P. (1997). *Teaching the commons: Place, pride, and the renewal of community*. Boulder: Westview.

Theobald, P. (2009). *Education now: How rethinking America's past can change its future*. Boulder: Paradigm.

U.S. Census Bureau. (2000). *State & county quickfacts*. http://quickfacts.census.gov

Willis, G. B., Royston, P., & Bercini, D. (1991). The use of verbal report methods in the development and testing of survey questionnaires. *Applied Cognitive Psychology, 5*, 251–267.

Establishing Mathematics Classroom Culture: Concealing and Revealing the Rules of the Game

Christine Knipping, David Reid, and Hauke Straehler-Pohl

Abstract In this chapter we will propose several mechanisms in classroom interactions that give rise to disparity in learning opportunities. We will use terminology derived from the work of Basil Bernstein to describe these mechanisms in four episodes of classroom discourse. We conclude that efforts to enable "less able" students by incorporating elements of everyday "horizontal" discourse into school mathematics, which is oriented towards "vertical" academic discourse, can instead deny these students learning opportunities that are available to more privileged peers.

Introduction

In the first weeks of a school year, teachers and students establish the basic rules for classroom interactions in that classroom, for those students, in that year. These rules can be more or less flexible, more or less tacit, and more or less negotiable, but in every classroom there are rules that govern what is appropriate teacher behaviour, student behaviour, and mathematical behaviour. These rules taken together constitute the culture of the classroom. The rules apply differently to each person in the room. Most clearly the rules for the teacher are different from the rules for the students. For example, the teacher has much more influence in the establishing of the rules. Individual students also are assigned specific roles through the interactions in the first weeks. Among these are the roles of being "good at math" or "bad at math". Because these roles are associated with different rules, opportunities, and status, the assigning of these roles amounts to a stratification of the students according to perceived ability in mathematics, and a disparity in learning opportunities.

C. Knipping (✉) • D. Reid
Universität Bremen, Bremen, Germany
e-mail: knipping@math.uni-bremen.de; dreid@math.uni-bremen.de

H. Straehler-Pohl
Freie Universität Berlin, Berlin, Germany
e-mail: h.straehler-pohl@fu-berlin.de

In this chapter we will make use of terminology derived from the work of Basil Bernstein (2000) to describe several mechanisms in classroom interactions that give rise to stratification of students and hence to disparity in learning opportunities. We will discuss how these mechanisms can arise out of efforts to enable "less able" students by incorporating elements of everyday discourse into school mathematics, but which instead deny these students learning opportunities that are available to more privileged peers. In other words, attempts to establish supportive classroom culture for all students instead result in a classroom culture that, like the wider culture outside the school, assigns individuals specific roles that come with different opportunities and status.

Mathematics Classroom Culture

Mathematics classroom culture became a topic of discussion in the 1990s, indicating a change of perspective in education generally and in mathematics education more specifically (see for example Cobb and Yackel 1998; Nickson 1994; Seeger et al. 1998). Researchers coming from socioconstructivist and sociocultural perspectives considered recognising and acknowledging the social enterprise of teaching and learning mathematics to be both essential and under-researched. While the socioconstructivists emphasised the crucial function of human interaction and cooperation and the idea of "colearning" (Bateson 1975), the sociocultural perspective called for consideration of the cultural-historical dimension, relying on Vygotsky's work and thought.

As Lerman (2000) points out, the inclusion of the social in mathematics education progressed from research on interactions of individuals in social contexts to research on the social contexts themselves, including classroom cultures. Valero (2004) adds consideration of the sociology of mathematics to this mix, so that the social origins of mathematics are recognised as significant for what is going on in classrooms. Just as it is important to consider the origins of mathematics in examining classroom cultures, we feel it is also important to consider the origins of the students and the developmental trajectories that are made available for them through mathematics education. Who gets access to what? And with which outcomes? Addressing these questions calls for using a sociological framework. This implies approaching classroom culture not just as a "culture of the classroom", but to incorporate how this culture is embedded within a realm of culture that exceeds the walls of the classroom and includes broader social and political discourses that shape the school system. That is, in looking at the social process that produce and reproduce classroom cultures we consider also the social context outside that classroom. In order to do so, we will compare concrete instances of classroom culture in Germany, where the secondary school system is traditionally oriented towards selection and allocation of students for different segments of the labour-market, and Canada, where the secondary school system is oriented towards the ideal of inclusion and diversity.

Paul Dowling's (1998) seminal work "The Sociology of Mathematics Education" suggests that, in schools, different forms of mathematics are made available to different social groups of students. Analysing common British textbooks for different ability-groups, he concludes that an abstract form of mathematics is imbued with status, while a contextualised form of mathematics is subordinated. He reveals how these two different forms of mathematics are connected and oriented towards two different forms of labour, intellectual and manual, and how a stratification and hierarchisation of students is socially constructed along these lines.

These findings have been reproduced by several studies that investigated classroom interactions instead of textbooks and hence showed how this stratification is manifested in what we have referred to above as classroom culture (see Atweh et al. 1998; Hoadley 2007; Straehler-Pohl et al. 2014). All these studies compare mathematics classrooms in different—but in each case relatively homogeneous—contexts that have already been stratified socio-economically. They suggest that the stratification of students that precedes the teaching and learning of mathematics—be it in the form of ability-streaming, single-sex schooling, or geographical segregation—is reflected in the micro-interactions of the classroom culture and hence all these studies identify a tendency towards the reproduction of the given social division of labour, including its stratifications and hierarchies.

Such findings are not surprising in a selective school system that is aligned to allocating students to segments of the labour-market. However, can such mechanisms also be found in an inclusive school system, where significant efforts are made to provide the same form of mathematics to everyone?

Lubienski (2000), who was involved in the reform of NCTM standards at the end of the twentieth century, describes how efforts to improve the teaching of problem solving in ways that were expected to help students with lower socio-economic status (SES) in particular have exactly the opposite effect. She notes that the initial assumption was that "instruction centered around open, contextualized problems might seem particularly promising for lower SES students" (p. 456), because research has shown that such students have less exposure to open problems and their "families tend to be more oriented towards contextualized language" (p. 456). In her research she found that a focus on teaching through open, contextualised problems had the potential to "improve both lower SES and higher SES students' understanding of mathematics", however this potential went along with a tendency of "increasing the gap in their mathematics performance" (p. 478). Using Bernstein's sociological theory to reanalyse her data, she concludes:

> Lower SES students seemed to be more fearful of saying or believing the wrong thing; they desired more specific direction from the teacher and texts. Higher SES students were more comfortable with the open pedagogy, feeling confident to make sense of ideas being debated in discussions. Higher SES students moved more often from a focus on specific contextualized situations towards the intended generalizable, mathematical principles. (Lubienski 2004, p. 119)

Lubienski identifies her own classroom culture as characterised by an *invisible pedagogy*. Despite her own efforts as a teacher to empower students from lower socio-economic backgrounds in particular, the limited visibility of her mathematical intentions—that is, a lack of clarity about the desired form of mathematical activity—further marginalised the already underprivileged students.

As mathematics serves as a gatekeeper for access to university studies leading to high status professions (Martin et al. 2010; Stinson 2004), we see understanding the mechanisms that are involved in the production of differential access to high status mathematical knowledge as an issue of social justice. In our research, we study how this stratification occurs within the first weeks of school in contexts where the students are together for the first time. Specifically, we observe interactions in mathematics classrooms at the beginning of secondary schooling, and the process of establishing rules and roles in those classrooms. We attempt to analyse those interactions in order to identify mechanisms that give rise to disparity in learning opportunities. In this chapter we will propose several mechanisms, and illustrate them through two episodes of classroom activity.

We have found Basil Bernstein's (2000) concepts of *classification*, *framing*, and *implicit pedagogy* as especially useful in describing these mechanisms in classroom discourse. In our analysis of the episodes we will both describe and apply these concepts. Bernstein's work will also be used to deepen our discussion of the episodes and to show how the different mechanisms that we have found in the different contexts are embedded in the broader socio-cultural and socio-political context, in which schooling takes place.

The Episodes

The two episodes we will describe and compare below come from a research programme focussing on the emergence of disparity and the stratification of achievement in mathematics classrooms. This research is collaborative and comparative, involving researchers from Canada, Germany and Sweden. The classroom interactions were observed at the beginning of the first year of secondary school, in classrooms chosen for their variety. The episodes used here come from an urban selective, upper stream school (*Gymnasium*) in Germany, and a nominally inclusive rural public school in Canada. Links to further analysis of data from these classrooms and other aspects of the research project can be found at http://www.math.uni-bremen.de/~knipping/sd.

Our descriptions below will focus on how aspects of the classroom discourse contributed to the emergence of disparity in learning opportunities in the classrooms. We are interested in exposing the problems inherent in certain teaching practices, not in teacher knowledge or competencies. Therefore, we would like to emphasise that we do not consider the teachers involved to be negligent. In fact, as we have discussed elsewhere (Knipping et al. 2011), their concern for the educational success of the students and their embeddedness in a broader socio-cultural context may account for some of the choices they make. The actions of these teachers may contribute to the emergence of disparities in their classrooms and reproduce socio-cultural differences from outside the classroom, but this is not the teachers' intent. We maintain that teachers do not do this with the intention of limiting learning for some, but rather in the hope of promoting all their students' learning (cf. Morgan 2012, p. 188). As Bourne points out, teachers are produced by the institutions and practices they are embedded in.

> It is important to be clear that I am not here ascribing conscious intent to teachers as agents in some vast conspiracy to maintain the status quo insofar as power relations are concerned. ... As Sheridan (1980) explains with respect to Foucault's theories of power, through the numerous institutions and practices of socialization 'the forces of the body are trained and developed with a view to making them productive' (p. 219). Power produces, rather than simply reproduces. (Bourne 1992, p. 233)

In the following, we discuss two classroom episodes in which we identified mechanisms that provoke a stratification of students. In our conclusion, we will discuss the differences in relation to the broader institutional cultures in which the episodes are embedded. The first episode we discuss below, from Germany, depicts a well-known didactic situation: the Race to Twenty. Later we will make use of a didactical description of the situation (Brousseau 1978, 1997) in order to reveal significant aspects of the actual classroom episode. We will then report and analyse a second classroom episode from Canada, in order to make cross-cultural and cross-contextual comparisons. Throughout the discussion of the episodes, we will introduce our analytical tools step by step in order to gradually deepen and refine our analytical gaze.

Episode 1: The Race to Twenty and Classification

This episode is from an inner-city 5th grade (age 10 years) mathematics classroom in Germany (see Gellert 2010, 2012; Gellert and Hümmer 2008 for other analyses). It is the very first lesson after the summer holidays and it is the students' first day at a *Gymnasium* (upper stream secondary school). Mr. Black is a specialist mathematics teacher and his 5th graders come from different primary schools. The teacher and the students do not know each other and in fact Mr. Black was assigned to the class at the last minute and so had no opportunity to research the students' backgrounds. However, when he addresses the class, he talks about the "infamous class 5d, that I have already heard about". Rubbing his hands, Mr. Black starts the lesson by announcing that he wants to test if the students know how to count to 20, as this is supposedly a "basic condition" (his words) for being accepted at that school. Mr. Black asks "Who dares to count to twenty?" Some students giggle. Transcript 1 begins at this point.

3.1	Mr. Black:	Nicole, okay. So you think you can count till 20. Then I would like
3.2		to hear that.
4>	Nicole:	Okay, one two thr-
5.1>	Mr. Black:	-Two, oh sorry, I forgot to say that we alternate,
5.2		okay?
6	Nicole:	Okay.
7	Mr. Black:	Yes? Do we start again?
8	Nicole:	Yes. One.
9	Mr. Black:	Two.
10	Nicole:	Three.
11.1	Mr. Black:	Five, [*short silence*] oops, I also forgot another thing. [*students'*
11.2		*laughter*] You are allowed to skip one number. If you say three, then I
11.3		can skip four and directly say five.
12	Nicole:	Okay.
13	Mr. Black:	Uhm, do we start again?
14	Nicole:	Yeah, one.
15	Mr. Black:	Two.

Transcript 1: Is it counting or is it a game?

Both continue 'counting' according to the teacher's rules. In the end Mr. Black states "20" and Nicole loses. In a joking way he declares "Nicole, you still have to try a little bit, you did not make it to 20". While her peers giggle, Nicole appears a bit puzzled: "okaaay".

Classification and Recognition Rules

The students are in a mathematics classroom and the first task set by the teacher is to count to twenty. In this kind of situation, students are probably prepared to engage in a mathematical activity, but they most definitely do not expect to be asked to count to twenty, as that is something that they know how to do since first grade. However, it turns out pretty quickly that the apparent "test" of the ability to count to twenty was meant as a joke, while *actually* what students were supposed to do is to compete in a game with the teacher, which involves saying a sequence of numbers leading to the goal of saying "twenty". The nature and the context of the activity are at first unclear: Are the students meant to be counting, which is clearly a mathematical activity but inappropriate for their grade level, or are they playing a game, which may be appropriate on the first day of school but not necessarily mathematical? The context of Mr. Black's class is a fifth grade mathematics class, but neither counting to twenty nor playing a game are normally part of the grade five mathematics context.

As *classification*, Bernstein (2000) refers to the principle regulating the boundaries between contexts: "Classification [refers] to the relations between categories, these relations being given by their degree of insulation from each other" (p. 99). In the context of schooling, classification can refer to the categorising of areas of knowledge within the curriculum or what is taught (p. 6). Here we are specifically interested in the classification of discourses, including everyday discourse, the discourse of school mathematics, and the discourse of academic mathematics.

Strong classification means that strong boundaries between topics and subject areas are maintained or that strong boundaries between a subject and the everyday are maintained. For example, a traditional mathematics curriculum has strong classification as few connections are made to other disciplines or to everyday contexts. A project-based mathematics curriculum, on the other hand, has weaker classification, as mathematics teaching is integrated with other subject areas inside or outside school. This means that everyday knowledge and subject knowledge are less insulated from each other.

In Mr. Black's classroom the classification seems to be weak. While there is some mathematics involved (dealing with numbers), this mathematics is blended with an everyday context (playing a game). After the initial confusion, however, the students do become aware that they are playing a game. This indicates that they have what Bernstein calls a *recognition rule* that allows them to identify the seemingly weak classification and participate in the game context.

Subsequently, eight other students play the game but lose against the teacher whilst an atmosphere of students-against-the-teacher competition develops. Finally, the tenth student, Hannes, manages to win. Hannes reads his numbers from his notebook.

226.1	Mr. Black:	Yeah, well done. [*students applaud*] Did you just write this up
226.2		or did you bring it to the lesson? Did you know that today you would–
227	Hannes:	I observed the numbers you always take.
228.1	Mr. Black:	Uhm. You have recorded it, yeah. [*silence*] Did you [*directing*
228.2		*his voice to the class*] notice, or, what was his trick now?
229	Torsten:	Yes, your trick.
230	Mr. Black:	But what is exactly the trick?
231	Cecilia:	You have to say certain numbers

Transcript 2: Explaining the trick

Hannes wins the game and his peers applaud, which is appropriate in the context of game playing. Mr. Black reacts rather differently. He asks whether Hannes has written the numbers in his notebook in advance. This seems to be an odd assumption. It turns out that Hannes did not bring his notes from home. Instead he observed what Mr. Black was doing to win the game and copied it. This is a good strategy in unfamiliar game playing contexts, but Mr. Black calls it a "trick" which is another indication that the context might not be game playing after all. In fact, the context derives from a well-known didactical situation, the Race to Twenty. But Mr. Black's students do not know this, and so continue to struggle to recognise the context, due to the weak classification.

Dual Recontextualisation

In order to be able to describe this ambiguity of the activity, being at the same time oriented towards game playing and mathematics, we have to elaborate our concept of classification. In order to do this, we have to take a step back and make some general reflections on what school mathematics actually is. According to Bernstein, any school subject comes into being through a process of recontextualisation.

> As a principle, pedagogic discourse is the process of moving a practice from its original site, where it is effective in one sense, to the pedagogic site where it is used for other reasons; what [Bernstein] calls the principle of recontextualization. In relation to work practices he offers the example of carpentry which was transformed into woodwork (in UK schools), and now forms an element of design and technology. School woodwork is not carpentry as it is inevitably separated from all the social elements, needs, goals, and so on, which are part of the work practice of carpentry and cannot be part of the school practice of woodwork. Similarly, school physics is not physics, and school mathematics is not mathematics. (Lerman 2001, p. 100)

In this process "other discourses are appropriated and brought into a special relationship with each other, for the purpose of their selective transmission and acquisition" (Bernstein 2000, p. 32). For the purpose of making the highly abstract meanings of academic mathematics learnable and teachable for novices, school mathematics draws heavily on everyday discourses. Thus, school mathematics is not only to be understood as a recontextualisation of academic mathematical knowledge and discourse, but further as a recontextualisation of everyday knowledge and discourse. Jablonka and Gellert (2010) have therefore coined the term *dual recontextualisation*.

The recontextualisation that occurs from academic mathematics can be described as a *transposition didactique* (Chevallard 1985), in which mathematics is transformed in order to adapt it to the school context. In particular, school mathematics must be both teachable and testable, which means that some kinds of mathematical knowledge that are emphasised in academic mathematics are not emphasised in school mathematics, and vice versa. In school mathematics, theoretical concepts and techniques may be related differently than they had been originally. The dominant principle for structuring meanings in school mathematics is the effective transmission and acquisition. School mathematics is structured in order to make learning and teaching effective and *not* in order to make mathematics itself (as a knowledge-producing science) effective.

The second dimension of recontextualisation occurs from everyday discourses. The game-playing in Mr. Black's classroom can be seen as such a recontextualisation of an everyday discourse, just like the references to everyday contexts in typical school mathematics word-problems. However, again the dominant principle for structuring meanings is the effective transmission and acquisition *of mathematics.* Thus, considerations that would be privileged in the original cultural segments of everyday discourses *are subordinated* to considerations deriving from the effectiveness of learning and teaching mathematics (e.g., Dowling 1998; Gellert and Jablonka 2009).

From Dual Recontextualisation to Dual Classification

Each recontextualisation implies a re-ordering of meanings that determines what counts as school mathematics and what does not. This means that each recontextualisation gives rise to a classification principle. Because these classification principles are specific to the institutional frame of school mathematics, Bernstein's general principle regulating the boundaries between contexts becomes more specific in these principles, referring to the boundaries of school mathematics.

In order to describe the classification that derives from the recontextualisation of everyday discourses, we will use the concept of:

Classification of content: the insulation between school mathematics and the everyday that results from the degree to which a school mathematical activity makes sense outside of school mathematics, e.g. in the students' everyday-lives (cf. Straehler-Pohl and Gellert 2013).

The nature of strength of classification changes as the frame of reference of this principle is school mathematics. The more sense an activity makes in an everyday context (e.g. shopping, game-playing), the weaker the classification of content; the less sense it makes in an everyday context (adding fractions, calculating modulo 3), the stronger the classification of content.

We will refer to the principle of classification that derives from the recontextualisation of academic mathematics (in the form of a *transposition didactique*) as:

Classification of praxeological organisation: the degree of the insulation of school mathematics that results from the extent to which school mathematical activities develop a specialised internal structure by integrating new knowledge in higher order principles (cf. Straehler-Pohl and Gellert 2013).

The more an activity makes use of the specialised structures and higher order principles of academic mathematics, the stronger the classification of praxeology. An activity that makes less use of such structures and principles, e.g. through remaining on the level of computation routines, is weakly classified in terms of praxeology.

The definition of this second dimension of classification requires us to expand the concept of *praxeology*, which we borrow from the Anthropological Theory of Didactics (Chevallard 1999). The word praxeology binds together the two words *praxis* and *logos* within one concept and thereby stresses the inseparability of a practice and its reflection and discursive legitimation. Chevallard (1999) considers any school mathematical activity as the study of a type of problem. In order to solve a problem, a *technique* must be developed, so that the problem can be solved through practice (praxis). Each technique can at least potentially be described, justified or explained. There are two fundamentally different forms of such justification. Chevallard calls the first form the *know-how* (*savoir-faire*) in which a technique may be simply legitimated by the correctness of a solution (e.g. through situational adequacy or teacher authority). However, each activity can further be considered as a part of a discursive environment, the *know-why* (*savoir*). When activities contain this second dimension, this implies the theorisation of techniques, which Chevallard refers to as *technology* (technique + logos). An activity which systematically develops technologies determines a coherent, explicit, and principled knowledge structure and therefore separates a discourse from other discourses by producing a specialised form of legitimating knowledge. In other words, activities that involve the systematic development of technologies are strongly classified in terms of praxeology. Without systematically developed technologies an activity is weakly classified in terms of praxeology.

From these two classification principles, classification of content and classification of praxeology, we can derive two recognition rules that students have to possess in order to be able to effectively follow classroom discourse. Students have to recognise the legitimate relationship between the everyday and mathematics, that is the extent to which they have to subordinate the former to the latter (recall that subordination is postulated as a necessity deriving from recontextualisation). Furthermore, students have to recognise the extent to which they are required to reflect techniques on the level of technology.

With this differentiation in mind, we want to return to the classroom, where we have already observed a weak classification of content, as the border between game playing and school mathematics is unclear. What can be observed about the classification of praxeology?

Mr Black says, "Cecilia says that you have to say certain numbers. But tell me, which are the important ones?" Cecilia calls out the numbers Hannes used, 2, 5, 8, 11, 14, 17, and Mr. Black writes them on the blackboard. One student shouts out that "seventeen is actually the most important number," but Mr. Black's only response is to remind her that she should raise her hand before commenting. He then asks whether eighteen is important, and Dana replies that eighteen is not important as saying eighteen allows the other player to say twenty. She goes on to say that if the other player says seventeen, then you have already lost, and this is true also for fourteen.

255.1	Mr. Black:	So, fourteen is impo- is 'the most important number,' she says;
255.2		you have already lost at fourteen
256	Students:	Yes, yes
257.1>	Dana:	When you say fifteen then the other says seventeen and
257.2		when you say, uh seve-, no uh sixteen, then the

257.3		other also says seventeen and then you have actually already lost
258.1>	Mr. Black:	So, you must in any event, the fourteen
258.2		is the most important number of all these on the board

The discussion goes on and eleven is suggested (by Lena) as the most important number with an explanation similar to that of Dana. A student proposes "eight" when Jan shouts out that he has something "completely different."

286>	Mr. Black:	You mean something <u>completely</u> different from that
287>	Jan:	Yes, yes you have actually already won when you say <u>two</u>
288	Mr. Black:	Yes, why
289.1	Jan:	Because, when you say two, then the other has to say,
289.2		the other can only say three or four, then
289.3		you can immediately say five, then the other again has to
289.4		say six or seven, then you can say eight,
289.5		the other has to say nine or ten, then it comes to the eleven
289.6		and it goes on like that
290.1	Mr. Black:	Yes, perfect [*students giggle*] but that is not something completely
290.2		different but, okay, the complete explanation.

Transcript 3: What actually matters in the game, finally.

Here it becomes clearer that what is *actually* expected from the students is neither to count to twenty nor to win against the teacher but rather to analyse the game from a mathematical point of view. The teacher's idea of "the complete explanation" seems to be oriented towards developing a *know-why* of the Race to Twenty. There is a strong classification of praxeology, but up to now it has been hidden. It is revealed through Mr. Black's evaluation of the students' answers.

Hannes was the first to develop a technique to win the game, which his peers, who shared his recognition of the context as game playing, applauded. Within the weak classification of content, this recognition of the context made sense. However, Mr. Black's dismissive reaction to Hannes "trick" is indicative of the strong classification of praxeology that becomes clear in the interaction with Dana, Lena and Jan.

We see here two different classification principles at work: one of weak classification concerning content and one of strong classification concerning praxeology. It appears that it is the latter that is the dominant of these two principles. Thus, the students need to not only recognise the weak separation between the everyday activity of game playing and this mathematics class, but also recognise the strong importance of *knowing-why* (derived from academic mathematics) in Mr. Black's classroom. Furthermore, they need to recognise the hierarchy between these two classification principles. Those who recognise the two principles and their relation to each other early have an advantage, both because they recognise the purpose of the activity and have the opportunity to engage in it appropriately, and because they receive positive feedback from the teacher, giving them status.

Framing and Implicit vs. Explicit Pedagogy

As we noted above, the strong classification of praxeology in Mr. Black's lesson is revealed (for the observer) through his evaluation of students' answers. Even though it provides a window into the *structural* characteristics of the discourse (classification), in classroom practice, evaluation is an *interactive* feature. What becomes visible for the observer, whose distance allows a sensitivity for structure, may remain invisible for the participants, whose attention is bound to the flow of ongoing interaction. Bernstein therefore assigns evaluation to *framing*. Framing regulates—in the interactional dynamics—what kinds of communication and action are considered legitimate.

> The principle of classification provides us with the limits of any discourse, whereas framing provides us with the form of realisation of that discourse; … Classification refers to *what*, framing is concerned with *how* meanings are to be put together, the forms by which they are to be made public, and the nature of the social relationships that go with it. (Bernstein 2000, p. 12)

Like classification, framing can be strong or weak.

> Where framing is strong, the transmitter has explicit control over selection, sequence, pacing, criteria … Where framing is weak, the acquirer has more *apparent* control (I want to stress apparent) over the communication. Note that it is possible for framing values—be they strong or weak—to vary with respect to the elements of the practice, so that, for example, you could have weak framing over pacing but strong framing over other aspects of the discourse. (Bernstein 2000, p. 13)

Selection, sequence, pacing and evaluation criteria for knowledge are four instructional aspects of discourse considered by Bernstein when analysing framing. In Mr. Black's lesson the framing of *selection* is strong; Mr. Black selects what is studied. He also controls the order in which topics are studied; the *sequence* is strongly framed. The *pacing* is framed ambiguously. Mr. Black waits for the students' contributions, thus gives them some control of the pacing, but he also moves on to the next topic as soon as one or two students have responded appropriately. The contributions of these students are taken as an indication that the whole class is ready to move on, and the students have no option to return to earlier topics. This means that overall Mr. Black controls the pacing, which becomes very quick as the students can accelerate it (by answering quickly or correctly) but not decelerate it. And Mr. Black controls the evaluative criteria for a legitimate contribution, so framing over this aspect is also strong. However, as we have seen, this does not imply that it is clear to all students exactly what is expected from them. We will discuss this apparent opposition further below.

Additional to the four instructional aspects of framing, framing also includes regulative aspects, namely "expectations about conduct, character and manner" (Bernstein 2000, p. 13). Bernstein calls this the framing of the *social order*. However, Bernstein makes us aware that any instructional moment always includes a certain degree of social regulation.

In Mr. Black's lesson the social order is strongly framed. For example, Mr. Black is quite clear about who should speak when, that the students should introduce themselves the first time they speak, and should raise their hands when they wish to speak. So in general, the framing is strong and it appears to be clear to the students how they are meant to act. Theoretically, under these conditions students should be able to recognise what it means to act appropriately and be able to align their behaviour accordingly. In Bernstein's terminology, students should be set in a position to acquire the *realisation rule* that allows them to act in a legitimate way. Just as recognition rules allow students to correctly read the classification of a context, realisation rules allow students to conform to the framing.

However, just *some* students seem to have acquired such a realisation rule with regard to the evaluation criteria. In spite of the strong framing the evaluation criteria appear to remain hidden for the others. To describe such a situation, where the actual criteria for producing and evaluating a legitimate text are not made evident to all students, Bernstein uses the concept *implicit pedagogy* in contrast to *explicit pedagogy*.

> Explicit or implicit refers to the visibility of the transmitter's intention as to what is to be acquired from the point of view of the acquirer. In the case of explicit pedagogy the intention is highly visible, whereas in the case of implicit pedagogy the intention from the point of view of the acquirer is invisible. (Bernstein 2000, p. 200)

Explicit and visible pedagogy, on the one hand, and implicit and invisible pedagogy (see Lubienski 2004, as quoted above), on the other hand, are used synonymously throughout the literature. When the framing is strong and the pedagogy is explicit, all students are given the opportunity to conform to the requirements. When the pedagogy is implicit, at least some students will not be given this opportunity. Strong framing in a context of an implicit pedagogy becomes observable when a teacher explicitly communicates what she demands from the students, but her evaluation behaviour, however, does not match her explicit communications. This can be the case, for example, when a teacher is not really conscious of her own intentions, or when the sequencing of a lesson requires a dynamic development, where the intentions of the teacher are continuously changing. The *problématique* of this combination is that the teacher sends out the message that students can rely on the criteria that she communicates, while *actually* students have to be aware that they continuously need to look for the teacher's real intention autonomously.

The initial game between Nicole and the teacher (transcript 1) is paradigmatic of an implicit pedagogy combined with strong framing. The teacher first initiates the interaction and in the course of it, he states the criteria *retrospectively*. In such a situation the students have to be aware that they will not be given a full set of criteria from the beginning, but that they will have to be attentive for explications. However, the relatively strong framing over all aspects of discourse suggests that the students can (and must) heed the teacher's words, as he is in control.

The case of Hannes makes us aware that it is still not enough to just meet the criteria. When he pursued a strategy that was aligned to the *hitherto* available criteria, and furthermore appropriate to everyday game playing, Mr. Black deprecatingly labelled it "a trick". In contrast, the contributions of Dana, Lena and Jan were fully acknowledged.

There appears to be a mismatch between the communicated criteria and the teacher's intentions, as Dana, Lena and Jan violate the framing, but still appear to match the teacher's intentions. Instead of answering the teacher's question of why *eighteen* is important, Dana responds by giving the reasons for why it is *not* important and finally describes why *seventeen* is important (she ignores the teacher's framing of sequencing). The same applies for Lena, who instead of providing the requested explanation for the importance of fourteen, explains the importance of the number eleven. Jan is even allowed to take turns without raising his hand and to interrupt Mr. Black (line 287). While Hannes's behaviour is consistent with the framing, it received a less positive evaluation than that of Dana, Lena and Jan, who violated the framing. However, Dana, Lena and Jan answer in ways that are appropriate to the strong classification of praxeology, while Hannes does not. Here it seems that the classification principle is more important than the framing.

To summarise, it appears that there is a discrepancy between what the teacher openly communicates as criteria for a legitimate contribution (framing) and the more subtle structural demands of the discourse (classification). While the former provides a clear frame to act within, this frame can be undermined by structural principles that remain hidden. A valued contribution seems to require an alignment to the teacher's invisible intentions derived from the classification principles (a strong classification of praxeology that dominates a weak classification of content) rather than an alignment to the overt criteria provided by the teacher (strong framing).

This example makes us aware that strong framing may well be a basis for preparing all students to make *a* contribution. However, in order to prepare all students to make *an entirely valued* contribution, it requires an explicit pedagogy, where framing is in harmony with the classification principles. Otherwise, it is not only that *some text is privileged* in the mathematics classroom, but that an unequal access to the principles for the privileging of such text translates into a *privileging of some students* and a *marginalisation of others*. This is particularly the case for an implicit pedagogy in the context of strong framing, as strong framing suggests that students *can rely* on the rules that seem apparent to them, while the actual rules remain implicit. Under these conditions the mathematics classroom will favour those who already possess the recognition rules and further discriminate against those who need the teacher's help to acquire them.

In such a case, the combination of implicit pedagogy and strong framing becomes the mechanism which can mislead students; the knowledge they are meant to acquire is not in harmony with the teacher's interactions.

Classification and Framing in the Race to Twenty: Brousseau

The Race to Twenty is well known in mathematics education and it is interesting to use the concepts of classification and framing to compare what happened in Mr. Black's classroom with the canonical description by Brousseau (1978, 1997). Recall that we have described Mr. Black's lesson as showing weak classification of content but strong classification of praxeology, and showing a generally strong framing which, however, is misleading concerning the structure of knowledge. In contrast, we would claim that Brousseau describes the Race to Twenty in a way that shows strong classification of both content and praxeology, but weaker framing of sequencing and pacing than in Mr. Black's class.

In Brousseau's description, the Race to Twenty is played in several phases. We will briefly describe them before analysing them in terms of classification and framing. In the first phase, the teacher explains the rules and plays the game with one student in front of the whole class, until it is clear that the rules are understood, at which point she has a second child take her place. In the second phase, students play in pairs, keeping an open record of the numbers they have chosen. In the third phase, the class is divided into two teams, which play against each other, discussing each move and their overall strategy within their team. In the fourth phase the focus shifts to identifying propositions that help in winning the game.

In the first phase the teacher introduces the students into how to play the game.

> The teacher talked about the game; she communicated a message which contained the rules of the game so that the students could internalize and apply them. This message did not contain any new words; it is assumed that the children understand the terms and their organization (that is to say, the phases). The statement of the whole set of rules might have been too long for some children. So the communication of the message is accompanied by an action of the child (by making her play). (Brousseau 1997, p. 6)

Here we see strong framing. The goal is to communicate the rules of the game as clearly as possible, and the teacher controls the selection, sequence, pacing and criteria of the knowledge in order to do so. It is very important that the rules for the game are understood for the next phase, and the teacher does not rely only on a verbal communication of the rules, but also plays the game with the child as a way of checking understanding. "At the same time and while she is playing, the teacher comments on the decisions and illustrates the rules of the argument. She talks about these rules by matching them with the circumstances of the situation" (p. 7). Evaluation is directed towards the situation, not towards the student who is playing. Hence, it has solely the function of explication, not of assessment.

In the second phase, the children play the game against each other in pairs. Here the framing of pacing is much weaker. As the children play with their peers and not against the teacher, they themselves can choose how quickly to play. In this phase the classification of content becomes stronger; the game is presented as a mathematical activity, not as way of having fun. However, the classification of praxeology is, for the moment, still relatively weak. Brousseau explicitly allows that the student's strategies in this phase may be far from mathematical, for example, choosing a number because it is lucky, or worked previously (p. 9). Students evaluate their own actions by playing, hence the framing of criteria is weak. However, Brousseau makes clear that the purpose of the phase is to create a shared situation that can be collectively referred to later and not so much the transmission of mathematical criteria.

In the third phase (group playing against group), the group discussion creates a new context, in which the framing is generally weaker. The group members decide among themselves the pacing while making their decisions, its structure (taking turns, all speaking at once), and the criteria for deciding on a move. Brousseau describes the criteria as "feedback" derived from the situation and describes two forms:

> an immediate feedback (at the time of formulation) from the people with whom she has had the discussion, who show her that they do or do not understand her suggestion ... a feedback from the milieu at the time of the next round played, if the formulated, applied strategy is a winning one or not. (p. 11)

The classification of praxeology in this phase becomes stronger. Although strategies may remain non-mathematical, they must be articulated, and so become objects to be theorised rather than remaining implicit. Implementing this obligation of articulating strategies leads to an explication of criteria. Learning to articulate strategies occurs through what Brousseau called the "dialectic of formulation". He expects that children who are not able to articulate their strategies will learn to do so in this context.

> If some of the children whom we are addressing haven't sufficient mastery of this language (perhaps because they are too young), the dialectic of formulation will function for them; they will have to construct an efficient vocabulary (this will therefore be a learning exercise only for those who need it). (p. 12)

The fourth phase is quite different from the first three. Rather than playing the Race to Twenty game, the students are engaged in a "theorem competition" in which the goal is to propose conjectures about the Race to Twenty. Points are awarded when a conjecture is accepted or refuted, but the number of points depends on how exactly the two teams agree to accept the conjecture (without argument, by an empirical test of playing the game, through a mathematical argument). Brousseau's explicit intent is to shift the praxeology to the theoretical. In our terms, the "theorem competition" is focussed on a strong classification of praxeology. The name visibly announces the game as of mathematical and not of leisure nature: this determines a strong classification of content.

Summarising, the Race to Twenty, as intended by Brousseau, is characterised by a gradual progression towards strong classification of both content and praxeology, but relatively weak framing of sequencing and pacing. While criteria are not explicit from the beginning, they become more and more explicit along the gradual progression of classification as the students are given time (weaker framing of sequencing and pacing) to develop an understanding of the criteria. In the literature, a pedagogy as intended by Brousseau that contains elements of explicit and implicit pedagogy can be described as a *mixed pedagogy* (e.g., Lubienski 2004; Morais 2002).

The Race to Twenty: Mr. Black and Brousseau

Let us now compare in more detail the classification and framing in Mr. Black's lesson with Brousseau's description. In Mr. Black's classroom we observe a mechanism in the classroom interaction that gives rise to disparity in learning opportunities. Some students, like Dana, Lena and Jan, are able to engage in the intended mathematical activity because they arrive with the necessary recognition and realisation rules. Others, like Hannes, cannot participate legitimately until they acquire these rules. This mechanism involves the combination of:

- weak classification of content,
- strong classification of praxeology,
- strong framing,
- implicit pedagogy.

As the strong classification of praxeology dominates the weak classification of content, this privileges students who already have the right recognition rules to identify this hierarchy from the beginning. These students can concentrate on theorising about what is going on, instead of putting their efforts into playing against the teacher.

We can now compare Mr. Black's pedagogy with the ideal pedagogy as described by Brousseau:

Mr. Black (interaction)	Brousseau (canonical description)
Weak classification of content	Progression towards strong classification of content
Strong classification of praxeology	Progression towards strong classification of praxeology
Strong framing	Mixed framing
Implicit pedagogy	Mixed pedagogy

The Race to Twenty in Brousseau is dynamic. It appears that the situation is designed to gradually become more and more complex with the students' growing awareness of the situation. In the case of Mr. Black, the pedagogy appears quite stable; classification and framing values do not change throughout the lesson. The criteria for evaluation, however, change continuously. While the aim of Brousseau's Race appears to be the *transmission of criteria*, so that the students can autonomously apply them, the aim of Mr. Black's Race rather appears to be checking who is *already* in possession of the criteria. While it appears that Brousseau systematically conceptualises the Race to Twenty as the means to the goal of learning, in Mr. Black's case it appears more that the Race is an end in itself: Within a very short time, the "good students" are distinguished from the average; they are literally the winners of the Race to Twenty.

Episode 2: T-Tables

We now want to shift our attention to an episode that comes from the classroom of Mr. White, a Canadian Grade 6 teacher in a rural, *inclusive* secondary school. The school is inclusive in the sense that all children from the region attend the same school and are placed in heterogeneous classes. The students are in their first year in secondary school and they have attended several different elementary schools in the area. Some of them have been in the same elementary class, but not many. Others might know each other as they have gone to the same elementary school, but for many their new peers are not familiar. Mr. White is the home room teacher for this class. This means he is teaching them not only mathematics but most other subjects, for example, language arts, science and art. Mr. White is with his home class for more than 70 % of his weekly teaching. Mr. White may have met some of the students at an orientation day and he had the opportunity to examine their records, but his personal "philosophy" is that he prefers not to have "preconceived ideas" of his students. However, he spent the first few classes testing the students on basic arithmetic skills, which might have provided him with some expectations before the first lesson.

In the first mathematics lesson of this school year, Mr. White starts a unit on T-tables. When textbooks are handed out to the students, Mr. White shows a T-table in the textbook to them (see Fig. 1) and asks them to copy this table into their notebooks. They are to try to "fill in the blanks" for homework and Mr. White comments, "If you can fill those in you're well on your way to understanding T-tables. How many of you will be able to fill this in I wonder?" He makes no mention of the context given in the textbook relating Kevin's age to Alice's grade.

This is a "real-world" context found in school mathematics textbooks, relating to an unlikely question (Kevin wondering how old his sister will be when he is in grade *n*) and an unlikely method of addressing it (using a table of values). But in recontextualisation this is typical, as the presentation of the "real" context is subordinate to the requirements of the school mathematics context. While the textbook task mixes

Kevin's Grade	Alice's Age
6	4
7	5
8	6
9	7
?	?
?	?
?	?

Kevin uses a pattern. He predicts how old his sister will be during each of his school grades.

Fig. 1 Task from the textbook (Mathquest 2000, p. 8)

Fig. 2 The T-table as copied by the teacher on the blackboard

Kevin	Alice
6	4
7	5
8	6
9	7

elements of school mathematics: a table, blanks to be filled in and the word "pattern", with everyday elements like the names Kevin and Alice, their ages and school grades, it is clear that this content belongs to school mathematics. No real child would make such a table and look for patterns in it as part of their everyday activity.

At the start of the lesson Mr. White organises the students into groups of four and tells them to share the results of their homework with their group members. He asks them: "What are your three answers? How did you get them?" While they are working he sketches a T-table on the board (see Fig. 2). Note that in his table, almost all the context information has disappeared. Kevin and Alice remain as headers, but their meaning as grades and ages, or their connection as two siblings, is not mentioned.

After the students have worked in their groups for a few minutes, Mr. White asks for the attention of the class. He calls for a volunteer to fill in the blanks of the T-table he has sketched on the board. Alicia volunteers to go to the board and fills in the left column down to 12 and the right column to 10. While Alicia is writing on the board the other students in the class are attentive; they watch what Alicia is writing. Transcript 4 begins when Alicia finishes.

58.1	Mr. White:	Is there anybody from her group, as well as Alicia, who can tell us how
58.2		those numbers fit in the way they do? What did you do? [*waits*]
59	[*Max raises his hand*]	
60	Mr. White:	OK.
61	Max:	Added one on each time.
62	Mr. White:	Which side are we talking about? The left side or the right side?
63	Nick:	Both sides.
64.1	Max:	Either or both of them. Because Kevin, in one year he's in grade six and
64.2		Alice is four years old. So the next year, he's going to be in grade seven
64.3		and she is going to be five years old. So you add one on both groups.
65.1	Mr. White:	So in other words, you're adding down, adding one. Is this what you
65.2		mean? If you started, if you started here you just add one to get to ten.
66	Max:	Yeah.

67.1	Mr. White:	You just add one to get eleven. You just add one to get to twelve?	
67.2		Is that what you did?	
68	Max:	Yes.	
69	Mr. White:	So what did you did, what did you do over here?	
70	Max:	The same thing. I had the number seven because I knew she was two-	
71	Mr. White:	This one here?	
72	Max:	Yeah.	
73	Mr. White:	Yes.	

Transcript 4: How those numbers fit in the way they do.

Classification of Communication

In the video it is evident that most students lose interest once Alicia has completed the table on the board. For them the "filling in the blanks" task is complete. They do not recognise that something more is expected in Mr. White's question in line 58.

What can we say about the classification and framing here? As we have already noted, the classification of content is strong; filling in T-tables and looking for patterns in them is far from an everyday activity. The classification of praxeology is harder to gauge at this point, and we will postpone our discussion of it. The framing is also generally strong; Mr. White clearly controls the dialogue with Max. However, in two respects the framing is weaker: Mr. White does not require the other students to pay attention to what he and Max are saying, and his questions "Who can tell us how those numbers fit in the way they do? What did you do?" (line 58) are very open. The criteria for acceptable answers are invisible. At this point, Mr. White appears to apply an implicit pedagogy.

There is an interesting difference in the way Max talks about the numbers and the table and the way Mr. White does. Max refers to the context for the T-table given in the textbook: Kevin's grade and Alice's age. He uses his knowledge of the everyday in line 64 to say why both sides increase by one: "Because Kevin, in one year he's in grade six and Alice is four years old. So the next year, he's going to be in grade seven and she is going to be five years old. So you add one on both groups." He also begins to provide a context-based explanation in line 70, "because I knew she was two-" but Mr. White interrupts him. Mr. White, however, makes no references to the context. He refers to the columns as "the left side or the right side" and the numbers in the table as numbers, not grades or ages. Even when he copied the table from the textbook, he changed the column labels from the meaningful (in context) numerical labels "Kevin's Grade" and "Alice's Age" to the non-numerical labels "Kevin" and "Alice".

We describe this difference in Max's way of talking and Mr. White's in terms of a third principle of classification:

Classification of communication: the insulation between contextualised and decontextualised language use that results from the degree to which the discourse refers to contexts that are at least potentially perceptible (cf. Straehler-Pohl and Gellert 2013).

The distinction between contextualised and decontextualised language refers to *actual* and *virtual* contexts and their linguistic role of meaning orientation as discussed by the linguist Ruqaiya Hasan (2001). While actual contexts, either in the here-and-now or anytime-anywhere, are at least hypothetically accessible via the senses, virtual contexts are exclusively constituted by the means of language. They would not exist without language.

If an utterance refers to a virtual context, then its meaning has to be derived from specialised registers as the meaning can not be manifested by sensation or experience. For example, the word relation has a completely different meaning when used in a "family register" or in a "mathematics register". Decontextualised use of language thus implies strong classification of communication. If the discourse refers to potentially perceptible contexts, that is, it mainly makes use of contextualised language, then it is weakly classified in terms of communication.

In Transcript 4, the classification of communication is strong. Mr. White expects the way the numbers "fit" to be communicated using language that is decontextualised. Max has not recognised this, however, and while actually not meaning any perceivable persons, he expresses his arguments in contextualised language, referring to Kevin's grade and Alice's age. This is consistent with the references to the everyday in the original task. However, such contextualised language was dismissed by Mr. White and immediately replaced by more decontextualised formulations (line 65).

In the next transcript Mr. White's practice turns more and more towards an explicit pedagogy, and Max recognises this strong classification of communication.

74.1	Max:	Because I knew she was two years younger than the grade he was in.
74.2		So then I just added one on [the numbers?] from there.
75.1	Mr. White:	I have a question. This can come, the answer may come from any
75.2		group. You may look at the T-table here or you may look at the one
75.3		you've created in your notebook. Can anybody figure out or tell me the
75.4		relationship between the left side of this T-table and the right side
75.5		of the T-table.
76		[*Max is the only student who raises his hand.*]
77	Mr. White:	Okay.
78.1	Max:	The difference between the numbers, there's a difference of two on
78.2		each number.

79	Mr. White:	A difference of two. How do you mean difference?
80	Max:	There is, one is two higher.
81	Mr. White:	So in other words, this one is two higher.
82	Max:	Yes.

Transcript 5: In other words.

Recognition and Realisation Rules: Finding the Correct 'Other Words'

In line 74.1 Max completes his thought from line 70 and describes the relationship between the two columns, again making reference to the everyday context: "she was two years younger than the grade he was in". Mr. White's response (line 75) is interesting: He asks the class if anyone can tell him the relationship between the two columns. As this is what Max has just done, one might expect Max to interpret this as a signal that there was something wrong with what he said. In line 78 Max tries again "there's a difference of two on each number". He reformulates the relationship he stated in line 74 in a way that leaves out anything perceptible completely, expressing the relationship in the language of an entirely virtual context. He has now recognised the strong classification of the communication.

Let us now return to the question of the classification of praxeology. Mr. White's question "Can anybody figure out or tell me the relationship between the left side of this T-Table and the right side of the T-Table." (line 75) refers to the concept of a functional relationship between two sets. This suggests a strong classification of praxeology. Max's answers, especially in line 74.1, refer to the functional relationship, but as we have seen, they fail to do so in a decontextualised way. His decontextualised answer "there's a difference of two on each number" (line 78) also refers to the functional relationship, but less clearly. When Mr. White asks him to clarify, Max adjusts his description, replacing the static description "there's a difference of two" with one that gives priority to one column over the other "one is two higher". Mr. White replies "So in other words, this one is two higher" making sure that the target of the functional relationship is clear (it is the left hand column in this case).

Max is the only student who attempts to engage in this interaction with Mr. White. As we mentioned above, some students seem to not recognise that this interaction is important. Others may recognise the importance of Mr. White's questions, but might be unable to respond as they lack the necessary realisation rules. Still others may be as able to participate as Max is but choose not to until the framing of the criteria becomes stronger (Knipping and Reid 2013). By observing Max they may be able to copy the way he has learned to describe relationships, using formulations like "this one is two higher". However, as his earlier explanations failed to recognise the strong classification of communication, they were marked as inappropriate, and it is unlikely that anyone will copy that discourse.

83	Mr. White:	I have a question. How do you go from this number to this one? Remember you said that we added down or you folks added down. How do we get from this side if you were looking at these numbers and if you say they sort of, they sort of seem to match up in a way? How do we get from this side to this side?
…		[*Mr. White erases the numbers 8, 9 and 10 on the right side of the table and asks explicitly for a relationship between 6 and 4, 7 and 5, 8 and 6, 9 and 7. More students now volunteer to provide an answer.*]
104	Eric:	You subtracted two each time.
105.1	Mr. White:	We have a response here that says you subtract two each time.
105.2		Let's assume that we are going to give a letter to each of these numbers.
105.3		Let's assume we say each of these numbers represents X so any time we
105.4		want to put a number in we replace X with the number. And if we say
105.5		equals a number, [*bell*] whoops.
106	Larry:	That's the ten fifteen senior high bell.
107.1	Mr. White:	Can anybody else, can anybody figure what we would put in here
107.2		for a little tiny equation? To help fill this out? [*Mr. White points to Eric*]
107.3		What did he just say, what did he say we do?
108	Multiple Students:	Subtract two.
109	Mr. White:	Let's put it in. Subtract two equals the number.
110		[*Mr. White writes X - 2 = N on the blackboard.*]
[…]		
112.1	Mr. White:	Now, let's see if we can make it work. Because guess what, a lot
112.2		of T-tables work in a pattern something like this where you can fill in a
112.3		little tiny equation. If you understand that with this one you'll understand
112.4		most of what happens in most of the rest of the T-tables. See this little
112.5		equation here? It gets a little bit harder but they work basically the same
112.6		way. Now, let's see if we can—show you how this works. If we say X
112.7		minus two equals the number that we want, which is over here.
112.8		If we say six minus two equals
113	Student:	Four.
114.1	Mr. White:	Four. Now all of a sudden you are into real simple arithmetic.

114.2		You did this ages ago so guess what? We made the math look a little bit
114.3		hard, now we're trying to make it look easy. What would we do now if
114.4		we are down at ten, using the same pattern? Ten, minus, what are we
114.5		going to put in here?—Anybody? Yes sir.
	[*Emily, Ben and Larry raise their hands*]	
115	Ben:	Two.
116.1	Mr. White:	Right. Because we are always going to use the same one,
116.2		the same pattern that is. Ten minus two equals—
116.3		Wayne, are you with us son? What's ten minus two buddy?
117	Wayne:	Eight

[*Taking turns, Mr. White and several students go through more examples some of which he writes on the blackboard.*]

Transcript 6: How do you go from this number to this one?

Classification of Praxeology: A Little Tiny Equation

Mr. White's next question reveals something interesting about the classification of praxeology, which we had tentatively identified as strong, based on Transcript 5. Max's answer "one is two higher" (line 80) describes a relationship between the two columns, which is what Mr. White asked for in line 75, but it does not satisfy two criteria that become clear in Mr. White's next question, "How do you go from this number to this one?". With each question the framing of the criteria for answers has become stronger, and now it is clear that Mr. White wants a *procedure* that will change the numbers in the left hand column into the numbers in the right hand column. Max's answers were all expressed as relationships, not as procedures and either worked in both directions (difference of two) or went from right to left (one is two higher). Eric's response "You subtracted two each time" is positively evaluated. He has recognised that the classification of communication is decontextualised, and that the classification of praxeology is not as strong as Mr. White's mention of "relationship" suggested.

An interesting shift occurs here. The framing of the social order becomes stronger in that Mr. White now expects other students to become involved, and even calls on Wayne by name (line 116). Mr. White also controls more of the discourse, only allowing students to respond to simple arithmetic questions. At the same time the classification of praxeology becomes weaker. The mathematics of T-tables is reduced to writing the equation and plugging in values. The process of identifying the relationship between the numbers, which mathematically is the basis of the equation, was done entirely by Max and Eric with Mr. White's guidance. There was no sign that the rest of the class should be involved. Mr. White appeared to be satisfied with the rest of the class being able to answer his procedural questions.

Does Everyone in Mr. White's Class Have the Same Access?

What most clearly distinguishes Mr. White's class both from Mr. Black's and from the description of Brousseau is that it ends with a weak classification of praxeology. This difference of classification appears to happen for the sake of inclusiveness: "We made the math look a little bit hard, now we're trying to make it look easy." In this first lesson the emphasis is on adding more values to the table, and plugging numbers into the equation. It appears that this is sufficient for a legitimate contribution and Mr. White's final strengthening of framing guarantees that his students—in contrast to Mr. Black's—can rely on the criteria he communicates. These are apparently the final criteria. In the end Mr. White seems to apply an explicit pedagogy.

However, the generation of the equation has been done entirely by Mr. White, Eric and Max. And crucially, the combination of weak classification of praxeology and strong classification of communication implies that it is impossible for others to question and discuss this process. In particular, Max's use of the context to identify the relationship that underlies the equation becomes invisible for the other students. In the process of generating a meaningful mathematisation for a contextualised table, it is a crucial first step to verbalise relations in a way that is accessible to all. This allows students to gradually increase the strength of classification (of content, communication or praxeology), but without an abrupt change. However, as Max appears to be able to make such an abrupt change on his own, the process that underlies this change completely disappears from the classroom discourse and hence remains completely implicit for the other students, who still need to acquire the recognition and realisation rules in order to make such changes themselves. In later lessons Max continues to be able to identify relationships and find equations, but he never again describes what he is doing in terms of the contexts provided in the textbook. Max seems to have recognised that the contexts provided in the textbook are useful, at the same time he has learned in this first lesson that mentioning those contexts aloud is not useful. As Bernstein (2000, p. 17) notes:

> Certain distributions of power give rise to different social distributions of recognition rules and, without the recognition rule, contextually legitimate communication is not possible. It may well be, at the more concrete level, that some children from the marginal classes are silent in school because of the unequal distribution of recognition rules.

Max has learned to be silent about context as he has learned to recognise the strong classification of content and communication in Mr. White's classroom. However, he still has (from some prior experience at home or in primary school) the recognition rule that allows him to use contexts to find equations. Because of the strong classification of communication those students who do not already have this recognition rule are unlikely to acquire it in Mr. White's classroom. Similarly, Max has recognition rules related to a strong classification of praxeology, but as the classification of praxeology in Mr. White's classroom is weak, Max is unlikely to make explicit use of those recognition rules, and so other students will not observe them.

Our discussion shows how Mr. White did more than making "the math look a little bit hard" and then "making it look easy" again. There also appear to be two different kinds of mathematics going on. Both kinds of mathematics are characterised by a strong classification of content and communication. However, what remains invisible behind these strong classifications is that the two kinds of mathematics differ fundamentally concerning the strength of classification of praxeology.

In Mr. Black's class (Episode 1) everyone is measured on the same but implicit criteria. The ones who meet them are exposed as winners, the ones who fail against them are left feeling like losers. In Mr. White's class on the contrary, everyone is abled to feel that they have succeeded according to the given criteria. Mr. White sets out criteria that are attainable by everyone. However, what is concealed from most of the students is that there is a whole set of alternative criteria that are only applied to those who can autonomously qualify themselves as being able to match them. This is the other face of differentiation of criteria: it is not that the criteria are implicit, but the very fact that there is a differentiation going on.

In Mr. White's class, the mechanism that leads to a stratification of access for different students is much less characterised by an implicit pedagogy than in the case of Mr. Black. Instead we describe this mechanism as an *implicit differentiation*, that is covered by an apparently explicit pedagogy.

Conclusion

What we have seen in the episodes above are different classroom cultures evolving in the first weeks of a school year. Terminology developed from the work of Bernstein allowed us to identify mechanisms in classroom interactions that gave rise to these different classroom cultures and to disparity in mathematical learning opportunities. We also showed how this culture was embedded in a context that went beyond the walls of the classroom: a selective culture of the German Gymnasium where students are traditionally oriented to allocation of students for higher segments of the labour-market, and a Canadian secondary classroom where the system is oriented towards inclusion and diversity.

What both episodes have in common is a recontextualisation from everyday discourse to school mathematics. This recontextualisation of what Bernstein calls "horizontal" discourse is a typical feature of school mathematics and other school subjects. "As part of the move to make specialised knowledges more accessible to the young, segments of *Horizontal discourse* are recontextualised and inserted in the contents of school subjects" (Bernstein 2000, p. 169). But as we have seen above, inserting segments of the everyday into school mathematics does not necessarily make the subject more accessible, at least not for everyone.

In Mr. Black's lesson the weak classification of content and the implicit pedagogy related to the strength of classification of praxeology (and communication) seemed to allow the students to engage in the everyday discourse of game playing. It was only when a student's answer, corresponding to the strong classification of praxeology, was positively evaluated that the legitimate orientation in the discourse became gradually clearer. While some students were able to recognise the strong classification of praxeology early and became the true winners of the game, others might be left wondering why exactly these contributions were valorised, and in the end they are left out of the real "race".

In Mr. White's classroom, the classification of content and communication was strong, but the classification of praxeology was weak, and the former appears to dominate the classification of praxeology, so that contextual explanations are ignored even when oriented toward theorising the problem. Despite the weak classification of praxeology, theorising is yet not absent in Mr. White's class. However, it appears that not everyone is expected to take part in it. Although the framing is generally strong, during Mr. White's exchange with Max and Eric the rest of the class is not expected to participate. At that moment, when Max is able to engage with the task more theoretically, the framing also weakens, so that not everyone has to attend, and there is no indication that what is happening is important. Instead, the activity closes with an emphasis that everyone has successfully acquired the necessary means to legitimately participate in Mr. White's class.

In interviews Mr. White stated clearly that he was deeply concerned for his "less able" students to be successful in his class. This was visible also in his comments to the class. To do this he weakens the classification of praxeology and reduces mathematics to a procedural or technical level. In attempting to enable his "weak" students he denies them access to important features of high status mathematical knowledge.

Recontextualisation comes with a danger of reducing what Bernstein calls "vertical" discourse, the discourse of a subject like mathematics, to a set of procedures.

> Recontextualising of segments [of horizontal discourse] is confined to particular social groups, usually the 'less able'. This move to use segments of *Horizontal discourse* as resources to facilitate access, usually limited to the procedural or operational level of a subject, may also be linked to 'improving' the student's ability to deal with issues arising (or likely to arise) in the students' everyday world ... *Vertical discourses* are reduced to a set of strategies to become resources for allegedly improving the effectiveness of the repertoires made available in *Horizontal discourse*. (Bernstein 2000, p. 169)

What Bernstein describes here and what we have identified in Mr. White's class can be seen in relation to an inclusive school system that opts for a differentiation of learning opportunities, trying to meet the students, "where they are". This intentionally inclusive gesture always bears the danger of nobly (but nevertheless falsely) aligning pedagogy to lowered expectations and hence producing self fulfilling prophecies (Straehler-Pohl et al. 2014). As suggested by the Bourne quote in the introduction, the institution of education produces the teachers who participate in it, and Mr. White's teaching is not (only) a reflection of his own choices as a teacher, but also (and perhaps mainly) a reflection of the institution of education.

The way that Mr. Black has employed an everyday context—or in Bernstein's words, segments of horizontal discourse—is very different. He is not limited "to the procedural or operational level" and not making mathematics a "resource for allegedly improving the effectiveness" of the everyday. Quite the opposite, he employs the game for theorising and for introducing the students into the values of the *Gymnasium* in the selective German secondary school system: The *Gymnasium* is oriented to intellectual forms of labour, hence the orientation to theorising and the privileging of abstract thought. It is oriented towards a strong specialisation of students, hence the selectivity. Finally, the *Gymnasium* is supposed to identify and categorise the emerging member of the elite. The implicit pedagogy, that resembles a race rather than a learning opportunity, makes sense within this realm. However, despite the striking selectivity and competitiveness, one also has to acknowledge how rapidly Mr. Black's pedagogy brings *some* students to produce quite sophisticated theoretical insights in the structure of the Race to Twenty. As with Mr. White, we want to emphasise that Mr. Black's teaching is a reflection of the institution he is embedded in, the German *Gymnasium*.

Hence we see, in two quite distinct contexts, mechanisms that produce disparity in learning opportunities arising out of both sides of the dual recontextualisation that produces school mathematics. Classroom interactions mirror outside social and structural characteristics in ways that make the classroom culture in a German *Gymnasium* very different from the classroom culture in a Canadian school. However, there are also commonalities, one of which is the emergence of disparity. This is to be expected in a selective school like the *Gymnasium* but it occurs also in inclusive schools, and, as we have seen here, can be produced out of an attempt to ensure success for all. By understanding better such disparity producing mechanisms, and how they operate in different contexts, we establish a basis for teaching that can reduce or avoid them, so that inclusive schools can be more truly inclusive.

References

Atweh, B., Bleicher, R. E., & Cooper, T. J. (1998). The construction of the social context of mathematics classrooms: A sociolinguistic analysis. *Journal for Research in Mathematics Education, 29*(1), 63–82.
Bateson, G. (1975). Some components of socialization for trance. *Ethos – Journal of the Society for Psychological Anthropology, 3*(2), 143–155.
Bernstein, B. (2000). *Pedagogy, symbolic control and identity: Theory, research, critique* (Rev. ed.). Lanham: Rowman & Littlefield.
Bourne, J. (1992). *Inside a multilingual primary classroom. A teacher, children and theories at work*. Unpublished PhD thesis, University of Southampton.
Brousseau, G. (1978). Étude locale des processus d'acquisitions scolaires. *Enseignement élémentaire des mathématiques, 18*, 7–21.
Brousseau, G. (1997). *Theory of didactical situations in mathematics: Didactique des mathématiques, 1970–1990*. N. Balacheff, M. Cooper, R. Sutherland & V. Warfield (Eds. & Trans.). Dordrecht: Kluwer.
Chevallard, Y. (1985). *La transposition didactique*. Grenoble: La Pensée Sauvage.

Chevallard, Y. (1999). L'analyse des pratiques enseignantes en théorie anthropologique du didactique. *Recherches en Didactique des Mathématiques, 19*(2), 221–266.

Cobb, P., & Yackel, E. (1998). A constructivist perspective on the culture of the mathematics classroom. In F. Seeger, J. Voigt, & U. Waschescio (Eds.), *The culture of the mathematics classroom* (pp. 158–190). Cambridge: Cambridge University Press.

Dowling, P. (1998). *The sociology of mathematics education: Mathematical myths/pedagogic texts*. London: Routledge.

Gellert, U. (2010). Modalities of local integration of theories in mathematics education. In B. Sriraman & L. English (Eds.), *Theories in mathematics education: Seeking new frontiers* (pp. 537–550). New York: Springer.

Gellert, U. (2012). Pedagogic device: Ein Instrument für die Analyse impliziter Prinzipien mathematischer Unterrichtspraxis. In U. Gellert & M. Sertl (Eds.), *Zur Soziologie des Unterrichts. Arbeiten mit Basil Bernsteins Theorie des pädagogischen Diskurses* (pp. 166–190). Weinheim: Beltz Juventa.

Gellert, U., & Hümmer, A. M. (2008). Soziale Konstruktion von Leistung im Unterricht. *Zeitschrift für Erziehungswissenschaft, 11*(2), 288–311.

Gellert, U., & Jablonka, E. (2009). "I am not talking about reality": Word problems and the intricacies of producing legitimate text. In L. Verschaffel, B. Greer, W. Van Dooren, & S. Mukhopadhyay (Eds.), *Words and worlds: Modelling verbal descriptions of situations* (pp. 39–53). Rotterdam: Sense.

Hasan, R. (2001). The ontogenesis of decontextualised language: Some achievements of classification and framing. In A. Morais, I. P. Neves, B. Davies, & H. Daniels (Eds.), *Towards a sociology of pedagogy. The contribution of Basil Bernstein to research* (pp. 185–219). New York: Peter Lang.

Hoadley, U. K. (2007). The reproduction of social class inequalities through mathematics pedagogies in South African primary schools. *Journal of Curriculum Studies, 39*, 679–706.

Jablonka, E., & Gellert, U. (2010). Ideological roots and uncontrolled flowering of alternative curriculum conceptions. In U. Gellert, E. Jablonka, & C. Morgan (Eds.), *Proceedings of the sixth international mathematics education and society conference* (pp. 23–41). Berlin: Freie Universität Berlin.

Knipping, C., & Reid, D. (2013). Have you got the rule underneath? Invisible pedagogic practice and stratification. In A. M. Lindmeier & A. Heinze (Eds.), *Proceedings of the 37th conference of the international group for the psychology of mathematics education* (Vol. 3, pp. 193–200). Kiel: PME.

Knipping, C., Straehler-Pohl, H., & Reid, D. (2011). "I'm going to tell you to save wondering": How enabling becomes disabling in a Canadian mathematics classroom. *Quaderni di Ricerca in Didattica, 22*(Supplemento 1), 171–175.

Lerman, S. (2000). The social turn in mathematics education research. In J. Boaler (Ed.), *Multiple perspectives on mathematics teaching and learning* (pp. 19–44). Westport: Ablex.

Lerman, S. (2001). Cultural, discursive psychology: A sociocultural approach to studying the teaching and learning of mathematics. *Educational Studies in Mathematics, 46*(1–3), 87–113.

Lubienski, S. T. (2000). Problem solving as a means towards mathematics for all: An exploratory look through a class lens. *Journal for Research in Mathematics Education, 31*(4), 454–482.

Lubienski, S. T. (2004). Decoding mathematics instruction: A critical examination of an invisible pedagogy. In J. Muller, B. Davies, & A. M. Morais (Eds.), *Reading Bernstein, researching Bernstein* (pp. 108–122). London: Routledge.

Martin, D. B., Gholson, M. L., & Leonard, J. (2010). Mathematics as gatekeeper: Power and privilege in the production of knowledge. *Journal of Urban Mathematics Education, 3*(2), 12–24.

Morais, A. M. (2002). Basil Bernstein at the micro level of the classroom. *British Journal of Sociology of Education, 23*(4), 559–569.

Morgan, C. (2012). Studying discourse implies studying equity. In B. Herbel-Eisenmann, J. Choppin, & D. Wagner (Eds.), *Equity in discourse for mathematics education* (pp. 181–192). Dordrecht: Springer.

Nickson, M. (1994). The culture of the mathematics classroom: An unknown quantity? In S. Lerman (Ed.), *Cultural perspectives on the mathematics classroom* (pp. 7–35). Dordrecht: Kluwer.

Seeger, F., Voigt, J., & Waschescio, U. (Eds.). (1998). *The culture of the mathematics classroom*. Cambridge: Cambridge University Press.

Sheridan, A. (1980). *The will to truth*. London: Tavistock.

Stinson, D. W. (2004). Mathematics as "gate-keeper"(?): Three theoretical perspectives that aim toward empowering all children with a key to the gate. *The Mathematics Educator, 14*(1), 8–18.

Straehler-Pohl, H., & Gellert, U. (2013). Towards a Bernsteinian language of description for mathematics classroom discourse. *British Journal of Sociology of Education, 34*(3), 313–332.

Straehler-Pohl, H., Gellert, U., Fernandez, S., & Figueiras, L. (2014). School mathematics registers in a context of low academic expectations. *Educational Studies in Mathematics, 85*(2), 175–199.

Valero, P. (2004). Socio-political perspectives on mathematics education. In P. Valero & R. Zevenbergen (Eds.), *Researching the socio-political dimensions of mathematics education* (pp. 5–23). Dordrecht: Kluwer.

Part II
Working with Adults

Learning Mathematics In and Out of School:
A Workplace Education Perspective 99
Gail E. FitzSimons
University of Melbourne, Australia

Mathematical Modelling and Bank Loan Systems:
An Experience with Adults Returning to School 117
Vera Helena Giusti de Souza, Rosana Nogueira de Lima,
Tânia Maria Mendonça Campos
Anhanguera University, Brazil
Leonardo Gerardini
Colégio Dante Alighieri, Brazil

Working with Adults: A Commentary 131
Javier Díez-Palomar
University of Barcelona, Spain

Learning Mathematics In and Out of School: A Workplace Education Perspective

Gail E. FitzSimons

Abstract This chapter reports on the mathematical training of operators in the pharmaceutical manufacturing industry in Australia. Based on an analysis of workplace observations, the pedagogic design of the training addresses tensions and contradictions between the mathematically informed workplace practices of the operators and their formal mathematical competences. A Bernsteinian approach is used in this chapter to discuss issues of enhancement at the individual level and of social, intellectual, cultural, and personal inclusion.

Democracy and Pedagogic Rights

Basil Bernstein (2000) uses the concept of boundary as a metaphor to signify the realisation of distributions of power through various and often invisible means. Equally important is the way these boundaries are relayed by various pedagogic processes. Utilising the training of operators in the pharmaceutical manufacturing industry in Australia as a case study, this chapter will demonstrate how low-level manufacturing workers, who might have been excluded from democratic participation by invisible boundaries relayed through the official vocational curriculum, can overcome this exclusion. Through pedagogic processes informed by socio-cultural research—which acknowledges the value of human activity, including critical workplace interactions—workers may be empowered mathematically. In fact, all stakeholders have the potential to gain something from the process.

Bernstein (2000) advocates three pedagogic rights which enable democracy to function: enhancement, inclusion, and participation. At an individual level, the right to enhancement—experiencing boundaries as tension points between past and future options—"is the right to the means of critical understanding and to new possibilities" (p. xx) and a condition for confidence. At the social level, the right "to be included, socially, intellectually, culturally, and personally" (p. xx), possibly also

---

G.E. FitzSimons (✉)
Melbourne Graduate School of Education, University of Melbourne,
Parkville, VIC, Australia
e-mail: gfi@unimelb.edu.au

requiring the right to be separate or autonomous, is a condition for *communitas*. At the political level, the right to participate is more than a discourse and must be about practice which has outcomes; it is the condition for civic practice or discourse. Two of Bernstein's main themes are that (a) the distribution of knowledge carries unequal value, power, and potential; and (b) the visual and temporal features of images and voice in the learning situation project a hierarchy of values. In other words, education is necessarily a value-laden activity; curriculum and pedagogy cannot be separated from values.

In formal pedagogic situations, such as vocational mathematics or workplace numeracy education, arbitrary selections are made from the discipline of mathematics according to distributive rules which provide different forms of consciousness to different social groups, and differential access to unthinkable knowledge in the possibility of new knowledge, as distinct from thinkable knowledge which takes the form of official knowledge. However, in the workplace, as elsewhere in society, problems are ever-evolving and the development of new knowledge—locally new if not always universally new—is an essential requirement for completing the task at hand within constraints of time and/or money, so that workers often find themselves in "unthinkable" territory, creating new knowledge.

The Economic Realities of Pharmaceutical Manufacturing

This industry is highly regulated. The Pharmaceutical Manufacturing Licence depends on adequate *Good Manufacturing Practice* (GMP) standards: there is a requirement to demonstrate control over equipment, personnel, and processes. Formal training records are also required, and both are reviewed in regular and random GMP and *Quality Assurance* (QA) audits. The pharmaceutical industry requires that all activities are documented: Records of production, equipment maintenance, and laboratory testing are considered as legally-binding documents. Failure to comply with government regulations can mean loss of operating licence.

Production has to be streamlined, attuned to considerations such as meeting seasonal demands while remaining cognisant of the limited shelf lives of various products. Underlying every action in the pharmaceuticals production process is the mandatory requirement for accountability and traceability: Everything has to meet exact specifications as detailed in continuously updated *Standard Operating Procedures* (SOPs), and to be checked and rechecked. All operators (warehouse, production, and packaging workers) are acutely aware of the importance of this to their jobs.

Beyond initiation into routine procedures, the need for training is predicated on the ability to respond to unexpected or unforeseen situations which may be encountered in the workplace. Production delays can be caused by something as serious as equipment malfunction, or as simple as failure to access the specified quantities of raw or packaging materials. More serious losses are incurred through reworking, necessary as a result of error detection. Public recall of products can result in loss of sales as the company reputation is damaged. A lag in innovation, in product development or work organisation, can reduce market edge. As with industry in general, workplace structures are continually changing in this enterprise as a result of local or head office decisions. Training beyond day-to-day requirements is essential to work organisation predicated on new techniques, both in the engineering sense of tools and machinery, and in the socio-cultural sense of communication and participation in workplace decision-making.

Although there was a strong training ethos emanating from company headquarters in Germany, the decision by the Melbourne manager to participate in a formal, externally provided and accredited course was not taken lightly. Whereas in Germany all new operatives are assessed at recruitment and offered mainly in-house training, in Australia qualifications frameworks and recognition of skills for so-called unskilled and semi-skilled workers are a relatively recent development dating back to the 1990s. On the basis of the availability of the new vocational education credentials, and looking to increase productivity, management decided to offer a base-level program (Certificate I) to selected and volunteer workers already employed as operators. Although these workers were highly experienced, they lacked formal certification in this area, and it was intended that formal academic recognition and development of their skills would enhance the workplace culture.

I was employed to "deliver" in 20 h the competency-based education and training (CBT) modules of *Calculations A* and *Basic Computer Skills* to about 12 pharmaceutical operators, both on-site and in my institution. After negotiations were completed, production workers were informed of the availability of training and offered encouragement through a combination of paid and unpaid study schedules. Organising these schedules to fit in with production runs and rostered-days-off proved quite a demanding task for the company. Showing an insightful empathy for the workers' broader educational horizons, the manager requested that, while many classes were to take place on-site, others were to be held in the institutional setting of my vocational university, in order to provide familiarity and agency within this educational context (e.g., organising ID cards, using the library, locating classrooms in complex buildings). Unlike many others in Australia, this company showed its workers respect and encouraged a sense of belonging: It valued and celebrated outstanding contributions, recognising that difficult working conditions are sometimes unavoidable (e.g., building refurbishments, or economic downturns elsewhere).

The official mathematical knowledge took the form of a *Calculations A* curriculum. This was comprised of selections from the vertical discourse of mathematics (i.e., systematic, explicit theoretical knowledge found in general education curricula), which offers conceptual understanding and integration of meanings. This is in contrast to knowledge which is developed through experience via a collection of specific, unrelated contexts, described by Bernstein (2000) as horizontal discourse (see FitzSimons 2008; FitzSimons and Wedege 2007, for further discussion of vertical and horizontal discourses in mathematics). However, as illustrated below, the curriculum mostly resembled the content knowledge expected of children before they reach secondary education, but without offering the workers opportunities for higher level thinking and access to democratic participation.

Calculations and Computing in the Pharmaceutical Manufacturing Industry

The accredited curricula for pharmaceutical operators specified generic core modules (e.g., calculations, quality assurance, and industrial communication), together with a range of pharmaceutical core modules (e.g., good management practice, basic computer skills) and, finally, specialised electives in production, packaging, and materials handling. Operators found these groupings to be of ascending relevance to their everyday work. The following are examples of *Learning Outcomes* from Certificate I. Of the three Calculations modules, only Calculations A was core. Typically, workers were extremely unlikely to progress voluntarily along this mathematical path of study—discussed further below. Consistent with vocational education policy of the time, and since, all curricula had to be seen as relevant to work, so that this framed the pedagogical discourse.

Calculations A: Estimate, Calculate, and Record Workplace Data

Within the context of this calculations module, *estimate* refers to the ability to form an approximate judgement which is closely related to the formally calculated result.

- Estimate results from basic information used in typical workplace situations
- Calculate results involving whole numbers, simple fractions, & decimals used in typical workplace situations
- Record estimates and calculations on standard workplace forms/documents accurately and legibly

Calculations B: Use Routine Measuring Instruments.
Complete Routine Arithmetic Calculations. Chart Data

- Explain SI measurements for mass, volume, temperature, length
- Measure product weight, volume, temperature, length, and associated variations
- Record data on standard charts

Calculations C

- Calculate performance measures
- Convert imperial to SI measures
- Calculate percentages, ratios, and proportions
- Use imperial and SI measures to calculate performance
- Record data on standard charts

Basic Computer Skills

This module provides the learner with a basic understanding of computers and computing systems used in the pharmaceutical manufacturing process. The emphasis is on the ability of the person to manipulate the keyboard, respond to simple commands, and input data in their immediate work area.

- Explain the purpose of computers and their components and the impact of computers on society
- Describe the different computer and computerised control systems used in the pharmaceutical manufacturing/production industry and the differences between the systems
- Input, store, and retrieve data following computer menus and commands, using a range of input/output devices.

Quality Assurance A

- Identify and monitor critical control points at the work station
- Sample product for off-line testing
- Perform inspections and tests of own work

Industrial Communication A

- Express views verbally
- Read non-routine text
- Prepare written information to support groups and teams

It is difficult to justify the curriculum for Calculations A—where the arithmetic processes have been isolated and artificially separated from their historical genesis in the measuring activities of people—in terms of how it might contribute to the survival of the enterprise (or the industry for that matter) in times of rapid economic change. Except for recording of estimates and calculations, it is certainly not a reflection of the work I observed, as described below in the handling of raw materials and finished goods, or in production processes. In fact, the workers were routinely involved in a wide range of measuring activities, such as those listed in Calculations B & C, as well as planning and locating activities in 3-dimensions. Importantly, nobody actually performed these mathematical activities in the manner that would be expected and assessed in the situational context of formal schooling. Notably, the pedagogy implied by the Calculations A curriculum, and recommended in the teacher support materials, was restricted to lower level procedural activities of symbolic representation and manipulation, ignoring or deliberately avoiding mathematically higher level activities (or competences) of problem solving, reasoning, justification (EACEA 2011). Neither was there any reflection on, or evaluation of, the solution process—nor even of the discipline itself. Accordingly, it cannot be remotely considered as setting the foundation for advanced mathematical thinking. By contrast, *Basic Computer Skills* is noteworthy for its inclusion of the social dimension and its practical orientation to actual work processes. Yet, somewhat surprisingly, *Quality Assurance* appears to be completely dissociated from mathematics, even though the concept of representative samples is mathematically complex: It is likely to be reduced to a set of rules and procedures, restricting the possibility of meaningful communication between operators and technical staff in case of errors or breakdowns in understanding (cf. Kent et al. 2011).

One unintended effect of the Calculations A curriculum is that it actually acts as a barrier to pharmaceutical workers at the operator level, because they are forced to revisit the school mathematics that was so alien in years gone by. Fractions and decimals have suddenly reappeared, even though vulgar fractions are seldom used, if at all, in this workplace. In fact, these workers are likely to be using and understanding different, broader, even more sophisticated mathematical and statistical ideas, simply because they are contextually significant. Yet, they may be unable to satisfy competency requirements on a decontextualised basic skills test without the use of a calculator. Adults are positioned as incompetent in an educational system based on abstraction—as educational infants, recalling personal images of prior dependency on the teacher. Lack of agency was noteworthy among the workers on first viewing the Calculations A curriculum: burly, highly competent men were almost reduced to tears. In order to participate meaningfully in workplace decision-making, a sense of mathematical empowerment needs to be engendered, including higher order thinking as well as basic skills. However, it is questionable whether the accredited Calculations A curriculum could have made any contribution to the higher order thinking skills, to say nothing of day-to-day work practices.

The workers, many of whom had been out of formal education for many years, needed to be convinced of the value of further study and that it would not replicate their negative experiences of school mathematics education. For some, the inclusion of technology that they had not previously encountered in formal education was of initial concern. For others, the opportunity to gain recognition of their (plentiful) skills and to learn more through an officially accredited course provided a real morale booster. In the following section I illustrate some of the complexities of operators' daily practices as a stark contrast to the Calculations A curriculum.

Workplace Observations

With permission from the manager, in my own time I observed and discussed the activities taking place on a daily basis with the operators and other workers employed at all stages of the production process, from entry to exit. It was possible to frame the mathematical activities within the set of six universals that Bishop (1988) characterised as being an essential part of all cultures (including the workplace culture): counting, locating, measuring, designing, explaining, and playing. Each was found to be an integral part of the workplace taken in its entirety of space and time over the range of workers' shifts. In this way, my discussions with, and observations of, the operators and their colleagues enabled me to particularise and concretise the abstract official curriculum. I now outline and discuss relevant aspects of the work done by operators at each of the six major stages of production, namely: (a) inwards goods, (b) raw materials store/warehouse, (c) production: compounding, (d) production: packaging, (e) distribution: finished goods warehouse, and (f) outwards goods: packing/stock control.

Inwards Goods The operator's tasks were to record deliveries (raw materials, packaging, office supplies, etc.), unload the deliveries, check them, and sign them off. Where necessary, he had to redirect incorrect deliveries to a nearby plant in the same company which manufactured fertilisers and other agro-chemicals. Unloading could involve pallets of numerous boxes which he counted and labelled, on the spot if possible, processing his own labels using a computer and waiting overnight for unloading of the pallet to complete the labelling of those boxes not initially visible. Keeping a mental note of boxes hidden in the interior of a large delivery is certainly not a trivial mathematical skill.

Counting boxes required finding a part-box delivery notice in the case of a box not completely full. For example, the operator might receive 54 boxes, comprised of 53 with 1,925 plastic bottles in each and 1 with 1,375. The total number of bottles was then calculated using a calculator, which also required an estimation to ensure the reasonableness of the answer. The operator then had to photocopy the paperwork for stock control. After checking that the labels were correct he would forward the certificate of analysis to the QA department. At the end of each day he had to complete the daily goods manifest on all deliveries, accurately for traceability purposes.

After accepting the delivery of a pallet, this operator had to allocate a suitable place for immediate storage prior to transporting it to the next destination, the Raw Materials Store or the Warehouse. This required communication with the driver about the correctness of the order and where to offload the pallet. Large pallets require transport by forklift, and this is another skill requiring a range of mathematical competences such as reasoning, problem posing and solving, logical and spatial thinking (EACEA 2011), as well as effective verbal and non-verbal communication skills, in order to avoid mishaps which could damage products and even endanger human life. This is demonstrated in forklift sales and training websites which highlight the importance of inter-related contextualised mathematical measurement concepts, such as: (a) the weight and size of typical loads, (b) the height required to lift the loads, (c) interior or exterior uses influenced by the quality of the driving surface and the kinds of fuel types legally permitted, and (d) the average number of hours used per day. Any mathematical (or other) error has potentially fatal consequences, as shown on videos of "funny forklift accidents" found through web-based search engines. The mandatory possession of a valid forklift driver's licence for such operators highlights the generally unacknowledged range of workers' tacit mathematical knowledges.

Raw Materials Store/Warehouse The duties here were to record incoming stock, allocate appropriate locations, transfer stock as required to the next destination, and condense the remaining stock. In the store, in order to accommodate production process needs, operators were required to make conversions, using a chart to convert pre-made tablets from kilograms (as they were received) into actual numbers, and to convert huge rolls of foil from their length in metres into their weight. In the warehouse, the operators had to plan strategies for location and sub-location of goods other than raw materials. They had to allow for a small tolerance (or margin of error) in the quantities transferred. To allocate the locations, including sub-locations, they had to follow the 'first-in, first-out' or FIFO rule which requires meticulous attention to data on the labels as well as logical planning skills. The transferred stock was recorded using a special transfer screen on one of the in-house computer systems. SOPs provided explicit instructions for all transfer processes.

Production: Compounding This area of the plant is where tablets, creams, medicines, testing kits, etc. are manufactured. Because it is a restricted area I was not granted access to the actual site. In terms of mathematics, the operators were required to complete production and packaging order sheets, and to calculate the actual yield (usable quantity divided by total quantity, expressed as a percentage) at the completion of each process.

Production: Packaging Tubes of cream were packaged by hand, while small bottles of liquid used in testing kits and tablets embedded in foil blister packs were machine-packaged. The operators were required to complete production and packaging order sheets, and to calculate the yield, read data from the instrument used for torque testing of bottle caps (to ensure that they were not too tight or too loose), count the reject items, check the codes, and estimate the amount of glue needed on the day for the labels. They were also required to generate the labels using a computer. Every half hour, data was recorded for the tubes of cream and tablets: the codes, the number of leaflets used and the number still available; the number of boxes likewise, in order to ensure that they were 100 % accurate. If not, production was halted while the discrepancy was investigated. For the bottles of testing kits, samples were taken at the beginning, the middle, and the end of the process; the torque tests for caps (12–20 lb) were completed every 15 min. After production was completed, products were transferred to the Finished Goods Warehouse.

Distribution: Finished Goods Warehouse Here the operators were required to check that the company paperwork corresponded with their records. They would then store the finished goods, locating them according to turnover rates, and later pick the goods required for despatch. Picking goods on order required that the operators check the quantities. This meant calculating the total number of items by multiplying the contents of each box by the total number of items per box, again taking into account the part-boxes. For each individual order, there was to be one tray. For supermarkets, there was to be one pallet per product. The operators needed to check that the paperwork on the stock location sheet corresponded with the quantity, batch, and expiry date; they also needed to check the FIFO date. For chilled products they needed to check the temperatures on refrigerators by monitoring a 24-h graph, and a logarithmic graph on the deep freeze unit for storing goods.

Outwards Goods: Packing/Stock Control After the orders of picked goods arrived, they were packed up for despatch; chilled and frozen goods were always to be packed at the last minute. This section also dealt with the organisation and recording of transport arrangements. Here, operators were required to check that the pick slip data corresponded with what they received, do the packing (machine assisted where appropriate), and weigh the package to the nearest kilogram. Based on the length and weight of the order, they had to organise transport according to the need: they used a pictorial guide of different forms of delivery vans according to these weights and lengths, ranging from small taxi trucks to very large heavy vehicles used for supermarkets and shipping containers. Especially in the case of chilled or frozen goods, the transport services had to be phoned in order to confirm that they would meet purchaser deadlines. In terms of record keeping, all the data or codes, locations, pick slips and job numbers, and carrier information had to be entered onto the computer. Records of packing included entering the con-note [consignment note] number for proof of delivery and to follow up any subsequent queries. Finally, the company would expect to receive a fax to confirm delivery.

An additional observation was of a whole-staff meeting where management reported on the previous 12 months' results, using tables including percentages, charts, and graphs, comparatively, to elaborate on their claims, both positive and negative, about the year's productivity. Following Bernstein (2000), I believe that workers have the democratic right to understand and evaluate claims based on mathematical objects such as these.

At first glance the operators' work might appear fairly banal, focusing on counting, estimating, completing and checking forms, reading of measurements, storing and locating materials and finished goods, and using technology as a form of shared record keeping. This kind of work was typical of many work sites I have visited across a range of industries (e.g., FitzSimons 2005; FitzSimons et al. 2005). However, here there was an even more acute emphasis on avoidance of errors, and the workers had not only internalised this but took great pride in their work, encouraged and supported by management in forms of personal recognition at the annual Christmas function. Their levels of functioning were far more sophisticated than the curriculum of Calculations A would suggest. Many of the operators were new immigrants to Australia and, to my surprise, I learned that their educational backgrounds were quite varied, ranging from early school leavers to university graduates. Cohorts of students as varied as this are commonplace in adult education, and the challenge for teachers committed to maximising learning opportunities for each person is to make them feel respected and valued for what they bring to the learning-teaching situation.

Pedagogic Design: Addressing the Tensions and Contradictions

In order to overcome what I perceived to be the manifest weaknesses of the accredited curricula, I took the opportunity to combine Calculations A and Basic Computer Skills, and integrated, where possible, important aspects of other modules listed above. This offered a way around the constraints of teaching the decontextualised or pseudo-contextualised skills which would follow from a strict interpretation of the Calculations A module. It also enabled me to build on and share my observations of the whole process, from start to finish, enabling the workers to feel valued for their personal contribution, and to appreciate the needs of others, upstream and downstream.

As with the impoverished curriculum, I found the published student support materials to be unlikely to reflect the complex realities of production. For example, when a breakdown in production occurs for a given length of time, there needs to be line clearance, where the machinery must be cleaned out, as is normal at the end of any production run, and then the process started anew. However, the learners' guide ignored all of these important characteristics of actual work practice. In one worksheet, after conveniently choosing the breakdown time as 12 noon, it assumed that production could immediately resume at the point of stoppage once the breakdown had been fixed. Depending on the circumstances, this could be quite illegal! Another worksheet encouraged rote learning without any conceptual foundation: Multiplication and division by powers of 10 is "simply" accomplished by moving decimal points so many places to the left or right! But which way?

On the other hand, posing the question: "What can go wrong in your work that might be related to numbers or measurement?" provided operators with rich opportunities to demonstrate their depth of knowledge about the work processes they were involved in and to make their own connections with the official curriculum. In reality, workers everywhere may be confronted with problems arising from the lack of appropriateness, quality, and maintenance of tools and equipment of production for the specific task(s) at hand, not to mention other social, environmental, and communication problems. However, these are not generally the focus of training programs; many of my other vocational students have reported a general lack of interest by supervisory staff to the reporting of such problems in their workplaces, including air traffic controllers! Fortunately, the pharmaceuticals manufacturing industry is highly regulated, precisely because of the social, health, and environmental risks to workers and to the general public as end-users.

Technology was a major focus of the pedagogy. This meant both learning about the electronic and other technologies utilised at the site (e.g., various computer networks; processing, packaging, and weighing machines; storage and refrigeration facilities), and learning through these technologies as well as scientific calculators. It was possible for the operators to practise computer skills on various types of low-tech computerised machinery used for printing labels, for example; or to learn about the different internal and external communication systems and databases in use at this site. However, meeting the assessment requirements for both calculations and computing meant designing further learning activities.

The school-mathematical activity of calculating statistical means after physically rolling three dice at a time was a starting point, which eventually led to discussion of probability and the concepts of randomness and variability. Using a professional statistics package, *Minitab*, we reproduced the individual and combined data on the computer, and then conducted simulations of this work on a large scale. Plotting the results as bar charts and then as a quality control [QC] chart, with $n=3$, elicited the surprising information that these were the kinds of charts with which the operators were regularly confronted in production meetings. These charts were produced from data based on each area's work which operators were expected to understand in a general sense, and for which they were held accountable. Now, for the first time, the operators had some understanding of where these charts came from, and began to appreciate what happened to the values they had entered personally or which had been entered automatically. Even at the relatively unsophisticated level of statistical understanding made possible by our activity, operators were more likely to participate in discussions about the causes of defective production, without fear of showing their ignorance.

Throughout the pedagogic activity, calculations were performed in a natural manner, embedded in contexts meaningful to the workers, who were then able to locate themselves personally in the activity and to see the vital importance of accuracy in meeting official standards. Discussions reflecting on social issues such as gambling, uses and abuses of computer technology, together with industrial issues of quality management, took place. (Sadly, no discussions reflected on the uses and abuses of mathematics.) Formal computer activities took place in the university—familiarising the workers with educational institutions, including using the computerised catalogue system in the library, and locating books, magazines, etc. through putting the Dewey decimal system of cataloguing to practical use. Our guided tours of the plant, which integrated workplace calculations and computer usage, enabled the workers to learn about sections upstream and downstream that they had never previously visited, highlighting their interdependence. It also enabled the remaining staff to learn about the work of these colleagues, as well as the nature of the training program itself.

Assessment was integrated within learning activities and workers were given the opportunity to share their mathematical knowledge with others. Utilising the concepts of measurement and unit conversion from the non-core units of Calculations B & C—which were extremely unlikely to be taken up voluntarily—enabled the workers to locate these tasks within contexts with which they were familiar, allowing them to judge the reasonableness of their answers. For some, this meant entering new territory allowing them to develop more confidence should they take on these tasks in the future.

The *Calculations A Assessment Task* was designed using actual activities, actual product names, and the workers' own names, drawn from each of the six major production stages discussed above. It also included an open-ended question: "Sometimes things go wrong in the workplace. Think about one thing that can go wrong in your area, such as counting, measuring, or locating (finding something). Use your knowledge of mathematics to explain what went wrong and how it could be fixed." The *Basic Computer Skills Assessment Task* asked workers to show how much they knew about workplace processes which were related to computers. It included describing, explaining, and using computer technology; also discussing the purpose of computers and their impact on society. In the following section I discuss some of the workplace competencies needed in modern industrial economies and that are integral to the activities performed by workers.

Workplace Competencies

Techno-mathematical Literacies (TmL)

Noss (1997) claimed that sophisticated mathematical skills are required for interpretation of results as well as error detection or retrieval from catastrophic technological breakdown situations. He argued that in many work situations there is less reliance on traditional school mathematics skills which can be carried out more efficiently by computers, and greater reliance on an ability to think in a mathematical way. Workplace decisions are based on an interplay of complex relations between professional, vocational, and mathematical knowledge. When these become contested or problematic, a workplace mathematics far broader than basic numeracy is required. According to Hoyles et al. (2010), changes arising from the introduction of information technology (IT) into the workplace have resulted in a shift "from fluency in doing explicit pen and paper calculations, to fluency with using and interpreting output from IT systems in order to inform workplace judgements and decision-making" (p. 7).

Hoyles et al. (2010) also elaborated on the notion of boundary objects (see, e.g., Star 2010), such as paperwork commonly shared between managers and operators, where breakdowns in communication may occur due to misunderstandings. (Hoyles et al. added the term *symbolic* to indicate the link with TmL.) As discussed above, paperwork plays a crucial part in the pharmaceutical manufacturing industry in terms of records for accountability purposes. Not only do all the records of serial numbers, quantities, and locations of materials used in the warehouse, production, and packaging have to be scrupulously correct, but it is likely that at least some operators will be faced with complex tables, charts, and graphs. As explained above, QC charts were commonly used by managers at production meetings, and offer an example of a boundary object which, prior to our computer workshop, was virtually meaningless and in the realm of unthinkable knowledge (Bernstein 2000) for operators. Even though it was impossible for them to fully comprehend the high level statistical detail in our limited time frame, the operators at least began to have some appreciation of their meaning and use.

It may be argued that the case study described above is relatively low-tech and possibly dated in comparison to the very high-tech pharmaceutical manufacturing company researched by Hoyles et al. (2010). That company had a phenomenal output of generic drugs, utilising the cutting edge technology of automated packing machinery, supported by minimal technical staffing; all extensively electronically monitored by a system of *Overall Equipment Effectiveness* (OEE). However, because of their lack of appreciation of the system's complexity and how each variable worked, together with possible resistance to workplace surveillance and control, technicians were reluctant to engage with this information. Hoyles et al. recommended that, in order for this system to act as a symbolic boundary object, facilitating communication between workers and management, the technicians be given access to the whole production process—a critical feature of our case study.

Even though the mathematics needed in the pharmaceutical manufacturing plant in this case study appears, at first glance, relatively low level, a deep conceptual understanding of skills such as accurate counting, reading, and recording of numbers and codes is absolutely essential. In addition, understanding the implications of errors and how to redress them is fundamental. According to Lindenskov and Wedege (2001, p. 12):

> One does not simply count. There is a work-related aim in counting, and a certain precision is demanded. There are certain limits to the time consumption. Often one already knows the items that are to be counted. Often the shape of the items and the arrangement of the workplace will call for a special way of counting. Finally it is the organisation of work that determines who counts, controls and documents, whether it takes place individually or in co-operation, and who can suggest changes. Counting in a work context is not only counting.

As noted previously, any error can result in tragedy for end-users and/or withdrawal of the company's right to operate—apart from disastrous public relations.

Work Process Knowledge

Boreham (2004) describes work process knowledge as "a systems-level understanding of the work process in the organization as a whole, enabling employees to understand how their own actions interconnect with actions being taken elsewhere in the system" (p. 209). Boreham adds that:

> Work process knowledge is a synthesis of theoretical and experiential knowledge. While codified knowledge in the form of theory or written procedures might not be sufficient to guide action by itself, when it is synthesized with personal knowledge of the work situation the resulting construction—work process knowledge—allows people to make sufficient sense of the situation to enable them to act. (pp. 214–215)

Holistic conceptions of workplace learning, such as work process knowledge, do not appear to have any place in any atomistic curricula such as those described above. Work process knowledge is an essential part of a successful and safe workplace, and the teaching/learning program described here contributed to developing such knowledge in an informal and enjoyable way, as the worker-students learned with and from their workplace colleagues around the work site.

The study by FitzSimons et al. (2005) into the highly regulated chemical spraying and handling industry identified several components of numeracy learning. These include:

- having an awareness and understanding of the problems and risks
- having the confidence and knowing when to seek and gain information and confirmation from other workers, manuals, package labels, historical records; also the internet
- being able to cope with the complexity of information potentially available
- having the ability to learn from experience
- developing the teamwork skills of joint planning and problem solving.

In small-scale workplaces, such as plant nurseries or wine grape production, supervisors allowed novices restricted parameters for decision making, always under guidance, until they had a proven record of safe practice. In this way serious mistakes could be avoided, yet opportunities for reflection on errors could be provided as an explicit learning experience, in addition to the implicit learning taking place. This kind of learning experience is in direct contrast to the narrow focus of the learning outcomes described above.

Communication

The ICMI Study on *Educational Interfaces between Mathematics and Industry* (Damlamian et al. 2013) highlights on a broader scale the vital importance of accurate communication in sending and receiving information, online or offline, whether orally, in textual form, through gesture, etc. Not only do workers need to communicate about mathematically relevant issues with people in the workplace hierarchy who are above them, below them, or on their own level; they also need to have the confidence to question the mathematical assertions and assumptions of others in their workplace contexts. In addition, they may have to answer questions from consumers, and even to educate them. When people such as team leaders or supervisors, become responsible for the on-the-job training of apprentices and other newcomers, high level communication skills are vital, especially in science and technology.

What Are the Implications for Institutionalised Mathematics Education?

Even with year 12 or higher qualifications, many school leavers of today are likely to enter similar kinds of industrial workplaces. Their possible employment foregrounds could include paid full-time, part-time, or casual work; they could be employed by others, self-employed, or unemployed, or unpaid volunteers. They may wish to join or return to the paid workforce, get a promotion, reduce their workload, change jobs, become multi-skilled, return to study, and so forth. They will have important roles as citizens, partners, parents, carers of relatives or other significant people as members, even officials, in social, sporting, or other special interest groups, where mathematically-based decision making is required, possibly involving legal and moral responsibilities. They will also have domestic tasks, including home maintenance, budgeting, caring, and so forth. Notice that these foregrounds include aspects of life well beyond the normal assumptions of ongoing full-time paid employment. Many young people still in education are, or have been, already employed, paid or unpaid, and bring these experiences to learning situations; some they may wish to share and some not.

For me, the most important aspects of the pharmaceutical manufacturing workplace study included contextualised learning which valued learners' knowledges and their workplace experiences, treating them as respected partners on our mathematical and technological journey. This journey was made as holistic as possible, developing work process knowledge encompassing the whole plant, with contributions from other workers along the way, thereby enabling the operators to link their everyday knowledge with scientific knowledge, and integrating their assessment with their learning experiences. The inspired request from the manager led to an expansion of their learning horizons into, for many, the unthinkable territory of a large university where they could begin to feel at home; for some this was the first library that they had ever entered. How might school students be supported to broaden their learning horizons beyond the classroom?

In formal educational settings, it may be possible to identify and incorporate knowledge gained from formal and informal workplace or other out-of school experiences into mathematics activities (Bonotto 2013). It may be possible to give students practical experience and to develop confidence in using context-specific language to clarify understanding and resolve mathematical problems across different communities of practice (i.e., across boundaries). It may be possible to use their work or life experience to understand and see the potential of mathematics as a conceptual tool for critically appraising existing work or social practices, and to take responsibility for working with others to conceive and implement alternatives (see, e.g., FitzSimons 2011, 2012).

Revisiting Democracy

I now return to the earlier discussion of democracy and pedagogic rights. The implied assumption of failure in school mathematics achievement, together with the restricted levels of mathematical thinking apparent in the Calculations A curriculum, are consistent with Bernstein's (2000) theorisation that the workers will be "positioned in a factual world tied to simple operations" (p. 11). Socialisation into this kind of pedagogic code is likely to deny workers their pedagogic rights to democratic participation.

In relation to Bernstein's (2000) concept of boundary, I have tried to show how the boundaries seemingly established through what appeared as a very limiting and limited core curriculum could be overcome: First, by recognising that (often invisible) mathematical skills actually constitute critical components in a highly regulated workplace that can affect human lives and livelihoods, and where errors can be fatal; and second, by respecting the full integrity of the work actually done by operators on a daily basis to show that it has far more depth than a superficial reading of the official curriculum would suggest. Every single code or number carries a great depth of meaning, and all discrepancies between expected and actual values must be accounted for. Simplistic pedagogical practices such as instructions based on rote learning to move decimal places to the left or right, or "convenient" breakdowns do not do justice to the responsibilities, and the codified as well as tacit knowledges of the workers.

Bernstein's (2000) first pedagogic right is that of enhancement at the individual level—experiencing boundaries as tension points between past and future options, thereby developing the means of critical understanding and awareness of new possibilities, which he described as a condition for confidence. This chapter has argued that physical and intellectual boundaries can be transgressed, both in the workplace where workers were able to appreciate the contribution of their own role in the greater scheme of production, and through the dual location of studies in the comfort zone of their own plant, and at the university where they could gradually develop a new confidence and agency. In addition, by drawing on the applications of the non-core units of Calculations B & C as offering meaningful contexts for the mathematical work they were already doing, or might be expected to do, operators transgressed the boundaries of the official curriculum. No doubt they would have been more confident to attempt mathematical further study had these units been offered—or perhaps even to contemplate a different academic pathway.

Pedagogic activities such as those described above contributed to Bernstein's (2000) second right, that of social, intellectual, cultural, and personal inclusion. As a reflection of the importance of this principle, the workers' knowledge and experience, as well as their artefacts and work practices, were acknowledged and valued as their voices and stories in the activities were incorporated into our texts. Knowledge was distributed among us all: I had more experience and expertise in the mathematics/statistics field, whereas the workers had years of practical experience and know-how. Utilising teaching and learning places and spaces within and around the plant, including the canteen, was where the workers felt comfortable; whereas, initially, the university was where I felt more comfortable. But we each crossed our own boundaries to learn new ways of thinking and to appreciate and value the differences.

The right to democratic participation, as a condition for civic practice or discourse, must be about practice which has outcomes, according to Bernstein (2000). Access to unthinkable knowledge, going far beyond the official knowledge, was enhanced through the activities which removed the veil of secrecy, however unintentional, surrounding artefacts such as the production and printouts of QC charts and other official documents, tables, and graphs which had acted as boundary objects between workers and supervisors. The operators were also invited to critique the uses of computers and the social issues surrounding gambling. As it happened, one of the manager's final comments to me was to remark on his delight that the workers were actually beginning to pose questions about their daily work, rather than merely accepting outputs without question. From my perspective, the increased knowledge and awareness of previously opaque artefacts at least gave the workers a chance to challenge or "answer back" to criticisms, direct or implied—and to have the confidence to do so. For some operators, even setting foot inside a modern day education institution—let alone a university—or library had also previously been unthinkable. As Bernstein (2000) noted, education is necessarily a value-laden activity; curriculum and pedagogy cannot be separated from values.

Acknowledgement This paper is written as part of the *Adults' Mathematics: In Work and for School* research project awarded to Tine Wedege and led by Lisa Björklund Boistrup supported by the Swedish Research Council, 2011–2014.

References

Bernstein, B. (2000). *Pedagogy, symbolic control and identity: Theory, research, critique* (Rev. ed.). Lanham: Rowman & Littlefield.

Bishop, A. J. (1988). *Mathematical enculturation: A cultural perspective on mathematics education*. Dordrecht: Kluwer.

Bonotto, C. (2013). Artifacts as sources for problem-posing activities. *Educational Studies in Mathematics, 83*(1), 37–55.

Boreham, N. (2004). Orienting the work-based curriculum towards work process knowledge: A rationale and a German case study. *Studies in Continuing Education, 26*(2), 209–227.

Damlamian, A., Rodrigues, J. F., & Sträßer, R. (Eds.). (2013). *Educational interfaces between mathematics and industry: Report on an ICMI-ICIAM-Study*. New York: Springer.

Education, Audiovisual & Culture Executive Agency [EACEA]. (2011). *Mathematics in education in Europe: Common challenges and national policies*. Brussels: EACEA.

FitzSimons, G. E. (2005). Numeracy and Australian workplaces: Findings and implications. *Australian Senior Mathematics Journal, 19*(2), 27–40.

FitzSimons, G. E. (2008). Mathematics and numeracy: Divergence and convergence in education and work. In C. H. Jørgensen & V. Aakrog (Eds.), *Convergence and divergence in education and work* (pp. 197–217). Zurich: Peter Lang.

FitzSimons, G. E. (2011). A framework for evaluating quality and equity in post-compulsory mathematics education. In B. Atweh, M. Graven, W. Secada, & P. Valero (Eds.), *Mapping equity and quality in mathematics education* (pp. 105–121). Dordrecht: Springer.

FitzSimons, G. E. (2012). Family math: Everybody learns everywhere. In J. Díez-Palomar & C. Kanes (Eds.), *Family and community in and out of the classroom: Ways to improve mathematics achievement* (pp. 25–46). Bellaterra: Servei de publications de la UAB.

FitzSimons, G. E., & Wedege, T. (2007). Developing numeracy in the workplace. *Nordic Studies in Mathematics Education, 12*(1), 49–66.

FitzSimons, G. E., Mlcek, S., Hull, O., & Wright, C. (2005). *Learning numeracy on the job: A case study of chemical handling and spraying*. Final Report. Adelaide: National Centre for Vocational Education Research. http://www.ncver.edu.au/publications/1609.html. Accessed 15 May 2013.

Hoyles, C., Noss, R., Kent, P., & Bakker, A. (2010). *Improving mathematics at work: The need for techno-mathematical literacies*. London: Routledge.

Kent, P., Bakker, A., Hoyles, C., & Noss, R. (2011). Measurement in the workplace: The case of process improvement in the manufacturing industry. *ZDM – The International Journal on Mathematics Education, 43*(5), 747–758.

Lindenskov, L., & Wedege, T. (2001). *Numeracy as an analytical tool in mathematics education and research*. Roskilde: Roskilde University, Centre for Research in Learning Mathematics.

Noss, R. (1997). *New cultures, new numeracies. Inaugural professorial lecture*. London: Institute of Education, University of London.

Star, S. L. (2010). This is not a boundary object: Reflections on the origin of a concept. *Science Technology Human Values, 35*(5), 601–616.

Mathematical Modelling and Bank Loan Systems: An Experience with Adults Returning to School

Vera Helena Giusti de Souza, Rosana Nogueira de Lima, Tânia Maria Mendonça Campos, and Leonardo Gerardini

Abstract We chose two Brazilian bank loan systems to discuss with a class of 20–38 year-old students from a last year in Brazilian "Education of Youth and Adults", aiming to verify which kind of discussion—mathematical, technical or reflexive—emerges in a mathematical modelling environment. We divided the class in seven groups of three students, and organized four 90-min meetings during their mathematics classes, all of them in the evening. In each of the first three sessions, we presented three guiding questions to be discussed in the small groups. In the fourth, the teacher-researcher promoted a whole class discussion. In this article, we present some of the discussions raised during the sessions in one of the groups, which has shown interaction among them. We realised mathematical discussions were almost absent in their discourse and they were concerned with solving the questions as mathematical exercises.

Introduction

Education of Youths and Adults (EYA) in Brazil aims to provide education to those who did not finish basic school at the foreseen age and return to school later, most of the times coerced by working conditions. In EYA, each semester is equivalent to 1 year of the regular education, and this means that EYA High School in Brazil can be completed in 18 months instead of the three regular years. Students in EYA courses are adults and do not return to school to learn what they have missed from school when they were young, but to learn abilities they currently need in their

---

V.H. Giusti de Souza (✉) • R. Nogueira de Lima • T.M.M. Campos
Post-Graduate Program in Mathematics Education,
Anhanguera University, São Paulo, Brazil
e-mail: verahgsouza@gmail.com; rosananlima@gmail.com; taniammcampos@hotmail.com

L. Gerardini
Colégio Dante Alighieri, São Paulo, Brazil
e-mail: leogerardini@yahoo.com.br

© Springer International Publishing Switzerland 2015
U. Gellert et al. (eds.), *Educational Paths to Mathematics*,
Advances in Mathematics Education, DOI 10.1007/978-3-319-15410-7_6

professional lives, as suggested by Kooro and Lopes (2007, p. 2), when they state that "… the adult is not a child. The learner brings a life experience and an apprenticeship that is not usually considered in learning experiences". Hence, we chose a mathematical modelling approach with two bank loan systems to work with an EYA group of students, taking into account the words of Fonseca (2005) that this kind of education prioritises "the possibility to a more democratic access to literate culture, and mathematics teaching must engage in this task, considering its proper resources and opportunities" (p. 26).

Freire (2005) calls "banking education" not the study of financial mathematics, which is our aim in this research study, but the model in which a student is considered as a depository of information and contents (either mathematical knowledge or the knowledge concerning any other subject). This concept is meant as a criticism against the Brazilian educational system as a whole, in the same way that Moretto (2003) calls students who are faced with this kind of teaching "information accumulators". In our study, we aim at educating citizens and not information accumulators, meaning that these students should be able to critically participate in their social environment. With EYA, this characteristic is even stronger, since the students are older than the students from regular schools, have a broader experience in life and are already engaged in their workplaces.

We chose to discuss two Brazilian bank loan systems with distinct premises with a group of last semester EYA students, aiming at analysing what kind of discussions—mathematical, technical or reflexive (Barbosa 2001)—are raised in a mathematical modelling environment using activities involving two bank loan systems and the advantages and disadvantages for the ones in need of financial help. In this article, we briefly describe the EYA groups' discussions regarding their work to understand two Brazilian most used bank loan systems: Price System (PS) and Constant Amortisation System (CAS).

Adult Mathematics Education

Considering research studies outside Brazilian EYA, the main concern presented in various research studies regarding formal adult mathematics education is numeracy (Wedege 2009), especially related to developing personal empowerment (Hassi et al. 2010) and basic skills in mathematics at work (Wedege 2009). In our research, we are not directly discussing numeracy nor the kind of personal empowerment mentioned in those research studies, but we chose a content that is taught in formal adult mathematics education in Brazil: simple and compound interests, that we chose to discuss by the use of bank loan systems in a mathematical modelling environment. In this way, we believe, such subject can provide basic skills useful for work as well as for the personal empowerment to deal with situations in real life, as, according to Niss (1996, p. 13), one reason why it is necessary to teach mathematics is "providing *individuals with prerequisites which may help them to cope with life* in the various spheres in which they live: education or occupation; private life; social life; life as a citizen" (italics original).

FitzSimons (2008) argues that "In order to develop adults who are considered numerate in contemporary society, there needs to be a convergence between teaching of the vertical discourse of mathematics and the horizontal discourse of numeracy" (p. 10). In the way they were dealt with in our study, we consider bank loan systems as a horizontal discourse inserted in the vertical discourse (Bernstein 2000) of school mathematics. We are convinced that Brazilian EYA courses are restricted to vertical discourse of mathematical contents, which are extracted from the regular curriculum although taught in different rhythm, since the length of the courses is half the length of regular ones. Also, in these courses, the teacher usually plays a central role in the process. In our research study, we decided to use a mathematical modelling environment in order to give voice to the adult learners, as we believe their lifelong learning should be considered when they are in a learning situation (Hassi et al. 2010).

Considering research studies in Brazilian EYA, Araújo (2007) conducted semi-structured interviews with EYA students from secondary and high schools, solving problems involving elementary mathematical contents. Through these interviews, the author realised that those students had gaps in their reading abilities and trouble in interpreting the information presented in the problems. In addition, they lacked knowledge in elementary mathematics. The students who were able to solve the proposed problems applied mathematical knowledge they regularly used in their own lives. Fonseca (2005) also emphasised this, claiming that in EYA classrooms mathematical contents that are intimately related to students' needs should be taught by using resources that the students themselves bring to the classroom and which they have acquired through their social and/or professional experience.

Furtado (2008) describes Brazilian high scores of school failure, and most students that look for EYA courses have not been successful in their regular school lives. In the light of such a conclusion, he suggests the need to use special teaching strategies for EYA courses regarding students' interests and expectations in this modality of learning. This is what Corôa (2006) evidenced in his research study. He interviewed teachers who work in EYA in order to understand how they introduce mathematical contents that need to be developed according to EYA's program. He found that those teachers adapt contents to bring them closer to students' daily lives.

Considering these findings, we used mathematical modelling as a special teaching strategy for EYA and bank loan systems, as they stand in close relation to real life situations.

Students' Profile

This research study was carried out in a public school in a "favela" of São Paulo's outskirts, with four classes of EYA students in the last stage of the course. The students attend school in the evening. Our focus was on the students of one of those classes who agreed to participate in this research study. In order to build their profile, we administered a questionnaire with nine questions regarding gender, age, birthplace, profession, previous education, period of time out of school and reasons for that, as well as their feelings about mathematics. We assumed that these characteristics might interfere with the group's motivation to work with a mathematical modelling approach and with bank loan systems.

Analysing the questionnaire, we observed that the group had 21 students, 17 female and 4 male, with ages varying from 20 to 38 years old, most of them coming to São Paulo from the Northeast States of the country without having graduated from High School at the expected age (in Brazil, 17 years old). They have been away from school for 6 years or more and expect from EYA to prepare them adequately for working demands in Brazilian States different from that of their birth. Most of them work in commerce and finishing school is essential and necessary, since they are already working, and they cannot take the risk to be replaced by more qualified workers. They generally find it difficult to understand mathematics but consider mathematics a necessity for and in their lives.

In conclusion, we considered that students with such a profile are more mature and experienced than regular high school students, and that they work with interest. Furthermore we assume that the topic of bank loan systems might be important and motivational for them, for example with regard to the acquisition of their own house.

Methodological Choices

We decided to do a research study with a qualitative analysis of data. Therefore, we divided the class into seven groups of three students and organized four 90 min meetings during their mathematics classes. During the first three meetings, we presented PS and CAS tables of values and three guiding questions to be discussed in the small groups. In the fourth meeting, the teacher-researcher promoted a whole class discussion. Meetings were tape-recorded in agreement with all the participants and records were analysed.

In order to organize the activities, we chose Mathematical Modelling Case 1 (Barbosa 2001), in which the teacher presents a problem, gives data to solve it, and the students have to solve the problem using the given data. To work with the tables (PS and CAS), we chose a R$ 100,000.00 loan to be paid in 120 months, considering these data viable for the research study participants, since there was a government incentive program to make it easy for the population to borrow money to buy their own house up to R$ 100,000.00.

These students were not used to work autonomously with mathematics questions nor were they familiar with reading and analysing amortisation systems' tables. To help them to work in those ways, we thought it was necessary to give them some guiding questions to analyse the given tables. In each of the group meetings, we provided three guiding questions and the amortisation tables. In the whole class discussion meeting, one question for discussion was given.

For the first meeting, the guiding questions were intended to help students with the reading of the table:

> In each table, locate the value of the loan and the number of instalments.
> Observe if there is a column or line with constant values in each table. Does it happen in the CAS table? In PS? What does that mean?
> Observing the instalment columns, what can you grasp from each table? Discuss your point of view with your group.

On the second day, the questions incited a more critical reading, aiming to compare the nature of the two systems:

> Choose three random lines in each table and analyse what information you can get from them.
> Observing amortisation and interest columns and their relation to the instalments, what can you make from the CAS table? The PS table? What does that mean?
> Is the interest value to calculate the first instalment the same in both amortisation systems? Explain.

In the third meeting, the questions were designed to raise the mathematical content with which the tables were constructed:

> If you have already understood both amortisation systems, can you tell how much the total value of the amortisation column is without calculating it?
> Discuss with your group members whether you can explain how the instalments are calculated in each system.
> How do you calculate amortisation values using the CAS system?

Finally, for the discussion meeting, the question was mainly to incentivize the discussions:

> Give reasons for choosing one of the amortisation systems to get a loan to buy a house, depending on the social, economical and political scenario of the country.

For this article, we chose to present two kinds of data: dialogues from one small group, with the kind of discussions (mathematical, technical and reflexive) they presented; and some of the arguments given in the fourth meeting regarding the choices of an amortisation systems. For many reasons, only three of the seven groups participated in all meetings. The small group A was chosen for closer analysis, because they have participated in all meetings, have shown more interaction among them than the other two groups, and their interactions presented more characteristics of the three kinds of discussion we intended to look for. In addition, they were also chosen due to the kind of doubts they raised in the first meeting, when they showed less knowledge than the other groups about the tables, and even so, their development on the subject was higher than that of all other groups.

For the analysis of the discussions, we used Barbosa's (2006) ideas that in a mathematical modelling environment three kinds of discussion may occur: mathematical, technical and reflexive. Mathematical discussions are connected to mathematical content; technical, to the translation of the situation in mathematical language; and reflexive ones, to characteristics of the studied model and its consequences.

Mathematical Modelling

We chose to work in a mathematical modelling environment because we agree that it is beneficial to discuss theoretical concepts connected to real problems and to bring the mathematics classroom closer to students' real lives. Mathematical modelling environments also make possible a change in the teacher's role, since s/he is no longer considered the only one who holds the knowledge (Barbosa 2001), and the pedagogical process can be based in a dialogical relationship between teacher and student (Skovsmose 2004), allowing the teacher to respect the knowledge those students bring to school.

Among the possible mathematical modelling perspectives described by Barbosa (2006), we focused on the "socio-critical", as it seemed to be the most adequate for the students we worked with. We believe that social aspects are relevant because, in this environment, students may discuss the concepts involved in the model at hand in a way to understand such a model and use it appropriately in their real lives. Regarding the critical aspect, knowing the model and how to use it makes the students capable of criticizing its premises, and we believe that this is a way to foster critical citizenship.

According to Barbosa (2001), there are three cases to be considered in a mathematical modelling perspective taking into account the length in time in which the activities will be developed and the kind of tasks due to teacher and to students. In Case 1, the teacher presents a problem with all information needed to solve it, and students must solve it. In Case 2, the teacher presents a real problem, and the students must collect information needed to solve it. In Case 3, the students choose a non-mathematical theme, formulate a problem and are responsible for finding all information needed to solve it. For this research study we used Case 1, since students were not used to such an approach in mathematics classrooms, especially not to finding useful information to solve a problem.

The Price System and the Constant Amortisation System

In the PS, the monthly instalments are constant for the whole period of payment, and each instalment is composed of two parts: the interest and the amortisation. The interest is calculated according to a fixed rate over the amount due, and the amortisation is the amount subtracted from the amount due.

In the CAS, the monthly amortisation is constant and obtained by dividing the amount due at the beginning by the number of months of the loan. The instalment is also composed by interest and amortisation, interest calculated according to a fixed rate over the amount due.

By using these two kinds of systems, our goal was that students would understand and be aware of the differences between them, and be able to decide which system would be best for him/her in real-life situations.

Table 1 Excerpt of the PS table used by students

Price system (PS), interest rate per year: 12.79 %

Instalment number	Value of instalments (R$)	Interest (R$)	Amortisation (R$)	Amount due (R$)
0				100,000.00
1	1,480.68	1,065.80	414.88	99,585.12
2	1,480.68	1,061.38	419.30	99,165.82
3	1,480.68	1,056.91	423.77	98,742.05
4	1,480.68	1,052.39	428.29	98,313.76
5	1,480.68	1,047.83	432.85	97,880.91
...	...	...	...	...
55	1,480.68	745.23	735.45	69,187.14
56	1,480.68	737.40	743.28	68,443.85
57	1,480.68	729.47	751.21	67,692.65
...	...	...	...	...

Table 2 Excerpt of the CAS table used by students

Constant amortisation system (CAS), interest rate per year: 12.79 %

Instalment number	Value of instalments (R$)	Interest (R$)	Amortisation (R$)	Amount due (R$)
0				100,000.00
1	1,899.13	1,065.80	833.33	99,166.67
2	1,890.25	1,056.92	833.33	98,333.34
3	1,881.37	1,048.04	833.33	97,500.01
4	1,872.49	1,039.16	833.33	96,666.68
5	1,863.60	1,030.27	833.33	95,833.35
...	...	...	...	...
27	1,668.21	834.88	833.33	77,500.09
28	1,659.33	826.00	833.33	76,666.76
29	1,650.44	817.11	833.33	75,833.43
...	...	...	...	...

In Tables 1 and 2 we present an excerpt from the tables of both systems, PS and CAS, given to the students, in an attempt to clarify the description of data. In each table we chose to depict the first few lines of the tables and also three lines in which the interest is smaller than the amortisation.

Kinds of Discussions

According to Barbosa (2006), neither mathematics nor modelling are ends, but ways to question daily life situations intrinsically related to the society wherein the individual lives. This means that it is not expected that the students come out with explicit mathematical contents at play in the proposed situations, but that they discuss and criticize such contents.

Discussions emerging from an activity with mathematical modelling approaches can be categorized, according to Barbosa (2006), as mathematical, technical and/or reflexive. The mathematical discussions refer to concepts and ideas solely related to mathematics. This kind of discussion is raised in a mathematical modelling environment when the proposed problem demands the use of mathematical concepts or ideas either known to the students or new to them. Students discuss about the suitability of the concept in the situation, agree to use it, and solve the problem. If they do not know the mathematical content that seems to be necessary for solving the situation, they have to discuss how to search for ways to learn, understand and apply it. For instance, Oliveira (2009) presented two graphs to students, one representing the number of retired workers in time, and the other the number of active workers in time. When trying to find out when the number of retired workers would be smaller than the number of active ones, one student said "This is a first degree polynomial function. We can work with simultaneous equations in order to find out the values". He discussed with peers which mathematical idea would be the best to solve the problem, and this characterises a mathematical discussion.

Technical discussions are about strategies, procedures and techniques, related to the mathematical concept students have agreed to use in a situation and that can be applied to solve it. Again, if they do not know those strategies, procedures and techniques, they need to discuss how to achieve them. For example, we believe that a possible technical discussion that can be raised in the situation of Oliveira's (2009) study would be students attempting to find out how to solve simultaneous equations and actually solving them.

Reflexive discussions regard the understanding of the mathematical model, such as the premises used in building it, the results proposed by it and the usefulness of these results for the society. It may appear when students try to analyse the nature of mathematical models. In the same context of Oliveira's (2009) study, we believe that a possible reflexive discussion would be twofold: the discussion about the meaning of linear models, from the mathematical point of view; and the discussion about what more retired workers than active workers or vice versa implies for the society, from a socio-economical, critical point of view.

In the data analysis obtained from records of the seven groups, we searched for mathematical, technical and reflexive moments in the discussions in order to find out whether the study of PS and CAS with this group of students allowed the development of all three categories expected in the proposed approach.

Results

Considering first the dialogues of all seven small groups, we observed that students were not familiar with the idea of discussing a given situation to learn mathematics. For them, our questions represented exercises that needed to be solved and presented to the teacher. Consequently, students divided the tasks between them, and wrote their answers without much discussion. For instance, one student of Group F, in the first meeting, said:

F1: Do not worry, I will do it. If I have doubts, you help me, okay?

In the second meeting, the initial dialogue between students of Group F was:

F1: Let's do the following: you answer the questions about the CAS table, and I do it for the PS. Afterwards, we organize the solutions in order to give them to the teacher.
F2: No problem!
F1: In this way we do not waste time.

There was no further dialogue on the record of this group. In addition, Group E turned the tape recorder off before starting their discussions, so we do not know what their dialogues and conclusions were. We believe that this happened for some reasons. First, most of the students were women, and, in lower economical classes, many women are not used to making decisions about employment of family money in situations in which a large account of money is involved, such as a loan to buy a house. Hence, they may lack interest in the topic of our didactic activities. Of course further research would be necessary to prove this conjecture. Second, considering the kind of activity we proposed, involving discussion and analysis of the situation which the students are not used to, they may have been shy to let the teacher listen to their doubts and weaknesses regarding the subject.

Group A was one of the few that showed some argumentation among them and where there were reasons for their answers to the guiding questions provided by the teacher. Taking a closer look at the dialogues, we realised, however, that even while discussing among them students mainly clarify what is asked in the questions. They do not go further into the analysis of the systems. On the basis of the following dialogues, we present some categories of discussions this group realized in the meetings.

During the first meeting, students made an effort to understand both tables. For that, they engaged in technical discussions:

A3: The value of the loan is the same as the amount due, because we owe a hundred thousand Reais to the bank, which will be used to pay the house.
A2: Did you understand? If you borrow money, you will be owing to whom you asked it from, okay?
A1: Okay, I understand. But why didn't he [*the teacher*] ask us to locate the value of the amount due, since it was already written?

We interpret this segment as a technical discussion because the students describe how the table is presented, the content of each column and the relations between columns and lines.

After that, the students observed some characteristics of each system. We categorised this as a reflexive discussion, given that they analyse the assumptions of the bank loan systems.

A1: I understood that the amortisation column is constant and the instalments are decreasing in CAS table.
A3: In mine [*PS table*], I saw that amortisation is increasing and the instalments are constant.
A2: Good! The difference between them is that, in one of them, the amortisation column is constant [*CAS*] while in the other [*PS*] it is increasing; in one of them, the instalments are decreasing and in the other they are constant.

Another example of a reflexive moment is when a student associated the name of the system with what he saw in the amortisation column:

A1: Dude, this column is the amortisation, that's why it says "Constant System of Amortisation" on the top.

We did not notice the presence of any mathematical discussion in the first meeting, just technical and reflexive ones. It is possible that our questions did not stimulate students to raise them. In fact, the only mathematical discussion we found in the discourse of Group A occurred at the second meeting:

A1: I think we need to write down that the second column is equal to the third plus the fourth, in both tables.
A2: How did you notice this?
A1: Look, the value is R$ 1,480.68; the interest is R$ 1,029.08 and the amortisation is R$ 451.68, if you add the interest and the amortisation you have R$ 1,480.68, take the calculator.
A3: It is true, and also in lines 50 and 100, it is the same, and in the other table too.
A2: Dude, it is the same, the line doesn't matter.

We consider this a mathematical discussion because it entails a mathematical understanding: the amount you pay monthly is amortisation plus interest.

After this dialogue, the group members realised that the interest rate is 12.79 % per year in both systems, which we considered a technical discussion, because, although they realised the rate is the same, they did not show that they know how to apply such rate, as we can see in the following dialogue.

A1: In the interest column both values are R$ 1,065.80, but the instalment values are not the same.
A3: But he [*the teacher*] does not want to know the instalment value.
A2: We must explain why this occurs.
A1: I know! Here in the beginning a interest rate is the same for both tables, and it is 12.79 % per year.
A2: Okay, just write it down.

For the second meeting students in Group A had mathematical and technical discussions, with no evidence of reflexive discussion.

In the third meeting, when students were asked to give the total value of the amortisation column without any calculation, their utterances were as follows:

Adults Returning to School

A1: We are supposed to give the total value of the amortisation column.
A2: Look! The amount due decreases according to the amortisation, hence the total value must be the same as the initial amount due.
A3: I agree, that's the reason why yesterday he [*the teacher*] asked us to analyse the amortisation column.

This is a technical discussion, as the students understood that the amortisation is how much the debt decreases, and therefore, the total of amortisation column must be the initial amount due. Although they realised this fact, they did not go much further in their analysis of both tables. Their dialogues show that they were concerned with solving the questions as mathematical exercises, and that they were hardly thinking reflexively about the premises that rule the systems.

A1: How is the amortisation calculated in the CAS system?
A2: I have no idea.
A3: If the value is the same, there must be a formula.
A2: How can we make the calculation if we don't know the formula?
A1: If we needed a formula, he [*the teacher*] wouldn't ask.
A2: It must be some kind of division, because the answer is not an exact number.
A1: It is true, but it may not be something about the interest or the instalments.

In these last utterances, we see that the students believe that if there is a calculation to be made, there must be a formula for that, and the teacher is responsible for giving it to the students. If a number is not an "exact number", meaning a non-integer number, student A2 believes that it must come from divisions and the remainder of the group agree with him. This illustrates that although the students were immerged in a mathematical modelling environment, they do not show autonomous behaviour. They expect the teacher to give them all the information they need to do the task.

In the fourth meeting, the whole group discussed the guiding question with the teacher-researcher: "Give reasons for choosing one of the amortisation systems to get a loan to buy a house, depending on the social, economical and political scenario of the country." The whole-group utterances are interesting as they show that, although there are no mathematical discussions, students' critical senses were activated. In the future, if they need a bank loan, they will be aware of the existence of at least two kinds of systems, and the fact that they will have to decide which one is more appropriate for them. In the dialogue below, we present students utterances without connecting them to members of the previous groups.

K1: Teacher, the main difference of the tables is that in one of them the amount to be paid monthly is constant, and in the other, it decreases, so we must have different scenarios
K2: I believe that in our present scenario the ideal would be CAS, because here the amount to be paid monthly decreases, which would make our lives easier each month
K3: What is the problem with the amount being constant, in our current political scenario?
K4: Because I know every month what I must pay and if anything happens to the financial banking system, nothing changes to me

K5: So, Teacher, what is the best?
K6: But interest is higher in the price table
K7: But the amount to be paid in CAS is higher
K9: I believe that it depends on what kind of job you have
T: How do you explain that to us?
K9: Teacher, the job is a main factor, because if you have a regular job, the CAS table is better because I know that the amount will decrease and my wage won't. If you work on your own, Price is better, because you know that you have to run for that amount of money
K10: I agree with K9, but the best choice is when you have money!
K11: That's it!

During the study of these activities, students' reflexive discussions showed that they realised that the premises of each system are different. In PS, the instalments are a fixed value to be paid monthly, while in CAS, it is the amortisation that is a constant value. In addition, reflexive discussions in the fourth meeting evidenced that they realised that there is a better system to be chosen depending on their financial situation and what kind of job they are in—a regular job or an "autonomous" one. On the other hand, they did not discuss any of the other features of both systems. For example, in 120 instalments, the interest paid is higher than the amortisation up to the 55th instalment in PS (see Table 1), and up to the 27th in CAS (see Table 2); in PS, instalments are constant, and smaller than the first 48 instalments of CAS, which means 4 years of a higher monthly instalment if one chooses CAS. In our view, this means lack of mathematical and reflexive discussions during the work with these activities.

Conclusions

Literature regarding formal adult mathematics education has numeracy as its main subject of research as we discussed in section "Adult Mathematics Education". In this chapter, we presented a study with EYA students, analysing what kind of discussion—mathematical, technical or reflexive—is raised in a mathematical modelling environment. Therefore we used activities involving two bank loan systems, which we do not consider to be strictly related to numeracy as defined by Lindenskov and Wedege (2001, p. 5, displayed in Evans et al. 2013, p. 206), because we do not believe that bank loan systems can directly be related to "functional mathematical skills and understanding that in principle all people need to have". However, the mathematical content and interests are of course a topic here. Like Hassi et al. (2010), we also believe that curriculum items for EYA should consider students' environments, social and professional lives, and their needs. Accordingly, we chose to work with bank loan systems in order to discuss interests and to develop a horizontal discourse within the vertical one of school mathematics (Bernstein 2000). For that, we selected a mathematical modelling environment as a different teaching strategy, as Furtado (2008) suggests. We designed some guiding questions for students to work in groups during three meetings and to discuss with the whole class and the teacher in one last meeting. We used data from one of these groups—Group A—to exemplify the kinds of mathematical, technical and reflexive discussions they had when dealing with this theme.

As a result we found that technical discussions dominated. We assume that this is so because the students are more used to solve mathematical exercises than to reflecting about a situation presented to be discussed. Consequently, they rarely engage in mathematical discussions and do not verbally express their analysis of patterns that can be raised in the situation at hand. When asked about these patterns, students do not attempt to answer the guiding questions, which could guide them to mathematical, and hence, to reflexive discussions, relating the mathematical concepts in the activities to their own lives.

Such a behaviour may be due to the students' profile. Being in EYA and having a memory of an old school, in which the students were not encouraged to give their own opinion and where the teacher was the central holder of knowledge, may be an explanation to that. Even using a different strategy, that proved to be useful in Corôa (2006), was not completely successful in our study, since most students did not fully engage in the activity.

In a mathematical modelling environment, the teacher is not the central holder of knowledge, but the one who mediates students' interactions in small groups or in the whole group. This environment can raise positive aspects in an innovative way to apply mathematics in contextualized situations, which can be useful for these students' lives. Among these positive aspects there are: interaction between group members, which is not usual in mathematics classrooms; discussions about mathematics contents related to real life; and the need to take decisions in order to solve different activities in the classroom (the students' role might change from passive to more active).

In the particular case of this research study, with Constant Amortisation System and Price System, we point out as a positive aspect the arousal of conscience about different bank loan systems with different properties. Students should be aware of those differences in order to choose the system that is more suitable for his or her financial situation. However, as Araújo (2007) pointed out, the observed students lack basic mathematical skills to read the tables provided by us in order to understand the bank loan systems and raise reflexive discussions about their premises.

For future research, we suggest studies that consider more horizontal discourses to EYA students such as interests, both simple and compound, and even percentages. In addition, a stronger development of a mathematical modelling environment should be taken into practice with EYA students, in an attempt to help them to be more autonomous and more aware of mathematical patterns in contextualized situations. Type 3 of mathematical modelling, which uses contents chosen by the students themselves, should play a bigger role, as this might be a way to make them even more active during activities. Regardless of which case is being used, our conviction is that mathematical modelling environments are the most suitable way to work with EYA students in mathematics classrooms.

References

Araújo, J. L. (2007). Relação entre matemática e realidade em algumas perspectivas de modelagem matemática na educação matemática. In J. C. Barbosa, A. D. Caldeira, & J. L. Araújo (Eds.), *Modelagem matemática na educação matemática Brasileira: pesquisas e práticas educacionais* (pp. 17–32). Recife: SBEM.

Barbosa, J. C. (2001). *Modelagem Matemática: concepções e experiências de futuros professores.* Unpublished PhD thesis, Universidade Estadual Paulista (UNESP Rio Claro), Instituto de Geociências e Ciências Naturais, Rio Claro.

Barbosa, J. C. (2006). Mathematical modelling in classroom: A socio-critical and discursive perspective. *Zentralblatt für Didaktik der Mathematik, 38*(3), 293–301.

Bernstein, B. (2000). *Pedagogy, symbolic control and identity: Theory, research, critique* (Rev. ed.). Oxford: Rowman & Littlefield.

Corôa, R. P. (2006). *Saberes construídos pelos professores de matemática em sua prática docente na educação de jovens e adultos.* Unpublished Master thesis, UFPA, Belém.

Evans, J., Wedege, T., & Yasukawa, K. (2013). Critical perspectives on adults' mathematics education. In M. A. Clements, A. J. Bishop, C. Keitel, J. Kilpatrick, & F. K. Leung (Eds.), *Third international handbook of mathematics education* (pp. 203–242). New York: Springer.

FitzSimons, G. E. (2008). A comparison of mathematics, numeracy, and functional mathematics: What do they mean for adult numeracy practitioners? *Adult Learning, 19,* 8–11.

Fonseca, M. D. (2005). *Educação de jovens e adultos: especificidades, desafios e contribuições* (2nd ed.). Belo Horizonte: Autêntica.

Freire, P. (2005). *Pedagogia do oprimido* (45th ed.). Rio de Janeiro: Paz e Terra.

Furtado, E. D. (2008). Políticas públicas de EJA no campo: do direito na forma da lei à realização precária e descontinuidade. In *Anais do XIV Encontro Nacional de Didática e Prática de Ensino.* Porto Alegre: Nexus Soluções e TI.

Hassi, M. L., Hannula, A., & Saló i Nevado, L. (2010). Basic mathematical skills and empowerment: Challenges and opportunities, Finnish adult education. *Adults Learning Mathematics–An International Journal, 5*(1), 6–22.

Kooro, M. B., & Lopes, C. E. (2007). *Uma análise das propostas curriculares de matemática para a educação de jovens e adultos.* Paper presented at XI Encontro Nacional de Educação Matemática, Belo Horizonte, UFMG.

Lindenskov, L., & Wedege, T. (2001). *Numeracy as an analytical tool in mathematics education and research.* Roskilde: Centre for Research in Learning Mathematics, Roskilde University.

Moretto, V. P. (2003). *Construtivismo: a produção do conhecimento em aula* (3rd ed.). Rio de Janeiro: DP&A.

Niss, M. (1996). Goals of mathematics teaching. In A. J. Bishop, K. Clements, C. Keitel, J. Kilpatrick, & C. Laborde (Eds.), *International handbook of mathematics education* (pp. 11–47). Dordrecht: Kluwer.

Oliveira, M. K. (2009). *Cultura e psicologia: questões sobre o desenvolvimento do adulto.* São Paulo: Hucitec.

Skovsmose, O. (2004). *Educação matemática crítica: a questão da democracia* (2nd ed.). Campinas: Papirus.

Wedege, T. (2009). The problem field of adults learning mathematics. In G. Griffiths & D. Kaye (Eds.), *Proceedings of 16th adults learning mathematics* (pp. 13–24). London: Adults Learning Mathematics–A Research Forum.

Working with Adults: A Commentary

Javier Díez-Palomar

Abstract The commentary on the chapters of FitzSimons and of Giusti de Souza et al. reconsiders the distinction between 'horizontal discourse' and 'vertical discourse' to reflect on possible relationships between mathematics and meaningful contexts. By drawing on various sociological traditions the importance of these relationships for the design of 'democratic mathematical activities' is emphasized.

I want to start my commentary on the chapters authored by FitzSimons and by de Souza, Lima, Mendonça and Gerardini acknowledging my pleasure to discuss the significant contributions of their work to the field of adults learning mathematics.

Many years ago I met Dr. FitzSimons for the first time. She has a long professional career marked by her research in the field of numeracy, adult learning and mathematics in the workplace. Fourteen years ago, she edited one of the first reviews in adults learning mathematics research that I remember reading, with Diana Coben and John O'Donoghue (2000): *Perspectives on Adults Learning Mathematics, Research and Practice*. On that occasion, FitzSimons (2000) introduced the section on 'Adults, Mathematics and Work', contributing with a seminal work in which she compiled the main inputs made up to that time on the discussion about the relationship between teaching mathematics to adult learners and the analysis of how mathematics is present in a large range of tasks that workers face at their workplace. Fourteen years later, FitzSimons has just co-edited a special issue of the journal *Educational Studies in Mathematics* (Bakker 2014) about 'vocational mathematics education'. This special issue draws on discussions at the ICMI study on mathematics education and industry (Damlamian et al. 2013).

Throughout these years, the research in adults learning mathematics (ALM) has greatly evolved, thanks mainly to the emergence of socio-cultural and dialogical theories, whose theoretical concepts and epistemological basis opened new horizons of analysis previously unthinkable. Current research seems to indicate a move towards consolidating the use of concepts such as 'situated learning' (Greeno 1997; Lave and

---

J. Díez-Palomar (✉)
Department of Mathematics and Science Education, University of Barcelona,
Barcelona, Spain
e-mail: jdiezpalomar@ub.edu

Wenger 1991), 'contextualized mathematics' (Jablonka 2010) in local—vocational and non-vocational—settings. The proceedings of previous ALM annual conferences put on the table topics such as (professional) development of adult numeracy in nursing (Coben and Weeks 2012), recycling (Kane 2012), toolmakers (Mills 2012), using flyers to teach and learn reasoning about proportions (Abbott 2012), financial literacy (Beeli-Zimmermann and Hollenstein 2011), image of mathematics in advertising (Evans 2011), mathematics in engineering (Goold and Devitt 2011), etc. FitzSimons (2013) provides an updated picture of the state of the art in this field, acknowledging the irruption of the sociocultural theories (in particular sociocultural activity theory) to push forward the research field. The chapter written by de Souza and colleagues in this book is another example of that evolution. Their large experience working in the context of mathematics modelling and teacher training brings a fresh standpoint to the discipline, introducing new arguments to discuss a, perhaps, 'old' question in ALM: the problem of 'transferability' of knowledge in mathematics into everyday practice (Evans 1999; Kelly 2011; Keogh et al. 2012). Can concepts such as 'mathematization' open up the field to develop further the analyses on the links between mathematics and numeracy? Does starting from 'mathematics modelling' inspire innovative methodologies to improve adults' mathematics learning curricula? Can we re-read previous work in ALM drawing on this starting point? Is there any chance to democratise mathematics learning in adult education?

In their chapters, both FitzSimons and de Souza and colleagues use theoretical notions developed by Bernstein (1996) in order to explore the boundaries between everyday practices (manufacturing and loan calculations) and mathematics, particularly 'vertical and horizontal discourse'. These concepts are a more sophisticated extension of Bernstein's theory on elaborated and restricted codes, which Bernstein developed in the four volumes of *Class, Codes and Control*. FitzSimons and de Souza et al. elaborate the argument that mathematics (as a vertical discourse) may be 'invisible' in everyday practices, although those practices may content 'heavy' and 'sophisticated' mathematics (Wedege 2010). In the next paragraphs, I comment on this reading of FitzSimons's and de Souza et al.'s chapters.

A Sociological Look at the Mathematics Skills Used in the Workplace

In the decade of the 1990s, Mogen Niss popularized the idea of invisibility of mathematics in society (Niss 1994, 1995). Despite the (objective) social importance of mathematics, Niss discussed that, actually, its presence goes unnoticed for most people. He called the discrepancy between the objective transcendence of mathematics and its subjective invisibility the 'relevance paradox'. According to Niss, mathematics is a meaningful part of the world around us. We find it in virtually every single human action. However, despite being embedded in human actions, often mathematics goes unnoticed and actions are not identified as being mathematical. This poses three major problems for educators: the problems of justification, of possibility and of implementation (Niss 1995).

The 'invisibility' of mathematics has been a recurring theme in the literature about the connection between mathematics and adults' literacy skills, particularly in all that relates to the skills developed in the workplace. Harris (1994) draws on interviews with young and unskilled workers in the inner London area during the late 1980s, for instance with hairdressers who denied using ratio and proportion in their workplace. However, when Harris asked them to describe the mixtures they used to dye hair, often they used the word 'proportion.' According to her analysis, "the research responses also seemed to suggest that when mathematical skills caused the workers no problem, then they were regarded as common sense. They only became 'mathematics' when the worker could not do them" (Harris 1994, p. 19).

This type of evidence (Díez-Palomar 2004; Harris 1994) seems to suggest that there is a social representation (Moscovici 1988) of mathematics as an "elaborated code" (Bernstein 1971). Bernstein introduced the concepts of elaborated and restricted codes, and the classification and framing principles, to create a theoretical model for what happens in the classroom. He argued that the principles of classification and framing of instructional practice explain the difference between the academic performances of children of different social class background: children from privileged social classes do better in school because they master the elaborated code legitimated by the school, while children from working-class contexts use restricted codes with limiting effects on gaining access to academic knowledge.

Years later, Bernstein developed his theory drawing on the notions of "vertical and horizontal discourse". In an attempt to distance himself from Bourdieu's perspective on the sociology of education (who was criticized as reproductionist because his approach had a low value to improve children's learning), Bernstein in 1996 published for the first time the idea of "vertical and horizontal discourse",[1] developed further in 1999. Until that time Bernstein had analysed the different principles of pedagogic transmission and acquisition as code modalities, over a social basis. However, as he says: "the *forms* of the discourses, i.e. the internal principles of their construction and their social base, were taken for granted and not analysed" (Bernstein 1999, p. 157). It seems that Bernstein was aware of the international discussion around 'expert' and common-sense types of knowledge. He quotes Bourdieu's work, who distinguishes between symbolic and practical discourse, according to the functions the discourse plays. He also refers to Habermas's distinction between the 'life world' of the individual and the other as the source of instrumental rationality (Habermas 1984, 1987). And he cites the concept of 'expert

---

[1] Bernstein (1999) defines 'horizontal discourse' as "everyday or 'common-sense' knowledge. Common because all, potentially or actually, have access to it, common because it applies to all, and common because it has a common history in the sense or arising out of common problems or living and dying" (p. 159). In contrast, "vertical discourse takes the form of a coherent, explicit, and systematically principled structure, hierarchically organised, as in the sciences, or it takes the form of a series of specialised languages with specialised modes of interrogation and specialised criteria for the production and circulation of texts, as in the social sciences and humanities" (p. 159).

systems' proposed by Giddens, following Habermas's thinking (Giddens 1991). According to Bernstein, these authors were not able to differentiate between school(ed) knowledge and everyday common-sense knowledge, which is what Bernstein does with 'vertical and horizontal discourse'.

In her chapter, FitzSimons uses these notions to analyse her observations of workers' mathematical skills in a pharmaceutical manufactory where she was responsible for facilitating a course about competency-based education and training modules of *Calculations A* and *Basic Computer Skills*. She starts from Bernstein's approach to define mathematics as a vertical discourse because it is a discourse with a structure that is "coherent, explicit, and systematically principled" (FitzSimons and Wedege 2007, p. 51). In contrast, numeracy is an example of horizontal discourse (FitzSimons 2004; FitzSimons and Wedege 2007), because it is "embedded in on-going practices, usually with strong affective loading, and directed towards specific, immediate goals, highly relevant to the acquirer in the context of his/her life" (Bernstein 1999, p. 161; quoted in FitzSimons and Wedege 2007, p. 51).

The mathematics found in the curriculum of the courses does not correspond to the mathematical knowledge workers have after many years of experience working in the factory. FitzSimons talks about 'lack of agency' to describe the process of alienation that such type of courses mean. From a sociological point of view, we can understand the 'lack of agency' as a consequence of a process of bureaucratization (in Weber's terms), that converts the teaching of mathematics into a set of rules and procedures isolated from their (historical) meaning. FitzSimons's proposal of integrating contextualized elements, when possible, in the curriculum of the *Calculations A* module is a way to transform this process of anomie and lack of meaning. To do this, FitzSimons draws on her observations and the discussions with the workers, to design *contexts meaningful to the workers, who were able to locate themselves personally in the activity and to see the vital importance of accuracy in meeting official standards.*

De Souza and her colleagues decided to use another strategy: start from a 'real' problem and use mathematics modelling to recover the 'agency' in the learning process. Let us see.

Mathematics Modelling Within Adult Education

The first time I read the concept of 'horizontal mathematization' was in Treffers and Vonk (1987), who gathered results of the *Wiskobas* project at IOWO (Institute for the Development of Mathematics Education). In the late 1970s, Adri Treffers created the notion 'horizontal mathematization' to talk about the transformation of a problem field into a mathematical problem. Treffers was influenced by Freudenthal, who had published in 1973 *Mathematics as an Educational Task*, criticizing 'New Math' with its idea of set theory and the abstract deductive structures of mathematics as the basis for school mathematics. In that book, Freudenthal claimed that teachers should encourage children to use 'mathematical tools' to model real situations in order to solve the problems embedded within them. Using observation, experimentation, inductive reasoning, etc., students should transform an initial problem (situation) into a mathematical problem. Freudenthal, using Treffers' (1978, 1987) concepts, wrote in 1991:

> Horizontal mathematisation leads from the world of *life* to the world of *symbol*s. In the world of life one lives, acts (and suffers); in the other one symbols are shaped, reshaped, and manipulated, mechanically, comprehendingly, reflectingly: this is vertical mathematisation. (pp. 41–42).

In the example that de Souza et al. introduce, adult learners face a situation of contracting a loan (which could be potentially a real situation) to buy a house. Learners have to decide what type of plan they would chose: one based on a constant amortisation system, or another one based on a price system in which instalments are fixed. Students struggle to understand the problem and to take the right decision based on their mathematical analysis. In order to do so, they need to use the elaborated code of mathematics.

Here is where the authors of the chapter propose an innovative approach using Bernstein's notion of 'horizontal discourse' to reconceptualise the 'traditional' mathematics modelling approach based on Freudenthal (1973) and Treffers (1978, 1987). I have rarely seen the use of Bernstein's ideas to analyse the discourse that emerges from mathematics modelling. Jablonka and Gellert (2007) did something similar when they explained that "horizontal mathematisation as used by Treffers (1987) can be described, within Bernstein's theoretical framework, as a transfer between horizontal and vertical discourse" (p. 3).

In that book, Jablonka and Gellert address an interesting point, which has been controversially discussed (specially among authors working about numeracy): the transferability of mathematics knowledge across contexts. In the 1980s, many studies somehow questioned the possibility to have 'direct' transfer between academic (formal) knowledge of mathematics and knowledge linked to everyday situations (Carraher et al. 1985; Lave 1988; Vithal 2003). Lave (1988) claimed that it is not evident that we can transfer the mathematics learnt in school to everyday life situations outside school. Later, Evans (1999) and Noss and Hoyles (1996) were still questioning if we can make a link between academic mathematics and everyday life situations. It is not clear that we can generalize mathematics skills, even the more basic ones.

However, studies like these draw our attention to the fact that school needs to recognize (and include) other types of mathematical discourses, admitting that there are discourses (codes, arguments) with the same validity as the formal ones (vertical discourse, in Bernstein's terms), because these non-academic discourses contain valid, true and consistent explanations, too. Hence school must recognize (and legitimate) them. This statement coincides with contributions from contemporary sociological theories such as the dialogic learning theory (Flecha 2000) or the theory of communicative action (Habermas 1984, 1987). What matters is not the type of code used by learners (elaborated, restricted), as long as this code is used in order to achieve valid knowledge underpinned in a task; what matters is the type of argument they use. When learners explain a task using arguments based on validity claims, then (no matter which code they use) they are making an effort to understand the mathematics embedded in the task (and they use those arguments to navigate from horizontal to vertical mathematics).

Further Comments: Mathematics and Democracy

I am going to end this brief comment going back to FitzSimons's final words about mathematics and democracy. Understanding mathematics as a form of liberation, inclusion, and struggle for democracy, inevitably pushes me to think of the efforts of authors such as Paulo Freire (1996, 1998), Ramon Flecha (2000), Jesús Gómez (2004), and many others, who have devoted and dedicated their lives to overcome the social and educational exclusion of those people who never had the same learning opportunities as us. All of them struggle(d) to legitimate the knowledge that people without a university degree have, because it is as valuable as the one that 'experts' (in Giddens's terms) hold. Having lived my last 17 years working in 'CREA-Centre of Research in Theories and Practices that Overcome Inequalities' (founded by Flecha and colleagues in 1991) has given to me the opportunity to understand why what Gramsci called 'common-sense' and 'popular culture' is really valuable (Díez-Palomar 2011); because it comes from the accumulation of knowledge from hundreds of people observing and experimenting the world that surrounds us along the history. Concepts like 'horizontal discourse' are theoretical tools that try to account for what other people already realized long time ago. García Lorca organized 'La Barraca', a theatre company aiming to bring the knowledge of the classics of universal literature to people living in rural areas, to the families who never had the opportunity to attend a play in a theatre to see those classics that should be available for free to everybody. Having the chance to work at CREA has taught me to appreciate the tremendous democratic value of knowledge, not only regarding mathematics; all knowledge that we as human beings have accumulated over centuries of civilizations.

Both FitzSimons and de Sousa et al. draw on Bernstein's (1999) theoretical approach to support a democratized view of adults learning mathematics. Using those concepts, they move from the realm of mathematics (vertical discourse) to mathematics 'within the real world' (horizontal discourse). In order to learn mathematics, we must be able to solve problems in everyday life situations, and we also need to learn how to use abstract and sophisticated mathematical tools. Otherwise, we do not have the required repertory of 'savoirs' to give valid answers to such situations (in dialogical terms). As Greeno wrote in 1997:

> In the situative perspective, use of abstract representations is an aspect of social practice, and abstract representations can contribute to meaningful learning only if their meanings are understood. To the extent that instruction presents abstract representations in isolation from their meanings, the outcome can be that students learn a set of mechanical rules that can support their successful performance on tests requiring only manipulation of the notations, not meaningful use of the representations.
>
> On the other hand, it is perfectly consistent with the situative perspective that abstract representations can facilitate learning when students share the interpretative conventions that are intended in their use. (p. 13)

It seems that to learn mathematics we need teachers using both real and contextualized tasks (better if they do that with authentic situations) as well as artificial and abstract tasks (Sierpinska and Kilpatrick 1998) based on the vertical discourse of mathematics. But if those are different codes according to what we have learnt from the comments to FitzSimons's and de Souza et al.'s chapters, then how can we mix them? Is there a way to promote democratic teaching in mathematics including both elaborated and restricted codes in the pedagogic discourse of mathematics? Habermas's (1984, 1987) and Flecha's (2000) theoretical approaches may give us some hints to answer these questions satisfactorily.

Habermas (1984) introduces the principles of a theory of argumentation. Arguing, as Habermas says, can be done using two different types of arguments, which are based on very different kind of claims: we can use arguments drawing on power claims to impose our point of view (using language with a clear perlocutionary effect, in Austin's (1971) words). Or we can use arguments based in validity claims to reach consensus and agreement on true knowledge. That is when we use language in an illocutionary way, as a means to reach knowledge, to understand, etc. Trustworthiness does not depend on the power position occupied by the speaker, nor on the kind of code s/he uses. Trustworthiness depends on how a statement fits in with reality—the truth. What is important here is the use of validity claims.

This kind of language use requires the creation of dialogic spaces (as mentioned by de Souza et al.). However, these dialogic spaces, in order to be truly 'dialogic', must be based on what Flecha (2000) calls the principle of egalitarian dialogue. With this notion, Flecha is recognizing the value of the arguments of all people (both academic and not academic), according to the validity of their claims and not to the status that the person holds who is using them.

Pointing out that there are different types of codes (as Bernstein claims), does not imply that this difference cannot be overcome. Instead, drawing on Habermas's and Flecha's contributions, it is obvious that we have strategies available to break the gap between 'expert knowledge' and 'common-sense', which allow us to design more democratic mathematical activities and open up the universe of mathematics knowledge to a broader range of people. We need both activities extracted from everyday situations, to model them (as suggested by Barbosa 2006, and by de Souza et al.), as well as learning situations to lead learners (either adults or children) to also engage in decontextualized mathematical activities, interacting with other people (teachers, peers, volunteers, etc.) who must know how to solve them, using egalitarian dialogue. To me, this is what means democratizing the access to the mathematics knowledge.

FitzSimons's chapter provides us with an example of that. I cannot stop having the feeling that behind her statements there is a hint that makes me think of her work as a clear contribution to struggle against the negative consequences of what Weber called 'the iron cage', that is: the process of losing meaning in modern societies. The references used by FitzSimons make me think in the process of reification developed by Marx when he described the loss of meaning ('anomie' in Durkheim's terms) that workers experience when they are no longer the 'owners' of their own work; when the workers become simple 'pieces' in the production chain. With her proposal of democratizing mathematics teaching and learning by designing a meaningful curriculum, situated in context, open to dialogue, FitzSimons has found a way to break the 'iron cage'; and this is profoundly democratic. We need more contributions like this, more efforts like the one she and de Souza et al. make in this book, to really achieve a fairer and more responsible education for everybody. We need an education with more sense and provided in a more democratic way, thinking that learning opportunities should be available for everyone, not only for those people who are lucky because they were born with them.

References

Abbott, A. (2012). It's your money they're after! Using advertising flyers to teach percentages. In A. Hector-Mason & D. Coben (Eds.), *Proceedings of the 19th international conference on adults learning mathematics* (pp. 175–179). Hamilton: The National Centre of Literacy & Numeracy for Adults, University of Waikato.

Austin, J. L. (1971). *Palabras y acciones: cómo hacer cosas con palabras*. Barcelona: Paidós.

Bakker, A. (2014). Special issue: Characterising and developing vocational mathematical knowledge. *Educational Studies in Mathematics, 86*(2), 151–156.

Barbosa, J. C. (2006). Mathematical modelling in classroom: A critical and discursive perspective. *ZDM – The International Journal on Mathematics Education, 38*(3), 293–301.

Beeli-Zimmermann, S., & Hollenstein, A. (2011). Financial literacy of microcredit clients: Results of a qualitative exploratory study and its implications for educational schemes. In T. Maguire, J. J. Keogh, & J. O'Donoghue (Eds.), *Proceedings of the 18th international conference on adults learning mathematics* (pp. 26–36). Adults Learning Mathematics – A Research Forum.

Bernstein, B. (1971). *Class, codes and control. Volume 1: Theoretical studies towards a sociology of language*. London: Routledge.

Bernstein, B. (1996). *Pedagogy, symbolic control and identity: Theory, research, critique*. London: Taylor & Francis.

Bernstein, B. (1999). Vertical and horizontal discourse: An essay. *British Journal of Sociology of Education, 20*(2), 157–173.

Carraher, T. N., Carraher, D. W., & Schliemann, A. D. (1985). Mathematics in the streets and in schools. *British Journal of Developmental Psychology, 3*(1), 21–29.

Coben, D., & Weeks, K. (2012). Behind the headlines: Authentic teaching, learning and assessment of competence in medication dosage calculation problem solving in and for nursing. In A. Hector-Mason & D. Coben (Eds.), *Proceedings of the 19th international conference on adults learning mathematics* (pp. 39–55). Hamilton: The National Centre of Literacy & Numeracy for Adults, University of Waikato.

Coben, D., O'Donoghue, J., & FitzSimons, G. E. (Eds.). (2000). *Perspectives on adults learning mathematics: Research and practice*. Dordrecht: Kluwer.

Damlamian, A., Rodrigues, J. F., & Sträßer, R. (2013). *Educational interfaces between mathematics and industry: Report on an ICMI-ICIAM-study*. Cham: Springer.

Díez-Palomar, J. (2004). *La enseñanza de las matemáticas en la educación de personas adultas: un modelo dialógico*. Unpublished PhD thesis, Universitat de Barcelona.

Díez-Palomar, J. (2011). Being competent in mathematics: Adult numeracy and common sense. In T. Maguire, J. J. Keogh, & J. O'Donoghue (Eds.), *Proceedings of the 18th international conference on adults learning mathematics* (pp. 90–102). Adults Learning Mathematics – A Research Forum.

Evans, J. (1999). Building bridges: Reflections on the problem of transfer of learning in mathematics. *Educational Studies in Mathematics, 39*(1–3), 23–44.

Evans, J. (2011). Students' response to images of mathematics in advertising. In T. Maguire, J. J. Keogh, & J. O'Donoghue (Eds.), *Proceedings of the 18th international conference on adults learning mathematics* (pp. 118–128). Adults Learning Mathematics – A Research Forum.

FitzSimons, G. E. (2000). Section III: Adults, mathematics and work. In D. Coben, J. O'Donoghue, & G. E. Fitzsimons (Eds.), *Perspectives on adults learning mathematics: Research and practice* (pp. 209–227). Dordrecht: Kluwer.

FitzSimons, G. E. (2004). *An overview of adult and lifelong mathematics education*. Keynote presentation at Topic Study Group 6, 10th International Congress on Mathematics Education. http://www.icme10.dk

FitzSimons, G. E. (2013). Doing mathematics in the workplace: A brief review of selected literature. *Adults Learning Mathematics: An International Journal*, 8(1), 7–19.

FitzSimons, G. E., & Wedege, T. (2007). Developing numeracy in the workplace. *Nordic Studies in Mathematics*, 12(1), 49–66.

Flecha, R. (2000). *Sharing words: Theory and practice of dialogic learning*. Lanham: Rowman & Littlefield.

Freire, P. (1996). *A la sombra de este árbol*. Barcelona: El Roure.

Freire, P. (1998). *Pedagogy of freedom: Ethics, democracy, and civic courage*. Lanham: Rowman & Littlefield.

Freudenthal, H. (1973). *Mathematics as an educational task*. Dordrecht: Reidel.

Freudenthal, H. (1991). *Revisiting mathematics education: China lectures*. Dordrecht: Kluwer.

Giddens, A. (1991). *Modernity and self-identity: Self and society in the late modern age*. Stanford: Stanford University Press.

Gómez, J. (2004). *El amor en la sociedad del riesgo: una tentativa educativa*. Barcelona: El Roure.

Goold, E., & Devitt, F. (2011). The role of mathematics in engineering practice and in the formation of engineers. In T. Maguire, J. J. Keogh, & J. O'Donoghue (Eds.), *Proceedings of the 18th international conference on adults learning mathematics* (pp. 134–154). Adults Learning Mathematics – A Research Forum.

Greeno, J. G. (1997). Or claims that answer the wrong questions. *Educational Researcher*, 26(1), 5–17.

Habermas, J. (1984). *The theory of communicative action, vol. 1: Reason and the rationalization of society*. Boston: Beacon.

Habermas, J. (1987). *The theory of communicative action, vol. 2: Lifeworld and system: A critique of functionalist reason*. Boston: Beacon.

Harris, M. (1994). *Mathematics in Denman textile courses. Three reports for D College*. National Federation of Women's Institutes. Unpublished manuscript.

Jablonka, E. (2010). Contextualised mathematics. In H. Christensen, J. Díez-Palomar, J. Kantner, & C. M. Kinger (Eds.), *Proceedings of the 17th international conference on adults learning mathematics* (p. 3). Adults Learning Mathematics – A Research Forum.

Jablonka, E., & Gellert, U. (2007). Mathematisation – Demathematisation. In U. Gellert & E. Jablonka (Eds.), *Mathematisation and demathematisation: Social, philosophical and educational ramifications* (pp. 1–18). Rotterdam: Sense.

Kane, P. (2012). Spatial awareness and estimation in recycling and refuse collection. In A. Hector-Mason & D. Coben (Eds.), *Proceedings of the 19th international conference on adults learning mathematics* (pp. 56–68). Hamilton: The National Centre of Literacy & Numeracy for Adults, University of Waikato.

Kelly, B. (2011). Learning in the workplace, functional mathematics and issues of transferability. In T. Maguire, J. J. Keogh, & J. O'Donoghue (Eds.), *Proceedings of the 18th international conference on adults learning mathematics* (pp. 37–46). Adults Learning Mathematics – A Research Forum.

Keogh, J. J., Maguire, T., & O'Donoghue, J. (2012). A workplace contextualization of mathematics: Visible, distinguishable and meaningful mathematics in complex contexts. In A. Hector-Mason & D. Coben (Eds.), *Proceedings of the 19th international conference on adults learning mathematics* (pp. 32–38). Hamilton: The National Centre of Literacy & Numeracy for Adults, University of Waikato.

Lave, J. (1988). *Cognition in practice: Mind, mathematics and culture in everyday life*. Cambridge: Cambridge University Press.

Lave, J., & Wenger, E. (1991). *Situated learning: Legitimate peripheral participation.* Cambridge: Cambridge University Press.

Mills, K. R. (2012). Some correspondences and disjunctions between school mathematics and the mathematical needs of apprentice toolmakers: A New Zealand perspective. In A. Hector-Mason & D. Coben (Eds.), *Proceedings of the 19th international conference on adults learning mathematics* (pp. 69–83). Hamilton: The National Centre of Literacy & Numeracy for Adults, University of Waikato.

Moscovici, S. (1988). Notes towards a description of social representations. *Journal of European Social Psychology, 18*(3), 211–250.

Niss, M. (1994). Mathematics in society. In R. W. Scholz, R. Sträßer, & B. Winkelmann (Eds.), *Didactics of mathematics as a scientific discipline* (pp. 367–378). Dordrecht: Kluwer.

Niss, M. (1995). Las matemáticas en la sociedad. *UNO: Revista de Didáctica de las Matemáticas, 2*(6), 45–57.

Noss, R., & Hoyles, C. (1996). The visibility of meanings: Modelling the mathematics of banking. *International Journal of Computers for Mathematical Learning, 1*(1), 3–31.

Sierpinska, A., & Kilpatrick, J. (Eds.). (1998). *Mathematics education as a research domain: A search for identity.* Dordrecht: Kluwer.

Treffers, A. (1978). *Wiskobas Doelgericht.* Utrecht: IOWO.

Treffers, A. (1987). *Three dimensions: A model of goal and theory description in mathematics instruction – The Wiskobas project.* Dordrecht: Reidel.

Vithal, R. (2003). Teachers and street children: On becoming a teacher of mathematics. *Journal of Mathematics Teacher Education, 6*(2), 165–183.

Wedege, T. (2010). People's mathematics in working life: Why is it invisible? *Adults Learning Mathematics – An International Journal, 5*(1), 89–97.

Part III
Working with Pre-schoolers

'Number in Cultures' as a Playful Outdoor Activity: Making Space for Critical Mathematics Education in the Early Years 143
Anna Chronaki, Georgia Moutzouri, and Kostas Magos
University of Thessaly, Volos, Greece

Fairness Through Mathematical Problem Solving in Preschool Education 161
Zoi Nikiforidou
Liverpool Hope University, UK
Jenny Pange
University of Ioannina, Greece

How Do Fair Sharing Tasks Facilitate Young Children's Access to Fractional Concepts? 173
Julie Cwikla
University of Southern Mississippi, Ocean Springs, USA
Jennifer Vonk
Oakland University, Rochester, USA

Working with Pre-schoolers: A Dual Commentary 191
Michaela Kaslová
Charles University, Prague, Czech Republic
Sixto Romero
Universidad de Huelva, Spain

'Number in Cultures' as a Playful Outdoor Activity: Making Space for Critical Mathematics Education in the Early Years

Anna Chronaki, Georgia Moutzouri, and Kostas Magos

Abstract The chapter discusses an attempt to design and implement a playful outdoor activity for a culturally diverse group of young children that emphasizes a critical perspective on mathematical learning. Critical mathematics education is taken here to embrace identity and learning as interrelated processes that are not simply analytical categories but deeply rooted in collective action and human subjectivity. Based on this premise, "Number in Cultures" can be conceived as a counter event that constitutes an open space for young children and adults to play, explore and question issues concerning the cultural underpinnings of number and its relatedness to their lives. Specifically, it has aimed to create awareness around diversity in number-words and number-symbols rooted in the context of Greek, Arabic and Romany languages and cultures. Number in cultures has been implemented as part of a playground workshop where Roma children could potentially become active participants along with adults and children of a non-Roma background.

Introduction

During the last three decades, varied efforts have been placed over raising a sociopolitical agenda for mathematics education. The importance of reforming curricula, creating counter pedagogic practices, and establishing educational policy and practice for social justice and critical citizenship has been stressed. Such efforts are often rooted within theoretical and empirical research that helps to sensitize public audiences on issues concerning the political, cultural and social dimensions of mathematics, mathematical learning and mathematical knowledge transmission/communication. Towards this direction, the perspective of critical mathematics education as epistemology and pedagogical praxis could pave the ground that allows us to imagine and weave alternative acts to what one might call established or dominant formal and/or informal practices. The interest towards embracing such an

---

A. Chronaki (✉) • G. Moutzouri • K. Magos
Department of Early Childhood Education, University of Thessaly, Volos, Greece
e-mail: chronaki@uth.gr; mountgeo@ath.forthnet.gr; kmago@tee.gr

© Springer International Publishing Switzerland 2015
U. Gellert et al. (eds.), *Educational Paths to Mathematics*,
Advances in Mathematics Education, DOI 10.1007/978-3-319-15410-7_8

endeavor at the levels of theory, policy and pedagogy is rooted through an expressed impetus for battling and resisting social injustice, racism and stereotyping that still exists and thrives in modern multicultural and technological societies and for considering the complexity of cultural diversity as it is embedded in particular contexts, where mathematics education becomes implemented from the early nursery years through tertiary education (Apple 2004; Atweh et al. 2011; Brown 2011; Stinson and Bullock 2012; Valero and Zevenbergen 2004; Walshaw 2011).

Although over the years the re-organization of curricular practices either in terms of policy or classroom practice has become the center of attention (e.g., Gutstein 2006), the actual re-contextualisation of a sociopolitical agenda concerning mathematical content (what to teach) and process (how to teach) is almost at its birth. Using Michael Apple's distinction amongst functional (i.e. developing competencies and skills) and critical mathematical literacy (i.e. encouraging reflection, critique and dialogue over knowledge production and learning) it could be easy to claim that most national or local curricula for mathematics, from early childhood up to secondary and tertiary education, follow mainly a functional paradigm (Apple 2004). The above is true not only for semi-peripheral countries such as Greece but also for metropolitan states in both Europe and USA. Despite the odds, a recent reform of the mathematics curriculum in Greece has placed attention on the integration amongst issues concerning child development, society and learning. Especially for the early years, there are claims for the need to pursue a holistic approach of children's academic competencies along with skills that support a double process of self and social development, or, in other words, children's identities as learners and their socialization as citizens (see the new national curriculum in Greece: http://ebooks.edu.gr/). Although such claims are not carefully analyzed in terms of the potential (and conflicting) discourses that they might serve, one needs to appreciate that such an emphasis encourages the pursuing of children's learning as learning interrelated with re-constructing identity and weaving subject positions.

In our earlier work, the issue of 'learning identity' was interwoven with children's spaces of learning—where children learn not only to perform but also to value mathematical activity (Chronaki 2005). Roma children, their families and communities, by and large, resist formal schooling, and remain marginalized or 'voiceless' (Chronaki 2009). Bertau (2007) argues that 'voice' is related to changing and positioning self simultaneously in diverse positions. Identity was, thus, discussed as a complex, fluid and hybrid process that does not develop in isolation but in direct interaction and participation within open communities and practices. Such openness might foster dialogicality through troubling essentialist conceptions of identity (Chronaki 2009, 2011).

Critical mathematics education is taken here to embrace identity and learning as closely interrelated processes that are not simply analytical categories but deeply rooted in collective action and human subjectivity. Based on the above, 'number in cultures' was designed as an activity focused on numbers valued in different cultures. The event constitutes an open space—a 'space of appearance'—for young children and adults to play and explore issues concerning the cultural underpinnings of 'number' and its relatedness to their own lives. Specifically, it has aimed to create awareness through memory work around cultural diversity of number-words and number-symbols in the context of Greek, Arabic and Romany languages. 'Number in cultures' was implemented as part of a playground workshop, where Roma children became active participants (as they took a leading role in the activity) along with adults and children of a non-Roma background.

We created a pedagogical space where children (and adults) can both play with numbers and begin to explore some complex and silenced stories about numerals as an integral part of their culture. In the following sections, we first account on what might mean a critical perspective for the learning of mathematics, second we move towards analyzing the design for a critical perspective on number, third, the playful activity is described in detail, fourth, the implementation of the activity is outlined and finally, the whole experience is discussed.

A Critical Perspective for the Learning of Mathematics

Gert Biesta (1998) argues about the impossibility of critical pedagogy as a political project in the context of education from an instrumental modernist point of view. As he claims, "critical pedagogies are in one way or another committed to the imperative of transforming the larger social order in the interest of justice, equality, democracy, and human freedom" (p. 499) and he interrogates the extent to which this humanitarian mission can be really fulfilled within the complexities, uncertainties and risks embedded within our globalized post-modern societies. Critical pedagogy has been, by and large, considered as a democratic and emancipatory educational practice that often becomes the main means and laboratory in the human struggle for liberation, justice and democracy. It is based on the assumption that such virtues can only be cultivated, developed and sustained when people develop the capacity for critical reflexivity that enables them to create an adequate understanding of their situation, to imagine alternatives and to become agents for social transformation. Yet, he claims that this very 'impossibility' is what makes critical pedagogy itself possible.

In Europe, Ole Skovsmose and his colleagues have emphasized a philosophical perspective on critical mathematics education that engages us with an alternative epistemology of mathematics and its role in society and education (Skovsmose 2011). Their theoretical endeavors are rooted in critical pedagogy as proposed by Paulo Freire, in critical theory with strong influences from the Frankfurt school and more recently in contemporary post-modern accounts of critique, ethics and knowledge. They try to emphasize mathematics education as an open, complex and yet uncertain space of critical activity. This perspective has been embraced by colleagues in Latin America and South Africa who have invested towards the creation of interventionist programs in communities that encounter problems of poverty, injustice and illiteracy (Knijnik and Wanderer 2010, Vithal 2003). In the USA, the devoted work of Marilyn Frankenstein and more recently Rico Gutstein, based on Freire's work has developed with a clear interest towards the conception of a curriculum that encourages critical mathematics education as praxis (Frankenstein 1989; Gutstein 2006). Given the particular context of the USA multicultural society, their emphasis has been on battling injustice and exclusion due to class, gender and race difference by means of promoting critical literacy.

Skovsmose (1994) has argued about 'mathemacy' as a way to turn our attention to critical mathematical literacy as a process that involves not merely the development of skills and competencies but also a critique of a naive dependence on mathematics as 'ideology of certainty' and as a 'formatting power' in society at large. This ideology serves to operate as a gatekeeper of social order, rationality and regulation through a passionate and unquestionable trust in political arithmetic as the ultimate tool for safeguarding democracy. Mathematics education can mean both empowerment and disempowerment and thus serves as a double role over history. On the one hand, it has served colonizers and modernists to conquer the world by enculturating people into a western sense of logic and rationality by means of mathematical literacy. On the other hand, it has also contributed as a means towards the creation of critical citizenship and democratic ideals (Skovsmose et al. 2013). Taking this dual role into account, one realizes that the sociopolitical positions and potentialities of mathematics and mathematics education are neither fixed nor determined. Skovsmose et al. (2013) encourage us to consider its relevance and connectivity to a number of concerns. (1) Globalization and ghettoizing. (2) Basic assumptions of modernity where scientific and epistemic transparency are constructed as the mere path to social progress and democratic order, arguing that a critical perspective on mathematical knowledge production needs to question (and create alternative) trajectories to this utopia. (3) Application of knowledge in specific contexts (e.g., problem solving, modelling) but also as closely related to knowledge, power and ethics. (4) Disempowerment potential through mathematics education and in particular through imposition or appropriation of stereotypes concerning race, gender, ability and language use. (5) Empowerment by means of cultivating ethics of responsibility as integral components of mathematics education (i.e. programs where mathematical literacy is closely related to issues of social justice and critical citizenship).

Taking the above into consideration, we might be able to revisit discussions in the field of mathematics education related to whose knowledge counts as 'official' and whose knowledge and culture gets labeled as being merely 'popular', 'exotic' or 'weak' in status and, hence, not seen as legitimate for formal schooling. According to Gutstein et al. (1997), a critical mathematics education needs to encounter an intercultural perspective on mathematical learning that relies and builds on children's informal knowledge. A critical perspective cannot develop without espousing children's social and cultural backgrounds as well as their personal ways of knowing and learning as they develop and grow through experiencing the particular political contexts of living and work. However, at the same time it needs to 'resist' an 'ideology of certainty' that permeates 'formal' mathematical knowledge (Skovsmose 2011; Skovsmose et al. 2013). In our view, there is an additional challenge that can be captured around identifying ways for 'troubling' essentialist discourses and identities as they are related to mathematics education (Chronaki 2011).

Nowadays, it is by and large recognized that mathematics can be considered both as a tool and space for analyzing and interpreting social activity as an integral part of diverse social and cultural experiences. An example of this might be to undertake memory work concerning 'number' in varied cultures and to embrace the possibility of a co-existence amongst multiple ways in talking and symbolizing number. In this way, mathematics and mathematical activity can work out as a means for intercultural interaction and as a potential tool for developing awareness of mathematics as embedded in culture, everyday life and potential mutual exploration of 'difference' and acceptance of cultural codes and resources. On the one hand, mathematical learning needs to be connected, with children's cultural roots including their linguistic affordances and, on the other hand, with their potential to develop critical awareness about mathematics and its role in society as a genre of knowing. Both of these two dimensions are closely connected to their potential to develop a sense of hybrid identity and in particular a sense of 'who they are', 'who they want to become' and 'how they position themselves' as non-fixed but fluid in the midst of multiple knowledge hierarchies. This critical dimension of mathematics education can work towards supporting an entry to dialogicality. As it has been argued elsewhere, "Dialogicality is a basic Bakhtinian concept that serves to critique formalism in language and particularly in literary texts … [It] is embedded in all cultural products, of which educational practices and curricula are examples" (Chronaki 2009, pp. 127–128).

Although, there are currently efforts to design (and implement) formal and informal curricular activities that support critical practices, such attempts address mostly the needs of older learners. We focus on the very young learners in order to promote and develop their self-image and views of their potential participation. Views, images and ways of acting become mediated to younger ones through varied levels of informal education and contact as part of their family, school and community. There is evidence that young children -even at the age of three- adopt appropriate stereotypical views of the 'other' as ethnic, cultural, race and gender identities (Derman-Sparks and Ramsey 2006; Nieto 2004).

At the level of educational policy, the structure and content of curricula, teaching methods, textbooks and educational materials become the backbone that mediates the reproduction of stereotyping and the diversity amongst self and other. Discourse can be communicated subtly and through collective practices not only with eldest students via school-based and formally-structured curricula activity but also with younger ones through everyday talk and playful experiences. As such, adults mediate an unjustified 'trust in numbers' (Porter 1953/1996) through both curricula and playful materials and stories that serve to represent 'mathematics' as a certain and universal body of knowledge. We now turn to the following two sections towards outlining our design for a critical perspective on number as a playful outdoor activity for young children and its implementation as part of an intercultural festival co-organized with our undergraduate students in the city of Volos.

Designing for a Critical Perspective on Number

Over the centuries, number symbols and number operations have had a long and turbulent history of development accompanying the social, cultural and political circumstances in the respective communities where mathematical knowledge emerges, grows and transforms (Bartolini-Bussi and Boni 2009; Radford 2001). Specifically, it is a fact that the ten 'Western' digits (0, 1, 2, 3, 4, 5, 6, 7, 8, 9) are also known as Arabic, Hindu or Hindu-Arabic numerals. Although mentioned in formal curricula, it does not often become a theme for investigative teaching.

Anthropological studies in remote communities of Oksapmin in Papua New Guinea have revealed how people up until now resort to an embodied way of numbering where numbers are indicative of certain parts of the body, specifically hands, arms and the head. This genre of numbering makes use of 27 points of the upper part of the human body and seems to be efficient for the everyday exchanges amongst people within the community. Geoffrey Saxe (1981) and his colleagues have studied the importance of bilingual education of number use in this community, where Western number symbols and number words are used next to body related numerals. Other studies of a similar focus have revealed that illiterate people use personal strategies for undertaking complex operations necessitated for solving particular problems in street commerce (Crump 1990; Saxe 1981).

Mathematical knowledge situated in 'other' cultures (often non-Western ones) has become marginalized as certain voices and bodies in multicultural classrooms have become silenced, marginalized and, even worse, ignored. Children, at large, learn to naturalize 'number' as universal and static, and to appreciate the numerals of 0, 1, 2, 3, 4 … 9 as universally and uniformly used everywhere all over the world—a 'common sense' view of 'what is number'. This has enormous consequences on children's learning identities, particularly for those children who feel they cannot identify themselves with the dominant 'Western' culture and realize a huge gap amongst their own cultural resources and the formal culture of numbering. An example of the above is the case of Roma children (Chronaki 2005, 2009, 2011) who resort on oral practices for transmitting knowledge from generation to generation. For them, number sense and the practice of number operations is based more on using number words than on symbols in a written genre, and also more on mental strategies for identifying ways to enact number operation and problem solving than on algorithm use.

School experiences for number and numbering can closely be related to a functional literacy perspective through which children's learning is strongly equated with developing certain abilities and skills. Functional literacy refers, by and large, to the fulfillment of a variety of competencies needed to function appropriately within a given society. A curriculum based on such values serves the reproductive purposes (i.e. maintaining the status quo) of the dominant interests in society which is often stressed by ensuring a close interaction amongst education, market and workplace.

From a functional literacy perspective, children have the opportunity to develop the necessary competences (i.e. handling number, symbolizing number, arithmetic problem solving, reasoning) needed to function appropriately within society as literate or numerate citizens. This optic seems the only alternative and promotes the view that the learning of 'number' happens as if number words and symbols are universally similar and uniformly used in every part of the world, in every culture and in every language. This viewpoint becomes hegemonic since adults including teachers, educators and parents are neither aware nor sensitized towards number diversity and its relative connection to promoting cultural diversity and exclusion. Specifically, children from an early age and particularly those who belong in the Roma community resort mainly to oral genres of communicating. They have cultivated a sense of number that is deeply situated in their confidence to use number-words and to perform in-situ strategies for number operations. However, current mathematics curricula practices ignore the importance of sensitizing young students, educators and parents in such an issue. Being sensitive about knowledge, diversity means accepting knowledge (and mathematical knowledge) as fallible and deeply situated within local cultural contexts and knowledge politics. This, consequently, may mark an important gesture towards respecting the tacit links amongst knowledge, cultural diversity and identity work, which is particularly pertinent in multicultural classroom settings (Chronaki 2005, 2011).

Number in Cultures: Imagining a Critical Counter Practice

The notion *'numbers are the same everywhere'* hides certain concerns; first that numbers, as we currently know them in the West, are based on the convention of using the so-called Arabic symbols; second that Arabic symbols are not currently used universally (e.g., in a number of Arabic countries they are not formally used); and third the convention of using the Arabic symbols today is historically and semantically situated. If we ignore the above, we tend to teach mathematics within a restricted focus on developing merely functional literacy (i.e. learning how to name numbers, how to operate with numbers and how to use current numerals in realistic contexts). In this case, the chances of developing critical literacy skills become rather slim. Emphasizing a critical literacy perspective on number might also mean providing opportunities towards expanding children's learning to include awareness of number as non-static and fixed and to consider number as multicultural and diverse in essence. Such awareness can only develop in a pedagogic context that allows children to engage simultaneously in two interrelated dimensions; *first* engagement with relevant information about number in cultures and *second* engagement with opportunities that support critical engagement and reflection over the possible connections amongst number and language and amongst self and other. The fulfillment of this double aim is not a straightforward and easy task. On the contrary, developing awareness and critical reflection demands focused design and sensitized implementation. We will now turn towards describing the particular steps undertaken to meet these demands in terms of the design and the implementation of the playful activity 'Number in Cultures'.

'Number in Cultures', as was outlined above, has been organized around three parts in our design; (a) a narrative about numbers and their role in our lives, (b) exploring and discussing the number-symbols used today in three cultures in a pre-prepared 'Magic-Board' where contemporary Arabic, Romany and Greek symbols and number-words were somehow interrelated, and (c) introducing the game 'my name-my number'. In this chapter, we discuss the above design experience as a critical counter-practice. A critical counter practice is, according to Ilan Gur-Ze'ev (1998), a perspective of critical pedagogy that emphasizes the need to educate participants' "decipher reality" and aims to engage them in the struggle of developing the "reflective potential of human beings and their ability to articulate their worlds as a realization of their reason" (Gur-Ze'ev 1998, p. 485). In addition, it strives for conditions under which everyone will be able to become part of the human dialogue in order to work towards the possibility that the human subject will be able to stand up and confront Heidegger's 'forgetfulness of being'.

Based on the rational outline in the preceding sections and especially on the ideas that (a) the curriculum is not a 'neutral' space, and (b) a counter or non-majority practice that strives for human dialogue potentially might be a space for critical education, we have moved towards designing an alternative approach of teaching number. As such, the 'Number in Cultures' event is deeply related to a core urge for supporting children to become part of a dialogue about diverse social and personal ways of thinking and talking about number. Through this, they may become able to get involved in memory-work about number as fallible and culturally situated. They can also become part of a context where the demands for 'forgetting' number as a fluid entity that develops over time can be opposed and confronted. Memory-work about number is a way of sensitizing children about cultural diversity and fallibility embedded within mathematical knowledge and it supports the aim to approach number as non-static and culturally situated entity.

The activity design was undertaken by the first two authors, Anna and Georgia, and its implementation took place in an outdoor workshop as part of an intercultural festival organized by Kostas. The intercultural outdoor festival called *'One Volos— One Color'* is a yearly event that takes place in a playground nearby the university. It aims to bring students and teachers, children and adults together in a playful and relaxing atmosphere, where they can all interact by developing relations and by exchanging ideas, views and perspectives. Although the word 'color' is used metaphorically to denote an anti-racist political agenda geared towards inclusion of ethnic minorities, it indicates the importance of this very basic feature (i.e. color) on which exclusion practices produce diversity at the same time. The aims of the festival and workshops were: (a) to co-create a temporary open 'space' for children and adults to interact and communicate by means of playing in varied organized 'games' in order to realize that diversity, at varied levels, is a creative part of our lives, (b) to engage student-teachers to design such processes as part of their participation in specific courses, where they can problematize the complexities involved, and (c) to provide opportunities for human dialogue as an entry to dialogicality. During these workshops a variety of games were organized and they were related to literature, visual arts, puppet shows, music, and geography.

A Story, a 'Magic Board', and the Game 'My Name: My Number'

The playful outdoor activity 'Number in Cultures' comprised a story, a 'Magic Board' and a game. Initially, children and adults were gathered together in a circle and participated in a story narrative. The story was about the presence of numbers in our life and the values associated to each one of the single digits from 0 to 9. The presence of numbers was discussed in other cultures and languages such as Greek, Arabic and Romany focusing on questions such as; 'Do you think that all cultures use the same words and the same symbols for numbers?' 'Would you like to know more? Based on these questions, an open discussion was held with the participants (children and adults), where they could express ideas and questions but also could talk about any possible emotional investment on number (e.g., Do you have a favorite number? What does a number might mean to you?).

Participants were then introduced to the 'Magic Board' -a board made out of paper that represented the number words and symbols of the 10 digits (0, 1, 2 till up 9) in three languages -Greek, Arabic and Romany. It served the needs of a visual display or a graphic organizer for them to explore and come in contact with differences and similarities amongst number words and symbols in these three languages. The 'Magic Board' was accompanied by a game entitled 'My name: my culture', which was focused on transforming one's own name into a number and then the number into number symbols across languages (see Fig. 1). The rules for the game were as following;

- write down your name
- count the number of letters in your name and write it down
- identify the number-word and number symbol in Arabic or in Romany
- create a head ribbon with that number symbol and wear it!
- walk around and introduce yourself by using your own symbol

For all their trials, children (and adults who accompanied them) could use the 'Magic Board' for assistance in order to identify exact number words and symbols across languages. The final tasks of constructing the ribbon and introducing themselves as number-word were opportunities for active interaction amongst members of the workshop and for getting to know each other in an informal and playful way. Both also meant to provide a relaxing closure to the game. As such, the game became a means to use and investigate the 'Magic-Board' deeper and more extensively. This was their first game and introduction to connections across various symbols.

'Number in Cultures': Experiencing a Counter Event

As explained above, 'Number in Cultures' was designed as a way to imagine and materialize a critical counter practice based on a playful activity aimed to address a diverse audience including young children and adults who did not necessarily

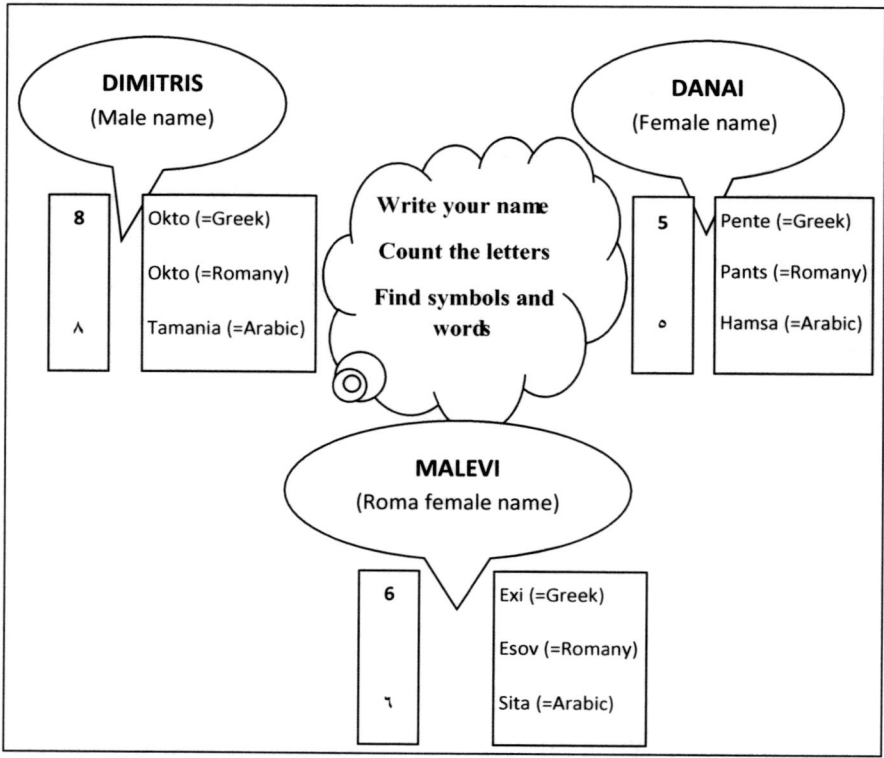

Fig. 1 Anchoring the game 'My name: my number'

possess a mathematical background and who happen to meet at a public space such as the playground. The activity was implemented at the intercultural festival described above in year 2009. The audience, who participated in this outdoor workshop, was a mixed group of people in terms of age (young children and adults who accompany them) and of ethnic backgrounds. Most of the children were of Romany origin but there were also a few immigrants from Albania or Bulgaria. Along with them, a number of non-Romany Greek children came to join together with their parents, thus creating a multicultural and multi-age audience. The age of children varied from 2 to 12 year olds and the age of the adults who accompanied them and participated varied from 20 to 80 year olds.

Taking into account that the workshop took place outdoors in a playground situation, and that certain features concerning the profile of the participants such as their age, gender and ethnicity were not known till the last minute, our design had to consider an emphasis on the creation of short span and playful activities. Doing outdoor activities in large groups, children and adults have difficulties keeping their attention on the meticulous completion of tasks that demand complex problem solving skills or investigational work. The work we described above was organized as an outdoor creative workshop that could take place in a public space. It was experimental as well as explorative and ethnographic in nature and came close to a 'teaching experiment' methodology (Hedegaard and Chaiklin 2005) that emphasized a systematic ethnographic observation of how an experimental activity becomes implemented with humans and facilitates qualitative data collection (Chronaki 2011). In the following sections we will describe some indicative events of experiencing 'Number in Cultures' as a critical counter-practice as part of its implementation process.

Narrating a Story: Opening up 'a Space of Appearance'

During the first phase of implementing 'Number in Cultures', our attempts were geared towards involving participants to sit around a circle and to participate around the oral narration of a story concerning 'The King of Numbers' (see Fig. 2). The story evolved in an imagined school classroom, where all children were called to discuss about their beloved number by accounting its value and importance in life. Apparently, although all numbers seemed to be assigned a distinct value in children's eyes, number zero (0) was left outside as if it was completely insignificant. Zero needed to defend itself by explaining what might be its own distinct potentialities and affordances (e.g. zero can be added at the end of every number and can create a different number as in 10, 100, 1,000 etc.) (Fig. 3).

Participants, children and adults had the opportunity to talk about number in their life and to find simple ways of identifying themselves with number (e.g., what is my favorite number). This discussion was followed by questions trying to investigate how competent the small children were with number through questions such as, 'Do you know how to read and write symbols for all numbers? Can you write them down?' It became obvious that children in early primary and late nursery knew

Fig. 2 Narrating a story about number

Fig. 3 Introducing the 'Magic Board'

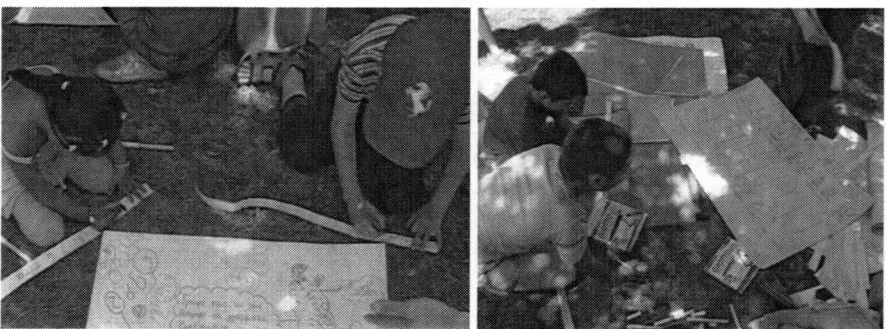

Figs. 4 and 5 Playing the game & investigating number words and symbols

how to recognize and write Western number symbols well. Younger children knew how to handle the use of number words in oral speech and expressed confidence with counting a small number of objects (see Figs. 4 and 5).

Introducing Number in Cultures: Expanding Possibilities for Memory-Work

In consequence, the idea that numbers are not the same in all cultures was introduced by posing a critical question such as: 'Do you think that everybody in this world has the same way to symbolize number?' Here, the response of the vast majority was a unanimous *'yes!'* This exemplifies the dominance of the discourse that mathematics is culturally uniform and unchangeable in terms of a public understanding of number. Such a simple response reflected how number has become stereotypically fixed as a common language for public communication and Western symbols as the norm to the extent that its sociocultural roots have become forgotten and lost. The fact that different number symbols are currently used in some countries (e.g., Egypt) came as a huge surprise to all participants and provided a terrain for discussion and reflection. It is essential to emphasize that this 'surprise' element was instrumental towards shifting participants' attention (both children and adults) from an essentialist view of number as a universal signifier towards the idea of exploring number symbols' development and use across Western and Arabic cultures.

At this point, the 'Magic Board' was introduced as a way of introducing them systematically towards the related differences and similarities amongst Arabic and Western symbols and for providing a resource to undertake their coherent reading. To begin with, all numbers and their symbols were read aloud and children were given the chance to point out specific number symbols and to check out the relatedness with number-words or vice versa. Several questions arouse concerning sameness and diversity across number symbols and number words. At the same time, children were seriously engaged into checking the correspondence amongst number words and symbols in three languages: Greek, Romany and Arabic. Using the 'Magic Board' was an important affordance to their efforts. A few 'Magic Boards' were left around to facilitate children's attempts to search corresponding words and symbols as they were investigating what might be their own secret number. In all, the 'Magic Board' served to substantiate the questions around 'Number in Cultures' and also provided a source for information to be taken further (see Figs. 4 and 5).

Playing the Game: Creating a Space to Meet and Act with the 'Other'

As described before, the game *'my name: my number'* created the opportunity to explore the information provided through the 'Magic Board' more deeply. Children turned towards playing the game with great interest. At first, they enjoyed telling and repeating their own names, spelling it out and counting its letters. This was an additional and wonderful opportunity to introduce themselves to children sitting next to them in the same group who had met at the playground for the first time. Performing little acts such as telling their names, counting the number of letters in their names, and identifying that particular number in a language different to their own became steps of a magic process that released excitement and enthusiasm. This number became a secret 'code' that enabled them to disclose themselves, to become present, recognized and valued.

This game created curiosity, surprise and engagement—qualities found in processes that can capture children's attention and motivation. The game *'my name: my number'* also created affordances for children towards making further explorations into number symbolization and developing awareness of number words and diversity of symbols. Children observed that numbers could be written in different symbols—an idea that they did not have before. In parallel, they expressed their admiration of Arabic symbols commenting that number symbols are like calligraphy. Via the 'Magic Board' they could systematically perceive that some numbers may have similar symbols or number words and some others do not. Such observations created the impetus to ask more questions and the motivation to learn more about number and the history of symbol transformation. Moreover, this event provided the chance for the Roma children to identify themselves with the roots of a disciplinary form of literacy that was mostly denied. Their eager participation and happy faces showed evidence of how proud they felt about being able to use Romany number words next to Western and Arabic ones. To give one example, a gypsy girl called Malevi counted the number of letters in her name and observed (i.e. by searching in the 'Magic Board') that the number word 'esov' meaning 6 in Romany echoes the number word 'sita' meaning 6 in Arabic. Malevi invested emotionally in this as it provided her the opportunity to experience presence and recognition by means of having her own cultural heritage becoming recognized and present (see also Chronaki 2011).

As an ending of the game, the children were asked to create head ribbons using their own beloved color in both paper and pens. They were excited about the idea that their name could transform into a number and said that they now have a secret code to accompany them for life. Moreover, this excitement was communicated with the other participants and as they were walking around wearing the head ribbons, everybody was involved in a social sharing of names that were transformed into number words and number symbols. This event created surprise, curiosity, interest and emotion. The use of the three languages (via number words and symbols) had embedded both a functional aspect (i.e. learning to read aloud the number word and symbol) and a critical aspect (i.e. becoming aware of epistemic and cultural diversity and of positioning as potential learners and active participants of epistemic history). Experiencing both within the context of such a playful activity meant safeguarding the possibility of creating a space for dialogue as they do and share things and ideas collectively in the context of playing together (Figs. 6 and 7).

Concluding Remarks

The present study is rooted within the belief that political, cultural and historical circumstances where we all live and construct notions of 'self' and 'other' simultaneously provide the context and the motivation for our learning. They frame and reframe our will and potential to participate in formal educational practices. Thus, identity work and learning practices are interrelated. 'Number in Cultures' has here been approached as a counter event that has provided us with the potential to make space for critical mathematics education for young children. The urge to act within a perspective that sensitizes children (and adults) towards troubling essentialist

Number in Cultures 157

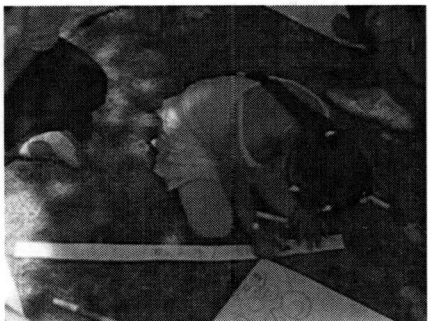 

Figs. 6 and 7 Children playing the game: creating their head ribbons

views of mathematical knowledge is a political act. It recognizes the need to pave the way for encountering alternative discourses of mathematics knowledge use and of mathematics education. The enactment of experiences with mathematics as formal or informal activity in Greece, but also worldwide, represents knowledge, by and large, as a universal, unchangeable and static set of abstract entities addressing number and geometry as ideal structures. At their best, current curricula try to model real-life activity through word-problems or 'thematic contexts' where mathematics can be applied as a set of tools that instrumentally facilitate problem solving. By and large such practices dominate the available learning and educational experiences for children. The present study encourages us to consider how the design of counter events that promote human dialogue can pave trajectories for re-articulating discourse(s) and encourage new power constellations to appear as possibilities. As seen above, experiencing a counter event can: (a) open a space of appearance for marginalized children, (b) expand possibilities for memory-work related to mathematical knowledge and silenced identities, and (c) create a space to meet and act with the other as entry to dialogicality.

Such educative experiences have been, by and large, discussed as potentialities with older children and adults. The present study has indicated that purposefully designed mathematical activity that makes space for critical education has the potential to also address the needs for younger children who belong in marginalized and dominant groups alike. The counter event of 'Number in Cultures' presented here can be considered as an opportunity for memory-work around number that strives to create an open space for individual children to enact alternative learning identities. Specifically, Roma children from being undervalued as marginalized students, come to appear as potential owners (and ancestors) of a past in the history of number development. In parallel, non-Romany Greek children can engage in a critical dialogue with the 'other'. For both of them, appreciation of number diversity (i.e. number words and symbols are not written and uttered similarly in every culture), encourages them to realize the multiplicity of mathematical knowledge itself and to re-consider how they identify and position themselves.

According to Maaluf (1998), a static and monolithic view of identity that focuses merely on color, religion, race or gender can be considered as 'murderous'.

He argues how a monolithic and static view on identity leads to 'murderous identities' that becomes disastrous and catastrophic for both self and community. The complexity, uncertainty and precariousness of everyday life in today's world should sensitize us towards addressing ways that encourage and support the individual subject to identify his/her own ways of positioning himself/herself in the plurality and multiplicity offered in educational practices. Having children, even at such early years, become introduced to a perspective of learning identities as fluid, multiple and hybrid becomes a crucial element for embracing diversity as a resource and not as a hazard.

References

Apple, M. (2004). *Ideology and curriculum*. New York: Routledge.

Atweh, B., Graven, M., Secada, W., & Valero, P. (Eds.). (2011). *Mapping equity and quality in mathematics education*. New York: Springer.

Bartolini Bussi, M., & Boni, M. (2009). The early construction of mathematical meanings: Learning positional representation of numbers. In O. A. Barbarin & B. H. Wasik (Eds.), *The handbook of child developmental and early education: Research to practice* (pp. 455–477). New York: Guilford.

Bertau, M. C. (2007). On the notion of voice: An exploration from a psycholinguistic perspective with developmental implications. *International Journal of Dialogical Science, 2*(1), 133–161.

Biesta, G. (1998). Say you want a revolution … Suggestions for the impossible future of critical pedagogy. *Educational Theory, 48*(4), 499–510.

Brown, T. (2011). *Mathematics education and subjectivity: Cultures and cultural renewal*. London: Springer.

Chronaki, A. (2005). Learning about "learning identities" in the school arithmetic practice: The experience of two young minority Gypsy girls in the Greek context of education. *European Journal of Psychology of Education, 20*(1), 61–74.

Chronaki, A. (2009). An entry to dialogicality in the maths classroom: Encouraging hybrid learning identities. In M. César & K. Kumpulainen (Eds.), *Social interactions in multicultural settings* (pp. 117–143). Rotterdam: Sense.

Chronaki, A. (2011). Troubling essentialist identities: Performative mathematics and the politics of possibility. In M. Kontopodis, C. Wulf, & B. Fichtner (Eds.), *Children, development and education: Cultural, historical and anthropological perspectives* (pp. 207–227). Dordrecht: Springer.

Crump, T. (1990). *The anthropology of number*. Cambridge: Cambridge University Press.

Derman-Sparks, L., & Ramsey, P. G. (2006). *What if all the kids are white: Anti-bias multicultural education with young children*. New York: Teacher College Press.

Frankenstein, M. (1989). *Relearning mathematics*. London: Free Association Books.

Gur-Ze'ev, I. (1998). Toward a non-repressive critical pedagogy. *Educational Theory, 48*(4), 463–486.

Gutstein, E. (2006). *Reading and writing the world with mathematics: Toward a pedagogy for social justice*. New York: Routledge.

Gutstein, E., Lipman, P., Hernandez, P., & de los Reyes, R. (1997). Culturally relevant mathematics teaching in a Mexican American context. *Journal for Research in Mathematics Education, 28*(2), 26–37.

Hedegaard, M., & Chaiklin, S. (2005). *Radical-local teaching and learning: A cultural-historical approach*. Aarhus: Aarhus University Press.

Knijnik, G., & Wanderer, F. (2010). Mathematics education and differential inclusion: A study about two Brazilian time–space forms of life. *ZDM – The International Journal on Mathematics Education, 42*(3), 349–360.

Maaluf, A. (1998). *Les identités meurtrières*. Paris: Grasset & Fasquelle.

Nieto, S. (2004). *Affirming diversity*. New York: Pearson.

Porter, T. (1953/1996). *Trust in numbers: The pursuit of objectivity in science and public life*. Princeton: Princeton University Press.

Radford, L. (2001). The historical origins of algebraic thinking. In R. Sutherland, T. Rojano, A. Bell, & R. Lins (Eds.), *Perspectives on school algebra* (pp. 13–63). Dordrecht: Kluwer.

Saxe, G. (1981). Body parts as numerals: A developmental analysis of numeration among the Oksapmin in Papua New Guinea. *Child Dev, 52*, 306–316.

Skovsmose, O. (1994). *Towards a philosophy of critical mathematics education*. Dordrecht: Kluwer.

Skovsmose, O. (2011). *An invitation to critical mathematics education*. Rotterdam: Sense.

Skovsmose, O., Yasukawa, K., & Ravn, O. (2013). Scripting the world with mathematics. In P. Ernest & B. Sriraman (Eds.), *Critical mathematics education: Theory, praxis and reality* (pp. 255–281). Charlotte: IAP.

Stinson, D. W., & Bullock, E. C. (2012). Critical postmodern theory in mathematics education research: A praxis of uncertainty. *Educational Studies in Mathematics, 80*(1–2), 41–55.

Valero, P., & Zevenbergen, R. (Eds.). (2004). *Researching the sociopolitical dimensions of mathematics classroom: Issues of power in theory and methodology* (pp. 107–123). Dordrecht: Kluwer.

Vithal, R. (2003). *In search of a pedagogy of conflict and dialogue for mathematics education*. Dordrecht: Kluwer.

Walshaw, M. (2011). Identity as the cornerstone of quality and equitable mathematical experiences. In B. Atweh, M. Graven, W. Secada, & P. Valero (Eds.), *Mapping equity and quality in mathematics education* (pp. 91–102). New York: Springer.

Fairness Through Mathematical Problem Solving in Preschool Education

Zoi Nikiforidou and Jenny Pange

Abstract This chapter reports on a study of fairness in young children. Its aim is to examine how preschoolers respond to mathematical problem-solving contexts that imply fairness and the claim of sharing and distributing. The results suggest that by the age of four, children are cognitively and socially capable, at a certain level, to transmit from their own state of mind to that of the others, overcome self-interest and understand, expect and prefer fairness, while distributing justice.

Democracy learning in the preschool context appears, directly or indirectly, through many levels and forms. Fairness and the notions of sharing, owing and giving characterize children's cognitive, social and moral development during the preschool ages. The aim of this study is to investigate whether young children ($N=40$), aged 4–6, express fairness through problem-solving situations. Scenarios were used in two conditions with three identical mathematical problems engaging processes of correspondence of uneven sets, equal distribution and fair open-ended switching; in the 1st Condition children were addressed in the 1st person, whereas in the 2nd Condition they were asked to assist a neutral hero. Results imply that while children at this age develop their Theory of Mind (ToM), they may accomplish division and correspondence in a fair way independent of the person who is directly affected. Concerning the swapping process, children would propose an 'identical other' item to be considered as a fair solution in the process of giving and taking, mainly when they were the direct recipients. Such findings indicate educational and methodological aspects in encouraging fairness-related activities through mathematical problem solving in preschool education.

Z. Nikiforidou (✉)
Department of Early Childhood, Liverpool Hope University, Liverpool, UK
e-mail: nikifoz@hope.ac.uk

J. Pange
Department of Preschool Education, University of Ioannina, Ioannina, Greece
e-mail: jpagge@cc.uoi.gr

Theoretical Background

Fairness and Early Years

Values, norms and concepts of democracy and justice underlie everyday activities and engagements formally and informally at an individual and collective level. From a philosophical approach, Rawls's theory of justice (1971) attempts to unify all the principles and ethical criteria in a coherent system, through reasonable ordering and scoping. Under 'justice as fairness' there are two principles to be met in a just and fair society: the principle that everyone has the same basic liberties and the principle of fair equality of opportunity (Rawls 2001). "A decision process is fair to the extent to which all those concerned are well informed and have their interests and perspectives expressed with equal force and effectiveness" (Barry 1995, p. 110). Even if there is no 'perfect' justice, being fair implies a 'comparative exercise' (Sen 2009) in the sense that in order to identify justice we need to compare different social states.

However, from a developmental point of view, fairness is mainly addressed not only as a value, attitude, virtue or social norm, but also as a skill, a perception or a cognitive process that emerges through age and growth. Notions like the allocation of resources, sharing, taking into account the welfare of others, proportional in/equality, equitable resource distributions, egalitarianism, reward-dispensation contexts (i.e. Sloane et al. 2012; Takagishi et al. 2010) are examined through child-friendly problem-solving situations or dilemmas. Children, and even infants, get exposed to problems that imply their understanding of fairness; either in relation to the other, the experimenter, another player, a neutral hero, a hypothetical actor and/or in relation to themselves and their personal interests. Is what is fair for me, fair for the other/s too?

Fairness in Early Years contexts can be linked with cognitive, logico-mathematical processes on one hand (developmental approach), and on the other hand, with issues of moral, personal and social aspects (philosophical approach). Thus, in every case there is a common point: Fairness is not only about oneself. Am I fair? Fair towards what or towards whom? Fairness always involves someone else and/or others as well as a specific context and a particular situation. So, in order to be fair someone needs to be able to manipulate the given information and to take the perspective of the other and understand their intentions, desires, true beliefs, hidden emotions and perceptions through the development of one's Theory of Mind (i.e. Wellman and Liu 2004; Wu and Su 2013).

In the classroom, children participate in a shared reality, while developing prosocial behaviors and moral reasoning (Miller 2006) and while 'being' and 'doing' democracy. Fairness and the ideas of sharing, distributing, turn taking, giving and owing undergo multiple activities and realities. Some examples include free play or thematic learning activities, digital games or group games, peer interactions or individual tasks, scenario-based and inquiry-based learning, where children face uncertainties, conflict situations, novel circumstances and have to negotiate, debate, make decisions, consider what is just and right for themselves and for the others. Through these everyday interactions children learn to make their own choices, take initiatives, solve problems, take the perspective of others, cooperate, take risks and above all develop democratic attitudes, skills and strategies (Emilson 2011).

The Origins of Fairness in Young Children

From a classical piagetian approach, children undergoing the pre-operational stage, at the ages of 2–7, have difficulty in viewing the world and diverse situations through the point of view or the feelings of others, as their thinking is characterized by egocentrism and centration (Damon 1975; Piaget 1972). Under these lines, preoperational children primarily focus on their own needs and consider objects and causality to be analogical to their own actions or activities and "everything is judged from the individual's point of view" (Piaget 1972, p. 215). Accordingly, at this stage they reveal self-interest, they view the world through their own lenses and are about to gradually mediate between the awareness of their own mental states and that of the others, under their Theory of Mind (ToM) (Hale and Tager-Flusberg 2003).

More recent research argues that the Piagetian tasks to explore children's egocentrism (i.e. the three-mountain task, Piaget and Inhelder 1956) were highly verbal, closely linked to advanced cognitive mechanisms such as memory, 3-D representations and language acquisition, and not very child-friendly in the sense of not being meaningful and purposeful for young children (Flavell 1988; McCrink et al. 2010; Yost et al. 1962). Furthermore, in the last 35 years there is an ongoing interest in how egocentrism and selfishness are linked with the ToM, first introduced by Premack and Woodruff (1978), who, through their studies with chimpanzees, defined ToM as the ability to impute mental states to oneself and others.

ToM enables "children to predict and explain actions by ascribing mental states, such as beliefs, desires, and intentions, to themselves and other people" (Astington 1991, p. 158). Depending on the situation, young children and even infants can take into account the perspective of the other and reason as 'the other' by showing empathy, altruism, compassion and other key pro-social characteristics (i.e. Callaghan et al. 2005; Sally and Hill 2006; Sloane et al. 2012; Wellman et al. 2001). While developing their ToM children start to differentiate their thinking from the others' in order to split the first-person and the third-person angle (Rochat et al. 2009).

Other-regarding preferences and the mental concepts of false-belief, perspective-taking, sharing and inequity aversion are mainly studied through fairness-related behavior games (Fehr et al. 2008; Sloane et al. 2012; Takagishi et al. 2010; Wu and Su 2013), implying mathematical concepts like division, fractions or probabilities (Watson and Moritz 2003). So, an important factor in the development of children's understanding of fairness is their ability to infer the mental states of others. How can I be fair to myself and to others if I don't conceptualize what others want, feel, believe, see, expect, desire? How can I understand if a game is fair if I don't relate to particular information?

Precisely, the fairness-related behavior can be facilitated by ToM, according to Sally and Hill (2006), and the perspective-taking abilities play a significant role in normative behavior like fair distribution (Fehr et al. 2008). If a child can take into consideration the views, beliefs, desires and intentions of the others, then, in his/her allocation of resources or distribution of goods, s/he might be more equal, more just, more fair and less egocentric or self-centered. Takagishi et al. (2010) found in their study that preschoolers who had more advanced ToM would make less selfish offers and would propose a fairer division of candies, whereas preschoolers who had not yet developed their perspective-taking abilities made offers on a more selfish basis.

Significant Factors in Studying Fairness in Young Children

One important aspect in examining young children's development of fairness is whether they are direct or indirect recipients of the problem-solving situation. A claim for subjective and objective fairness has to do with whether it relates or does not relate to one's personal interests, experiences and needs. Participants might show different preferences if they are personally affected by the result of their participation or if they call for decisions that involve other actors and protagonists. This is a reason why methodologically third-party and first-party tasks are used (Gummerum et al. 2010; Moore 2009; Sloane et al. 2012; Sommerville et al. 2013).

Along these lines, research on fairness has shown that children starting at the age of 3.5 demonstrate sensitivity towards fairness at least under some conditions when the recipients are others and not themselves, namely in situations where self-interest cannot intrude (e.g., Fehr et al. 2008, Olson and Spekle 2008). Harbaugh et al. (2003) underlined that with age, bargaining behaviors change towards greater fairness in distributive justice and participants overcome self-interest after the age of 7. In a cross-cultural study carried out by Rochat et al. (2009) 202 children demonstrated fairness in sharing even at the age of 3, through both differentiation and strategic coordination of first-person and third-person perspectives. In the same direction, Moore (2009) found that, among 66 preschoolers, the resource-allocation decisions depend on the recipient with a strong preference firstly to themselves, after to their friends and finally to the non-friends. Around the age of 5 children can show sensitivity to fairness in both cases where they are direct recipients (first-party tasks) or external players (third-party tasks) (e.g., Gummerum et al. 2010; Takagishi et al. 2010)

Many times, young children are aware of fairness standards, but when they are personally involved they might discard the hypothetical distributive justice and seek to favor themselves by allocating resources unfairly. As Smith et al. (2013) underline there is a judgment-behavior gap in children, in the sense that even though young children endorse fairness norms related to sharing, when given a chance to share they often act in contradiction to those norms. Smith et al. propose that a possible reason for this gap between fairness norms and actual resource allocation is connected to the tensions between desires (what I want to do) and norms (what

I have to do). Children at younger ages want to keep more for themselves despite the fact that they have a sophisticated understanding of fairness (Sommerville et al. 2013). They might know and be able to judge which is the fair choice or solution to a problem but instead prefer what is more advantageous for them and in turn 'unfair' in that particular case.

Language is one more factor linked with children's perception and understanding of fairness. The linguistic articulation of a particular mathematical or conceptual notion, like fairness, is crucial in children's judgment and reasoning in the context of what is fair and what is not fair. LoBue et al. (2009) found that the ability to talk about fairness has a developmental trend between the ages of 3–5. They support that children are sensitive to unequal distributions before they can explain why and before they can articulate their opinions; although they recognize and dislike inequalities, they gain the capacity to express so only gradually.

The aim of this study is to examine how preschoolers respond to mathematical problem-solving contexts that imply fairness and the claim of sharing and distributing.

(a) Do they express this democratic value orally and graphically while participating in scenario-based problem-solving situations?
(b) Do they link fairness to the mathematical procedures of correspondence (problem 1), equal distribution (problem 2) and open-ended fair exchanges (problem 3)? In this last problem, what happens when children have the option to choose as an incentive whatever they wish (problem 3), as there are no specified resources to choose from?
(c) Do they show the same strategy when they are personally involved as when they participate through the perspective of another person; in this study that of a young hero?

Materials and Methods

The research was conducted at the university kindergarten linked to the University of Ioannina. Forty children from two classrooms, aged 4–6 years old ($M=5.1$), participated in this within-subject study.

Children participated in groups of five, randomly selected from the cohort of their classrooms, in a separate classroom within their school. Children sat around a small table with the researcher, who after introducing herself moved on to explain the task to them. Children were assigned to one of the two Conditions (1st person vs 3rd person) and completed the three problems at once. Storytelling and problem-solving situations were used in order to engage the children in reasoning and thinking practices concerning fairness. Children could collaborate and exchange ideas while giving and recording individually their 'fair' answers to the tests.

In the beginning children were tested on their prior knowledge of the meaning of 'fairness'. They were asked to give, according to their opinion, a 'fair' or an 'unfair' example. Subsequently, participants listened to a narrative, funny rhyme produced

Table 1 The methodology of the study

	Condition 1 (1st person)	Condition 2 (3rd person)
1st problem (correspondence)		
2nd problem (distribution)		
3rd problem (exchange)		

by the researchers, where the hero, a young boy named Takis, experienced three adventures and dilemmas (Table 1).

In the 1st problem children were asked to distribute fairly four items between two characters (correspondence among two uneven sets; corns and chicken). In the 2nd problem they were asked to divide fairly one object in four parts (equal distribution; a pie in four equal parts). In the 3rd problem they were asked to swap fairly an item (a bicycle) with another item of their own preference considered as 'fair' (exchange). In the first two problems children were given the available data to be used fairly or unfairly, whereas, in the last problem, children had to complete the image of justice by proposing what they subjectively considered to be fair.

There were two conditions differing in the perspective. In Condition 1 the questions and dilemmas were personally addressed to the participants as if they were the protagonists (1st person); in this case the question was: '*What would you do to be fair?*' In Condition 2 the respective dilemmas and questions were referred to the neutral hero (3rd person); '*What do you think Takis should do to be fair?*' In both conditions identical colored pictures, sized 10 × 10 cm, were used to illustrate the key information of the problem solving situations.

Children had the opportunity to discuss and exchange ideas in order to contribute a fair solution to the three diverse situations, and in the end they recorded their answers individually on specially designed A4 sheets. Then the researcher would discuss with each participant 'why' they selected the precise response and keep notes on children's justifications. Children's personal recordings and explanations were used for further analysis.

Results

Concerning their prior knowledge on fairness, all children stated having heard and knowing the meaning of the term. However, only 60 % of the children could express their ideas semantically and link fairness to a precise example or a situation. Indicatively some responses were:

Child A: Fairness is when we don't cheat.
Child D: Fairness is when you can't draw something and the others help you.
Child K: It is not fair to play a game with cards and get few cards.

In the 1st problem 85 % of the children achieved equivalence among two uneven sets of objects correctly (i.e. two corns per chicken). In order to examine whether there was a difference in children's fair responses based on the 1st person or 3rd person perspective, a t-test was used. It was found that the perspective did not influence the fair attribution of items to recipients; t(38)=0.618, $p>0.05$. It is worth mentioning that there was one case where a group, in total five participants, argued and proposed to give one corn per chicken on that day and keep the remaining two corns for the following day. This response, apart from being a good example of collaborative learning, was scored as a 'fair' solution.

In the 2nd problem 70 % of the children recorded and explained how the distribution of a pie in four pieces would be fair. In this case, there were also some children who were able to explain orally the procedure but had difficulty in expressing such procedures graphically. In these cases, if children reasoned and explained orally the division 'correctly', they were recorded as fair. Again, the perspective, Condition 1 (1st person) vs Condition 2 (3rd person), did not play a significant role; t(33)=−1.08, $p>0.05$ and the null hypothesis of finding a difference in the two Conditions was rejected (Fig. 1).

In the 3rd problem, the answers were more subjective as the option of swapping one item with a non-presupposed item was open-ended and based on personal beliefs and estimations. In this test, item A (a bicycle) was taken away from the hero (in Condition 2) or from each participant (in Condition 1) while they were offered to get anything else in exchange: item B. 60 % of the responses indicated an 'identical other' item as a fair solution to the exchange; so, this would be a bicycle. Other responses were random and subjective like, 'a ball', 'a cat', 'a balloon', 'a doll house'. However, participants in Condition 1 showed a significant preference for the 'identical other' as a fair answer, compared to participants in Condition 2; t(33)=−2.56, $p<0.05$. This implies that when children were directly affected, they were more cautious with what they suggested as fair based on their personal interest. In this case, the majority strongly believed that the same item, a bicycle, was a fair solution.

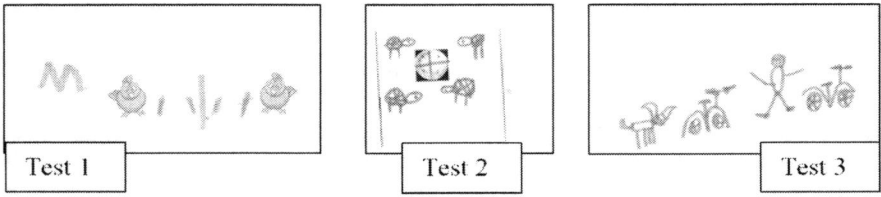

Fig. 1 Examples of the 'fair' responses

Discussion and Conclusions

Findings demonstrated that preschoolers can identify fairness in problem-solving situations involving mathematical procedures and fairness norms. They gave fair responses in the test of correspondence of uneven sets of items (problem 1) and in the test of equal distribution (problem 2), independent of the perspective of the actor. These responses support previous research (Callaghan et al. 2005; Fehr et al. 2008; Gummerum et al. 2010; McCrink et al. 2010; Moore 2009; Rochat et al. 2009; Sommerville et al. 2013) implying that no matter who was the recipient (themselves or the hero) they would pursue the most 'fair' solution. This suggests that by the age of 4, children are cognitively and socially capable, at a certain level, to transmit from their own state of mind to that of the others, overcome self-interest and understand, expect and prefer fairness, while distributing justice.

In the first two tests, preschoolers managed to make mathematically sound judgments in order to establish fairness norms. They were able to distribute two uneven sets of items fairly (4:2) and to divide an object in equal parts (1:4), no matter if they were personally addressed or if they had to assist the hero of the scenario. In both problems they knew what was normatively fair in the sense of 'what is expected to be done to be fair' (Fehr et al. 2008) and, in most cases, made mathematical and graphical connections. These results entail that, within the ages of 4–6, children can relate fairness to other mathematical notions (like probabilities; Watson and Moritz 2003), can take into account the perspective of the other and can advise another person on what to do in order to select a fair solution in a problem; they acquire these skills with the development of their ToM (Wellman et al. 2001; Wellman and Liu 2004).

Fairness is one of the attributes linked with the ToM and vice versa ToM encourages fairness-related behaviors (Sally and Hill 2006; Takagishi et al. 2010; Wu and Su 2013). Both statements were evident in the current study as children showed that they can take into account the hero's views and intentions while recommending to him fair solutions that would make everyone happy (Condition 2) in the same way they gave solutions to the problems when asked personally (Condition 1). The proposed distribution of goods in both tests, 1 and 2, were similarly independent of whether the children were helping the hero or not. Being able to see the same problem from a personal and an 'other' stance implies that children at this age within this meaningful context responded for justice allocation in a similar manner (Gummerum et al. 2010; Rochat et al. 2009).

However, this was not the case in the last problem of item exchange (problem 3). The nature of the design of this task is quite rare in the sense that it is open-ended with no predetermined payoffs. Participants in this test suggested as fair the 'identical other' significantly more often when they were directly affected rather than when the neutral hero was the recipient of their proposed solution, demonstrating a distinction between what is fair to them and what is fair to the other. In Condition 1, under the 1st person perspective, children were more cautious in selecting a 'fair' answer and agreed on the 'identical other' item as a fair exchange.

Here, they responded in a rather selfish manner, although this idea of selfishness may be discussed. In this last test, where the attributed value to be considered as fair was subjective and open to selection, most children chose to keep the 'identical other' item. They considered the exchange of a bicycle with a bicycle as fair, even though they were free to pick anything else. Despite this freedom of selecting anything, children gave responses with a normative and egalitarian basis (Fehr et al. 2008); they suggested in most cases the exact same-equal item. However, if they had been egocentric and had only sought to stimulate their own interest as the pre-operational characteristics (Piaget 1972) suggest, they could have asked for something more valuable than a bicycle, as they had this option. Someone might dispute that a bicycle is really valuable for preschoolers, thus again, if they were driven by their own desires and needs only, they could have asked for something different or bigger or more individualistic.

Some children, while dealing with dilemmas, depicted a judgment-behavior gap (Smith et al. 2013). This can be due to confronting a source of conflict or imbalance between their individual desires and the broader norms or expectations, between subjective and objective fairness; what I want to choose and what I have to choose. Thus, this gap could apply to all ages, not only children, and is not an indicator of limited perception or appreciation of fairness. Everyone, at any age, might choose an option which is not fair but rather more beneficial per se at a personal level. Further research in this direction could highlight the precise factors that interfere with this gap and how education can play an important role in supporting children from early ages to make wise and fair decisions within the wider context of judgment and decision-making.

In the current study, the problem solving scenarios were presented in a child-friendly participatory way; funny rhymes within a context that combined a semantic framework, real life data, useful visual input, a motive to engage (by asking them to help the hero to overcome his dilemmas or by asking them what they would do if they were in his place) with children's personal active participation, collaboration and reasoning. Children also had the opportunity to represent their ideas graphically and construct solutions through group discussions with the others. They could exchange ideas and discover problem-solving techniques instead of participating in a task that would make no sense to them, taking into consideration some of the criticisms on the methodology of Piaget's experiments (Yost et al. 1962).

Children revealed difficulty to verbally use the words 'fair' or 'unfair' in a sentence. This semantic difficulty complies with the study of LoBue et al. (2009) who supported that the ability for verbal articulation of unfairness and inequity is age related. Children admitted knowing what the word 'fair' means but not all of them could include it in an example or a situation. Or children were able to explain how to divide one pie in four pieces but had difficulties in representing it graphically. Such findings imply that children might be aware of unfair or fair conditions but might not be able to express the relevant linguistic connotations. This aspect of the role of language, oral and written, should be taken into account when designing teaching-learning activities.

In further research more factors, like the use of new technologies or the use of more scenarios linked to other mathematical notions, could be studied on the grounds of educational practice, as children's emergent democracy learning can be enabled and supported in the preschool context (Emilson 2011). Some methodological alterations could arise issues of individual participation or/and collaborative learning in problem solving. Scenarios more directly connected to children's personal interests or with real tokens or gifts or incentives could be taken into account, as this was not the case in the current study. Another limitation of the study was the small sample and the fact that there could be a developmental trend between the age ranges of 4–5 and 5–6.

Overall, the implications of this study may be constructive in an educational, didactic direction since fairness can be supported in the preschool classroom. Based on their cognitive, personal, emotional, moral and social development, preschoolers at the age of 4 may be inserted to the meaning of fairness through logico-mathematical meaningful activities. At this early age, children developmentally start to overcome their sense of ownership and self-interest, they start to develop a normative and egalitarian thinking, they build up their ToM, and mediate between their own and others' views, desires, intentions, beliefs. Through their formal education, children begin to develop reasoning and understanding of the world in a more organized, shared and goal-oriented way. Therefore, preschoolers should be given the opportunities to share, give, distribute, interpret, decide, assert, negotiate and last but not least be fair and just; fair to themselves and to others.

References

Astington, J. (1991). Intention in the child's theory of mind. In D. Frye & C. Moore (Eds.), *Children's theories of mind: Mental states and social understanding*. Hillsdale: Lawrence Erlbaum.

Barry, B. (1995). *Justice as impartiality*. Oxford: Clarendon.

Callaghan, T., Rochat, P., Lillard, A., Claux, M. L., Odden, H., Itakura, S., et al. (2005). Synchrony in the onset of mental-state reasoning: Evidence from five cultures. *Psychological Science, 16*(5), 378–384.

Damon, W. (1975). Early conceptions of positive justice as related to the development of logical operations. *Child Development, 46*, 301–312.

Emilson, A. (2011). Democracy learning in a preschool context. *International Perspectives on Early Childhood Education and Development, 4*, 157–171.

Fehr, E., Bernhard, H., & Rockenbach, B. (2008). Egalitarianism in young children. *Nature, 454*, 1079–1083.

Flavell, J. H. (1988). The development of children's knowledge about the mind: From cognitive connections to mental representations. In J. W. Astington, P. L. Harris, & D. R. Olson (Eds.), *Developing theories of mind* (pp. 244–271). Cambridge: Cambridge University Press.

Gummerum, M., Hanoch, Y., Keller, M., Parsons, K., & Hummel, A. (2010). Preschoolers' allocations in the dictator game: The role of moral emotions. *Journal of Economic Psychology, 31*, 25–34.

Hale, C. M., & Tager-Flusberg, H. (2003). The influence of language on theory of mind: A training study. *Developmental Science, 6*, 346–359.

Harbaugh, W. T., Krause, K., Liday, S. G., & Vesterlund, L. (2003). Trust and reciprocity: Interdisciplinary lessons from experimental research. In E. Ostrom & J. Walker (Eds.), *Trust in children* (pp. 302–322). New York: Russell Sage.

LoBue, V., Nishida, T., Chiong, C., DeLoache, J. S., & Haidt, J. (2009). When getting something good is bad: Even three-year-olds react to inequity. *Social Development, 20*, 154–170.

McCrink, K., Bloom, P., & Santos, L. (2010). Children's and adult's judgments of equitable resource distributions. *Developmental Science, 13*(1), 37–45.

Miller, J. G. (2006). Insight into moral development from cultural psychology. In M. Killen & J. Smetana (Eds.), *Handbook of moral development* (pp. 375–398). Mahwah: Lawrence Erlbaum.

Moore, C. (2009). Fairness in children's resource allocation depends on the recipient. *Psychological Science, 20*(8), 944–948.

Olson, K. R., & Spelke, E. S. (2008). Foundations of cooperation in young children. *Cognition, 108*, 222–231.

Piaget, J. (1972). *Judgment and reasoning in the child*. London: Routledge & Kegan Paul. (Original work published 1924)

Piaget, J., & Inhelder, B. (1956). *The child's conception of space*. London: Routledge.

Premack, D., & Woodruff, G. (1978). Does the chimpanzee have a "theory of mind"? *Behavioral and Brain Sciences, 4*, 515–526.

Rawls, J. A. (1971). *A theory of justice*. Cambridge, MA: Harvard University Press.

Rawls, J. A. (2001). *Justice as fairness: A restatement*. Cambridge, MA: Harvard University Press.

Rochat, P., Dias, M., Liping, G., Broesch, T., Passos-Ferrera, C., Winning, A., & Berg, B. (2009). Fairness in distributive justice by 3- and 5-year-olds across seven cultures. *Journal of Cross-Cultural Psychology, 40*, 416–442.

Sally, D., & Hill, E. (2006). The development of interpersonal strategy: Autism, theory-of-mind, cooperation and fairness. *Journal of Economic Psychology, 27*, 73–97.

Sen, A. (2009). *The idea of justice*. Cambridge, MA: Harvard University Press.

Sloane, S., Baillargeon, R., & Premack, D. (2012). Do infants have a sense of fairness? *Psychological Science, 23*(2), 196–204.

Smith, C. E., Blake, P., & Harris, P. L. (2013). I should but I won't: Why young children endorse norms of fair sharing but do not follow them. *PLoS One, 8*(8).

Sommerville, J. A., Schmidt, M. F. H., Yun, J., & Burns, M. (2013). The development of fairness expectations and prosocial behavior in the second year of life. *Infancy, 18*, 40–66.

Takagishi, H., Kameshima, S., Shug, J., Koizumi, M., & Yamagishi, T. (2010). Theory of mind enhances preference for fairness. *Journal of Experimental Child Psychology, 105*, 130–137.

Watson, J. M., & Moritz, J. B. (2003). Fairness of dice: A longitudinal study of students' beliefs and strategies for making judgments. *Journal for Research in Mathematics Education, 34*(4), 270–304.

Wellman, H., & Liu, D. (2004). Scaling of theory of mind tasks. *Child Development, 75*, 523–541.

Wellman, H., Cross, D., & Watson, J. (2001). Meta-analysis of theory-of-mind development: The truth about false belief. *Child Development, 72*(3), 655–684.

Wu, Z., & Su, Y. (2013). Development of sharing in preschoolers in relation to theory of mind understanding. In M. Knauff, M. Pauen, N. Sebanz, & I. Wachsmuth (Eds.), *Proceedings of the 35th annual conference of the Cognitive Science Society* (pp. 3811–3816). Austin: Cognitive Science Society.

Yost, P., Siegel, A., & Andrews, J. (1962). Nonverbal probability judgments by young children. *Child Development, 33*, 769–781.

How Do Fair Sharing Tasks Facilitate Young Children's Access to Fractional Concepts?

Julie Cwikla and Jennifer Vonk

Abstract The National Science Foundation funded a 2-year program in the US to investigate the cognitive and mathematics development of children aged 3–6. Interview and assessment data of children's strategies to solve fair sharing tasks involving fractional quantities are reviewed. Children were asked how snacks could be shared among differing numbers of friends (social condition) or to distribute items into containers (physical condition). Some children demonstrated a qualitative understanding of fractional unit and with increasing grade level we observed an increased quantitative understanding. In addition, we documented significant relationships between children's emerging causal reasoning, verbal IQ, and mathematics performance.

Introduction

The field of mathematics education research has well documented elementary students' and teachers' difficulty with rational numbers and fractions (e.g., Armstrong and Bezuk 1995; Bezuk and Bieck 1993; Gonzales et al. 2009; Hiebert and Wearne 1986; Mack 1998; Meagher 2002; Moss and Cass 1999; Tirosh 2000). It is possible that students and teachers develop and maintain a "whole number bias" because fractions and rational numbers are introduced too late in the curriculum (Mack 1990; Ni and Zhou 2005). There is limited evidence regarding the ability of preschool and kindergarten children to understand fractional quantities (Cwikla 2014). Even less work has focused on examining the manner in which fractions should be introduced to help facilitate early learning. Some have suggested that fractions are best presented in the context of fair sharing problems (Baroody and Hume 1991) and with the explicit use of metacognitive strategies (Pennequin et al. 2010). This

---

J. Cwikla (✉)
Director of Creativity & Innovation in STEM, Office of the Vice President for Research, University of Southern Mississippi, Ocean Springs, MS, USA
e-mail: julie.cwikla@gmail.com

J. Vonk
Associate Professor, Department of Psychology, Oakland University, Rochester, MI, USA
e-mail: vonk@oakland.edu

area of early human development has been explored to a limited degree and by very few researchers over the past several decades (Empson 1999, 2001, 2003; Empson and Levi 2011; Hunting and Davis 1991; Miller 1984; Pothier and Sawada 1983). Given the larger context and difficulty that children, pre-service, and in-service teachers exhibit in understanding fractions, it is clear that exploring young children's abilities and understanding prior to formal in-school curricular introduction, as well as predictive factors, could inform our understanding of childhood development and shape mathematics curricular improvements.

We were interested in investigating whether a social context (e.g., sharing with friends) that children might be more familiar with would facilitate their fraction manipulation compared to a nonsocial context (e.g., placing items in containers).

In addition we were interested in how solutions of fair sharing problems might be related to fair sharing preferences in a pro-social task where they could put their ideas about equity into action. Over the past 2 years (2011–2012), we have embarked on a program of research (funded by NSF FIRE Award #1043020) to investigate young children's understanding of fractional tasks prior to formal classroom instruction. They were between the ages of 3 and 6. We have interviewed and tested approximately 60 young children in each of three age groups. We have collected data on fraction problem solving when problems were framed both socially and physically, as well as measures of causal reasoning (including theory of mind), prosocial behavior, and intellectual ability.

Background

It has been posited that separate core domains of knowledge exist for reasoning within psychological and physical domains (Goswami 2008; Kinzler and Spelke 2007; Spelke 2000, 2008; Wellman et al. 1997; Wellman and Gelman 1998), with mathematical reasoning being subsumed within the physical domain (Feigenson et al. 2004). Thus, reasoning within a social or psychological domain might have facilitative effects that are limited to social problems. Alternatively, the benefits of such reasoning processes (i.e. theory of mind or ToM, metacognition) might be domain general and might enhance cognitive reasoning and problem solving even within the physical domains, such as mathematical reasoning. Therefore, we are curious if children's ability to solve fraction and fair sharing tasks are correlated with ToM, general intellectual abilities, verbal ability, executive function, and prosocial tendencies.

Core Domains

There has been extensive theoretical discussion about whether there are areas of core knowledge (see above) and Cosmides (1989) demonstrated that adults more readily solve problems when framed in a social context. Mack (1990) noted that children have difficulty solving math problems symbolically, even when they can do so when problems are framed within the context of their everyday lives. This work suggests that children will find social problems easier to comprehend. Once typically developing children achieve the developmental milestone of theory of mind around the age of four, they consistently pass false belief tests. However, when, queried with regards to changing physical states such as moving photographs, they find these problems slightly more difficult. However, researchers have since challenged the comparability of such tasks and pointed out that the executive processing and language demands may not be well equated in such comparisons (Apperly et al. 2005, 2007; Iao et al. 2011). These concerns highlight the importance of executive processing and verbal ability in problems of social and physical reasoning. It remains an open question whether placing problems in a social context provides a facilitative effect across all domains, and for all children, or only for those skilled in ToM. Will these facilitative effects transfer to mathematical problems? Will verbal ability, uniquely predict performance on fraction problems, or will performance be predicted by general intellectual ability alone? Lastly, will changes over time in social cognition, verbal ability and general intellectual function parallel changes in mathematical understanding? Addressing such questions will be informative in terms of understanding the nature of cognition and intelligence. Little empirical research has addressed the idea of domain general versus specific intelligence, despite common folklore about 'well-rounded' individuals versus those with specialized knowledge.

Mathematical knowledge is considered to be contained within the physical domain, separate from the domain of psychological reasoning (Feigenson et al. 2004), and is often taught only with reference to other mathematical constructs (Kreienkamp 2009). Brigham et al. (1996) suggest that fractions are best taught to those with learning disabilities by showing how problems of fractions, decimals and proportions are all related to a central mathematical concept—in this case, the idea of division of wholes. Their approach of integrating various mathematical concepts around organizing principles within a single domain runs counter to the strategy of integrating ideas across domains of learning. Others (Empson 1999, 2003) have focused on integrating mathematical concepts more firmly in the context of familiar life events. Given the intuitive connection between fraction problems and the practice of sharing and division of resources, it seems an ideal place for the intersection of two domains of knowledge to scaffold the developmental process and provide input that could enhance learning (see also Watson et al. 1999). Teachers who take advantage of how learning in one domain promotes learning in another could greatly enhance their students' understanding (see also Empson 1999). Here we focus on the framing of fraction problems in a social context and examine the contributions of social and physical cognitive processes.

Early Fraction Knowledge

Young children's knowledge of fractions will be explored in the context of their prosociality with both close friends and strangers. This study will allow us to explore the relationship among variables that we think can scaffold the learning process. The existing work on fraction learning in elementary school children, older than our population has tended to focus more strongly on proper versus improper fractions. We examine: (a) children's naïve understanding of division of quantities, and differences in strategies for solving fraction problems between 4 and 6 years prior to formal instruction and (b) what intrinsic capacities correlate with their abilities to solve these problems with (c) both proper and improper fraction problems. The findings could yield implications for how to best bolster students' learning, which will be highly informative both within the fields of mathematics education and developmental and cognitive psychology.

Wing and Beal (2004) suggest that children enter school with some knowledge about partitioning objects. They based this conclusion on the results of an experiment in which 5–7 year olds answered questions about the division of concrete or continuous quantities among friends, involving halves or thirds. Children performed better with concrete amounts, and their strategies did not appear to rely on counting. Although previous studies have attempted to understand individual students' processes in learning to comprehend fraction problems, these studies have tended to involve small numbers of students and approach the problem from a nonsystematic framework (Empson 1999, 2003; Hunting and Davis 1991; Pothier and Sawada 1983). There have been no longitudinal studies systematically controlling and comparing large groups of preschool aged children with varying abilities across the dimensions we have assessed. In addition, prior studies focus on proper fractions with little emphasis on ability to solve problems relating to improper fractions and mixed numbers in preschool children. Keijzer and Terwel (2003) demonstrated that children learned fractions better when placed within the framework of the number line than when given fair sharing problems, but these were older children (9–10 year olds), and the former group was invited to discuss their thinking and the latter group was not. Thus, this study compounded social context and its absence with opportunities for metacognitive reflection. Clearly, there is a demonstrated need for a systematic and scientific approach to the study of young children's learning of fractional concepts.

Young children might naturally vary in their propensity to solve fraction and fair sharing problems, based on innate and learned attributes such as prosociality, given that the division of resources relates to an appreciation of morality, equity, fairness. These traits are all foundations of distinctly cooperative human societies (Boyd 2006; Boyd and Richerson 2009). Furthermore, when such problems are framed in terms of giving to others (fair-sharing, Baroody and Hume 1991), it might be expected that children with greater abilities to take the perspective of others (ToM) will solve the problems more readily, or divide more equitably relative to those with lesser capacities. Therefore, our program of work tests children between the ages of 4–6 years of age on measures of ToM, and PSB and their ability to solve fraction problems varying in difficulty.

Perspective-Taking and Mathematics

Several studies lend at least some indirect support for the idea that metacognitive and ToM capacities may construe broad advantages across a wide variety of academic tasks, including those within the physical domain. Blair and Razza (2007) demonstrated a link between self-regulation and math and verbal abilities in kindergarten. In their study, false belief understanding measured in preschool was moderately related to math ability.

Prosociality and Fair Sharing

Because the fraction tasks involve the sharing of resources when framed socially, it is also of interest to determine whether children who are more prosocial than others find it easier to perform the divisions more efficiently and equitably. In addition, children before the age of 6 years tend to use strict equity as a principle for sharing, although initially, prior to 5 years, they are primarily concerned only with benefits to themselves (Larsen and Kellogg 1974; Lerner 1984). Slightly older children use principles such as merit and benevolence, but do not do so accurately until adolescence (Coon et al. 1974; Damon 1977; Hook and Cook 1979; Streater and Chertkoff 1976). Researchers have typically examined these changes in cross-sectional, rather than longitudinal designs, so it will be interesting to document changes in prosocial behavior in concert with changes in other aspects of social and mathematical cognition.

It has been posited that children younger than 5 years are unable to share equitably because they do not yet comprehend proportions (Damon 1975; Hook 1978; Hook and Cook 1979). Put another way, children may have difficulty comprehending proportions and divisions and, as a result, fail to engage in fair sharing. Alternatively, as with many of Piaget's foundational tasks, the seminal methods for studying early childhood concepts of proportions relied on verbal instruction (McCrink et al. 2010). McCrink et al. (2010) designed an elegant paradigm in which children engaged in a social game, evaluating the 'niceness' of clown characters who distributed desirable items. The researchers controlled either the absolute or proportional amounts of items shared. They found that 4 year olds focused exclusively on absolute amounts, while 5 year olds showed some sensitivity to proportions. Adults relied solely on proportions given. This recent study is one example of how framing a problem in a fun and social context might promote young children's ability to reason about proportions. Others (Goswami 1992; Mix et al. 1999; Singer-Freeman and Goswami 2001; Spinillo and Bryant 1991; Yost et al. 1962) have also observed early-childhood competence with respect to understanding proportion, and a rudimentary understanding of proportion has even been observed in infants, using a looking time procedure (McCrink and Wynn 2007). All of these findings lend credence to the idea that fractions can be taught as early as kindergarten, if not pre-school.

One might suspect that the use of fair sharing principles might vary as a function of prosocial tendencies. The children in McCrink et al.'s (2010) study reasoned about the 'niceness' of the character that did the sharing. Those who were more sensitive to the principles of fairness and kindness might have been more likely to attend to the distribution of resources, even when it involved difficult concepts, such as fractional divisions, but McCrink et al. did not test these relationships. The effects of prosocial characteristics, as well as the number of siblings and birth order, on children's strategies and comprehension of sharing are examined in our work. McCrink et al. demonstrated a developmental shift from focusing on what was given to self, towards what was being given to others. Their work reflects a child's transition from self to other-perspective taking, which occurs between the ages of 4 and 5.

In the comparative literature, more extensive work has been conducted on prosocial behavior (PSB), tolerated theft and inequity aversion. Chimpanzees, our closest living relatives, tolerate inequity better among cagemates than among less close associates (Brosnan et al. 2005). It is of interest to compare children's prosocial tendencies and division of resources in fraction problems when the problems are framed as giving to strangers versus family members and close friends. Using close relationships may help enhance their understanding of fraction concepts even further if they are relying on ToM and building the blocks for co-operative relationships—a fundamental aspect of human society (Boyd 2006). Moore (2009) found that children between 4 and 6.5 shared equally with friends and strangers when there was no cost to themselves, but treated strangers like non-friends when there was a cost. Even at this young age, children preferred equitable distributions with friends and behaved accordingly, but were sensitive to the identity of the recipient. Likewise, we initially detected different patterns of sharing in this study to the different potential recipients between 3, 4 and 5 year olds. As with ToM, prosocial tendencies might be expected to increase with age (Blake and Rand 2010 (ages 3–6); Lipscomb et al. 1983 (ages 5 and 6); Lourenco 1993 (ages 5, 7, 10)). Therefore, we should expect age related differences on this measure in our sample as well. We have also manipulated the relationship between the donor and potential recipient in our study as with Moore (2009).

To summarize, we have uncovered some evidence that children as young as five have demonstrated some rudimentary understanding of fractions and proportions. We have also indicated that a systematic investigation of the relevant variables that could inform early mathematics education is lacking.

Method

Participants

Pre-school and kindergarten children (N = 158) ranging from 3 years 6 months to 6 years 6 months participated in this study. The students were drawn from six different learning centers and schools along the Mississippi coast. None of the children had formal curriculum or activities that involved fractions or fair sharing concepts specifically. The PreK–3 students are required to be 3 years of age by September 1st of the academic year. PreK–4 students are aged 4 by September 1st and the Kindergarten students are 5 years of age by September 1st. Data were collected over one and a half years in a sequential design from March 2011 to November 2012. Each student was interviewed and assessed in three different sessions at two different time points approximately 1 year apart. The sessions consisted of (1) eight fraction and fair sharing tasks framed socially, and prosocial sharing tasks (2) eight fraction problems framed non-socially and theory of mind sequencing tasks, and (3) the WPPSI—111 IQ test (only in Year 1). Each session was 30–40 min in length depending on the student and was video recorded. Children completed the sessions in counterbalanced orders within each age group (3, 4, and 5 years). Participants' responses to each task were later coded from video by research assistants that were ignorant of the main hypotheses of the study.

Demographic Survey

A brief survey was sent home, along with the consent form to the child's primary caregiver to collect data on household income, number of siblings and their ages (birth order), whether the child was a product of a single or two-parent home, language(s) spoken in the home, and ethnicity.

Fraction Items

Fraction items were framed both socially and non-socially (see Table 1). In addition to their context, items were designed to allow us to examine both proper and improper fractional solutions. For example, one item was, "Jade wanted to share 6 carrot sticks with her 4 friends. How can she do this fairly?" This is a socially framed item in which the number of items was greater than the number of friends. Thus, this problem results in a mixed number of improper fractional solution. An example of a non-social task is, "There are 5 apples and 3 bags. The same amount of apple must be placed into each bag. How should the apples be placed into the bags?" Each child received the same items in the same order for each context, and

Table 1 Core of oral assessment protocol

	Social task	Physical task	Concept
1	Amanda wanted to share 3 muffins with her 3 friends. How can she do this fairly?	There are 3 oranges and 3 boxes. How should the oranges be evenly placed into the boxes?	1-to-1 correspondence
2	Chris wanted to share 3 crackers with his 3 friends. How can he do this fairly?	There are 6 sheets of paper and 3 notebooks. How should the papers be placed evenly into the notebooks?	Distributing 2 wholes Count by 2s Count by 1
3	Jade wanted to share 6 carrot sticks with her 4 fiends. How can she do this fairly?	There are 6 pretzels and 4 snack packs. How should the pretzels be put into the packs evenly?	Distribution of wholes Dividing into half Dividing into fourths
4	Sam wanted to share 5 oranges with his 3 friends. How can he do this fairly?	There are 5 apples and 3 bags. How should the apples be placed evenly into the bags?	Distribution of wholes 2 remain Dividing into half Dividing into thirds
5	Matthew wanted to share 7 pretzel rods with his 4 friends. How can he do this fairly?	There are 7 sticks and 4 fires. How should the sticks be put evenly into the fires?	Distribution of wholes 3 remain Dividing half, fourths Dividing into fourths Dividing into ¾
6	Tim wanted to share 3 grapes with his 6 friends. How can he do this fairly?	There are 3 cookies and 6 plates. How should the cookies be placed on the plates?	Less items than recipients Dividing into halves
7	Emily wanted to share 2 granola bars with her 6 friends. How can she do this fairly?	There are 3 cakes and 6 plates. How should the cake be placed evenly on the plates?	Less items than recipients Dividing into thirds Dividing into sixths
8	Tina wanted to share 2 cheese sticks with 8 friends. How can she do this fairly?	There are 2 loaves of bread and 8 baskets. How should the bread be placed evenly in the baskets?	Less items than recipients Dividing into fourths

they were presented orally by the investigator. The order in which the fraction items from different contexts (social, non-social) was presented was counterbalanced within the age groups.

Prosocial Tasks

The following materials were used in this study. Four Ziploc bags were used to hold the stickers for each child, friend, non-friend, and stranger for each participant. Sharpie markers were used to write names onto the plastic bags. Four plastic plates, two orange and two blue, were used to place the stickers during the game (see Fig. 1). Sets of identical stickers were used. A video camera was used to record the sessions. The trial orders were randomized for each child so that the experimenter followed

Fig. 1 Prosocial task

the order presented on a data sheet, where they indicated whether the child chose the prosocial option on each trial.

The children were tested individually by the researcher and a graduate student in either their school or day care center. The camera was positioned to not distract the student.

When the child was welcomed into the room, the researcher turned on the video camera, and recited the participant's identification number. The child was placed in front of a Ziploc bag labeled with their name. The researcher sat beside the child holding three other Ziploc bags. In the middle of the table, there were four plates. The child had one blue plate and one orange plate directly in front of them, and plates of the same colors directly behind these plates. In this way, the two plates on the left were of the same color and indicated one option (prosocial or non-prosocial option). The two plates on the right were of the same color (but differed from the plates on the left) and indicated the opposite option. Children were told that they could keep the sticker on the side that they chose that was closest to them, but the sticker on the same side as their choice, but farther away, would go to either a friend, non-friend, stranger, or no-one, depending on the trial. Identical stickers were placed on the plastic colored plates at the beginning of each trial. One of the far plates was always empty.

Each child was given two practice trials to explain the object of the task. In Practice Trial one, the child was given the option of choosing no stickers for themselves and one for the experimenter, or one for themselves and one for the experimenter. In Practice Trial two, the child was offered no stickers for themselves and one for the experimenter, or one for themselves and none for the experimenter. Children were allowed to keep the stickers exchanged.

After these initial trials, the experimenter asked the child to name a friend that they played with. Then the experimenter wrote the playmate's name onto one of the Ziploc bags. Then the child was asked for the name of a classmate (non-friend) with whom s/he did *not* play. Again, the experimenter wrote the name of the non-friend onto a Ziploc bag. The child was reminded that these bags belonged to them. The experimenter then labeled a third bag for an unknown child of the same sex as

Fig. 2 Theory of mind sequencing tasks

themselves, who they were told attended another school. The experimenter placed the child's own bag in front of them and the other bags that were not in play, off to the side. The appropriate bag was placed across the table for each trial. The counterbalancing schedule of each trial determined who the recipient would be, and thus which bag would be presented.

A total of 16 test trials were completed with each child. The counterbalancing schedule consisted of four trials of each of four types (friend, non-friend, stranger and no recipient, as recipients). Within those four trials of each type the prosocial option appeared twice on the right and twice on the left.

The experimenter did not influence the child's response. If the child asked questions or attempted to engage the experimenter, their response was to say "Do whatever you would like to do" or "Do what you think is best." At the end of the session the experimenter indicated that they will distribute the bags to each recipient accordingly.

Theory of Mind Tasks

We modified Baron-Cohen et al.'s (1985) sequencing tasks—a sample of which is reproduced in Fig. 2, for use in the current study. There were six sets of four images belonging to mechanical and behavioral categories, which assessed the child's ability to reason about the actions of objects alone (mechanical) or interacting with humans (behavioral). There were three sets of four images, labeled as "intentional" designed to assess the child's ability to reason about the intentions of others. There was also a practice set of four images depicting a boy washing his hands at the sink, used to orient the child to the task. The experimenter scrambled the pictures belonging to each series and handed the child the mixed up pictures one sequence at a time, asking the child to look at each of the images and then laying them out in a line to express the proper order to "tell a story". The child was asked to indicate when they had finished creating their story and then the experimenter asked "can you tell me about your story". The child's verbal response was also coded to indicate whether they used descriptive or causal terms to describe the actions of objects and individuals depicted in the images. In addition, the child's sequence was scored as follows. They received 2 points for placing each of the first and last cards of the sequence in the correct order (there was only one plausible sequence of events for each set of images). They received one point each for placing the second and third cards in the correct sequence. The experimenter recorded their scores as they described their stories, before taking back the cards and shuffling the next set. Each child received the sequences in a randomized order, with the Intentional sequences always presented in the middle of the sets. Randomization orders were balanced within each age group.

We also administered the WPPSI: Wechsler Preschool and Primary Scale of Intelligence, 3rd Edition, which produces separate verbal, performance and overall IQ scores.

Results

Using grounded theory (Glaser and Strauss 1967), we examined students' solution strategies on both sets of fraction items (socially and non-socially oriented). Both students' verbal and written responses were considered together when coding each response. Each item was coded as correct or incorrect and the type of strategy the student used was also described.

The Incorrect/Unfair Strategies were categorized into five subcategories: (1) Distribute all the wholes but not fairly, (2) Distribute the wholes fairly but then partition the remaining items incorrectly, (3) Distribute pieces of the items only, (4) Distribute some of the wholes only (typically those for which one to one correspondence with the friends could be achieved), leaving the remaining unassigned, and (5) Change the task, in which students might add more friends or items, typically to make the ratio of friends to items one-to-one, or giving extras away or to the "hungry" or biggest friend drawn on their sheet, or keeping some for themselves. The Correct/Fair Strategies were categorized into (1) Distribution of the wholes then economical partitioning of the remaining; that is, partitioning into the fewest and largest necessary pieces in order to distribute fairly and (2) Distribution of the wholes and then non-economical partitioning of the remaining. For instance, instead of dividing into thirds, dividing into sixths, or instead of dividing two pretzels into halves and one into quarters—dividing all three remaining pretzels into fourths/quarters. (3) All items could be partitioned properly to allow for fair distribution, without any allocation of wholes. Examples of students' written work and the accompanying transcript that exemplifies how they solved the following task are provided in Table 2. Their work is organized by age from youngest to oldest, left to right. "Jade wanted to share six carrots with her four friends. How can she do this fairly?"

Table 2 Examples of students' drawings and narratives as they worked through Task 3

First year	First year	Second year	Kindergarten
4 Years, 1 Month	4 Years, 6 Months	5 Years, 0 Months	5 Years, 6 Months
E: How many did this friend get?	C: She can cut 'em up	Child drew picture without speaking	C: So we would give one to each person ... and then we have 2 more left. Cut this one into a half and cut this one into a fourth and then give one fourth to each person. That would make two and then we have one more. Cut it into a fourth. Give one to each person. And then they'll each have 3
C: One	E: How?	C: They both get two	E: And they'll get 3 what?

E: How about this one? (etc. all answered correctly)	C: *In 4 pieces right there?*	E: Are they each getting 2 wholes?
E: So is that fair?	E: Can you draw the lines for me?	C: *Nuh uh*
C: *Yes*	C: *One line, one line, and they can each have four more*	E: What are they getting?
E: So if you were one of these friends which one would you be?	E: So how many carrots would they get all together?	C: *One whole thing and one half!*
C: *I want to be this one. Because I like carrots*	C: *2*	
E: Would you want to be this one?	E: 2 whole carrots?	
C: *No, Because she only has one*	C: *Nope*	
E: So is that fair?	E: So how many carrots do they get all together?	
C: *No*	C: *Half*	
	E: All together they get ½ a carrot?	
	C: *Yup!*	C: *One whole one and two thirds,* (shaking head) *I mean fourths*

Discussion

Children can reason about fractions at even an early age, before first grade (Cwikla 2014) and formal introduction of such concepts. We documented differences in the strategies used by 3, 4, 5, and 6 year old children. This work addresses an important realm of cognition that sits at the intersection of mathematics education and psychology. The strategies and abilities documented here begin to imply that pre-school and kindergarten children might be capable of understanding and illustrating preliminary fractional concepts and fair sharing better than previously thought. And—equally important—this knowledge has developed informally, likely through play and life experiences, prior to curricular introduction.

Given the difficulties that students have with rational number concepts, is it possible that students *and* teachers might develop and maintain a "whole number bias" because fractions and rational numbers are introduced *after* whole numbers (Mack 1990; Ni and Zhou 2005)? There is little empirical evidence as to whether children can comprehend or acquire such fractional concepts before the whole number bias is ensconced (Ni and Zhou 2005); yet this study offers some evidence for that comprehension in these young learners' minds.

However, this is a testable hypothesis that, although it is beyond the scope of the current project, could build upon this work. Other limitations of the current study preclude these results from beginning to inform curricular decisions, but at the very least it does beg further investigation into the young child's intuitive and socially influenced development the understanding of fractions. Our ongoing data collection will expand these results and likely illuminate more areas requiring investigation into the young child's mathematical understandings and ability to problem solve with fractional quantities.

Acknowledgements This material is based upon work supported by the National Science Foundation under Grant No (#1262281). Any opinions, findings, and conclusions or recommendations expressed in this material are those of the author(s) and do not necessarily reflect the views of the National Science Foundation.

References

Apperly, I. A., Samson, D., & Humphreys, G. W. (2005). Domain-specificity and theory of mind: Evaluating evidence from neuropsychology. *Trends in Cognitive Science, 9*, 572–577.
Apperly, I. A., Samson, D., Chiavarino, C., Bickerton, W., & Humphreys, G. W. (2007). Testing the domain-specificity of a theory of mind deficit in brain-injured patients: Evidence for consistent

performance on non-verbal, "reality unknown" false belief and false photograph tasks. *Cognition, 103*, 300–321.

Armstrong, B. E., & Bezuk, N. S. (1995). Multiplication and division of fractions: The search for meaning. In J. T. Sowder & B. P. Schappelle (Eds.), *Providing a foundation for teaching mathematics in the middle grades* (pp. 85–119). Albany: SUNY Press.

Baron-Cohen, S., Leslie, A. M., & Frith, U. (1985). Does the autistic child have a "theory of mind"? *Cognition, 21*(1), 37–46.

Baroody, A. J., & Hume, J. (1991). Meaningful mathematics instruction: The case of fractions. *Remedial and Special Education, 12*, 54–68.

Bezuk, N. S., & Bieck, M. (1993). Current research on rational numbers and common fractions: Summary and implications for teachers. In D. T. Owens (Ed.), *Research ideas for the classroom: Middle grades mathematics* (pp. 118–136). New York: Macmillan.

Blair, C., & Razza, R. P. (2007). Relating effortful control, executive function and false belief understanding to emerging math and literacy ability in kindergarten. *Child Development, 78*, 647–663.

Blake, P. R., & Rand, D. G. (2010). Currency value moderates equity preference among young children. *Evolution and Human Behavior, 31*, 210–218.

Boyd, R. (2006). The puzzle of human sociality. *Science, 314*, 1553.

Boyd, R., & Richerson, P. J. (2009). Culture and the evolution of human cooperation. *Philosophical Transactions of the Royal Society B, 364*, 3281–3288.

Brigham, F. J., Wilson, R., Jones, E., & Moisio, M. (1996). Best practices: Teaching decimals, fractions, and percents to students with learning disabilities. *LD Forum, 21*, 10–15.

Brosnan, S. F., Schiff, H. C., & de Waal, F. B. M. (2005). Chimpanzees' (Pan troglodytes) reactions to inequity during experimental exchange. *Proceedings of the Royal Society of London B, 1560*, 253–258.

Coon, R. C., Lane, I. M., & Lichtman, R. J. (1974). Sufficiency of reward and allocation behavior: A developmental study. *Human Development, 17*(4), 301–313.

Cosmides, L. (1989). The logic of social exchange: Has natural selection shaped how humans reason? Studies with the Wason selection task. *Cognition, 31*, 187–276.

Cwikla, J. (2014). Can kindergartners do fractions? *Teaching Children Mathematics, 20*(6), 354–364.

Damon, W. (1975). Early conceptions of positive justice as related to the development of logical operations. *Child Development, 46*, 301–312.

Damon, W. (1977). *The social world of the child*. San Francisco: Jossey-Bass.

Empson, S. B. (1999). Equal sharing and shared meaning: The development of fraction concepts in a first-grade classroom. *Cognition and Instruction, 17*, 283–342.

Empson, S. B. (2001). Equal sharing and the roots of fraction equivalence. *Teaching Children Mathematics, 7*, 421.

Empson, S. B. (2003). Low-performing students and teaching fractions for understanding: An interactional analysis. *Journal for Research in Mathematics Education, 34*, 305–343.

Empson, S. B., & Levi, L. (2011). *Extending children's mathematics: Fractions and decimals*. Portsmouth: Heinemann.

Feigenson, L., Dehaene, S., & Spelke, E. (2004). Core systems of number. *Trends in Cognitive Sciences, 8*, 307–314.

Glaser, G., & Strauss, A. L. (1967). *The discovery of grounded theory: Strategies for qualitative research*. Chicago: Aldine.

Gonzales, P., Williams, T., Jocelyn, L., Roey, S., Kastberg, D., & Brenwald, S. (2009). *Highlights from TIMSS 2007: Mathematics and science achievement of U.S. fourth and eighth-grade students in an international context*. NCES 2009-001 Revised US Dept. of Education Report.

Goswami, U. (1992). *Analogical reasoning in children*. Hove: Lawrence Erlbaum.

Goswami, U. (2008). *Cognitive development: The learning brain*. New York: Psychology Press.

Hiebert, J., & Wearne, D. (1986). Procedures over concepts: The acquisition of decimal number knowledge. In J. Hiebert (Ed.), *Conceptual and procedural knowledge: The case of mathematics* (pp. 199–223). Hillsdale: Lawrence Erlbaum.

Hook, J. (1978). The development of equity and logico-mathematical thinking. *Child Development, 49*, 1035–1044.

Hook, J., & Cook, T. (1979). Equity theory and the cognitive ability of children. *Psychological Bulletin, 86*, 429–445.

Hunting, R. P., & Davis, G. E. (Eds.). (1991). *Early fraction learning*. New York: Springer.

Iao, L.-S., Leekam, S., Perner, J., & McConachie, H. (2011). Further evidence for nonspecificity of theory of mind in preschoolers: Training and transferability in the understanding of false beliefs and false signs. *Journal of Cognition and Development, 12*, 56–79.

Keijzer, R., & Terwel, J. (2003). Learning for mathematical insight: A longitudinal comparative study on modelling. *Learning and Instruction, 13*, 285–304.

Kinzler, K. D., & Spelke, E. S. (2007). Core systems in human cognition. *Progress in Brain Research, 164*, 257–264.

Kreienkamp, K. (2009). Exemplary project: Math fact memorization in a highly sequenced elementary mathematics curriculum. In R. A. Schmuck (Ed.), *Practical action research: A collection of articles* (2nd ed., p. 143). Thousand Oaks: Corwin.

Larsen, G., & Kellogg, J. (1974). A developmental study of the relation between conservation and sharing behavior. *Child Development, 45*, 849–851.

Lerner, R. M. (1984). *On the nature of human plasticity*. New York: Cambridge University Press.

Lipscomb, T. J., Bregman, N. J., & McAllister, H. A. (1983). The effect of words and actions on American childrens prosocial behavior. *Journal of Psychology: Interdisciplinary and Applied, 114*, 193–198.

Lourenco, O. M. (1993). Toward a Piagetian explanation of the development of prosocial behaviour in children: The force of negational thinking. *British Journal of Developmental Psychology, 11*, 91–106.

Mack, N. K. (1990). Learning fractions with understanding: Building on informal knowledge. *Journal for Research in Mathematics Education, 21*, 16–32.

Mack, N. K. (1998). Building a foundation for understanding the multiplication of fractions. *Teaching Children Mathematics, 5*, 34–38.

McCrink, K., & Wynn, K. (2007). Ratio abstraction by 6-month-old infants. *Psychological Science, 18*, 740–745.

McCrink, K., Bloom, P., & Santos, L. R. (2010). Children's and adults' judgments of equitable resource distributions. *Developmental Science, 13*, 37–45.

Meagher, M. (2002). *Teaching fractions, new methods, new resources*. Columbus: ERIC Digest.

Miller, K. (1984). Child as the measure of all things: Measurement procedures and the development of quantitative concepts. In C. Sophian (Ed.), *Origins of cognitive skills: The 18th Carnegie Symposium on Cognition* (pp. 193–228). Hillsdale: Lawrence Erlbaum.

Mix, K., Levine, S., & Huttenlocher, J. (1999). Early fraction calculation ability. *Developmental Psychology, 35*, 164–174.

Moore, C. (2009). Fairness in children's resource allocation depends on the recipient. *Psychological Science, 20*, 944–948.

Moss, J., & Case, R. (1999). Developing children's understanding of the rational numbers: A new model and an experimental curriculum. *Journal for Research in Mathematics Education, 30*, 122–147.

Ni, Y., & Zhou, Y. (2005). Teaching and learning fraction and rational numbers: The Origins and implications of whole number bias. *Educational Psychologist, 40*, 27–52.

Pennequin, V., Sorel, O., Nanty, I., & Fontaine, R. (2010). Metacognition and low achievement in mathematics: The effect of training in the use of metacognitive skills to solve mathematical word problems. *Thinking & Reasoning, 16*, 198–220.

Pothier, Y., & Sawada, D. (1983). Partitioning: The emergence of rational number ideas in young children. *Journal for Research in Mathematics Education, 14*, 307–317.

Singer-Freeman, K., & Goswami, U. (2001). Does half a pizza equal half a box of chocolates? Proportional matching in an analogy paradigm. *Cognitive Development, 16*, 811–829.

Spelke, E. S. (2000). Core knowledge. *American Psychologist, 55*, 1233–1243.

Spelke, E. S. (2008). La théorie du 'core knowledge'. *L'Année Psychologique, 108,* 721–756.
Spinillo, A., & Bryant, P. (1991). Children's proportional judgments: The importance of 'half'. *Child Development, 62,* 427–440.
Streater, A., & Chertkoff, J. (1976). Distribution of rewards in a triad: A developmental test of equity theory. *Child Development, 47,* 800–805.
Tirosh, D. (2000). Enhancing prospective teachers' knowledge of children's conceptions: The case of division of fractions. *Journal for Research in Mathematics Education, 31,* 5–25.
Watson, J. M., Campbell, K. J., & Collis, K. F. (1999). The structural development of the concept of fraction by young children. *Journal of Structural Learning and Intelligence Systems, 13,* 171–193.
Wellman, H. M., & Gelman, S. A. (1998). Knowledge acquisition in foundational domains. In W. Damon (Ed.), *Handbook of child psychology* (Cognition, perception, and language, Vol. 2, pp. 523–573). Hoboken: Wiley.
Wellman, H. M., Hickling, A. K., & Schult, C. A. (1997). Young children's psychological, physical, and biological explanations. In H. M. Wellman & K. Inagaki (Eds.), *The emergence of core domains of thought: Children's reasoning about physical, psychological, and biological phenomena* (pp. 7–25). San Francisco: Jossey-Bass.
Wing, R. E., & Beal, C. R. (2004). Young children's judgments about the relative size of shared portions: The role of material type. *Mathematical Thinking and Learning, 6,* 1–14.
Yost, P., Siegel, A., & Andrews, J. (1962). Nonverbal probability judgments by young children. *Child Development, 33,* 769–781.

Working with Pre-schoolers: A Dual Commentary

Michaela Kaslová and Sixto Romero

Abstract The commentary on the chapters of Chronaki et al., of Nikiforidou and Pange and of Cwikla and Vonk sets the topic of working with pre-schoolers in a historical context. It draws on contributions of Comenius, Froebel, Pestalozzi and others and traces the key developments in developmental psychology. Finally, it points to the importance of research on pre-school mathematics as a condition for well-thought conceptualisations of pre-school education practice.

Sixto Romero Sánchez: A Brief Historical Sketch on the Teaching and Learning of Mathematics in Preschool

The history of education is not made of violent revolutions, such as Copernicus's revelations caused in astronomy for example, but has developed in a logical sequence of phases. From Greco-Roman times to the sixteenth century, no concern for consciously teaching a "child" can be detected, whereas teaching "children" had been a custom. How did the concern for "the child" arise? In my view, it was a sign of a pedagogical necessity, even in adverse situations, that has never ceased to encourage the process of the secular education of humanity. It is in this sense that one needed to create new concerns in and new forms of education.

The concerns for "the child" and for education in general, and mathematics education in particular, gained in importance in the late nineteenth century. These concerns were, at the time, traced back to Socrates and Plato, pointing at their dialogic

---

Editors' note: The commentary collects historical and theoretical reflections on early childhood (mathematics) education, made separately by the two authors of the commentary before discussing the three chapters of the section briefly.

M. Kaslová (✉)
Faculty of Education, Charles University, Prague, Czech Republic
e-mail: michaela.kaslova@pedf.cuni.cz

S. Romero
Escuela Técnica Superior de Ingeniería, Universidad de Huelva, Huelva, Spain
e-mail: sixto@uhu.es

heuristic techniques. By drawing on these 'classics', Montaigne, Rabelais, Comenius, Locke, and others emphasized the realistic experience of each single student. Unquestionably, Rousseau was significant: for his naturalistic philosophy, for his insights about evolutionary psychology, for the perception of a qualitative difference between an adult and a child. As precursors of primary education in the late eighteenth century and early nineteenth century, Condillac and Itard should be cited.

In the nineteenth century, Froebel and Pestalozzi have been influential and can truly be referred to as educators, innovators and reformers of traditional teaching. Froebel created kindergartens in 1837 and made significant contributions to pedagogy for improvements in teaching, with the aim of providing students with a good education, and Pestalozzi, already in 1810, indicated that education should adhere to a harmonic law of nature, from which the need for freedom in the child's education was derived.

As Beales (1956) suggests, a particular point in the historical development of education is the development of educational institutions, in which the principles of theoretical knowledge on education were systematically exposed and discussed. It is important to note that in the late nineteenth century schools in Switzerland, France, England, Germany, Belgium, Spain, Portugal, United States, Italy and elsewhere implemented what has been called 'active education'. As a summary, the following points characterized the new proposal:

- Education should respect each child's individuality.
- Academic studies and, more generally, life skills should open the way for the child to pursue his/her innate interests.
- Each age has its own physiognomy. There is a necessity of both individual and collective discipline leading to reinforcing the concept of responsibility.
- Cooperation versus selfish competition.
- Co-education as an instructional tool.
- Education should prepare children for the future, to be good citizens and competent, which is conducive to living as active members of society (cf. Ernest 1994).

It needs to be remembered that a fundamental principle, on which many theoretical conclusions are based and upon which many practical applications have been designed and realized, is that the central pedagogical value lies irreplaceably in the child as the focus of the teaching-learning process. Hence, the proposal translates into a plan for action:

- A need to know the child deeply, on a scientific and humane basis, with affection.
- To determine the individual potential and the development of particularly deep interests of each child, and to guide him/her to the full actualization of its personality.
- To generically teach all disciplines, and particularly mathematics, according to an individualized program in which each student is to progress according to her/his abilities, by using appropriate teaching techniques (Tyminski et al. 2014).
- The teacher is subordinate to the activity, to the true interests, and ultimately to the true freedom of the child.

In most countries mentioned above, during the last few decades, educational reforms have urged governments to establish curriculum design bases. The sources of the curriculum based on active learning considerations are important here. Any curriculum attempts to answer the following fundamental questions: What to teach? When to teach? How to teach? and also Why, when and how to evaluate? Responses are generated from sources of different types and origins. Usually four types of sources are mentioned (Blömeke et al. 2013) while each of them makes a specific contribution and provides specific information:

- A sociological source refers to the social and cultural demands on the education system, the content of knowledge, procedures, attitudes and values that contribute to the socialization process of students, the assimilation of social knowledge and cultural heritage of society. The curriculum must include the purpose and social functions of education, trying to ensure that students become active and responsible members of the society to which they belong (Liman et al. 2013).
- A psychological source relates to the processes of development and student learning (Tuckman and Monetti 2010). Knowledge of the regularities of the evolutionary development at different ages and the laws that govern learning and cognitive processes in humans provides an indispensable framework curriculum about opportunities and ways of teaching, when learning what is possible to learn at all times, and how to learn it (Rico et al. 2014).
- A pedagogical source collects both the existing theoretical foundation and the educational experience of teaching. The experience accumulated over the years is an irreplaceable source of curricular knowledge. Specifically, curriculum development in the classroom, in the actual teaching of teachers, proves indispensable to the development of the curriculum in its design phases and further development elements (Appleton 2003).
- An epistemological source refers to scientific knowledge that makes up the corresponding areas or curricular materials. The methodology, internal structure and current state of knowledge in the various scientific disciplines and interdisciplinary links between them, also make a decisive contribution to the configuration and contents of the curriculum (Natthapoj 2012, Schoenfeld 1983; Schommer 1990).

These four sources play a role in all phases of development and implementation of the curriculum (Barnett et al. 2014) at different times: (a) base design curriculum, curricular projects and programs; (b) in the development of the curriculum in the classroom. Thus, both the education authorities, which establish a normative curriculum, and the teachers in their projects, programming and educational practice should refer to these sources, which display both the curriculum content and their legitimation.

The historical perspective, briefly outlined above, clearly shows that mathematics is a body of knowledge evolving, and that such an evolution often plays a major role with respect to its interrelationship with other knowledge domains and the need to solve certain practical problems. For example, consider the historical origins of geometry: part of it developed by responding to problems of agriculture and architecture. Statistics has its origin in the organization of the first population census. Different number systems evolved with the need to handle elementary calculations in different contexts. Probability theory developed in response to gambling problems. In more recent times, research in discrete mathematics and numerical calculus experiences a considerable rise as a result of the increasingly widespread use of new technologies. Moreover, to some extent mathematics is the frame in which scientific models are constructed, taking part in the process of modeling reality itself, and often have served as a means of validating these models. However, the evolution of mathematics has not only proceeded by accumulation of knowledge or application to other fields. Mathematical concepts themselves have been changing their meaning over time, extending or revising, acquiring relevance or, conversely, being relegated to the background.

Taking all these considerations into account, the child's development of logical-mathematical knowledge should be placed in the following frame:

- Emphasis on the child's performance on objects and on the relationships that through operation are established between them. By means of manipulations the child discovers what is hard and soft, which movements are possible, etc. He or she also learns about relationships between objects (e.g., it discovers that a ball rolls faster than a truck, the dummy is larger than a ball which is heavier than … etc.). These relationships allow organizing, grouping, comparing, etc. They are not objects as such but they are constructs of the child on the basis of the relationships found and detected.
- The relationships between a newly-discovered object and other objects: relationships develop from sensory-motor to progressively logical and intuitive (in elementary education). These relationships are finding expression through and in language. The child not only learns how to refer to objects, but also how to express relations between them.
- Expressions of these relationships will be first through the action, then through oral language and then through the mathematical language by making use of iconic representations and finally numbers.
- The development of the child's mathematical representations requires teachers who, from time to time, intervene by pushing the child's curiosity forward; by supporting the child's actions; by scaffolding the child's manipulation of representations, raising levels; and by guiding him/her to appropriate expressions of mathematical language.
- Thanks to the intervention of the teacher, the child will be able to learn to:

    (a) discover the features of the objects,
    (b) set different order relations,
    (c) perform collections of objects based on certain attributes,
    (d) use simple counting strategies properly and graph icons and figures. The child will also be able to learn about the convenience of measurements, to solve measurement problems and become familiar with measuring units of space and time. The child will learn to distinguish various geometrical shapes and to establish relationships between them, and between the shapes and the child him/herself.

The implemented curriculum should give priority to the practical activity of the child—the discovery of the properties and relationships between objects through active experimentation.

Michaela Kaslová: From Comenius to Developmental Psychology

Through human history, changes in the approaches to education, its philosophy, system, concept etc. have abounded. The history of early-childhood education, pedagogy focussing on the development of children up to the age of 7 years, is however not long. In the majority of cases it addresses the issue of equality of access to education. First discussions were about the question "What stimulates and in which domain?" Later debates focussed on the issue of "Who can be/should be systematically educated?" This second question is related to what we might call 'the democratisation of the educational system', and it emerged from developments within the socio-political system. One of the first movements, born in Europe in the fifteenth century, enforced and partially realised an education for all, independent of gender and social position. This movement addressed the problem of a basic education for all in a context that, in contemporary terminology, is called 'primary schooling'. During the next centuries this idea had been developed and enriched. There were different attempts to answer the upcoming questions of 'Why, how and under what conditions?' as well as whether education should be organised as compulsory.

School and pre-school education was presented systematically in the theoretical work of Comenius in the first part of the seventeenth century. The democratic principles of why and how to stimulate the pre-school child are described in detail in his book *Informatorium scholae maternae* (1632) (German version of 1633: *Informatorium der Mutterschul*; English translation of 1663: *School of Infancy*). Other ideas can be found in *Didactica magna*, *Orbis sensualis pictus* and *Pansofia*. Comenius wrote, among other things, that a new knowledge and a new recognition must be discovered by a child in the child's real world; that we have to stimulate cognitive abilities; that this process must respect the needs of the child, which means that we must offer different stimuli which are perceived by different senses; and that we discuss this with the child in his/her native language if we desire the child to understand the lessons well; that the adults must start child activities at that stage of a child's life in which they are crucial for her/his successful development. His avant-garde principles were seized by many pedagogues, teachers and didacticians. For example, Maria Montessori referred to Comenius in her conception of didactics of mathematics. Later constructivist approaches correspond with Comenius's principles, too.

When analysing different philosophies (especially European) of conceptions of pre-school education, we can see that there are mainly three different approaches: (1) children are similar to 'small adults', thus we should use similar methods as at school with similar types of communication; (2) children have specific characteristics, thus we stimulate them respecting their stages of development (with tailored approaches); (3) it is not necessary to stimulate the children because their instinctive interests help them automatically in their development. If we evaluate these approaches from the point of view of economic resources required, we contend that (1) and (3) are more economical for the suppliers of the resources (such as state, region, village, institution, association) than the approach (2). (2) needs research and professional teachers (including studies at specific universities focussed on respective teaching skills). However, we argue in favour of approach (2) and would like to add further that it is necessary to accept that the child's characteristics can change over the years, in the context of a changing society, and that education must respect specificities of the child's family, region, language etc. Research in the domain of pre-mathematical literacy is not focused only on mathematics and pre-mathematics, but also recognises that a child's development can be understood from the perspectives of pedagogy, special-needs pedagogy, cognitive and evolutional psychology, sociology and other disciplines. Economic constraints sometimes force the institutions to unify and simplify the pre-school system and its conditions, curricula and educational methods. For this reason it is necessary to identify what can count as crucial elements of pre-school literacy. Therefore, let us concentrate here on the subject matter: pre-mathematics.

Pre-mathematics is a special domain: we cannot speak of mathematics yet, because the developmental stage of the pre-school child, in a Piagetian sense, is characterised as a pre-operational stage, as a stage of concrete images and pre-logical thinking, while mathematics is characterised by operations with abstract notions that lean on logical thinking. The step from pre-mathematics to school-mathematics is not easy for a lot of reasons—so called specific pre-school phenomena: (a) the *egocentrism* typical for pre-school age colours the perception, imagination and evaluation of experiences: a pre-school child evaluates the information according to his/her needs and desires, and for this reason it is difficult to say that the child can be objective; the child's tendency to personify objects or identify them with other objects changes the character of the perception of this object; (b) in the majority of cases children *learn in plays/games*: the play is characterized as a process accompanied by a wealth of emotions, and we must accept that these emotions can block or distort rational thinking (these arguments are confirmed by neurological studies looking at the functions of specific parts of the brain); (c) the *presentism* and the *topism* determinate the condition of the process of generalisation: pre-school children live intensively 'now and here'—in present time and present space; they have difficulties to orient themselves well in time and in plane space; (d) *syncretism:* a dominant perception prevents the proper use of an analytico-synthetic approach; (e) the *vocabulary* is not yet well developed and the children have problems with recalling words in new contexts: we mark this stage as a period of "*mixed-communication*", because the pre-school children show a restricted ability to communicate in one special code, including oral communication; the pre-school children communicate alternatively by oral, manipulative, gestural, kinaesthetic, pantomimic or drawn and other codes of communication; (f) *fine motor skills* and *muscle coordination* are not yet sufficiently developed and the children need to concentrate more on the movement than on the object/base of their activity: they are more focussed on why rather than on what they are doing.

The three chapters of this section correspond, in certain aspects, to Comenius's general principles and especially to the strategy (2), particularly with regard to current work in the cognitive psychology of pre-school age. It is important to emphasize that it is not a simple undertaking to prepare a child for school mathematics.

Julie Cwikla and Jennifer Vonk underline in their article the need of continuity in education. The principal idea is to demonstrate how the concept of fractions starts to develop within a group of pre-school children. Table 2 of their chapter shows the important role of language in the communication of ideas about fractions. This topic is key in that it permits us to complete a mosaic of research in this domain and in that it facilitates us to distinguish the common characteristics of the developmental process. A second important aspect is related to the work of primary-school teachers. Insights into the obstacles at the level of the first encounters with the concept of fractions can be informative for the instructional design of subsequent mathematical lessons on the topic. For researchers, in addition, there is the possibility to repeat this experiment in a new context and to compare the new data with presented data.

Zoi Nikiforidou and Jenny Pange report in their research about the importance of codes of communication. The child's capability to "read drawn information" or to "transform the drawn code to the oral code of communication" influences the success in problem-solving activities at pre-school age. This outcome harmonizes with other research outcomes, and it points to the importance of professional development in the context of communication in kindergartens and with respect to the media used. The communication difficulties that some children experience are related to differently developed analytico-synthetic thinking abilities.

Anna Chronaki, Georgia Moutzouri and Kostas Magos introduce their research in activities with respect to the specific characteristics of pre-school age. The set of described activities in age-heterogeneous groups can serve as good examples of global/comprehensive activities suitable for practice at kindergarten, clubs or families. Each of the activities described serves a number of goals: development of communication, imagination, notion, orientation in space etc. The conception of each of the activities is based on a development of culture, or better still, on each pre-mathematical concept and pre-mathematical ability, as well as on part of the culture. Their strategy shows that the level of progress depends on the existence of a good social atmosphere as well.

All three articles indicate the necessary complexity of research of pre-school mathematics. The contributions offer substantial arguments for the continuance and enlargement of research in pre-school age and in pre-mathematics. They insist on a protection of pre-school children from overhasty and ill-thought conceptualisations of pre-school education practice.

References

Appleton, K. (2003). How do beginning primary school teachers cope with science? Toward an understanding of science teaching practice. *Research in Science Education, 33*(1), 1–25.

Barnett, J. H., Lodder, J., & Pengelley, D. (2014). The pedagogy of primary historical sources in mathematics: Classroom practice meets theoretical frameworks. *Science & Education, 23*(1), 7–27.

Beales, A. C. F. (1956). El desarrollo histórico de los "métodos activos" en educación. *Revista Española de Pedagogía, 14*(55), 254–270.

Blömeke, S., Hsieh, F.-J., & Schmidt, W. H. (2013). Introduction to this special issue. *International Journal of Science and Mathematics Education, 11*, 789–793.

Ernest, P. (1994). Social constructivism and the psychology of mathematic education. In P. Ernest (Ed.), *Constructing mathematical knowledge: Epistemology and mathematics education* (pp. 62–72). London: RoutledgeFalmer.

Liman, M. A., Salleh, M. J., & Abdullahi, M. (2013). Sociological and mathematics educational values: An intersection of need for effective mathematics instructional contents delivery. *International Journal of Humanities and Social Science, 3*(2), 192–203.

Nattapoj, V. H. (2012). Relationship between classroom authority and epistemological beliefs as espoused by primary school mathematics teachers from the very high and very low socio-economic regions in Thailand. *Journal of International and Comparative Education, 1*(2), 71–89.

Rico, L., Gómez, P., & Cañadas, M. C. (2014). Formación inicial en educación matemática de los maestros de Primaria en España 1991–2010. *Revista Española de Pedagogía, 363*, 35–59.

Schoenfeld, A. H. (Ed.). (1983). *Problem solving in the mathematics curriculum: A report, recommendations, and an annotated bibliography*. Washington, DC: MAA.

Schommer, M. (1990). Effects of beliefs about the nature of knowledge on comprehension. *Journal of Educational Psychology, 82*, 498–504.

Tuckman, B., & Monetti, D. (2010). *Educational psychology*. Belmont: Cengage Learning.

Tyminski, A. M., Land, T. J., Drake, C., Zambak, V. S., & Simpson, A. (2014). Preservice elementary mathematics teachers' emerging ability to write problems to build on children's mathematics. In J.-J. Lo, K. R. Leatham, & L. R. Van Zoest (Eds.), *Research trends in mathematics teacher education* (pp. 193–218). Cham: Springer.

Part IV
Taking Spaces and Modalities into Account

Digital Mathematical Performances: Creating a Liminal Space for Participation 201
Susan Gerofsky
University of British Columbia, Vancouver, Canada

Participation in Mathematics Problem-Solving Through Gestures and Narration 213
Luciana Bazzini and Cristina Sabena
Università di Torino, Italy

Considering the Classroom Space: Towards a Multimodal Analysis of the Pedagogical Discourse 225
Eleni Gana and Charoula Stathopoulou
University of Thessaly, Greece
Petros Chaviaris
University of the Aegean, Greece

Commentary: Semiotic Game, Semiotic Resources, Liminal Space—A Revolutionary Moment in Mathematics Education! 237
Peter Appelbaum
Arcadia University, Philadelphia, USA

Digital Mathematical Performances: Creating a Liminal Space for Participation

Susan Gerofsky

Abstract This chapter introduces performative mathematical modes of expression for students. Performance engages with aesthetic, spiritual, ethical and contextual facets of life. The chapter argues in favour of 'liminal spaces', of liminality and play, that allow for fuller participation in mathematical thinking and expression. Examples of digital mathematical performance spaces are presented and discussed.

Introduction: Disembodied, Antiperformative Traditions in School Mathematics Pedagogy

Even after more than 25 years of reform movements in school mathematics, it has been noted in many places (e.g., Boaler 2002; Taylor 1996) that 'traditional' mathematics classes are still prevalent in many schools worldwide. These classes offer little space for learners to participate and interact as they are introduced to mathematical patterns and ideas. The stereotypical 'traditional-style' mathematics class, particularly at the secondary school level, encourages learners to sit quietly, work individually with pen and paper, and copy notes obediently from the teacher's lecture. There is little opportunity to engage in discussion, to voice basic questions, to express surprise or wonder, doubt or fear. Students are generally required to remain physically still and quiet and to keep their thoughts and emotions to themselves.

The philosophical and pedagogical assumptions underpinning this kind of traditional mathematics instruction and students' role in this kind of class are antithetical in a number of ways to notions like 'performance' and 'participation', at least as so far as these describe the actions of students.

- Firstly, such traditional classes are based firmly on a *transmission model* of pedagogy, where teachers (and textbooks, and exams) hold knowledge, and students do not. In this model it is the responsibility of teachers to transmit knowledge to students, through verbal explanations and demonstrations, followed by student

---

S. Gerofsky (✉)
Department of Curriculum and Pedagogy, University of British Columbia,
Vancouver, BC, Canada
e-mail: susan.gerofsky@ubc.ca

exercises that (on the analogy of physical exercises in a gym, as elaborated in Thorndike's 'faculty psychology') are meant to 'build mathematical muscles' through repeated practice of particular mathematical skills thought to have been transmitted through the teacher's lecture-based lesson (Gamon and Bragdon 1998; Thorndike 1924).
- Secondly, these classes make the pedagogical assumption that knowledge is most efficiently transmitted by packaging it in *small, logically-sequenced packages* and doled out to learners a bit at a time. These bits of knowledge and their sequence are known, named and discussed by teachers, but for students, they are often 'given out' as just the next lesson, and the next lesson after that, without establishing any sense of the big picture or context (within mathematics or beyond it). There is no space and no encouragement for students to ask big or contextual questions, and no stopping the forging of the inexorable chain of logic comprised of these daily small links (Herbel-Eisenmann et al. 2006; Noss 1994; Schmidt et al. 1997).
- Thirdly, traditional mathematics classes are based on the philosophical premise of a *Platonic/Cartesian mind-body split*. The physical world, physical objects and the students' and teachers' own bodies are treated as an (unfortunately) necessary but base, coarse and primitive element in the learning situation. The disembodied minds of teachers and students are, in contrast, treated as fine, exalted, transcendent. Mathematics, following the Platonic doctrine of perfect, immaterial forms, is taken to be an entirely mental and symbolic activity, searching for eternal perfection and absolute truth (Borasi 1984; Plato 1961; Radford et al. 2009).

Based on this third assumption, students in mathematics class are required to sit very still and quiet at their desks, listening to the teacher and thus entering the disembodied realm of abstract thought. Note-taking may be encouraged, but these notes are prescribed to take the form of symbolic (numerical and algebraic) statements, with the occasional geometric diagram allowed into the mix. Any action that asserts the physicality and material presence of the full-fledged human beings in the classroom is discouraged; conversation, locomotion, gesture, the use of material manipulatives and physical model-building, or even the presence of colour, movement and references to human personalities and the world beyond the classroom are sternly discouraged, particularly as students approach adult age. Philosophical and contextual questions, discussion, physical animation, connections of mathematics with human nature and the more-than-human world, the use of objects to model mathematical structures are all typically derided as babyish and extraneous to secondary and sometimes post-secondary mathematics classrooms, where students and teachers are supposed to be involved in work of pure and immaterial abstraction, of mind divorced from body.

The resulting pedagogical situations are all too familiar to everyone who has suffered and/or thrived in secondary-school mathematics class: a plain, grey, unadorned room, students sitting still and silent in desks arranged in a grid pattern of rows, perhaps taking notes quietly, as the teacher stands at a chalkboard or overhead projector delivering an extended lecture, all followed by a stretch of time when students work on exercises to practice the small bundle of knowledge or skills presented by the teacher.

If this were a pedagogical strategy used occasionally or once in a while as a moment of calm reflection or a gathering of energies in math class, amongst other more engaging and engaged learning activities, it might be justifiable as a change of pace and a chance to revisit, regroup and summarize one's learning before moving on. As one pedagogical approach in a rich repertoire, this could have merit in offering a meditative pause. Unfortunately, though, this lecture-based teaching to minimally-participating students is the everyday norm in most secondary (and many elementary and post-secondary) mathematics classes and often the sole teaching method, day in, day out, for about 100 classes a year and for many years on end.

Gadanidis and Borba (2008) have noted that, while kids often come home and tell their families what they learned in literature, history, science or art classes, they often have a very difficult time saying (or even knowing) what they learned in mathematics class. When classes consist of the introduction of small increments of technical, algorithmic knowledge and then drills of the same, it becomes difficult even to name the topic that was raised, never mind have a conversation about it. I remember my own children answering my question, "What did you learn about in math class today?" with an exasperated, "We just DID MATH, Mom!" Mathematics becomes something that learners *just do*, without space for interrogating what that thing is, why it might or might not be important, or fascinating, or ethical, or beautiful. Without space for student voices, actions, questions and stories, mathematics becomes an esoteric study available only to those who can cope with demands for obedience in a sensorily-impoverished learning environment.

Performative Pedagogy

Performance as pedagogy is a very different, almost opposite approach to teaching and learning mathematics. Performance in the sense of an artistic, theatrical, musical and/or dance performance always involves embodied ways of being, interaction and expression. The performer ideally engages the whole self in the moment of performance, bringing all the resources of mind-body-spirit together to sing soulfully, act mindfully, dance intellectually and so on. There is no separation of aspects or levels of being in an engaged performative moment, even as technical skills and intellect are brought into play (Berliner 1994; Sawyer 2003). The material world and embodied ways of being are celebrated as facets of an integrated whole performer, rather than being ridiculed, devalued and shunned.

Note the special case of a High Modernist Western take on performance, in which a distinct separation is made between performer and audience. In this paradigm, while the performer is fully engaged, audiences are expected to sit quietly and still, and to show appreciation appropriately at the correct points, and in very physically constrained ways (through polite applause, for example). In fact, in the

Modernist model, the audience at a symphony concert, ballet or recital is expected to behave as if they were students in a secondary-school mathematics class. Even performers in minor roles in such performances are expected to restrain their movements and emotions; for example, instrumental players in a symphony orchestra are taught never to tap their feet, sway their heads or show emotion when performing. Only featured stars, soloists, principal dancers and superstar conductors are encouraged to express emotion through less-constrained bodily movement.

However, apart from the very short historical time period of High Modernism in the West in the mid-to-late twentieth century, human performance has been considerably more participatory and audiences less silent and still. The boundary between performers and audience has more typically been a porous one, in which everyone is an active participant at some level, although some may have prepared beforehand and be ready to take on special roles. The field of Performance Studies considers the performative in human cultures in this broader context, bringing together cultural examples from theatre and dance but also from the anthropology of religion and from performative rituals and community practices. In considering pre-Modern, non-Modern and our contemporary post-Modern societies, it becomes clear that most or all cultures include participatory ritual, playful, fully-participatory embodied practices that are involved in the community's education and entertainment, and in repairing rifts in the human-and-more-than-human world. Leading theorists in the field of Performance Studies include Schechner (2003), Turner (1986) and Conquergood (2002), and the New York University (NYU) Department of Performance Studies is a key international centre for studies in this field.

Performance necessarily engages with aesthetic, spiritual, ethical and contextual facets of life. Even the most abstract kinds of performances elicit emotion and thought, physical responses and empathy. Since performance engages people on multiple levels, the most integrative performances speak to people in many ways at once, and raise many kinds of further questions and speculations when the performance is finished.

In contrast, the very limited participation allowable in so-called 'traditional' mathematics classes discourages the integration of multiple aspects of the learner as a whole person, emphasizing instead a solitary emphasis on rationality, logic and/or the obedient following of algorithmic steps in solving set problems. Mathematics taught in this way is an impoverished mathematics, one in which learners are so severely restrained from active participation that very few can muster a sense of curiosity, in-depth questions or a sense that the subject could ever have meaning to oneself as a person.

Given the rather stultifying, anti-performative traditions of school mathematics, how might we try to open up mathematics education and pedagogy to performative ways of being? Is it even possible to bring performance into mathematics teaching and learning—apart from the very reduced sense of students' 'performance' on a test, or a teacher's 'performance' in front of a silent, passive audience?

In the contemporary world, digital means of expression and communication are transforming human life and society. Aspects of digital performance have had a profound effect on many of the performing arts, including theatre, music and dance (see, e.g., Auslander 1999; Dixon 2009; Farley 2002). Education too has begun to work closely with a variety of digital tools and media, although school systems have been slower to adapt to digital technologies than many other areas of society and culture. With these contemporary developments in mind, we may ask the following question: Beyond actual physical performative pedagogies, how might we also engage digital and online resources to open up a space for exploration in *digital* mathematical performances in our era, where so much of our lives takes place on the internet via digital media and text?

I claim that digital, performative mathematical modes of expressions for students can create a new liminal space that allows for fuller participation in mathematical thinking and expression. I will first discuss the concept of liminal spaces, then outline some of the ways that online performative sites for student mathematical expression might afford liminal spaces that invite participation by those who might not otherwise have access to mathematical questioning and meaning-making.

Liminality and Play in Liminal Spaces

The concept of liminal space comes from the field of performance theory, the contemporary inter-discipline mentioned earlier which draws from theatre studies, art theory and the anthropology of religion. The idea of performance is expanded beyond the conventions of mainstream theatre to include the performance of rituals, theatrical 'happenings', and participatory community events. In this conception of the performative, there is not necessarily a boundary line drawn between performer and audience. Rather than expecting to see expert, professional actors performing on a proscenium arch stage, with the audience seated quietly and appreciatively in darkened banks of seats, performance theory blurs categorical boundaries, so that everyone is a participant and potential contributor to the action. We are all rapidly becoming used to this way of thinking about participation through our experiences of the internet, where everyone is a participant in community actions. The metaphor of a net or web is particularly apt; movement by anyone anywhere in the web or net shakes everyone else, as we are all involved in intertwined action.

In this context, performance theorists have developed the idea of liminal space (Schechner 2003; Turner 1986; cf. Gerofsky 2006). A limen is a threshold, the boundary line between two places or (metaphorically) two states of being. For example, a limen may refer to the threshold of a house, the boundary line between 'inside' and 'outside', or it may refer to the boundary between two states of consciousness (for example, sleeping and waking), or to chronological boundaries (between past and present, summer and winter, etc.), or to the boundaries between people or peoples.

In performance, this liminal space is expanded and opened up, to become a living-space. In other words, performance pries open a much larger space on the boundaries or the margins. These kinds of liminal spaces allow for exploration of the contradictions, paradoxes, transitions and transformations that take place as we pass boundaries or live in the spaces on the thresholds of our experienced worlds.

I have written elsewhere (Gerofsky 2011) about such boundary or threshold limens as powerful transitional spaces in the history of human culture and thought. There is a sense in which a boundary between two opposites functions as a kind of portal between what we perceive as different universes: between the worlds of waking consciousness and the altered consciousness of our sleeping states; between our experiences of indoor and outdoor worlds; between one society and another on a national borderline; and the potent boundary between the world of the living and the dead. In world religious practices, these kinds of powerful liminal spaces of transition are the realm of priests, shamans, saints and other mediators who, through training and natural abilities, are seen to be able to safely negotiate these dangerous boundaries on behalf of others, linking spiritual realms with our quotidian material world, and facilitating communication between deities and spirits and the living.

Even in the secular and abstract world of mathematics, one can find traces of sacred liminal spaces and their bodily engagements that can never be erased completely from human presence in the world. As developed in Gerofsky (2011), even the axes of the Cartesian coordinate plane have bodily correlations, and connect with the symbolism of the Christian cross, the orthogonal patterns on traditional shamans' drums and the geometry of the World Tree and middle earth in Nordic and other cosmologies. As much as we might like to remove all extraneous references to bodily and spiritual ways of being to create a neutral mathematical symbolism, it is not really possible to do so; the power of these transitional spaces persists. I suggest a different approach: one in which we embrace, expand and explore liminal spaces as part of a performative (and potentially, digital performative) aspect of a mathematics pedagogy repertoire.

Liminal performative spaces allow for *play* in in-between spaces, in the sense of the 'play' between a wheel and its axle (McLuhan 2003), without which the mechanism would freeze up and become immobile. A space of play is a necessity for any movement, and in order for any new understandings to develop. The idea of a space for play might seem strange to those steeped in the traditions of the stereotypical school mathematics class described above, where efficiency, seriousness and a strict adherence to logical exposition might seem inimical to a playful attitude. Yet children's play, artists' playful explorations and the imaginative play that is part of intellectual and creative work have an aspect of seriousness and an affinity for logic. To treat play as something foolish or trivial is to misunderstand the importance of play in all its senses. A system without play is frozen and locked, lacking a key requisite for development and renewal.

Performance theorists conceive of a theatrical stage, or a religious altar, or any performance space as an expansion of a liminal or boundary space. For example, the site for performance of a religious rite is an expansion of the boundary between the material and spiritual worlds. A stage creates a kind of sacred space as well, whether the stage is an elaborately-lit raised platform in an architecturally-designed theatre, or a chalk circle drawn on the pavement in street performances. Similarly, a classroom (or schoolyard, or garden, or library) has the potential to function as a liminal space—a space between not-knowing and knowing, between childhood and adulthood, between the community of the classroom and the outside world, for example. When conceived of as a kind of liminal space (as a stage, altar, or more simply as a potent space of multiple possibilities), a school classroom is suddenly opened to reconfiguration and repurposing as a mutable, flexible space susceptible to new geometries, new rules, new dreams and potentiality. Immediately, teachers and students are invited to participate in making spaces of possibility together in this unexplored and significant boundary region, where past converses with future in the enculturation of young people.

Actions taking place in this special, liminal space do not necessarily follow the rules and conventions of everyday life. Performers can explore personas that are not their quotidian selves; actors can convincingly play in role, because the story space of the stage lies in the expanded limen between truth and falsehood, where fictional truths can be told through that which is not literally true. (Are there ways that mathematics education can explore the spaces between fiction and literal truth? Do we already do so at some level through mathematical conjecture, proof by contradiction/reductio ab absurdum, as Pimm (1993) suggests, and imaginative work with physically impossible or logically paradoxical spaces?)

Play in the spaces of paradox and contradiction, ambiguity and transition—the space of the limen—is vitally important for learners who are novices to a field like mathematics, and just as vitally important for experts who aim to discover fresh approaches by seeking to regain 'beginners' mind', to get beyond the habituation born of experience. Mathematics is woven through and through with ambiguities that can be very confusing for novices; for just a few examples, consider the deliberate ambiguities of elementary concepts like 'multiplication', 'the equal sign', 'zero', and 'infinity' as one applies them in different areas of mathematics (Davis and Simmt 2006; Thurston 1995). To discuss one of these examples in more detail, consider the concept of multiplication which brings together models as diverse as repeated addition, equal groupings of objects, number line hopping, multiple folds in a sheet of paper or cloth (literally, 'multi- ply'), the area of a rectangle or other figure, an array of rows and columns, etc. (see Davis and Simmt 2006). What these very diverse models and representation have in common is a shared structure that we have named 'multiplicative', and which we can symbolize as $a \times b$. Apart from that mathematical structure, these phenomena or representations have little in common with one another, and yet, to understand the concept of multiplication as fully as possible, we have to hold a number of these multiple meanings or models all at the same time. This simultaneous holding of multiple, in some senses contradictory, meanings is what I mean when referring to deliberate ambiguity in mathematics.

In the stereotypical, 'traditional' mathematics classroom, there is no space for the expression of confusion or wonder at the multiple meanings of mathematical terms, objects and operations, no place to play with the rich and complex, seemingly contradictory and paradoxical concepts that are central to mathematical understanding. In a performative space, these liminal places can be opened up to allow for play, and for engagement of full-bodied, emotional, mindful exploration, which may lead to deeper levels of understanding and appreciation for the aesthetic and intellectual aspects of mathematics.

Gee (2003) has written extensively about the affordances of online spaces, especially massively multiplayer online roleplaying games, as potentially liberatory educational spaces. Gee gives numerous examples of children from disadvantaged families and neighbourhoods who go online to find a congenial place for learning, creating, gaining status and respect, and even making money. In Gee's home in the USA, and increasingly around the world, more and more people have access to the internet and to computers or hand-held internet access devices. We have begun to live our lives more and more in the virtual (and liminal) spaces of the internet, so that the boundaries between actual and virtual worlds become blurred, and the two worlds affect one another in a seamless way. In this, our new reality, digital performative spaces become some of our primary spaces for play, where we can interact with one another, explore and learn.

Some mathematics educators have been thinking about digital mathematical performances as a liminal space for exploration, learning and asking big questions. I will introduce three examples of online digital mathematical performances I have been involved with or am familiar with: Mathfest.ca, Vi Hart's mathematics education videos, and my own and others' mathematics/dance videos. These are all interesting and laudable projects that bring together online video and performative expressions of big mathematical ideas for the sake of stimulating curiosity, inquiry and learning. But these are still early steps in work on digital mathematical performances as pedagogy. They can all be critiqued as still being 'shows' that present mathematics to an audience that may be quite passive. My question to readers is about interactivity: how to make digital performances of mathematics that blur the boundary between performer and audience, and that encourage active participation and learning?

Some Examples of Digital Mathematical Performance Spaces

Digital Math Performance

This is a Canadian/Brazilian project that offers space for an alternative way of learning and expressing mathematical queries, wonderment, questions, aesthetics and humour through digital mathematical performances. I will begin by describing this project, which has been underway since 2008. I will discuss the project's results in terms of the creation of liminal spaces, and make suggestions for further development of liminal spaces for mathematical participation using digital performances and the affordances of online cultures.

The Digital Math Performance project, and its website at www.mathfest.ca, were developed by a Canadian/Brazilian team of mathematics educators: George Gadanidis (University of Western Ontario), Marcelo Borba (Universidade Estadual de São Paulo) and Susan Gerofsky (University of British Columbia). This project grew out of the collaborators' interest in making a space for school mathematics as something that could be talked and wondered about, told as a story or poem, danced, acted, sung, sculpted—in a word, performed. By opening up such a space, we wanted to stimulate conversations about 'what we learned in math class today', among students, teachers and families.

The project centres around an online mathematics performance space, characterized as both a contest and a festival. Each year, school students, teachers and pre-service teachers submit videotapes of up to 3 min in length as contest/festival entries. Dozens of performances are uploaded to the website each year, including mathematical skits, poems, songs, animated films, video stories, pick-a-path stories, comics and music videos. Eight celebrity judges (including well-known writers, musicians, a filmmaker, broadcaster and mathematics educator Ubiratan D'Ambrosio) and a panel of mathematicians and mathematics educators pick their top choices based on criteria of richness of mathematical ideas, creativity and imagination, and quality of performance. Prizes are awarded, but more importantly, the performances are shared online, with the aim of creating a community of learners and teachers sharing their questions, musings and artistic work around mathematical ideas. In recent years, the project has expanded the online performance festival to include science performances as well.

The Math Performance Contest was conceived in part as supplementing the competitive mathematics problem-solving contests that are a well-established feature of Canadian school mathematics. Where the problem-solving contests reward students for *doing* mathematics (through solving challenging, non-standard problems in a timed individual written test), the performance contest rewards students for working collaboratively to express their understandings and queries about mathematical concepts, or meta-mathematical musings about the nature of mathematics itself.

Vi Hart's Mathematics Videos

Vi (Victoria) Hart is a young 'mathemusician' whose YouTube mathematical videos have gone viral and made her famous over the past 2 years. Vi, the daughter of mathematician/computer scientist and sculpture George Hart, has a background in music as well as mathematics, and a strong curiosity about the areas of mathematics that inspire passionate interest in their practitioners.

In 2012, Hart began making a series of videos called 'Doodling in Math Class', and posting them online on her own YouTube channel www.youtube.com/user/ Vihart linked to her blog www.vihart.com. These fast-paced videos with a speeded-up voice-over begin with Vi doodling, folding paper or making glitter-glue lines on pinecones, and advance quickly to sophisticated ideas and speculations about interesting mathematical topics (infinite series, Fibonacci sequences in nature, fractals and geometric conundrums, just to name a few). The videos are framed with a story of being bored and doodling in math class, a conceit which soon leads to a lively discussion of exactly the topic the teacher was explaining in such a boring way through a standard lecture and exercise drills.

Vi's videos were immediately a great success, and some of her videos have now had millions of views on YouTube. ('Hexaflexagons' is the top with 5,703,604 views to date, May 8, 2014, but others number in the millions as well.) Vi quickly gained a world-wide following, and her work is well-known to many mathematics teachers (and is used in many of those same math courses she paints as boring!) She was soon hired by the Khan Academy to "sit in a cubicle all day and make random doodles" (as cited on her Wikipedia page)—actually, she was hired to create more of her mathematics videos as part of the Khan Academy's huge, free, online curriculum resource. Unlike most of the Khan Academy's recorded lessons that explain mathematical concepts in a personal but typical tutoring lecture format, Vi Hart's videos have a quirky, fast-paced performative quality, and are full of generation-specific contemporary references to zombies and cute cats, work with mathematical foods like Fibonachos and hexaflexamexagons (hexaflexagons made of tortillas), and a Moebius strip music box.

The phenomenal popularity of her work has meant that Vi has been written up in Salon and the New York Times, is invited to speak at countless events (and has already retired from the lecture circuit), and has a huge online following. Her videos merit re-viewing, especially since they move at such a rapid pace, and are offering mathematics educators and students a repertoire of digital math performance resources full of energy, curiosity and lively interest about mathematical ideas.

'Mathematics Through Dance' Videos

In contrast to Vi's wildly popular videos, the small collection of mathematics-through-dance online videos is still a small, underground phenomenon. However, I want to mention these in terms of their potential for stimulating whole-body participatory embodied learning of a large number of mathematical ideas (see also Gerofsky (2013) for further discussion of dance as a learning modality for mathematics.)

The most active proponents of dance as a modality for mathematics learning are California dancers Dr. Schaffer and Mr. Stern and their dance company (at www.mathdance.org and www.schafferstern.org). Schaffer and Stern have a limited online presence in terms of showing videos of their mathematical dances since their focus is live, professional dance performances. However, they do offer a rich set of online resources to support teachers in using dance and movement as a modality for teaching a wide range of mathematical ideas in school classes.

Other examples of representing and exploring mathematical concepts through dance and bodily movement include Diana Davis' 2012 award-winning 'Dance Your PhD' video representing her doctoral research results in mathematics through dance www.news.brown.edu/pressreleases/2012/10/phddance, 'Dimensional Yearnings' on www.mathfest.ca (www.edu.uwo.ca/mathscene/mathfest2009/mathfest244.html), and my own videos on exploring mathematical patterns in traditional longsword dance (https://vimeo.com/66303546) and in Sarah Chase's modern dance (https://vimeo.com/68811119, password-protected site). These early works in mathematic education through dance are noteworthy in that they not only *show* finished works, but can also be viewed as *invitations* for teachers and learners to participate in similar activities to explore mathematical patterns in embodied, movement based kinaesthetic modes.

Discussion: Making Digital Mathematical Performance Spaces More Participatory

The liminal spaces for digital mathematical performances have only begun to be explored through pilot projects like those described above. It is up to us to take these ideas further, to make them more flexible, more broadly accessible and democratic, more international and more interactive.

The present Digital Math Performance site is run by adults in positions of authority (teachers, university researchers)—could it be opened up to be more self-organizing? Could there be more dialogue, more call-and-response among performers and viewers? How could digital mathematical performances potentially function as a student-generated, peer-mediated activity? Would learners be motivated to collaborate with teachers and mathematicians to create such activity? Do students have deep mathematical questions they want to address through mathematical online performative modes?

Sites like Vi Hart's YouTube channel and the math/dance sites are presently geared towards spectators who are not necessarily participants in mathematical inquiry, although skilled teachers could potentially use these videos as catalysts for more participatory activities with their students. Are there ways that these kinds of performances could engage participants with mathematical ideas in depth, using digital media?

Would it be better to host mathematical performances on more widely available social networking sites (like Facebook, YouTube or Twitter)? Would immersive gaming sites or online worlds like Second Life or World of Warcraft (or more protected educational versions of these worlds) offer better opportunities for participatory digital mathematical performances using avatars in a sophisticated sim environment? Could the idea of learner digital mathematical performances become more truly cosmopolitan, crossing boundaries of nationality, race, class and gender? I believe that the possibilities are already in our reach, and I encourage people to work collaboratively to develop these new, liminal learning spaces for participation.

References

Auslander, P. (1999). *Liveness: Performance in a mediatized culture*. New York: Psychology Press.

Berliner, P. (1994). *Thinking in jazz*. Chicago: University of Chicago Press.

Boaler, J. (2002). The development of disciplinary relationships: Knowledge, practice and identity in mathematics classrooms. *For the Learning of Mathematics, 22*(1), 42–47.

Borasi, R. (1984). Some reflections on and criticisms of the principle of learning concepts by abstraction. *For the Learning of Mathematics, 4*(3), 14–18.

Conquergood, D. (2002). Performance studies: Interventions and radical research. *TDR/The Drama Review, 46*(2), 145–146.

Davis, B., & Simmt, E. (2006). Mathematics-for-teaching: An ongoing investigation of the mathematics that teachers (need to) know. *Educational Studies in Mathematics, 61*(3), 293–319.

Dixon, S. (2009). *Digital performance: Doubles, cyborgs and multi-identities in digital performance and cyberculture*. https://smartech.gatech.edu/handle/1853/28545. Accessed 15 Jan 2014.

Farley, K. (2002). Digital dance theatre: The marriage of computers, choreography and techno/human reactivity. *Body, Space and Technology, 3*(1), 39–46.

Gadanidis, G., & Borba, M. (2008). Our lives as performance mathematicians. *For the Learning of Mathematics, 28*(1), 44–51.

Gamon, D., & Bragdon, A. D. (1998). *Building mental muscle: Conditioning exercises for the six intelligence zones*. New York: Barnes & Noble.

Gee, J. P. (2003). *What video games have to teach us about learning and literacy*. New York: Palgrave Macmillan.

Gerofsky, S. (2006). *Performance space and time*. Symposium discussion paper for digital mathematical performances: A fields institute symposium, University of Western Ontario, London, Canada, June 9–11, 2006. http://www.edu.uwo.ca/mathstory/pdf/GerofskyPaper.pdf.

Gerofsky, S. (2011). Ancestral genres of mathematical graphs. *For the Learning of Mathematics, 31*(1), 14–19.

Gerofsky, S. (2013). *Learning mathematics through dance*. Proceedings of Bridges 2013, Enschede: Mathematics, Music, Art, Architecture, Culture. http://archive.bridgesmathartorg/2013/bridges2013-337.pdf

Herbel-Eisenmann, B., Lubienski, S. T., & Id-Deen, L. (2006). Reconsidering the study of mathematics instructional practices: The importance of curricular context in understanding local and global teacher change. *Journal of Mathematics Teacher Education, 9*(4), 313–345.

McLuhan, M. (2003). *Understanding me*. Toronto: McClelland and Stewart.

Noss, R. (1994). Structure and ideology in the mathematics curriculum. *For the Learning of Mathematics, 14*(1), 2–10.

Pimm, D. (1993). The silence of the body. *For the Learning of Mathematics, 13*(1), 35–38.

Plato (1961). *The collected dialogues of Plato including the letters* (E. Hamilton & H. Cairns, Eds.). New York: Pantheon.

Radford, L., Edwards, L., & Arzarello, F. (2009). Introduction: Beyond words. *Educational Studies in Mathematics, 70*(2), 91–95.

Sawyer, R. K. (2003). *Group creativity: Music, theatre, collaboration*. Mahwah: Lawrence Erlbaum.

Schechner, R. (2003). *Performance theory*. London: Routledge.

Schmidt, W. H., McKnight, C. C., & Raizen, S. (Eds.). (1997). *A splintered vision: An investigation of US science and mathematics education*. New York: Springer.

Taylor, P. C. (1996). Mythmaking and mythbreaking in the mathematics classroom. *Educational Studies in Mathematics, 31*(1–2), 151–173.

Thorndike, E. L. (1924). Mental discipline in high school studies. *Journal of Educational Psychology, 15*(1), 1–22.

Thurston, W. P. (1995). On proof and progress in mathematics (reprint). *For the Learning of Mathematics, 15*(1), 29–37.

Turner, V. (1986). *From ritual to theatre: The human seriousness of play*. New York: PAJ.

Participation in Mathematics Problem-Solving Through Gestures and Narration

Luciana Bazzini and Cristina Sabena

Abstract On the one hand, recent results in neuroscience and communication bring the role of multimodal resources in cognitive and communicative processes to the fore. On the other hand, Bruner highlights the different role of narrative and logical thinking in human understanding. Before this background, we investigate the teaching and learning processes when children are introduced to the mathematical meanings by the use of narrative context and multimodality. We focus on a narratively-based problem-solving activity in the first year of primary school. We analyse the teacher-students multimodal interaction and discuss the potential and limits of what we call the "semiotic game" between teacher and students. The importance of the teacher being conscious of the multimodal aspects of teaching and learning mathematics, and of the complexity of narratively-based contexts comes to the fore.

Introduction

Problem-solving is the core of mathematical discovery and plays a crucial role in the processes of teaching and learning. However, there is evidence that students experience difficulties when trying to solve problems. Moreover, they often lack interest and their approach to problems is only determined by school obligations. They solve problems for duty, not for beauty. The dynamics that occur in the mathematics classroom are of great interest for mathematics education research. In particular means for promoting participation and inclusion are worth considering and discussing. Thus, the necessity of finding ways to improve students' interest in doing mathematics arises.

We will discuss here two boundary conditions that, among many others, we believe to be of great help in stimulating students to solve problems and check solutions. The first condition is the creation of meaningful contexts for problems;

L. Bazzini (✉) • C. Sabena
Dipartimento di Filosofia e Scienze dell'Educazione, Università di Torino, Torino, Italy
e-mail: luciana.bazzini@unito.it; cristina.sabena@unito.it

the second is the adoption of suitable working methods in the classroom, based on the use of linguistic and embodied resources as semiotic means for thinking and communication. In this perspective, we focus on two specific issues, i.e. the role of narration and the use of multimodality, which we consider relevant tools also for promoting interest and participation in mathematical problem-solving.

Our research hypotheses are grounded on Bruner's ideas on the role of *narrative thinking* in human cognition (Bruner 1986), and the *multimodal perspective* in mathematics education (Arzarello et al. 2009). In the following, we will briefly outline the theoretical premises of the research, and we will present a case study. The case study comprises a narratively-based problem-solving activity set in the first year of primary school. We will analyse the problem-solving processes, focusing on the role of the teacher, in a multimodal perspective. The use of gestures and of narrative contexts for problem-solving are afterwards discussed in the light of the results presented.

Theoretical Background

With respect to the complex relationship between thinking and speaking, Bruner (1986) identifies two different cognitive styles: *narrative* thinking and *logical (or paradigmatic)* thinking. Narrative thinking focuses on intentions and actions, and is strongly anchored in experience in context as the subjects perceive them. It constitutes the ordinary way through which human beings make sense of their experiences and of the world. In fact, Bruner argues that we show a natural attitude to organize experiences by giving them the form of the narratives we use when talking about them. In contrast, logical thinking is based on cause-effect reasoning and is guided by principles of coherence and non-contradiction. Thus it is in consonance with the typical character of mathematical argumentations and proofs.

Though irreducible, since based on different principles and procedures, narrative and logical thinking are to be conceived as complementary, rather than juxtaposed modes of thought. Indeed, Bruner claims that the human condition can only be understood by focussing on how human beings create their "possible worlds" through the inclusion of both kinds of thinking (*ibid.*). This claim is very relevant to mathematics educators, who face the double challenge of introducing students to deductive ways of arguing, and of including all of them in such a process (lower achievers not being excluded). In our view, the route leading to theoretical thinking in mathematics, which makes intense use of logical-scientific thinking, starts in and is deeply intertwined with the use of the narrative dimension. It is our conviction that this intertwining can be a fruitful premise for an inclusive teaching of mathematics at all levels.

With the aim of realising a synergy between narrative and logical thinking in mathematics activities, we think that it is fundamental to go beyond the purely linguistic plane and to include gestures and extra-linguistic modes of communication. This claim is in line with recent results in psychology and cognitive science about the role of the body for communication and cognition. In particular, our research is informed by the so-called *embodied cognition* perspective, which assigns the body a crucial role in the constitution of mathematical ideas (Lakoff and Núñez 2000). We also draw on studies on the role of *gestures* in communication and in thinking (Goldin-Meadow 2003; McNeill 1992, 2005).

The perspective of embodiment, criticizing the Platonic idealism and the Cartesian mind-body dualism, advocates that mathematical ideas are founded on our bodily experiences and developed through metaphorical mechanisms. Many limits have been recognized in this theory, in particular concerning the lack of social, historical, and cultural dimensions in the formation of mathematical concepts (for a discussion, see Sabena et al. 2012). However, the embodied cognition studies have the great merit of having directed the attention towards the role of the body in the process of knowledge formation, including mathematical knowledge. In accordance with historical-cultural perspectives, we assume that the embodied nature of thinking has to be related to the historically constituted cultural systems. In short, the two aspects (i.e. embodied and individual on the one side, and historic-cultural and social on the other) must be considered together.

Furthermore, recent results on *multimodality* in neuroscience and communication give us new suggestions when analysing classroom dynamics. Gallese and Lakoff (2005) use the notion of "multimodality" to highlight the role of the brain's sensory-motor system in conceptual knowledge. Their claim is in particular based on the recent discovery of multimodal mirror neurons, which fire both when the subject, first, performs and directly observes an action but also, second, when it imagines an action. This model entails that there is not a central "brain engine" responsible for sense making which controls the different brain areas devoted to different sensorial modalities (what would occur if the brain behaved in a modular manner). Instead, there are multiple modalities that work together in an integrated way, up to the point that they overlap with each other, like vision, touch, hearing, but also motor control and planning. On the other hand, in the field of communication design the term "multimodality" is used to refer to the multiple modes we have to communicate and express meanings to our interlocutors: e.g. words, sounds, figures, etc.

If we look at the mathematics teaching-learning processes from the multimodal perspective, we have implications at both, didactic and research levels. We are convinced that increasing participation and deeper learning will be obtained if children are left free to express their thinking through their body and non-verbal languages like gestures.

At a didactic level, keeping in mind the theory-practice dialectics (Bazzini 2007), this approach suggests that classroom mathematical activities are to be planned and managed in order to foster the students' multimodal participation in mathematical activities. As it is well known, students' participation is highly sensitive to the teacher's prompts and choices. The role of the teacher in the classroom becomes hence crucial. This applies for both domains: choosing the tasks for the students and establishing a didactical contract (Brousseau 1997) that allows and encourages the use and intertwining of gestures with words and written signs, in order to support the students to evolve from their personal meanings to the scientific ones (Vygotsky 1978).

At the research level, our approach calls for the enlargement of the focus of attention and for the consideration of variables that are usually neglected, especially those of extra-linguistic or extra-symbolic origins. We mainly refer to gestures, gazes, and inscriptions (drawings and sketches). These variables can be considered important semiotic resources in the mathematics classroom (for further theoretical analysis and analysis with a semiotic lens, see Arzarello et al. 2009).

These hypotheses are at the base of our on-going research study aiming at (i) the introduction of children to mathematical meanings by the use of narrative context and multimodality, and (ii) the study of teaching and learning processes therein. In this paper, we focus on the role of multimodal resources in a narratively-based problem-solving activity in a primary school. In particular, we highlight the role of the teacher against the outlined background.

Teaching Experiment

We will analyse a problem-solving activity carried out in a class of 6-year-old pupils (first grade of Italian primary school). The teacher is a member of our research group and shares the theoretical background of the study. In order to be able to capture the dynamics of the interaction, the research methodology includes audio and videotaped recordings, additional to the collection of written products by the students; the analysis of the processes is mainly based on the semiotic productions of students and teachers in a multimodal perspective (Arzarello et al. 2009).

The activity is based on a story taken from the adventures of Iride, a fanciful character that accompanies the children during their mathematics lesson. It is also linked to previous experiences of the children, i.e. the growth of small tomato plants in the school garden. The story is delivered to the children by the teacher. Here the text:

> Iride likes our idea of having a vegetable garden. She plants tomato plants like we did and observes their growth. Close to a small tomato plant she puts a "Lego" tower (tower composed of small cubes). As soon as the plant grows, she adds cubes. From the beginning of the month, Iride has added 18 cubes. Today the plant has the height of a 26-cube tower. How tall was the plant at the beginning of the month?

The story presents a subtraction problem where the initial data is unknown. It is the first time that the children face this type of problem. Under these conditions, the pupils cannot transform the initial data, as they usually did in previously encountered problems. Furthermore, the problem requires a temporal inversion in order to be fully understood. On the base of this analysis, the teacher recognizes that the problem could be difficult for the children and decides to organize the work in groups that work under her supervision. Each group is formed of students with the same level of mathematical abilities, according to the teacher's evaluation.

This setting allows us to closely observe the teacher-students interaction, in a narrative and multimodal context. In the following, we focus on a group of low-medium achieving students, with members Marco, Zoe, Matteo, and Alessandro.

Data Analysis

Children are around a table, where paper, coloured pencils and little cubes are at their disposal to solve the problem.

All children become soon engaged in the narrative side of the story, and Matteo provides immediately a hypothesis: *"In my view, the plant is high like this"*, while showing the height by means of his hands, one over the other (Fig. 1).

Matteo appears not to consider the information about measures provided by the story; he imagines a plant, probably referring to his experience and real data, i.e. the height of the tomato plants which are actually present in the garden of the school. All the children declare to agree with Matteo, even if they are not able to justify their hypothesis, and the solution process seems blocked up.

However, the teacher decides to profit from Matteo's proposal and makes an intervention using words and a gesture:

Teacher: *When we planted them in the garden, the plants were higher than zero (performing a gesture with two open hands one over the other, as shown in Fig. 2).*

The teacher's words and gesture highlight an important mathematical feature of the problem, i.e. the initial non nul height of the plant. After the experiment, the teacher disclosed to us the purpose of her intervention: she feared that the children did not consider that the initial height of the plant could be different from zero, and wanted to make it clear for them. We recall that for the first time, the children are facing a variation problem in which the unknown data is the initial step of the process, and not its final result (as in previous problems). Therefore the teacher is anxious to support the children in focusing on this unknown data.

To reach her goal, the teacher reproduces the same gesture as Matteo. She accompanies the gesture with words that point to and refine a specific mathematical

Fig. 1 Matteo shows with a gesture (one open hand over the other) the initial height of the plant

Fig. 2 The teacher' gesture, reproducing Matteo's previous gesture

property. This particular use of gesture and speech is quite frequent in the teaching experiments that we observed during our research: we have called it the *"semiotic game"* of the teacher (Arzarello et al. 2009). The semiotic game is a didactic phenomenon that can be grasped only if we consider the teaching-learning processes in a multimodal perspective. In fact, it typically involves embodied resources, such as gestures, and speech. A semiotic game may occur when the teacher interacts with the students, for example in classroom discussions or during group work. In a semiotic game, the teacher adapts to the students' semiotic resources (e.g. their words and their gestures), and uses them to develop the mathematical knowledge towards scientifically shared meanings. More specifically, the teacher uses one kind of sign in tune with the students' productions (usually gestures), and another one to support the evolution of meanings (usually language). For instance, the teacher repeats a gesture that one student just did, and accompanies it with appropriate linguistic expressions and explanations.

From our overall research, we observe that the students often appear to benefit from the semiotic game. In this particular case, however, this does not apply to all children. In fact, after the teacher's semiotic game, the children face a serious difficulty caused by the linguistic ambiguity between the two sentences: "the plant grows another 26 cubes" and "the plant grows up to 26 cubes". Some children even say: *"the plant grows 26 cubes"*, and do not clarify further what they exactly mean. To help the children distinguish these two usages, the teacher proposes to simulate the plant's growing with the cubic Lego blocks. By building the Lego model, she persuades the children that the number 26 refers to the final height of the plant (and not to its growth). However, there is still no agreement about the initial height of the plant: Alessandro is still convinced that the initial height is zero.

This analysis shows that not all children grasped the previous semiotic game performed by the teacher. Yet we cannot be surprised about this result if we consider the teaching-learning processes as a complex phenomenon, not reducible to simple cause-effect schemas. In the next section, we will deepen the discussion about the limits of the semiotic game.

Considering the following interaction, it is interesting to notice that the teacher does not continue along the same line (i.e. with another semiotic game, or recalling the previous one), but changes the strategy to tackle the problem. She guides the

Gestures and Narrations

children in checking the proposal of Alessandro, by building a tower: 18 cubes are used (and added to zero) and eventually all children agree that, in this last case, the plant's height would be of 18 cubes and not 26.

In order to support the children in getting a visual image of the situation, the teacher resorts to the Lego tower: she lifts the tower of 18 cubes up from the table (Fig. 3) asking:

Teacher: *"What do the 18 cubes represent?"*

In this way, she guides the children towards the realization that the 18 cubes have to be put on a pre-existing tower, which measures the initial height of the plant. Then, she points to the space at the top of the tower to stimulate the children to visualize the cubes that are needed to pass from 18 to 26 (Fig. 4).

Alessandro, who is attentive to the teacher's gesture, asks: *"How many cubes did we add to reach 26?"*

The children add cubes to the tower until it reaches 26 while Alessandro looks at them and checks the counting in his mind. At the end he claims: "8!". At this point,

Fig. 3 The teacher lifts the tower of 18 cubes

Fig. 4 The teacher points to the space above the tower

Fig. 5 The drawing made by one child to represent the solution

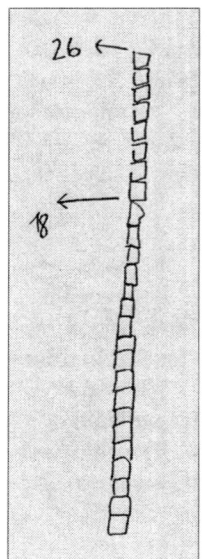

the teacher focuses the children's attention on what the numbers 18, 26 and 8 represent with respect to the height of the plant.

Finally, the solution procedure is clear for all and the situation is graphically represented with a drawing (Fig. 5).

The Lego tower is used as a concrete model for the growing of the plant, which was one of the didactical goals of the teaching experiment. Let us observe that in Fig. 5 the blocks corresponding to the number 8 are placed at the *top* and not at the *bottom* of the tower, as it would be coherent with Iride's story. On the contrary, the representation is coherent with the teacher's gestures reported in Fig. 4, where a 18-block tower is on the desk, and the teacher points on the space above it, asking how many cubes are needed to go from 18 to 26. This result is a confirmation of the importance of the non-verbal communication of the teacher with the students: not only do the children pay attention to the teacher's gestures, but also they rely on them in solving the problem.

Discussion

The analysis of the teaching experiment gives evidence of what we called the *multimodal* nature of the processes of teaching and learning. In order to tackle a mathematical problem, the children and the teacher used words, embodied signs (gestures), tools (the Lego blocks), and diagrams in synergic ways, confirming our hypothesis that teaching-learning processes develop in a multimodal way. This synergy was exploited in particular by the teacher when students met difficulties in modeling the situation, as in the case we have just described.

In the last decade, the studies on students' gestures and embodied communication in mathematics learning has gained a certain attention in mathematics education literature (see for instance Edwards et al. 2014). So far, little attention, however, has been dedicated to the teacher. The semiotic game notion goes into such a direction and intends to model a specific kind of teacher-students interaction.

The semiotic game can develop if the students produce something meaningful with respect to the problem at hand, using some signs (words, gestures, drawings, etc.). The teacher, monitoring the students' productions, can then seize the opportunity and enact her/his semiotic game. Even a vague gesture can indicate the level of comprehension of the student, who has not yet found the words to express it. In a Vygotskian perspective, the gesture documents that the student is in the Zone of Proximal Development for a certain concept (Vygotsky 1978). As a consequence, the teacher has the chance to intervene and to encourage the student to develop his/her intuition. In the case study analysed in this paper, we showed an example of an unsuccessful semiotic game, i.e. we showed evidence that not all students could benefit from the teacher's gesture-speech intervention.

This sort of "failure" of the semiotic game can be framed in a wider perspective that considers the teaching-learning processes as a complex phenomenon, not reducible to simple cause-effect schemas. Another example of a not-successful semiotic game is presented in Arzarello et al. (2010) and discussed in the light of the epistemological dimension of the teacher-students interaction. Here, we want to underline that the study of the potential and, above all, of the limits of the theoretical constructs of our research is of paramount importance both on theoretical and on practical planes.

The reasons for the limited success of the semiotic game in the given example are possibly to be found also in the complexity of the context proposed by the didactic situation. The mathematical problem is proposed to the students by means of a story (the story of Iride), which refers to the students' real experience of gardening. The narrative dimension was important to provoke the students' active involvement in the problem-solving activity and to enhance their participation in the mathematical work. However, differently from other cases we have previously analysed (Bazzini et al. 2009), it did not provide an essential support for the solution of the problem. Probably, the reference to the actual experience of the tomato plants in the school garden prevented the children from properly imagining Iride's situation and from activating the narrative thinking to deal with it. Furthermore, during the solving process, the teacher introduced the cube blocks to model the growth of the plant. We can therefore list three different contexts that the children had to make sense of and coordinate in their activity: Iride's story, the real plants in the school garden, and the cube tower. The resulting multi-layered context was probably too cognitively demanding for some children. At an older age, explicit discussions on the differences and relationships between mathematical and non-mathematical contexts (and mathematical and non-mathematical solutions) can promote authentic mathematical learning based on problem solving and modeling (Gellert and Jablonka 2009).

From a methodological point of view, it is worth noticing the synergy between teacher and researchers in analysing classroom episodes. Such synergy resulted very fruitful for both the teacher and the researchers: on the one hand, such an opportunity allowed the researchers a deeper insight and a more comprehensive analysis of the experiment. On the other hand, discussing the analysis of her own teaching activity, the teacher gained a great consciousness of many phenomena that did happen in her interaction with the pupils, but that she could not notice in the flow of the activity. For instance, the teacher herself highlighted some missed opportunities. As a result, the analysis methodology allowed the teacher to become more conscious of her own actions during her interactions with the children.

Finally, an important result concerns the possibility of widening the teacher's horizon with regard to the multimodal aspects of the classroom processes by means of an attentive analysis of the classroom processes. With a changed perspective, the teacher acquires new possibilities of intervention (e.g. by means of the semiotic game) to guide the students in their construction of mathematical meanings. To do so, the teacher has to learn how *to look* at the processes according to a multimodal perspective; that is, considering not only words and mathematical symbols but also other kinds of signs, such as gestures, gazes, and so on. Furthermore she also has to be sensitive to the relationships existing between the different signs and the different contexts involved in the activity. We are aware that acquiring such abilities may be a long and hard process. However, we are convinced that their impact on the educational practice may significantly improve school mathematics. This will be our concern in future research, and our challenge as well.

Acknowledgements This work is supported by grants from the Italian Ministry of Education (Progetto PRIN 2008PBBWNT "Teaching Mathematics: beliefs, best practices and teachers' formation"). We wish to thank the teacher Luisa Politano for her precious contribution to the study.

References

Arzarello, F., Paola, D., Robutti, O., & Sabena, C. (2009). Gestures as semiotic resources in the mathematics classroom. *Educational Studies in Mathematics, 70*(2), 97–109.

Arzarello, F., Bikner, A., & Sabena, C. (2010). Complementary networking: Enriching understanding. In V. Durand-Guerrier, S. Soury-Lavergne, & F. Arzarello (Eds.), *Proceedings of the 6th congress of the European society for research in mathematics education* (pp. 1545–1554). Lyon: INRP.

Bazzini, L. (2007). The mutual influence of theory and practice in mathematics education: Implications for research and teaching. *ZDM – The International Journal on Mathematics Education, 39*, 119–125.

Bazzini, L., Sabena, C., & Villa, B. (2009). Meaningful context in mathematical problem solving: A case study. In *Proceedings of the 3rd international conference on science and mathematics education* (pp. 343–351). Penang: CoSMEd.

Brousseau, G. (1997). *Theory of didactical situations in mathematics*. Dordrecht: Kluwer.

Bruner, J. (1986). *Actual minds, possible worlds*. Cambridge, MA: Harvard University Press.

Edwards, L., Ferrara, F., & Moore-Russo, D. (Eds.). (2014). *Emerging perspectives on gesture and embodiment in mathematics*. Charlotte: IAP.

Gallese, V., & Lakoff, G. (2005). The brain's concepts: The role of the sensory-motor system in conceptual knowledge. *Cognitive Neuropsychology, 22*(3), 455–479.

Gellert, U., & Jablonka, E. (2009). "I am not talking about reality": Word problems and the intricacies of producing legitimate text. In L. Verschaffel, B. Greer, W. Van Dooren, & S. Mukhopadhyay (Eds.), *Words and worlds: Modelling verbal descriptions of situations* (pp. 39–53). Rotterdam: Sense.

Goldin-Meadow, S. (2003). *Hearing gesture: How our hands help us think*. Cambridge, MA: The Belknap Press of Harvard University Press.

Lakoff, G., & Núñez, R. (2000). *Where mathematics comes from: How the embodied mind brings mathematics into being*. New York: Basic Books.

McNeill, D. (1992). *Hand and mind: What gestures reveal about thought*. Chicago: University of Chicago Press.

McNeill, D. (2005). *Gesture and thought*. Chicago: University of Chicago Press.

Sabena, C., Robutti, O., Ferrara, F., & Arzarello, F. (2012). The development of a semiotic frame to analyse teaching and learning processes: Examples in pre- and post-algebraic contexts. In L. Coulange, J.-P. Drouhard, J.-L. Dorier, & A. Robert (Eds.), *Recherches en Didactique des Mathématiques, Numéro spécial hors-série, Enseignement de l'algèbre élémentaire: bilan et perspectives* (pp. 231–245). Grenoble: La Pensée Sauvage.

Vygotsky, L. S. (1978). *Mind in society*. Cambridge, MA: Harvard University Press.

Considering the Classroom Space: Towards a Multimodal Analysis of the Pedagogical Discourse

Eleni Gana, Charoula Stathopoulou, and Petros Chaviaris

Abstract In an attempt to contribute to the discussion about the semantic potential of classroom space design in the teaching and learning of mathematics, this chapter reconstructs relationships and meanings from the use of space by teachers in two primary mathematical classes. Adopting a social semiotic perspective, our study reveals that semantic spatial configurations are functioning as the material forces, which are subject to and reflective of teacher's pedagogical conceptions, and are actively involved in constructing students' social experience in the specific teaching and learning environment.

Introduction

The physical setting, such as the arrangement of rooms and the way we use them along with the objects within them, conveys subtle and more overt socialization messages of which most people are typically unaware. Solomon (1992) and, more recently, McGregor (2004a) state that the spatial layout of schools, such as the organization into classrooms or desks in rows within them, is until today almost universally taken for granted so that the fundamental ways in which these spatial arrangements sustain ideologies about knowledge and power remain unseen. Several empirical studies from different fields of educational research demonstrated that classroom space is more than a neutral container for practices. In contrast, the studies suggest that classroom arrangements constitute pedagogical resources serving to transmit the pedagogical practices and the fundamental regulatory principles that govern the school. McGregor (2004b, p. 17) even claims that the classroom space itself represents "a hidden form of the curriculum".

E. Gana (✉) • C. Stathopoulou
Department of Special Education, University of Thessaly, Volos, Greece
e-mail: egana@uth.gr; hastath@uth.gr

P. Chaviaris
University of the Aegean, Mytilene, Greece
e-mail: chaviaris@rhodes.aegean.gr

Different forms of classroom arrangement facilitate or inhibit different forms of learning (Bissel 2002; Higgins et al. 2005). They construct, maintain or impede relationships between students and between students and teachers (Pierce 2009). Furthermore, they indirectly inform pupils about the teacher's expectations and prevalent power structures as teachers draw on space to assert their authority, of physical space per se that is crucial to the experience, but instead the ways in which the physical space is used by the participants. Specific arrangements support certain kinds of practices and power relations and consequently affect the distribution of knowledge. The metaphorical relationships established through the experience of such arrangements define, in practice, the space of the classroom as a specific place which supports expectations and assumptions about certain kinds of practices and relationships of knowledge and power.

In a similar line of thought, Massey, working in the field of geography, argues for conceptualizing space as "the product of interrelations" (2005, p. 9). She describes space as a sphere ranging from the global to the tiny, where many different bits and trajectories—physical, social, political or cultural—coexist and interact with each other and which is always open to transformation. In recognizing the social protagonist's agency, Massey allows for classroom space to be understood as a dynamic and multifocal construction, which can be constantly re-arranged and modified to meet different social or political requirements as well as particular discourse practices.

In an attempt to contribute to the discussion about the semantic potential of classroom space, this study explores the relationships and meanings which are generated by teachers' use of space in two mathematical classes of primary school (in Athens, Greece). The study draws on theories from the field of social semiotics and especially on multimodality for conceptualizing and analyzing communication in learning environments (Bezemer 2008; Flewitt 2006; Lim et al. 2012; O'Halloran 2005).

The basic assumption underlying this study is that space is significant as the material site where various semiotic resources of the teacher (language, movement etc.) are realised, and as such it is related to different conceptions and attitudes of his/her pedagogical attitude. The study addresses two key questions:

- How does classrooms space organize the interpersonal relations of the participants, that is amongst the students themselves and between the teacher and the students?
- Does the semiotic space configuration correspond with the semantic landscape established by the teacher's use of other communicative means (e.g. his/her discourse, movement) during the teaching-learning processes? In other words, do the semiotics of specific classrooms contribute to a coherent pedagogical discourse?

The analysis is performed through two main axes of focus: first, we consider the "perceptual spaces" (Scollon and Scollon 2003) established by the teacher's and the students' stationary position in the classroom. In a second step, we look for the potential reconfigurations of their semantics due to their interaction and integration with other semiotic resources used by the teacher during his/her teaching.

The analysis reveals that the observed mathematics teachers seem to understand that classroom space arrangement realizes pedagogies. They endeavor to modify and reconstruct the space layout of their classrooms in order to express discourses of progressive pedagogy. However, when overlapped with other semiotic resources used by the teachers, such as their movement in the classroom or their discursive choices when addressing the pupils, the meaning potential of such spatial configurations becomes subject to transformation and often reflects "authoritative" mindsets and methodologies. Overall, the multimodal experience of classroom space reflects an "interdiscursive dialogicality" (Scollon and Scollon 2003) between the different semiotic resources used by the teacher and an inherent heterogeneity in terms of pedagogical discourses.

The following section of this chapter considers classroom-space semiotics based on the research on multimodality of the pedagogical discourse. Thereafter, the focus is on the methodology of this study (the research questions, the collection and analysis of the data). Finally we discuss the results of the analysis and the implications of a multimodal approach to classroom studies and teachers' education.

The Semiotic Potential of the Classroom Space

Social semiotic theorizing (Blommaert and Huang 2010; Scollon and Scollon 2003) and research on multimodality acknowledge that space is integral to all acts of meaning-making (Bezemer 2008; Lim et al. 2012; Norris 2004; O'Halloran 2010). Scollon and Scollon (2003) developed a theory about the spatial scope of semiotic processes ("geosemiotics") revolving around notions such as "emplacement", which means the actual semiotic processes resulting from the specific location of signs in the specific material site. Furthermore, Blommaert and Huang (2010) state that the spatial arena always imposes its own rules, possibilities and restriction on (verbal) communication, and in that sense, space constitutes a real actor in sociolinguistic processes. Space is consequently understood as a sign of the social—with special attention to the meaning affordance of the particular structuring environment—and as a dynamic instance of the embodied experience of the social communication which takes place in that material site.

With regard to the classroom, Kress et al. (2005), in an ethnographic study conducted in three inner London schools, show that classroom space arrangements and displays are designed by teachers, on the one hand, in order to distribute a specific content such as a school subject (e.g. English, Science). On the other hand, however, the authors showed that classroom space arrangements also contribute to the implementation of fundamental regulatory principles of school discourse. In more detail, Kress et al. propose that the ideational meanings in classroom are actually realized through the interaction of three factors: (a) the teacher's movements, (b) the meaning of the space in which the teacher moves and (c) the students' position and how and where they eventually move. To give an example, the researchers describe a teacher's slow and deliberate movements in the classroom as "invigilating", which they term "a patrol".

Matthiessen (2010) states that the teacher's positioning in the different material sites in the classroom can realize "semiotic distance". In this way, social interpersonal relations between the students and the teacher are established due to the teacher's spatial position. In terms of social distance, Hall (1966) suggests four general categories of space, according to the typical distance in which they occur and the degree of physical contact experienced by the participants during the communication: (a) *the public space,* (b) *the social-consultative space,* (c) the *casual-personal space* and (d) *the intimate space* (Fig. 1).

Lim et al. (2012), referring to Hall's taxonomy, claim that in the context of the classroom most communication takes place in the social-consultative space, which creates a formal relationship between teacher and students. Therefore, in order to further investigate the semiotic potential of the social-consultative space, they develop a sub-division within this category, namely they propose to distinguish: (a) *the authoritative space,* (b) *the personal space,* (c) *the supervisory space* and (d) *the interactional space.* Their coding is based on the characteristics of the material site and the degree of distance between the teacher and the students as well as on the type of activities that are typically taking place there.

The authoritative space is located at the outer limit of the social-consultative space. It corresponds with the space in the front center of the classroom which is usually the furthest away from the students. The authoritative space is the space where the teacher's desk is typically located and where the teacher positions himself/herself to provide instructions. In accordance with Hall's (1966) hypothesis of social distance, Lim et al. specify that "the material distance in the Authoritative

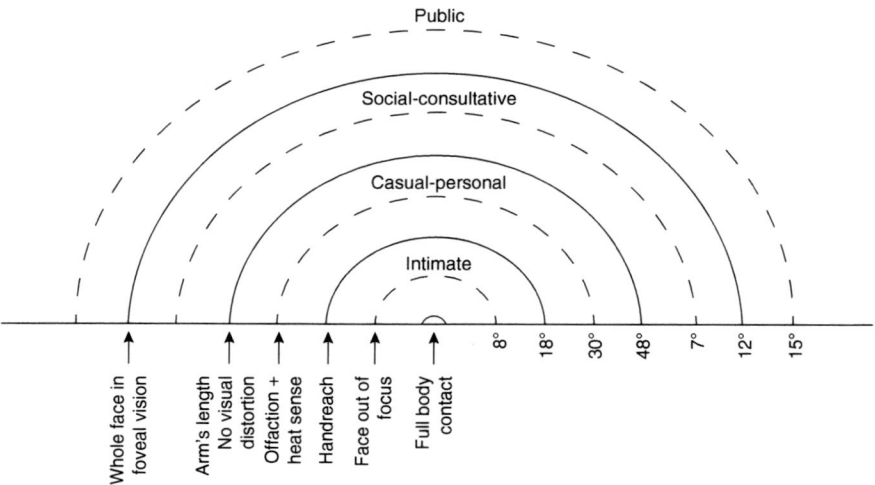

Fig. 1 Hall's (1966) distance sets (Reproduced from Lim et al. 2012, p. 237)

Space constructs a formal tenor in the relationship between teacher and students" (Lim et al. 2012, pp. 237–238).

The space behind the teacher's desk is described as the personal space where the teacher packs his/her items and prepares for the next stage of the lesson. However, the same classroom space can be transformed into an authoritative space when the teacher points or teaches from behind his/her desk. Lim et al. (2012, p. 238) explain: "physical spaces in the classroom may not only serve a single function". When the teacher moves up and down the classroom, or even alongside or between the rows of student's desks for the purpose of supervision, he is "transforming these sites into the supervisory space" (p. 238).

Finally, the location of the interactional space is defined "along the cline of Social-Consultative Space but inclined towards the Casual-Personal Space" where the "closer proximity between the teacher and the student(s) facilitates interaction and reduces interpersonal distance" (p. 238). Lim et al. call this patterning of movement and positioning as well as the intersemiotic correspondence between the use of space and the other resources used in the development of the teaching "spatial pedagogy" (p. 248).

The Study

The Setting of the Research: Place—Time—Participants

The data for this study include two one hour-long, video-recorded lessons from two different classes. The two lessons have several similarities, which make them comparable in terms of the data analysis. Both lessons take place in a 6th grade of a primary school, are on the same topic, comprise the same number of students (19 students per class), are conducted in classrooms of similar spatial capacity and are addressed to students with the similar mixed linguistic and mathematical ability. Looking at the teachers' profiles, there is a difference in gender but not in their teaching experience. Both teachers have 5 years of experience in teaching.

The Research Questions

The research questions underlying the study are the following:

- Which semantic configurations of space are established through the teachers' space management (students' desks arrangement/teacher's position in the classroom space)? What types of interpersonal relationships are established?
- In which way do the semantic configurations of space correspond with other semiotic resources used during the teaching-learning process? What kind of pedagogical discourse is thus sustained?

The Methodology

In exploring the use of classroom space as a semiotic resource, it is attempted to carry out:

- a statically-based analysis which considers the semantic relationships between the students' and the teachers' stationary positions in the specific locations of the classroom;
- a more dynamic analysis which, through the different stages of the lesson, is looking for the potential reconfigurations of spaces' semantics due to their interaction and integration with other semiotic resources used by the teacher during his/her teaching.

Therefore we first identify the "perceptual spaces" (Scollon and Scollon 2003) established through desk arrangements as well as the semantics of the different locations where the two teachers choose to position themselves based on the semantic categories for classroom space proposed by Lim et al. (2012). Then we map each teacher's functional use of different classroom sites during his/her teaching. In this part of the analysis, we make use of the methodological notion of "lesson-microgenre" as developed by the Curriculum Genre Theory (Christie 2005; O'Halloran 2004) in the context of social semiotics. O'Halloran (2004) has described lesson-microgenres in mathematics secondary classes as follows: "pre-lesson", "preliminary", "main lesson", "end of the lesson" and "interpolated disruptive microgenres". According to the Genre Curriculum Theory, the lesson-microgenres are realised through particular selections in the co-deployment of semiotic resources (e.g. discursive choices, positioning and movements, functional uses of pedagogical media resources like the whiteboard, the textbook, the digital media etc.). Consequently, the spatial semiotics of the classroom could be strongly influenced by such different realizations of the pedagogical discourse; in that case the meaning potential of a physical place is often changed and the specific location is reconfigured into "a new semiotic place" according to the different activities and interactions that occur in that space. In this study, we make use of Chaviaris et al.'s (2011) indicative description of mathematical lessons' microgenres in a primary school. Based on an ethnographic study it was found that the following lesson-microgenres sequence: "initial activity", "establishing relations", "establishing the new knowledge", "exploration-assessment of the new knowledge", and optional, "assessment of the previous knowledge". The use that the teachers make of each particular classroom location during the specific lesson-microgenre is coded in protocols in a table in 1 min intervals and along the following parameters: type of the proceeding task, specific language choices, conversational turns, paralinguistic markers, gaze, use of the pedagogic media which is used by both teachers, that is the whiteboard and the textbook (see Table 1 for an abbreviated protocol). The coding of the above parameters is followed by descriptions of the specific realizations and of their textual relationship with the spatial and the other modal configurations. The researchers viewed the lessons' videotapes many times, individually and collectively, and they discussed the

Table 1 Lesson-microgenre (initial activity)

Time	Teacher's position/ movement	Task proceeded	Linguistic/ paralinguistic markers	Artifact-tool	Gaze
13:25	Authoritative–front center of the classroom	Invitation to the whole class to notice a mathematical relation	Use of 2nd person plural *"What do you see?"*		To the students and for seconds to the whiteboard where a student is writing
14:10	Authoritative–a step towards students–left center of the classroom	Legitimizes talk between students	*"You can speak to each other about..."*		To the students
15:05	Authoritative–in front of the whiteboard	Asks for students' responses	Imperative/ personal pronoun indexing himself *"Tell me what you noticed?"*		To the students
16:05	Authoritative–front center–step towards students	Gives some hints- rephrases the invitation	*"Could you tell me what ..."*	Whiteboard	To students and to whiteboard
16:19	Authoritative–in front of whiteboard	Utilizing a reply-presents a mathematical procedure	*"I see/I take this and then I calculate ..."*	Whiteboard	To whiteboard, occasionally to students

multimodal analysis of the protocols in detail in order to establish the relational semantics provided by each lesson-microgenre experience as well as by the entire lesson.

Discussing the Results

In the class C1 the students' desks are united in pairs with four students sitting around them. In this way the potential collaboration between the students is facilitated. The teacher's desk is at the left side of the authoritative space. There is sufficient space for the students or the teacher to move between the group tables. The students, sitting around the united pairs of desks, are in personal social distance and direct visual contact with a group of peers. The vision towards the front of the class, where the teacher's desk and the whiteboard are located, requires the students to make the additional and perhaps more conscious movement of turning their heads. All in all, the semantics of the "perceptual spaces" created by the specific arrangement of the desks in the C1 classroom clearly support the methodological approach and hence the mindset of learning through participation in small groups.

The discreet location of the teacher's desk combined with the sufficient space left for him to move between the student's desks raises the expectation of a teacher functioning as a supporter for students' work as opposed to a teacher in control of knowledge. However, the teacher's (C1) stationary positions in the classroom during the teaching-learning process as well as the use of other semiotic resources he makes during the teaching-learning process does not correspond to the pedagogical discourses suggested by the design of the students' desk arrangement. In terms of the semiotic distance, teacher C1 positions himself mainly in the front space of the classroom where he conducts a lecture-type teaching. Although the design of the students' stationary positions provides adequate space between the students' desks, the teacher never uses this physical site for personal involvement and interaction with his students. Even when he moves into or close to the personal space, he uses this position to supervise the entire class. In fact, during the entire lesson the teacher C1 is standing: (a) in the front center of the classroom and in front of the whiteboard (authoritative space), (b) in front of his desk or the right of his desk (authoritative space/supervisory function), (c) behind his desk (personal space).

The teacher C1 starts the lesson by using the authoritative space for giving a formal type of teaching; he initiates and solves a problem by himself in order to show the students the new mathematical relation, which they have to learn. While he is modeling the solving procedures of this type of problem, his discourse is mainly monological as far as he answers by himself most of the rhetorical type of questions he poses during his writing in the whiteboard. Students, mostly silent, copy the proposed solution; some of them lose interest or become frustrated in trying to understand the modeled mathematical procedures, they drop the task and start looking around. When finishing the writing on the whiteboard, teacher C1 moves behind his desk, which is the personal space but functions as such only for 2 s, that is the time he needs to take and find the appropriate page in the textbook. After this very short time, the personal space turns into an authoritative space as the teacher stands up behind his desk, holding the textbook (motion as semiotic resource) and reading the guidelines for a typical procedural solution of the type of problems he has just taught (mediation resources). By his intonation, he emphasizes the attention the students should give to the "authority" of the book (paralinguistic resource). In addition, he keeps complete control over the conversation by initiating every communicative turn, and when questioning, his discourse is very regulative. The third lesson-microgenre observed in this class is the exploration-assessment of the new knowledge. Here, the teacher provides the students with a problem to solve, moves to the right of the desk and stands there, supervising the students work from the distance. While his posture may at some point suggest a sense of casualty, the phrasing of his discourse supports an authoritative teaching profile ("I want you to solve a problem for me", "I want to see who feels strong enough and certain of himself to be on the whiteboard", "don't forget to adopt the typical procedure I just spoke of").

In the class C2 the desks are arranged in the shape of a rectangle with the students sitting around all sides, so as to have direct visual access to the whiteboard without any other body movement being necessary. Those who sit at the two larger sides never have direct eye contact with each other. In terms of "perceptual spaces", the prominent role of the whiteboard within the design of students' positioning indicates that the social performance valued is the individual access to the content knowledge as it is presented in and elaborated through the whiteboard. Around the rectangle of the students' desks, there is enough space for circulation, but the pupils' positions seem to complicate their eventual movement. The teacher's desk is not placed at the front looking towards the students, but instead it is positioned as an extension to the short side of the rectangular. The teacher's desk touches also the students' desks, which creates a sense of informality. Another factor which contributes to the reduction of semiotic distance between the teacher and her students is the fact that the teacher positions herself in the interactional space for approximately half the time of the lesson. From the total of 42.46 min of the videotaped lesson's duration, she positions herself close to the students for 28.08 min. For 18.02 min teacher C2 is in the front space of the classroom (authoritative space), where she is teaching the new mathematical knowledge.

The teacher of C2 begins the lesson with the microgenre of "assessment of the previous knowledge" which, in the process of teaching practice, functions also as a "bridge" to the new information. During this lesson-microgenre she stands alongside the students' desks. This material site is typically recognized as the interactional space which, however in this scene, functions as a supervisory space: She often looks at the whiteboard making comments about a pupil's writing there (gaze/ linguistic resources). At the same time she is regulating the recollection of mathematical concepts of the whole class and the establishment of relational connections between them by an inquiry-response cycle (linguistic resources: relational clauses for the scaffolding of understanding). In this case the spatial proximity between teacher and students reduces the dominance of power and authority asserted by the action of supervision and establishes a "structured informality" (Lim et al. 2012) which evidently encourages the observed students to participate actively in the construction of knowledge. In the next lesson-microgenres there is an alteration in the use of space, with the authoritative space being used only for the establishment of new knowledge. The teacher uses the whiteboard to illustrate a particular mathematical method that she brings across to the students using the textbook for guidance. This latter instructional activity expands its semiotic meaning when the teacher moves towards the students and gives personal advice to some teaching points (interactional space). To summarize, the interpersonal relationship installed by teacher C2's positions in the different material sites, her minimal distracting movements, her linguistic choices (e.g. the frequent use of modality and adjuncts which display solidarity with her students) and her challenging of the textbook's authority (she points to a misstatement in the book and therefore builds the students' confidence in mathematic solutions as they are handled by the class as a team) construct a non-threatening learning environment where students feel comfortable enough to respond and speak up.

Conclusion

The study focused on the classroom space semiotics and particularly on the relationships and meanings generated by two primary teachers' use of classroom space during their teaching. By exploring the semantics of students' and teachers' positioning as well as how these meanings are contextualized in their co-deployment and interplay with other semiotic resources, the study argues that teachers' use of classroom space is an integral part of their pedagogy. Although applying the same curriculum standards, the intentional or unconscious use of the classroom space of the observed teachers as well as their use of other semiotic resources (discursive choices, gaze, use of media) constructed different learning experiences and therefore different representations of notions such as authority, power and participation. In fact, in the present study neither of the two primary classrooms had the "traditional" box-like structure with the teacher zone at the front of the classroom and the students' seating arranged in rows facing in one direction—an arrangement associated with a teacher-centered teaching. In both classes the teachers endeavored to recontextualize the semiotic potential of their classrooms' space and re-shape the spatial display according to the rhetoric of progressive pedagogy: the arrangement of the students' desks reflect teaching methodologies which suggest pedagogical notions and processes such as students' engagement and active participation, group collaboration, the teacher's movement into the students' placement for personal or group scaffolding, etc. However, in their co-deployment and interplay with the other semiotic resources used by the teachers during their teaching, like the teachers' positions and movements in the classroom, their discursive choices etc., the initial semantics of such space display often have to be reconfigured. The complex pedagogical discourses realized by the different semiotic resources delineate an "interdiscursive dialogicality" (Scollon and Scollon 2003) and reflect a hybridization in terms of pedagogical discourses employed by the teachers. The "classroom space experience" (Appelbaum 2008) emerging from the interplay of the varied semiotic resources appears to be constructed and reconstructed even during a one-hour lesson, reflecting very often conflicting and eventually confusing pedagogical meanings.

This study argues that the social semiotics and in particular the multimodal approach offers a lot for classroom studies and teachers' education. While many former approaches and methods in classroom research remain informative, the multimodal perspective broadens the definition of classroom discourse to include multimodal resources such as the use of space as part of the teachers' pedagogical practice. A classroom analysis, thus, which focuses on the meaning-making potential provided by the space as well as by each particular mode in use and which explores the semiotic landscape emerging from their constant interplay during learning-teaching processes will advance teachers' understanding of the dynamics and the complexity of the pedagogical semiosis (Lim et al. 2012). It would be productive for teachers to be able to reflect on their use of classroom resources and the way these contextualize pedagogical discourse. Such awareness would help them to design a more congruent co-deployment of the classroom resources which would reduce the semantic divergence and the conflicting and eventually confusing meanings arising through the use of different semiotic modes during their teaching. Lim (2011) proposes that a reconstruction of what actually goes on in the classroom, through video recording of classes and time-based transcription of the semiotic resources used by the teachers, would be productive for teachers' understanding of the semiotic complexity required by the pedagogical implementation.

References

Appelbaum, P. (2008). *Embracing mathematics*. London: RoutledgeFalmer.
Bezemer, J. (2008). Displaying orientation in the classroom: Students' multimodal responses to teacher instructions. *Linguistics and Education, 19*(2), 166–178.
Bissel, J. (2002). Teachers' construction of space and place: A study of school architectural design as a context of secondary school teachers' work. *Dissertation Abstracts International, 64*(02), 311A. (UMI No. 3082110).
Blommaert, J., & Huang, A. (2010). Semiotic and spatial scope: Towards a materialist semiotics. *Working Papers in Urban Languages and Linguistics, 62*. London: King's College.
Chaviaris, P., Stathopoulou, C., & Gana, E. (2011). A socio-political analysis of mathematics teaching in the classroom. *Quaderni di Ricerca in Didattica, 22*(Supplemento 1), 233–237.
Christie, F. (2005). *Classroom discourse analysis*. London: Continuum.
Flewitt, R. (2006). Using video to investigate preschool classroom interaction: Education research assumptions and methodological practices. *Visual Communication, 5*(1), 25–50.
Hall, E. (1966). *The hidden dimension*. New York: Doubleday.
Higgins, S., Hall, E., Wall, K., Woolner, P., & McCaughey, C. (2005). *The impact of school environments: A literature review*. http://www.ncl.ac.uk/cflat/news/DCReport.pdf.
Kress, G., Jewitt, C., Bourne, J., Franks, A., Hardcastle, J., Jones, K., & Reid, E. (2005). *English in urban classrooms: A multimodal perspective on teaching and learning*. London: RoutledgeFalmer.
Lim, F. V. (2011). *A systemic functional multimodal discourse analysis approach to pedagogical*. Unpublished PhD thesis, National University of Singapore.
Lim, F. V., O'Halloran, K. L., & Podlasov, A. (2012). Spatial pedagogy: Mapping meanings in the use of classroom space. *Cambridge Journal of Education, 42*(2), 235–251.
Massey, D. (2005). *For space*. London: Sage.
Matthiessen, C. I. M. (2010). Multisemiosis and context-based register typology: Registerial variation in the complementarity of semiotic systems. In E. Ventola & J. Moya (Eds.), *The world told and the world shown: Multisemiotic issues* (pp. 11–38). Basingstoke: Palgrave Macmillan.
McGregor, J. (2004a). Editorial. *Forum, 46*(1), 2.
McGregor, J. (2004b). Space, power and the classroom. *Forum, 46*(1), 13–18.
Norris, S. (2004). *Analyzing multimodal interaction: A methodological framework*. London: Routledge.
O'Halloran, K. L. (2004). Discourse in secondary school mathematics classrooms according to social class and gender. In J. A. Foley (Ed.), *Language, education and discourse: Functional approaches* (pp. 191–225). London: Continuum.
O'Halloran, K. L. (2005). *Mathematical discourse: Language, symbolism and visual images*. London: Continuum.

O'Holloran, K. L. (2010). The semantic hyperspace: Accumulating mathematical knowledge across semiotic resources and modalities. In F. Christie & K. Marton (Eds.), *Disciplinarily: Functional linguistics and sociological perspectives* (pp. 217–236). London: Continuum.

Pierce, J. L. (2009). *A co-construction of space trilogy: Examining how ESL teachers, English learners and classrooms interact.* Unpublished PhD thesis, Indiana University of Pennsylvania.

Scollon, R., & Scollon, S. W. (2003). *Discourses in place: Language in the material world.* London: Routledge.

Solomon, I. (1992). *Power and class in modern Greek school* (in Greek). Athens: Alexandria.

Commentary: Semiotic Game, Semiotic Resources, Liminal Space—A Revolutionary Moment in Mathematics Education!

Peter Appelbaum

Abstract The commentary on the chapters of Gerofsky, of Bazzini and Sabena and of Gana and Stathopoulou emphasises the importance of transforming the relationship between people and mathematics education. It synthesises the three chapters by pointing to semiotic perspectives as crucial to our understanding of mathematics education practices and research.

With the advent of the three chapters in this section, we are witnessing a truly revolutionary moment in mathematics education, a paradigm shift that cannot be ignored or denied, the dawn of a new era that will forever change the terrain of mathematics teacher education, mathematics teaching and learning, assessment of mathematics learning spaces, and the nature of research in and around mathematics education. I write this not because the chapters here are bizarrely new and provocative, but because I use these chapters as evidence that mathematics educators on different continents, in very different cultures and national contexts, trained themselves in incommensurate traditions of pedagogy, research, and the philosophy of mathematics, all share a common sense of semiotics as crucial both to our understanding of mathematics teaching and learning, and to the practices of teaching, assessment, and research.

The chapter on Narration and Gesture as characteristic of participation in mathematical problem-solving establishes the significant ways that embodied communication, story-telling contexts, models of mathematical concepts, and real-life experiences all come together in complex webs of meaning that are not evident in a surface description, but which contribute to our understanding of a particular moment. This is the classic idea of "thick description" referred to by the anthropologist Clifford Geertz (1973). His canonical example is the symbolic meaning of a person's wink that is only understood in an ethnographic encounter through familiarity with the possible interpretations in a particular culture, the relationships among the people involved in doing or witnessing the wink, place and time of the wink, and the event in which the wink transpires. Similarly, in this chapter, Bazzini

---

P. Appelbaum (✉)
Department of Curriculum, Cultures and Child/Youth Studies,
School of Education, Arcadia University, Philadelphia, PA, USA
e-mail: appelbap@arcadia.edu

© Springer International Publishing Switzerland 2015
U. Gellert et al. (eds.), *Educational Paths to Mathematics*,
Advances in Mathematics Education, DOI 10.1007/978-3-319-15410-7_15

and Sabena describe the ways that a seemingly simple gesture, a student moving his hands in a particular way, takes on a multi-layered set of meanings when brought in complex relation with a fantasy story, experience with gardening, and with Lego cubes used as a model of the growth of a plant in a garden. What we see is not that this is too complex to comprehend, but rather, that the teacher-researcher, who has been collaborating with the observer-researchers, is thinking about the semiotic meaning of the gesture, and thus uses the same gesture, informed by her Vygotskyian knowledge of scaffolding, to gently bring the student toward her own, possibly more mathematically sophisticated, thinking about the way that the gesture could be related to the mathematics question embedded in the fantasy story, and perhaps produce an "answer" to the question in the story.

Here is a term, coined by Bazzini and Sabena, which you will use routinely in a few years: "Semiotic Game." You saw it here first, folks! The teacher is skilled at playing this semiotic game, and also skilled at methods of enculturation and acculturation into the game that is part of the culture of her mathematics classroom. The students are getting better and better at this game if they are exhibiting evidence of success in this classroom, as determined by their mutually-negotiated didactic contract, which is simultaneously made clear as classroom events unfold.

The authors come to the conclusion that the semiotic practices of the teacher are crucial, yet not a simple instructional panacea. I will come back to my own interpretation later in my commentary. For now, note as well that their attention to gesture and narration brings to the foreground new ways to think about mathematics education as significantly about participation and interaction, as well as about the creation of a particular kind of place, the mathematics classroom, or the space of the mathematics lesson/unit, in which semiotics and symbolic meaning are or are not recognized—that is, where symbolic meaning is or is not seen, and thus, is or is not present for the participants.

We can layer on top of Bazzini and Sabena the interesting discussion contributed by Gana, Stathopoulou and Chaviaris, of the "space" in which mathematics teaching/learning encounters take place. On the one hand, they nicely describe the ways that the physical properties of the space—the chairs, tables, desks, windows, light, arrangement of the furniture, the placement and/or movement of the teacher and students—are all items that we need to seriously return to in our thinking about mathematics teaching and learning. In our rush to innovation and in our attention to curriculum development that will efficiently produce indicators of success on such things as international comparisons of standardized test performance, we sometimes forget the simple things, like whether or not the children in a primary classroom have a chair to sit in, or the kinds of authority that a teacher can communicate by standing or sitting in various physical arrangements of the classroom.

More importantly, though, is the way that these three authors help us to comprehend the space of teaching and learning: they describe the open space and the materials of that space as "semiotic resources," much in the same way that Bazzini and Sabena write of narration, gesture, manipulative materials used as models, and connections to real-life experiences. That is our second keyword of future mathematics education discussion: Semiotic Resources. You heard it here first, everyone! As

Gana, Stathopoulou and Chaviaris describe semiotic resources, their availability—bear with me here while I create a convoluted, seemingly self-referential sentence!—the availability of semiotic resources structures the possible ways that choices by the teacher structure the possibility of classroom structure to emerge. That is, semiotic resources are excellent examples of what the Sociologist Anthony Giddens (1993) termed materials of "structuration". Structuration theory prioritizes ontology over epistemology, so that production and reproduction of social practices—that is, the things and situations that we take for granted as the way things are and the way things unfold in social interactions, including the possibilities for what might take place—are, in a sort of behind the scenes way, deep behind multiple layers of curtains that obfuscate the sources of those beliefs, critical reflections of stasis and change, expectations by those agents who effect change in those situations, behaviors, creative, skillful and knowledgeable agents and solutions, and strategic thought that exists in those places in general. Like Giddens, these authors examine the beliefs that teachers hold, both implicitly and explicitly, as indicators of structuration—those things that structure the possibilities of structures and structural changes—including spatial arrangements, actions by individuals in the spaces, and so on.

When we consider semiotic meanings as crucial to the choices of the teacher in the room, as Bazzini and Sabena do, a teacher who is reflecting on the dynamic interplay of gesture, speech, narration, associations that people in the room are making with real-life experiences, and models of mathematical concepts, and, when we look at such a teacher in the context of structuration, informed by the notion that space is a critical dynamic influence on the structuration processes in that room, we might also think, as do Gana, Stathopoulou and Chaviaris, that the people in the room are also, in these moments, creating the symbolic meaning of the space in which they interact. That is, the ways that people behave, move about, when they speak and do not speak, the power relations supported or discouraged by the arrangement of the space, all contribute to the possible ways in which those present continue to interact and whether or not they might be able to sustain those ways of behaving, whether or not they might introduce new structural possibilities and so on. In other words, the people in the room are co-constructing by their behaviors and beliefs about what is possible in that space a particular *place* that they then refer to implicitly as their "mathematics classroom," or as "this mathematics lesson" or "unit" (Appelbaum 2008). I would go further to say that, in the micro-moments of a classroom event, the place is constantly recreated, over and over and over again, so that it appears to be the same place in that space, but in fact is always everrenewed. The place in that space thus appears to be a powerfully stable place to those present, since with each micro-moment initiated by agents in that space, a seemingly unchanged place is made even more evident as there "for them to be in."

I note two significant ways in which my own thinking is slightly different from that of the authors of these two chapters. First of all, I am not so focused on the teacher's choices. I see the choices made by the teacher as traditionally more evident than the choices made by the students—after all, in the typical place constructed by those present in mathematics education encounters, students do not make the choices that determine the nature of the place in which they find themselves.

Nevertheless, such a view erases the potential agency of the students, who in fact, in many school and out-of-school learning situations, do have a powerful role in the negotiation of that place in which they exist, and how successfully that place supports their learning of mathematics, their social interactions with each other, their opportunities to physically move in and out of that place, and so on.

Second, I find the concern about traditional definitions of "success" to be structuring the potential structures of research and thus structuring the potential structures of change in mathematics education in these chapters to be limiting our ability to open up some new ways to use the powerful concepts that are explicated through their examples of exciting research. For example, I disagree with the characterization of the teacher using gesture as having "limited success." For me, conflict, confusion and paradox are at the heart of the learning. The semiotic game worked perfectly, I say, in the sense that it allowed for the confusion to emerge, for the stop-making-sense aspect of the experience to bubble up to the surface so that it could become part of the experience. This taking of the experience of the problem-solving effort as something outside of oneself that can be examined in and of itself might also be identified as an important element of mathematics teaching and learning, i.e., as something also at the heart of a common, Hegelian and Piagetian notion, of the "cause" of learning.

The idea is that the conflict across at least four contexts of narrative story, the story problem, the school garden, the cube tower creates a cognitively demanding situation. But I maintain, instead, that it is the richness of interweaving these multiple stories, and thinking across them as comparative models for each other, that creates the exciting opportunity for engagement and learning. Why do the authors propose that Gellert and Jablonka's (2009) notion of relationships between mathematical and non-mathematical contexts is better experienced at an older age? Maybe it is classroom experiences such as this one that make the vibrant possibilities at the older age seem more successful, because of this enculturation/acculturation that begins in grade 1 and prior.

I agree that a teacher using the semiotic game concept to reflect on her and her students' experiences is surely able to approach her work in these terms, and to create a new way of thinking about teaching and learning, and a new way of thinking about the didactic contract!—that the children with the teacher are agreeing to this semiotic game, constantly challenging each other to figure out how to interpret their next move, then to discuss their strategies not only for obtaining answers to problems posed, but also to discuss their ways of making sense or coping with the as-yet non-sense of each others' gestures, drawings, words, and other representations.

I also agree that a teacher can make important applications of the types of social spaces that are described in the chapter by Gana, Stathopoulou and Chiavaris. A researcher, a curriculum designer, a teacher, a team of visitors, probably should describe for themselves a classroom along each dimension: first, Hall's categories, the public space, the socio-consultative space, the casual-personal space, and the intimate space; and then the subdivisions of the social-consultative space from Lim et al.: the authoritative space, the personal space, the supervisory space, and the interactional space. More important for reflection on teaching, for interpretation of success or efficacy in the space, and for understanding the co-created place, is the notion that these are all social spaces, and not necessarily about the physical structuration, which our authors here bring into play. Yes, the literally physical materials and characteristics of the space are important—any teacher who has been frustrated by the lighting, the overly warm room, the dream of air conditioning, the breakdown of technology, or the lack of open space for running around, surely knows the importance of the physical attributes of their classroom. What the authors point to for future research, I contend, is the symbolic meaning of the materials and particular choices made in the space, which determine the semiotic character of the place and thus the semiotic processes of structuration taking place; this will also enable us to think about all human beings in the place as potential agents of change, that is, as the initiators and makers of changes in structuration processes, and how these come about.

I see the two chapters discussed so far as helping us see how we can understand contemporary practices in mathematics education in new ways that lead to significantly new and insightful interpretations of those practices. They also cry out for further research into the ways that agents can be change agents, and not merely co-creators of social reproduction in mathematics education practices. How might we move beyond the status quo in mathematics education? How might we leap into new "ways into mathematics," and initiate important discourses into mathematics education practice, research, collaboration, and sharing? Gesture and Narrative offer us truly new conceptions of pedagogy which also highlight the notion of participation as critical for mathematics educators to consider and to explicitly apply in their work. Semiotic Space and, in general, Semiotic Resources, transform the relationship between people and mathematics education broadly conceived; these ideas make it clear that participants are caught up in processes of structuration, yet might also at the same time be the agents of transformation in mathematics education practice and research. Yet we seem, in the words of the third chapter, Gerofsky's "Digital Mathematics Performances," to live in "disembodied, antiperformative spaces," with little interactive, dialogic opportunity in this "place" we call a "mathematics lesson." Gerofsky's use of the term "Performative Pedagogy"—in the sense of an artistic, theatrical, and/or dance performance that always involves embodied ways of being, interaction and expression—echoes the previous two chapters' emphasis on "participation" in the space, which, in turn, constructs a particular kind of "place" (Appelbaum 2008). As Gerofsky describes it, there is no separation of aspects or levels of being in an engaged performative moment, even as technical skills and intellect are brought into play. The material world and embodied ways of being are celebrated as facets of an integrated whole performer, rather than being ridiculed, devalued or shunned.

Gerofsky's appreciation of participatory arts as the model for "Participatory Performance in Mathematics Education" is useful, given the attention given to ideas about participation in the various plastic and performative arts in the twentieth and twenty-first centuries. Once we conceive of mathematical activity as creative production of ideas and of the communication of those ideas, we can ask, "What does it mean for any or all of teachers, students, family members, and members of the broader community to collaborate on creative production?" (Appelbaum 2012) All forms of the arts involve participation in some sense—after all, experiencing art (observing, listening, watching, and so on) is a kind of participation in and of itself, and artworks have only rarely been created without the intention of connection in some way to a public or an audience. An important distinction when we translate the analogy from the creative arts to creative mathematics, however, is that between the participation of an audience, and the co-creation of the art with the audience. That is, there is a distinction between an artist who uses participation as part of the process of creating the art, and an art in which the participation itself *is* the art. Similar to participative democracy or participation management, it is not so much the fact that people participate that matters, but rather the fact that participation might be the main principle governing human interactions which makes the difference. A teacher, in other words, might focus on his or her work in a classroom as creative expression, and/or that same teacher might focus on facilitating the participation of the students as co-collaborators in artistic production.

Attention to Gesture and Narrative raises the specter of the teacher and students together as embodied performers of the "problem solver." A careful analysis of the Space and Place heightens our awareness of those spaces in which the place is one of integrative, holistic presence, and thus a contributor to the structuration of the possible events that might take place in this place at this time with these participants, in ways that are drastically different from traditional mathematics classrooms. Now, as Gerofsky notes, performance further transforms our understanding of the participants, the place(s) in which they may be simultaneously coexisting, and thus our understanding of performance itself: *Since performance engages people on multiple levels, the most integrative performances speak to people in many ways at once, and raise many kinds of further questions and speculations when the performance is finished*. This is a new thought about pedagogy as opening up speculation and questions rather than ending with answers! Is this, then, the playing out of this section of the book? Have we come to a significant understanding of how embodiment, multimodality, space, gesture, place, that is, semiotics and semiotic resources, become an abstract "structuration of mathematics education practice and research"? It may be the case that the implication of semiotic theories for mathematics education, if they are taken seriously, is that we are moving away from closure in teaching/learning events, toward openings, ambiguity, questions and provocations!

Gerofsky claims the digital provides some new possibilities and insights: It helps us focus on "Liminal spaces," those places where people are on the edge or border or simultaneously in more than one place at the same time. She posits that a liminal space is a "living-space"—the boundaries are portals between what we perceive as different universes, different places in which we simultaneously exist. Students and teachers are both learners and teachers in the best places of learning. They are at home in their classroom, and debating a mathematical conclusion across the globe via the Internet at the same time. They are members of their family, and at once members of other social groups, racial or ethnic groups, religious or gender groups, and so on. Now, I would add that existence in multiple worlds at once sometimes seems, at first, to be too complex, to be avoided, to make things so ambiguous as to make interpretation and decision-making impossible. This has been the case in mathematics education since it first became dominated by cognitive psychology in the nineteenth century: if we eliminate confounding variables, the thinking might suggest, we can draw conclusions and then act upon them. This is what I would describe as existing in the "uncanny valley" (Appelbaum 2011)—the uncanny valley is a term from robotics to explain why people are revolted in response to robots that are too much like humans (the Polar Express syndrome, referring to the failed animated film that people believe went awry when it made the animated characters "too human"). When we are living in living spaces that are liminal spaces, we are *not* in the uncanny valley: we might feel like we are mathematicians and not mathematicians at the same time, because we know a lot, yet at the same time we are having trouble figuring out what a question is asking of us. In a liminal space, this is exciting, it is enlivening, and it brings us to life. In a place where we cannot dwell in the amorphous boundary, our enthusiasm deadens. We are no longer present and no longer performing, no longer participants, but merely unpleasant robots who follow programming, no longer human. The antidote? Our third keyword of the future: Liminal Space.

Confusion and wonder are an essential part of participation as performance; in fact they contribute to the success of the play of the liminal space. Re-read the chapters in this section of our book; revisit them with the goal of becoming an agent of changes in structuration in mathematics education practice, or in mathematics education research, or in our unfortunate bifurcation of practice and research itself. Re-read them to collect a new vocabulary that will be a small component of your own, active, semiotic resources. Pay attention to the examples that Gerofsky provides of digital performance, and think, not how you might replicate these, but how you might translate the spirit of the liminal space into new places of collaboration, experimentation, aesthetic participation, and playful creation of new worlds to be in. These new worlds will be structuring the possibility of structuration processes, in a self-referential way. It will be fun, serious, challenging, unclear, open, and transformative. And, you will be welcomed by the many mathematics educators who have already begun this revolution, without realizing it perhaps—those who apply semiotic resources to make sense of the space they find themselves in, to recognize and embrace the embodied presence of teachers and learners in these places, through gesture, narration, analysis of the place of learning, and attention to the varieties of participation and performance, that are possible and that might be possible with a little bit of tweaking, in the new places that are co-creating with the spaces they find themselves in.

References

Appelbaum, P. (2008). *Embracing mathematics: On becoming a teacher and changing with mathematics*. London: Taylor & Francis.
Appelbaum, P. (2011) Carnival of the uncanny. In E. Malewski & N. Jaramillo (Eds.), *Epistemologies of ignorance and studies of limits in education* (pp. 221–239). New York: IAP.

Appelbaum, P. (2012, April 10–13). *The shape and feel of "participation": Flipping public spaces, performing democracy and playing with the exteriority of thought*. Vancouver: American Association for the Advancement of Curriculum Studies.

Geertz, C. (1973). *The interpretation of cultures*. New York: Basic Books.

Gellert, U., & Jablonka, E. (2009). "I am not talking about reality": Word problems and the intricacies of producing legitimate text. In L. Verschaffel, B. Greer, W. Van Dooren, & S. Mukhopadhyay (Eds.), *Words and worlds: Modelling verbal descriptions of situations* (pp. 39–53). Rotterdam: Sense.

Giddens, A. (1993). *New rules of sociological method: A positive critique of interpretive sociologies*. Stanford: Stanford University Press.

Part V
Criticising Public Discourse

Numbers on the Front Page: Mathematics in the News 247
Dimitris Chassapis
University of Athens, Greece
Eleni Giannakopoulou
Hellenic Open University, Greece

On the Role of Inconceivable Magnitude Estimation Problems to Improve Critical Thinking 263
Lluís Albarracín and Núria Gorgorió
Universitat Autònoma de Barcelona, Spain

Criticizing Public Discourse and Mathematics Education: A Commentary 279
Charoula Stathopoulou
University of Thessaly, Greece

Numbers on the Front Page: Mathematics in the News

Dimitris Chassapis and Eleni Giannakopoulou

Abstract This chapter explores the relationship between mathematics, politics and public media. It focuses on the headlines of leading Greek newspapers and analyzes the numerical discourse which appeared on the front pages between late October 2009, when the fiscal crisis of the country was first announced by the government, and June 2012, when, after a general election, the implementation of several policies seems to have been finally stabilized. The chapter discusses how by the choice of a particular numerical genre the reader is co-constructed and finally influenced in his/her interpretation of the numerical text.

Introduction

Since 2009, Greece has been submerged in a financial and social crisis unprecedented in its modern history, due initially to acute fiscal problems and, over the subsequent years, to the policies implemented on the occasion, which multiplied and transformed these problems. Since then, Greece has implemented structural adjustment policies, such as cutting down on social expenditures including pensions and raising taxes on personal incomes and sales. These have been accompanied by a dramatic attempt to reform Greece's economic system in the image of neoliberalism through massive privatizations of state-owned companies (e.g., power, telecommunication and water systems, highways, ports, etc.) and labor-market deregulation policies. In May 2010, Greece was given a bailout in loans from the European Union (EU) and the International Monetary Fund (IMF). This first bailout has kept Greece from defaulting on its loans; however, it soon became clear that additional financial support was necessary to continue servicing its debt-payments. Thus, on March 2012, the EU and IMF offered Greece a second bailout. The bailouts

---

D. Chassapis (✉)
School of Educational Sciences, University of Athens, Athens, Greece
e-mail: dchasapis@ecd.uoa.gr

E. Giannakopoulou
Educational Studies Unit, Hellenic Open University, Athens, Greece
e-mail: egian@tutors.eap.gr

invariably came with conditions in the form of austerity measures, mainly through large cuts of salaries and pensions, mass layoffs of employees in the public-sector, ever-widening labor-market deregulations leading to flexibility of working time and employees' firing conditions, as well as cutbacks in educational, health and most of social expenditures.

The austerity policies and the cutbacks of social expenditures were treated by employees and pensioners with objections and reactions leading to massive protests and strikes. In a general election in May 2012, a majority of Greeks voted for parties that reject the conditions imposed by the bailout agreements with the EU and IMF. However, since attempts to form a coalition government failed, a month later Greeks went again to the polls, and this time the pro-austerity parties got a narrow majority and formed a government.

All these years, every objection and reaction of the public and the parties opposing the implemented policies were repelled by the dominant political forces and interest groups with the argument (and the fears crafted around it and widely spread by the media) that if Greece does not comply with the EU and IMF demands, it will be forced out of the Eurozone, with the accompanying tragic consequences of defaulting on its debt and falling into bankruptcy. At the same time, the press and broadcasting media played a crucial role in legitimizing the austerity policies for the public, and thus helped to impose them upon those concerned, seeking on every occasion to mitigate their objections and reactions. In this respect, most newspapers' front pages as well as radio and TV news that announced or commented on governmental plans and decisions and, as a consequence, most of the related political debates, were flooded with numbers and numerical indices, interpreted and commented upon according to the case, thus producing a specific numerical discourse. Such a situation may be considered as a characteristic example of the use of numbers in political manipulations, given that, as Alonso and Starr (1987, p. 3) point out, acts of social quantification are "politicized" not in the sense that the numbers they use are somehow corrupt—although they may be—but because "political judgments are implicit in the choice of what to measure, how to measure it, how often to measure it and how to present and interpret the results".

The study reported in this paper focuses on the headlines and attempts to analyze the numerical discourse which appeared on the front pages of widely circulated daily Greek newspapers, issued from late October 2009, when the fiscal crisis of the country was first announced by the government, until June 2012, when, after a general election, the implementation of the aforementioned policies seems to have been finally stabilized.

Mathematics and Politics

There can be no doubt that,

> There is a constitutive interrelationship between quantification and democratic government. Democratic power is calculated power, and numbers are intrinsic to the forms of justification that give legitimacy to political power in democracies. Democratic power is calculating

> power, and numbers are integral to the technologies that seek to give effect to democracy as a particular set of mechanisms of rule. Democratic power requires citizens who calculate about power, and numeracy and a numericized space of public discourse are essential for making up self-controlling democratic citizens. (Rose 1991, p. 675)

On the other hand, it has been claimed that the referential meanings assigned to mathematical constructs, being properly manipulated, do not merely inscribe a pre-existing real world situation but constitute it. Techniques of inscription in numerical formats and accumulation of facts about aspects of the "national economy", the "public debt", "tax incomes" or "labor salaries" render visible particular domains with a certain internal homogeneity and external boundaries. The collection, manipulation and presentation of numerical data participate in each case in the fabrication of a "locus" within which thought and action can occur. Numbers delineate "fictitious spaces" for the operation of governments, and establish a "plane of reality", marked out by a grid of norms, on which governments can operate according to the case (Miller and O'Leary 1987; Miller and Rose 1990; Rose 1988). At the same time, every such fabrication of real world situations, ostensibly endorsed by the objectivity and neutrality of mathematics and enhanced by the publicity power of media, prevails or actually is imposed as the unique representation of reality and finally as the reality itself (Skovsmose 2010).

The relation between mathematics and politics has been addressed from different theoretical perspectives. Foucault (1979), for instance, has analyzed the relation between government and knowledge by considering "governmentality", or the mentalities of government that characterize all contemporary modes of exercise of political power in Western democracies. From within this perspective, Skovsmose (1998, 2010) has further exemplified the role of mathematics in the constitution of political, calculative, practices. The link between numerical information and a politics of calculated administration of a population is another perspective emphasized, for example, by Pasquino (1978). Latour (1987) has analyzed how, in modern societies, events and processes are inscribed in standardized forms, which can be accumulated in a central locale where they can be aggregated, compared, compiled and calculated, and how, through the development of complex relays of inscription and accumulation of "data", new conduits of power are created between those who wish to exercise power and those over whom power is to be exercised.

The Context of the Study

The study reported in this paper focuses on the headlines which appeared on the front pages of the main daily Greek newspapers, in selected editions published during the announcement of the initially fiscal, then economic and finally social crisis in the country, as well as the related political actions taken by the dominant political forces to supposedly confront these crises.

The front pages of the newspapers and especially the headlines acquire a prominence through diffusion since they reach an audience considerably wider than those who read the articles. Many more people than those who buy a newspaper glance, if only fleetingly, at the front pages of papers displayed on fliers, read on public transport, etc., or presented by the daily morning shows of main TV channels. Thus, the casual reader is lead to conclude on the importance of particular issues which have been given prominence in this way. At the same time, headlines orient the reader's interpretation of subsequent 'facts' contained in the article.

Therefore, front pages are put in focus in our study, since they present the most prominent and marked information from the news discourse and make up a summary of the news report, which "strategically serves as the expression of its macrostructure" (Van Dijk 1988, p. 226). At the same time, headlines perform, as Richardson (2007) explains, a double function: a semantic function, regarding the referential text, and a pragmatic function regarding the reader to whom the text is addressed. It is therefore important to take into account the participants, processes, and circumstances within each headline in order to identify patterns of issue framing. Since readers often read and recall only headlines, editors and reporters construct not only preferred meanings of the news reports for the audience, but also the most prominent ideological view of those news reports (Van Dijk 1988; Van Dijk and Kintsch 1983). The impact of headlines on the reader is likely to be all the stronger because certain linguistic features of titles make them particularly memorable and effective, as are, for instance, the use of puns, emotive vocabulary and other rhetorical devices as well as the use of numbers.

Any analysis of headlines in the print media poses a number of questions in relation to the constitution of the corpus. In particular, the following have to be answered at the outset: Over what period should the headlines be collected? Which newspapers should be included and what criteria should be used in the choice of headlines? The answers to these questions were provided in our study by its context, which, as already mentioned, was created by the related political actions taken by the dominant political forces during a certain period of time, in order to define, describe and confront the financial, economic and social crisis that broke out in the country.

In terms of Fairclough's framework for critical discourse analysis (see below), the context is crucial in the analysis of the processes of production and reception of a discourse; he distinguishes a "situational context" shaped by questions about time and place and an "intertextual context" constituted by additional texts and information about or from producers and their product as central for the process of interpretation (Janks 1997, p. 338). In other words, the function of discourse is twofold: to reflect on and to create context.

On this ground, the context of our study has been delimited by the following factors: the time period starting with the disclosure of the initial fiscal crisis (October 2009) up to the enforcement of the political actions introduced by the dominant political forces to confront the finally multifaceted crisis (June 2012); the widely-circulated newspapers which supported or opposed these political decisions and actions; and their headlines which utilize a numerical discourse.

The following examples of front-pages are illustrative of our corpus of headlines (Figs. 1, 2, 3, and 4).

2/11/2009 **TA NEA**

The new taxes for 2010
**They are looking for 4.5 bn €
in five ... wallets**

| 1.5 bn € abolishing separate taxation | 1 bn € collection of outstanding debts | Up to 1 bn € taxes on real estate | 700 m € taxes on cigarettes, spirits, fuel | 500 m € levy on bank profits |

Fig. 1 2/11/2009 TA NEA

15/01/2010 **ELEFTHEROTYPIA**

Stability program A proof for EU and markets
At any cost cutbacks 11.5 bn€

23.6 bn collection in 3 years	incomes increase (m €)		expenditure cuts (m €)	
	Tax increases	1,500	Social allowances	650
	Spirits, cigarettes, mobiles	1,110	Recruitments, contract employees	200
	Tax evasion	1,200	Military expenditures	457
	Levies evasion	1,200	Hospital supplies	1,400
	Levy by businesses	1,050	Subsidy to pension funds	540
	Investments, banking	1,680	Overtime, operating costs of public services	435
	Total	7,740	Total	3,682

Fig. 2 15/01/2010 ELEFTHEROTYPIA

10/06/2011   H AVGI

| Recession in the quarter -5.5 % | Economy is sinking from the first quarter of 2011 At 3.35 the inflation in May | The generic drugs are also high-priced (90 %) |

Coup de grace

| 38.2 % of the revenues should go to creditors for interest in 2015 from current 29.5 % | 140 % of GNP the national dept in 2015 on the basis of favorable assessments | 570 euro the minimum monthly income on which special tax will be imposed |

Fig. 3   10/06/2011 H AVGI

17/02/2012   ETHNOS

10 % reduction in all pensions over 1,300 €

primary pensions: cutbacks on all pensions aiming to save 45 m € per year

supplementary pensions: cutbacks 10 % on pensions amounting 200-250 €, 15 % on 251-300 € and 20 % on amounts over 300 €

allowances to families with more than 3 children: the bar for the beneficiaries is lowered from 55,000 to 45,000 euro

Fig. 4   17/02/2012 ETHNOS

The Method of Analysis

The method adopted for analyzing the numerical discourse that appeared on the front pages of widely read newspapers draws on a version of critical discourse analysis introduced by Fairclough (1989, 1995, 2003).

As summarized by Rogers et al. (2005, p. 371),

> Fairclough's analytic framework is constituted by three levels of analysis: the text, the discursive practice, and the sociocultural practice. In other words, each discursive event has three dimensions: it is a spoken or written text, it is an instance of discourse practice involving the production and interpretation of texts, and it is a part of social practice. The analysis of the text involves the study of the language structures produced in a discursive event. An analysis of the discursive practice involves examining the production, consumption, and reproduction of the texts. The analysis of sociocultural practice includes an exploration of what is happening in a particular sociocultural framework.

Janks (1997) provides an illustration of the connections between the three dimensions (Fig. 5).

The first dimension of the framework is textual analysis, which includes the study of

> the different processes, or types of verbs, involved in the interaction. The interpersonal functions are the meanings of the social relations established between participants in the interaction. Analysis of this domain includes an analysis of the mood (whether a sentence is a statement, question, or declaration) and modality (the degree of assertiveness in the exchange). (Rogers et al. 2005, p. 371)

The main goal of this analysis "is to describe the relationships among certain texts, interactions, and social practices" (ibid.). Analysis at the textual level involves use of Halliday's systemic functional linguistics and the three types of meanings that are created by the interface between language and the social context: textual, interpersonal, and experiential. Textual meaning is related to how a text is organized as a

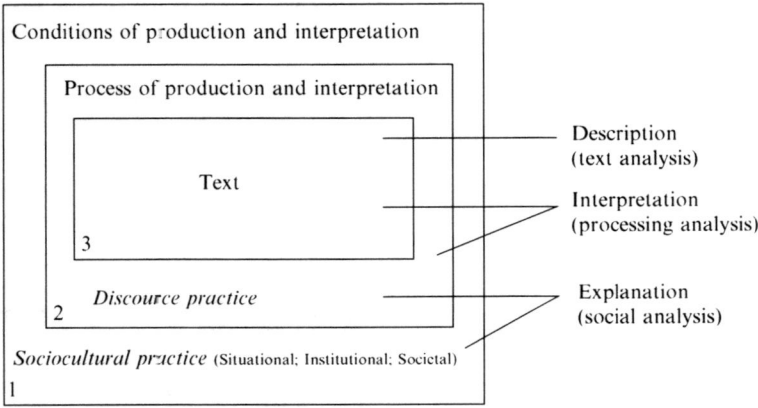

Fig. 5 Janks's illustration of Fairclough's discourse analysis (Janks 1997, p. 330)

coherent message; interpersonal meaning expresses the role of the relationships between participants; and experiential meaning deals with representing or constructing experience within language—the topic, subject matter, or content (Halliday 1975).

The second dimension, processing analysis, "involves analysis of the process of production, interpretation, distribution, and consumption. This dimension is concerned with how people interpret and reproduce or transform texts" (Rogers et al. 2005, p. 371) and the goal of analysis "is to interpret the configuration of discourse practices" (ibid.).

The third dimension, socio-cultural analysis, "is concerned with issues of power—power being a construct that is realized through interdiscursivity and hegemony. Analysis of this dimension includes exploration of the ways in which discourses operate in various domains of society" (ibid.) and is primarily aimed at using "description and interpretation to offer an explanation of why and how social practices are constituted, changed, and transformed in the ways that they are" (ibid.).

It must be underlined that critical discourse analysis in its various types has not gone without critiques. The three most common critiques, as summarized by Rogers et al.,

> are (a) that political and social ideologies are read into the data; (b) that there is an imbalance between social theory, on the one hand, and linguistic theory and method, on the other; and (c) that CDA [=critical discourse analysis] is often divorced from social contexts. (Rogers et al. 2005, p. 372)

These critiques and possibly many other concerns are part of the overall perspective in which we intend to place this study, which is a first attempt to investigate how numerical discourse is to be read.

The Numerical Discourse of Newspapers' Front-Pages

In the textual-analysis, we present issues regarding the choices and use of numbers in the newspapers' headlines, in order to investigate the context that the participants act in and the interaction between the participants, in our case between politicians, journalists and readers.

In our discursive practice analysis, we focus on the incorporation of numbers into sentences and in particular on the choice of words accompanying numbers, in order to examine the process of interpretation and reproduction. In the socio-cultural practice analysis we offer comments on macro issues related to the role of numerical discourse in political relations and its role on establishing social and political "realities".

The Choice of Numbers in the Headlines: Creating the Context

The main purpose of the newspapers that actively supported the governmental policies and decisions during the period under consideration was to persuade the public that the various consequences of the crisis that has broken out in the country have to be borne equally by all citizens regardless of their individual income and its sources, since they have all been involved through their various actions, behaviors and stances in the creation of national fiscal deficits and thus in the causation of the crisis. The Greek prime minister and leading persons of the government as well as the parliamentary opposition declared this thesis in a variety of verbal expressions.

Supporting this aim, the main numerical features incorporated in front-page headlines reveal a specific type and use of numbers. The numbers in each instance are whole numbers, decimal numbers, fractions, but mainly percentages. Whole numbers, mainly expressing monetary amounts, are in many instances abbreviated, thus require for their comprehension mental conversion and fluency, which many people may lack, e.g.: "4.5 bn new taxes on salaried and pensioners" (Vradyni, 21/11/2009).

The dominant numerical expression in the examined headlines is, as mentioned, percentages. Percentages, a seemingly simple and easily understandable numerical concept, are the main component in the creation and re-creation of a "reality" fabricated by newspapers. However, a percent does not express a quantity but a relation between two quantities; therefore, it is dependent on the change of the one or the other quantity or both and, at the same time, when it is written and re-written as a single numeral these quantities are not referred to at all. Percentages as relations have neither number properties nor may be subjected to number operations. Since they are numerical expressions but not numbers they are suitable for numerical alchemies and appropriate for political manipulations projecting—among other ideas—a false sense of equality, hence equity.

For example, the headline: "4 % reduction in salaries. *The Minister of Finance announced are cutback of 10 % in allowances and overtime pay of the public servants results in a loss of 4 % in their earnings*" (Vradyni, 19/12/2009),[1] uses percentages, emphasizing the equal treatment of employees regarding the reduction of their salaries but concealing the inequality of the amounts corresponding to each particular salary. This, presenting ostensibly and at the level of impressions the reduction in salaries by means of "equal percentages", obscures the inequality of differences that derives from the number operation of "subtraction" of amounts.

Whole numbers, decimals, fractions and percentages are, depending on the case, used to express:

- a situation of change, an increase or a decrease, e.g. "Breathing: 46 % lower deficit—Suffocation: 60,000 additional shops closed down" (Ta Nea, 13/07/2010) or "Reduction by 35 % in all salaries of employees in public enterprises" (Eleftherotypia, 25/11/2011);
- a part-whole relation in many cases expressed as a proportion, e.g. "Pensions: They lose three out of four" (Eleftherotypia, 28/06/2010), "The government plan for the employees in public enterprises: 60 % fired, 30 % change employment relationship, 10 % re-recruited" (Eleftherotypia, 09/09/2011) or "*Acknowledgment*

---

[1] In this and the following headlines reported in this paper the main headings are written in normal letters and the accompanying in italics.

of the Minister of Finance: *The application of the memorandum results in loosing 1 out of 4 jobs and,* In every 100€ they take away 30" (Eleftherotypia, 21/12/2011);
- a comparison involving one or more quantities, "Storm of measures, Less deficit at 8 bn euro in 2010" (To Vima, 15/12/2009).

According to cognitive analyses of mathematical problems, "change", "part-whole", "combination" and "comparison" are semantic structures that may be considered as describing many of the quantitative situations encountered in everyday life (e.g., Resnick 1992) and newspapers' editors capitalize on all these numerical structures. It is, however, interesting that in many cases the situations of change and comparison are expressed by positive or negative numbers and even positive or negative percentages, thus obscuring their comprehension.

The Choice of Numerical Genre: Interacting with the Readers

Genre is a typified rhetorical way of recognizing, responding to, acting meaningfully and consequentially within, and thus participating in the reproduction of, recurring situations. Genres both organize and generate kinds of texts and social actions, in complex, dynamic relation to one another. Genre is understood in our study as an ideologically active and historically changing force in the production and reception of texts, meanings, and social actions or as put by Bazerman (1997, p. 19) as "frames for social (…) locations within which meaning is constructed (…) [which] shape the thoughts we form and the communications by which we interact". On the front-pages of the newspapers studied here, numbers are included in different linguistic genres, the most common being declaration (primarily conveying information) and argument (primarily seeking to persuade), each one used in order to serve a different purpose according to the political interests promoted.

The editors organize their reports in headlines around numbers. Although the information provided may sound real to the reader, its textual structure is not clear because the numbers used are either not clearly comprehensible or may be misunderstood by most readers due to particular number properties and references which may be unfamiliar to them. In any case, however, the use of numbers gives a sense of objectivity to declarations and arguments, independent of the person who writes or reads them. As Porter (1995) has explained, the language of mathematics is well suited to embody objective judgments and it is adopted when claims to knowledge need to gain trust and credibility beyond the bounds of locality and society.

When readers engage in the reading of texts composed on the basis of numerical data, it appears to them that such texts are reporting factual "realities" or actual problems to be solved. However, they may feel overwhelmed and thus, although they may be possibly paying attention to the news, they become in fact inactive against the declared or argued situation. The editors strategically decrease the interaction between the two sides of participants, in our case, between politicians and readers. Numerical discourse seems to be an effective medium in defusing public resentment.

The Choice of Sentences: Influencing the Reader's Interpretation

The sentences always serve the main purpose of the text. Sentences composed around numbers acquire meaning by the referents of the numbers, the quantitative expressions related to the numbers and the modifiers of numbers (Chassapis 1997). The quantities expressed by the number on each instance are found to denote measures of collections or sequences of discrete instances, as, for instance, are money amounts or population groups. However, the referents of the numbers reported are in some cases individuals and in other cases population or employee groups. In the second case, the numbers reported in order to be made sense of must be traced back to individual cases, a calculation usually unfeasible due to lack of data.

For example, "1,000 days austerity, *Taxes:* elimination of tax exemptions *1.1 bn €*, real estate tax *400 m €*, taxes on cigarettes and spirits *710 m €*, tax on corporate profits 870 *m €*. Cuts: allowances of public servants – 650 *m €*, new recruitment – 80 *m €*, personnel reduction – 120 *m €*, reduction of subsidies – 540 *m €*," (Ta Nea, 15/01/2010). However, "Reduction up to 400 € in the salaries of the public servants, *Monthly allowances of a servant* at the Ministry of Finance 1,500 €, reduction 25 %, loss of pay 375 €" (Ta Nea, 21/01/2010) or "Up to 750 € loss of pay for public servants, *The protagonists of salary losses, the salary falls for military personnel, teaching staff, juridical officials and medical personnel*" (Vradyni, 24/01/2010).

Quantitative expressions related to a number are, depending on the case, verbal, e.g. "*New taxes for 2010:* They look for 4.5 bn € in five … wallets" (Ta Nea, 02/11/2009), "120 bn for three hard years" (Kathimerini, 29/04/2010) or numerical, e.g. "*Pensions reform:* Working 40 years for a pension amounting to … 35 years" (Eleftherotypia, 11/05/2010), "*The pension has been locked: 40 years work – 60 years age*" (Ta Nea, 02/06/2010).

Furthermore, it is interesting that the target number and the quantitative expressions to which it is related denote in many cases different referents, e.g. "*Pensions reform:* At 65 years and −30 % all pensions" (Eleftherotypia, 06/06/2010), thus forming a sentence rather difficult to grasp at a glance, in contrast to other cases with the same referent which is instantly comprehensible, e.g. "*The pension has been locked: 40 years work – 60 years of age*" (Ta Nea, 02/06/2010).

The Choice of Visual Displays: Reproducing the Numbers

Apart from the text, a variety of visual displays adds an elaborate visual dimension to the numerical expressions which influence the reader's interpretation. The visual displays that the newspapers under consideration utilize (coloring and/or highlighting the numbers) create a referential field for the numerical discourse in use which, as Potter et al. (1991, p. 343) have put it, function as "parallel commentaries" which reinforce textual numerical expressions. The following simple and complex depictions are illustrative (Fig. 6).

Recession	*Draft Budget for 2010*	*New taxes for 2010:*
in the quarter	aiming at lower deficit	They are looking for 4.5 bn €
-5.5 %	*Budget 2009-2010*	in five ... wallets
Avgi,	Eleftherotypia, 02/11/2009	Ta Nea, 02/11/2009
10/06/2011		

Fig. 6 Avgi, 10/06/2011; Eleftherotypia, 02/11/2009; Ta Nea, 02/11/2009

Moreover, the meaning assigned to these visual displays will influence the readers even before they begin to read the text, thus orient their interpretation of subsequent "facts" reported by the numbers. Numerical "facts" are produced and reproduced in the form of visual displays plus texts.

The Participants in the Newspapers' Front-Pages: Filtering Access to Numerical Discourse

In most newspaper headlines only the editors as language users or communicators have freedom in the use of special discourse genres or styles, or in the participation in specific communicative events and contexts, particularly in the cases where they include numerical information. In other words, many readers as participants in the communicative events based on numerical discourses are not in a symmetrical relationship with the other participant—the writer. For example, in the following headlines announcing the reform of the pension scheme, it is obvious that the author-editor is strongly against the increase of the retirement age: "*Pensions – Rollovers for 700,000 workers:* Five years 'penalty' to the heavy and arduous jobs *– From 55 to 60 (years old) the retirement with 35 (years* length of service*) – Twelve months per year from 2011 the rising of the retirement age*" (Ta Nea, 11/06/2010).

However, when reading these headlines, the readers are probably accustomed to the genre and the editor's strong voice in the text and this could be their only contribution to that communicative event, even if they disagree with the editor's view. Although this is one kind of communicative event, the reader may not freely communicate with the writer to express his/her opinion and may not change the discourse if they do not agree with it. Moreover, many people cannot understand and assign a meaning to the numerical data reported. That is, the use of numerical discourse determines that not everyone can have full access to it.

The Power Relations Behind the Headlines: Speaking for Whom to Whom

When we go through the front-pages of the newspapers under consideration, we find that the authors are trying to report the power struggle between the political forces in power, attempting to encounter and to overcome the causes and effects of the socio-economic crisis, against interest groups, ideologies and political parties opposing their approaches and policies.

In this struggle the numbers used are intentionally selected from the pool of available numerical information and properly reported in order to guide their interpretation.

The following examples are indicative:

- "*EU Commission's letter suggesting a shocking recipe* – Brussels is asking for a reduction of salaries by 7 %" (Ta Nea, 07/01/2010),
- "*A dramatic announcement by the Prime Minister* – A rescue package with painful measures: 0 % salaries in public sector are frozen, up to 20 % reduction in employees' allowances, 15 % increased taxes on fuel, 67 years of age the new limit for retirement" (Ta Nea, 03/02/2010).
- "*EU and IMF are uncompromised in shocking measures in order to provide the loan – The* package of un…happiness – Private sector: abolishment of 13th and 14th monthly salary and pension, freezing of salary increases, allowable personnel layoffs 4 % per month, the compensation for layoffs is questioned, compulsory arbitration between employers and employees is abolished – Public sector: abolishment of 13th and 14th monthly salary and pension – Other measures: Increase of VAT from 21 to 23 %, special tax 10–11 % on electricity bills. Cut offs: half of the social solidarity allowance to people with very low income, 1.2 bn euro from military armaments, grants to hospitals and social security funds" (Eleftherotypia, 30/04/2010).

In any case the government and the political forces supporting it are presented as exerting more power over its opponents, at least at the level of intentions. As a result of this, although many readers do not agree with the vigorously enforced approaches and policies, they accept them as inevitable.

On the other hand, the newspapers supporting political opposition assign quite different meanings to the same numerical data, not actually questioning the numerical "realities" created by these same numbers, which are accepted as neutral. For instance, a newspaper opposing the government interprets in its headline the announcement of the Minister of Finance for a cutback of public workers' salaries as "Cuts 10 % to the salaries of public workers, *Coordinated measures reduce by 10 % the salaries in the public sector and push for equal reductions in the private sector*" (Avgi, 19/12/2009). On the contrary, a newspaper which optionally supports the governmental policies reports on its same day headlines "4 % reduction in salaries, *The by the Minister of Finance announced cutback of 10 % to allowances and overtime pay of the public workers results in a loss of 4 % in their earnings*" (Vradyni, 19/12/2009).

Concluding Comments

In situations of economic and social crises, in which a plurality of political forces, interest groups and views are contesting, numerical discourse used by the media may produce a public rhetoric of interest or disinterest. Such rhetoric may play a crucial role in the creation of a public sphere where technical expertise dominates political debate excluding people not only from political debates but also from acting towards or reacting against political decisions and policies. This "technicization of politics", as called by Miller and Rose (2008, p. 77), emerges from the transformation of economic and social problems in problems demanding for their solutions technical manipulations of numerical data and thus the imperative of advanced technical expertise. In this account, numbers are not just "used" in politics, they help to configure the respective boundaries of the political and the technical and, furthermore, they are involved in establishing what it is for a decision to be based on interested or disinterest.

In these processes of technicization of politics, newspapers and media in general function as an important means for the production and dissemination of "truth", including, excluding or diverting, processing and representing, and therefore contributing to a large extent to the formation of a society's "general politics of truth": the appropriate political technologies of truth production and reproduction, the expressions of truth which are deemed acceptable or not, the acceptable mechanisms of judging true and false statements, the sanctioning of statements, and the valorization of claim-makers as truth-tellers (Foucault 1980). The particular role played by newspapers in the politics of truth concerns the production of "apparatuses of truth" which are discursive practices that are marked by "rules of formation" that define "concepts, procedures and objects", "limits and forms of the sayable" and "criteria of transformation" that condition discursive performance and change, and "criteria of correlation" that situate discourses among other discourses and non-discursive institutions (Foucault 1980, 1991).

In this context, the governing and dominant political forces and interests in a society organize the truth so that their versions of reality gain credibility over others and, as Rouse (1994, p. 112) remarks, "to make truth-claims is to try to strengthen some epistemic alignments, and to challenge, undermine, or evade others".

Of course, a complete and full account of the highly complex relationships between power, truth and news reports is outside the scope of an essay this size. Yet even the analysis undertaken here, of a limited sample of headlines, demonstrates the fact that numerical discourse, as utilized in the front-pages of newspapers published in Greece during the years when the economic and social crisis broke out, played a crucial role in manufacturing a certain "regime of truth", which served specific political views and policies as well as specific interest groups.

On the basis of the above conclusion, there emerge difficult questions and new challenges for critical mathematics educators, at least in Greece, who now have to face the following dilemma: they must teach the mathematical meanings of numerical concepts, e.g. percentage, and at the same time they have to select some of the referential meanings of these concepts and reject others, all of which are widely promoted and everywhere visible in public life, today.

References

Alonso, W., & Starr, P. (Eds.). (1987). *The politics of numbers*. New York: Russell Sage.
Bazerman, C. (1997). The life of genre, the life in the classroom. In W. Bishop & H. Ostrum (Eds.), *Genre and writing* (pp. 19–26). Portsmouth: Boynton/Cook.
Chassapis, D. (1997). The social ideologies of school mathematics applications: A case study of elementary school textbooks. *For the Learning of Mathematics, 17*(3), 24–26.
Fairclough, N. (1989). *Language and power*. London: Longman.
Fairclough, N. (1995). *Critical discourse analysis: The critical study of language*. New York: Longman.
Fairclough, N. (2003). *Analyzing discourse: Textual analysis for social research*. New York: Routledge.
Foucault, M. (1979). On governmentality. *Ideology and Consciousness, 6*(1), 5–22.
Foucault, M. (1980). Truth and power. In C. Gordon (Ed.), *Michel Foucault, power/knowledge: Selected interviews and other writings 1972–1977* (pp. 109–133). New York: Pantheon.
Foucault, M. (1991). The discourse on power. In C. Kraus & S. Lotringer (Eds.), *Michel Foucault: Remarks on Marx. Conversations with Duccio Trombadori* (pp. 147–187). New York: Semiotexte.
Halliday, M. A. K. (1975). *Learning how to mean: Explorations in the development of language*. London: Edward Arnold.
Janks, H. (1997). Critical discourse analysis as a research tool. *Discourse: Studies in the Cultural Politics of Education, 18*(3), 329–342.
Latour, B. (1987). *Science in action*. Milton Keynes: Open University Press.
Miller, P., & O'Leary, T. (1987). Accounting and the construction of the governable person. *Accounting, Organizations and Society, 12*(3), 235–265.
Miller, P., & Rose, N. (1990). Governing economic life. *Economy and Society, 19*, 1–31.
Miller, P., & Rose, N. (2008). *Governing the present: Administering economic, social and personal life*. Cambridge: Polity Press.
Pasquino, P. (1978). Theatrum politicum. The genealogy of capital: Police and the state of prosperity. *Ideology and Consciousness, 4*, 41–54.
Porter, T. M. (1995). *Trust in numbers: The pursuit of objectivity in science and public life*. Princeton: Princeton University Press.
Potter, J., Wetherell, M., & Chitty, A. (1991). Quantification rhetoric: Cancer on television. *Discourse and Society, 2*(3), 333–365.
Resnick, L. B. (1992). From protoquantities to operators: Building mathematical competence on a foundation of everyday knowledge. In G. Leinhardt, R. Putnam, & R. A. Hattrup (Eds.), *Analysis of arithmetic for mathematics teaching* (pp. 373–429). Hillsdale: Lawrence Erlbaum.
Richardson, J. E. (2007). *Analysing newspapers: An approach from critical discourse analysis*. New York: Palgrave Macmillan.
Rogers, R., Malancharuvil-Berkes, E., Mosley, M., Hui, D., & O'Garro Joseph, G. (2005). Critical discourse analysis in education: A review of the literature. *Review of Educational Research, 75*(3), 365–416.
Rose, N. (1988). Calculable minds and manageable individuals. *History of the Human Sciences, 1*, 179–200.

Rose, N. (1991). Governing by numbers: Figuring our democracy. *Accounting Organization and Society, 16*(7), 673–692.

Rouse, J. (1994). Power/knowledge. In G. Gutting (Ed.), *The Cambridge companion to Foucault* (pp. 92–114). Cambridge: Cambridge University Press.

Skovsmose, O. (1998). Linking mathematics education and democracy: Citizenship, mathematics archeology, mathemacy and deliberative interaction. *Zentralblatt für Didaktik der Mathematik, 30*(4), 195–203.

Skovsmose, O. (2010). Can facts be fabricated through mathematics? *Philosophy of Mathematics Education Journal, 25*, Online.

Van Dijk, T. A. (1988). *News as discourse*. Hillsdale: Lawrence Erlbaum.

Van Dijk, T. A., & Kintsch, W. (1983). *Strategies of discourse comprehension*. New York: Academic.

On the Role of Inconceivable Magnitude Estimation Problems to Improve Critical Thinking

Lluís Albarracín and Núria Gorgorió

Abstract In this chapter, we introduce inconceivable magnitude estimation problems as a subgroup of Fermi problems. The problems we use in our study require counting the amount of people in different situations. Based on the experience of a classroom activity carried out with 15-year-old students, we describe the process they went through to solve the problems, and discuss in which ways these problems provide knowledge to critically analyse the information that appears in the media.

Introduction

Diamond and Plattner (2006, p. 168) state that "democracy in its purest or most ideal form would be a society in which all adult citizens have an equal say in the decisions that affect their lives". In order to form their own opinion, which they need to make decisions, citizens need to understand their environment and be able to interpret a wide range of information. It would be desirable for the educational system to provide future adult citizens with interpretative tools which give them the appropriate interpretative skills to practise their rights and duties. Educational curricula have improved along these lines in recent years. However, activities that combine the usual mathematical content with global competence-related aspects such as decision-making are still rare.

It is necessary to have a wide range of activities to work on different aspects of critical mathematics in the classroom before the students reach the stage of having to take adult decisions. We regard compulsory secondary school (in Spain, the schooling between 12 and 16 years of age) as a suitable period in which to introduce realistic context activities to make sense of the real world and use mathematics critically. The contexts used in the classroom activities should be realistic in order to improve decision-making skills effectively. Moreover, increased realism will provide the activities with higher mathematical complexity. Yet, the reality to be

---

L. Albarracín (✉) • N. Gorgorió
Facultat de Ciències de l'Educació, Universitat Autònoma de Barcelona, Barcelona, Spain
e-mail: lluis.albarracin@uab.cat; nuria.gorgorio@uab.cat

© Springer International Publishing Switzerland 2015
U. Gellert et al. (eds.), *Educational Paths to Mathematics*,
Advances in Mathematics Education, DOI 10.1007/978-3-319-15410-7_17

worked on should be evaluated carefully as it must be ensured that the secondary school students are able to simplify the suggested situations without losing their essence.

For this reason we suggest Inconceivable Magnitude Estimation Problems (IMEP) as a means for improving critical thinking in secondary classrooms. IMEP confront the student with a situation that requires estimating the value of a considerably large real magnitude, well outside the range of their normal daily experience. By working on these activities, students will find that they need to confer meaning to values which they are not used to working with but that could be relevant in different contexts. These problems can be considered as a subgroup of Fermi problems, since they are solved in the same way. They also allow for different approaches to their resolution.

In this chapter, we present a classroom experience introducing some IMEP and discuss how critical thinking of 15-year-old students can be initiated didactically. The aim of this experience is for the students to estimate the number of people attending a demonstration and, in this way, develop a critical view on the information given by the media. During the activity different problems with real-life contexts are used, for which the students are asked to estimate amounts of objects on a surface.

According to Van den Heuvel-Panhuizen (2005), presenting a real context for problems can make them more accessible. Problems which involve a realistic context help to begin teaching mathematics within the realm of the concrete and then move on to the more abstract. Solving problems with a real context involves the construction of a mathematical model, the calculation of the solution, and transferring the result back to the real situation. Our experience has shown that the students elaborate their own methods to try to solve the IMEP they are confronted with, and that they introduce elements from the mathematical modelling processes of the situations dealt with. In this way they increase their confidence in their own methods and the solutions they come up with when initially working on problems that can be solved in their everyday environment. The results they obtain themselves enable the students to think about the validity of information given by the press or on different websites. Therefore, the work done in the classroom gives them the possibility to critically analyse published content and have their own views on certain events. Thus, the mathematical activity is centred on decision-making processes, since it allows us to understand and deal with the information the students obtain from their environment in problematic situations.

Real-Life Problems, Context and Authenticity

Mathematical problems posed in written form refer to concepts related to a particular context. This context may either be purely mathematical, such as in the classical problem of adding up the first hundred natural numbers; or it may have a real-life context, in the sense of being directly associated with a particular situation in the real world; or finally, it may have an imaginary context. Various studies on problem-solving in non-mathematical contexts have been conducted. Some of them are centred on understanding the way people solve problems in their work environment and focus on the differences in the use of mathematics in the classroom and at work (Jurdak and Shahin 1999, 2001). Other studies relate the use of mathematics in everyday environments to its use in the classroom (Jurdak 2006; Nunes et al. 1993). This research reveals that there is a significant distance separating the mathematics taught at school and the mathematics used in real life.

A considerable feature of problems based on real-life is their level of authenticity. Palm (2008) describes authenticity of coursework as the extent to which a task can be moved to a real-life situation. This works under the condition that the most important aspects of the situation are simulated in a highly realistic way. He presents a study aimed at determining the influence of the authenticity of a formulation of a proposed problem on the answers given by students. Palm shows that students who answer questions with a higher level of authenticity use knowledge from their everyday life and obtain more exact answers which are more consistent with reality than students who work with problems with a lower level of authenticity. Palm (2006) suggests various elements which should be included in real-life problems: He focuses his attention on the type of events which frame the problem, on the question, the information it contains, the type of formulation, the resolution strategies, the circumstances and determinants in the classroom, the prerequisites the answer should comply with, and the aim of the problem. Stocker (2006) adds the need to focus on how relevant the problems are to the students and the change-creating potential of the problem with the purpose of improving our environment.

Freudenthal (1983) states that discussing problems with real-life contexts in the classroom may be very enriching for the students. These problems offer various ways, which range from more abstract to more contextual, to use mathematical concepts thus avoiding to preproperate mathematical generalisation. However, Chapman (2006) observes that most teachers present real-life problems in a close-minded manner, so that the focus is on "de-contextualization in a mechanical way by fragmenting the context into a collection of words, phrases and sentences to be translated to mathematical representations" (p. 225). She suggests that the resolution should be centred on the meaning of the problem, with context-sensitive and particular explanations of the problem as a key point. Verschaffel (2002) states that the aim of introducing problems with written text and real-life contexts is approaching the reality of the mathematics classroom and creating opportunities to practise different aspects of problem solving without the drawbacks of direct contact with the real-life situation. Furthermore, the everyday context of problems may suggest different approaches to a resolution, and may also highlight aspects of the problem which might have been overlooked in a formulation lacking context (Arcavi 2002). Experiments from a critical mathematics education perspective on the use of mathematics show that the mathematical analysis of real situations allows the students to appreciate reality from a different viewpoint. An example is Camelo et al. (2010) in which the students conclude that the public media does not allow for adequate analyses of nutritional issues because of blurred boundaries of publicity and report.

According to Winter (1994), solving problems with real-life contexts involves mathematizing a non-mathematical situation, which implies constructing a mathematical model that respects the real-life situation as well as calculating a solution and transferring it from the model back to the real-life context. The most difficult step in this process is determining an appropriate model for the formulated real-life situation, since good knowledge of the context is required as well as of the involved mathematical concepts and a high level of creativity. Creating models to solve problems is not exclusively a resource of the higher mathematical levels; its use has been recorded in studies on mathematical production of students at different educational levels (English 2006; Esteley et al. 2010; Lesh and Harel 2003). Lesh and Harel (2003) define the concept of the mathematical model as a conceptual system designed to describe or explain mathematical objects, relations, patterns or regularities associated to solving problems. Therefore, modelling activities involve students in identifying the essential aspects of a studied reality and in creating an adequate representation with the mathematical tools available. Some real-life situations may be highly complex and the difficulty of the questions they raise should be previously assessed and adapted to the students' level of mathematical knowledge and comprehension of reality.

Estimation and Fermi Problems

Estimation involves making a judgement on the value of the result of a numerical operation or of the measurement of a quantity. We wish to obtain a quick result, specific to a certain context, in a simple way. An estimation is required when questions are posed such as: How long would I take to get to the metro station? Would a kilogram of potatoes be enough to feed eight people with this recipe? Or, would all of these clothes fit in the small suitcase or should I use a larger one? Hogan and Brezinski (2003) consider three areas of estimation: numerosity, measurement estimation and computational estimation. Numerosity refers to the ability to visually estimate the number of objects arranged on a plane. Measurement estimation is based on the perceptive ability to estimate lengths, surface areas, time, weight or similar measurements of common objects. Computational estimation refers to the process by which the value of calculations is attained, such as $2.7 + 4.4/2.5$. In this chapter we describe a sequence of activities in which students have to estimate large amounts of objects distributed on a flat surface. Each of the questions (problems) formulated has a specific context the students need to analyse and for which they should come up with a mathematical model. During this process the students choose the key aspects of each problem.

Following Ärlebäck (2009, p. 331), Fermi problems are "open, non-standard problems requiring the students to make assumptions about the problem situation and estimate relevant quantities before engaging in, often, simple calculations". What characterizes a Fermi problem is the possibility to break it up into smaller problems and to solve these separately, often by estimating. The most classical example of a Fermi problem involves estimating the number of piano tuners in Chicago. Based on estimating input data, such as the total population of Chicago, the proportion of families in Chicago that may own a piano or the amount of time needed to tune a piano, the number of piano tuners needed in the city can be estimated from the total hours of work needed in piano tuning (for more information and details, see Efthimiou and Llewellyn 2007). An essential characteristic of a Fermi problem is that it does not present all the necessary information that a solution would require. In addition, Fermi problems are resistant to simplistic transfers of mathematical methods used in other contexts. Ärlebäck (2011) states that working with Fermi problems may be useful for introducing modeling to the classroom. He characterizes them by:

> (1) their accessibility ... any specific pre-mathematical knowledge is not required to provide an answer; (2) their clear real-world connection ...; (3) the need to specify and structure the relevant information and relationships to be able to tackle the problem; ... (4) the absence of numerical data, that is the need to make reasonable estimates of relevant quantities; and (5) their inner momentum to promote discussion. (pp. 1011–1012).

Some examples of Fermi problems may be: How many jelly beans fill a one-litre bottle? How many cubic metres of biomass can be produced from fire debris? How long does it take to get from New York to Boston by bicycle? What fraction of the area of your city is covered by automobiles? A wide variety of Fermi problems of different difficulty levels can be found in online problem listings or otherwise in books such as Guesstimation (Weinstein and Adam 2008). Fermi problems might seem anecdotal and hardly relevant, but some of them may be directly connected to aspects of citizenship which are rarely addressed in the classroom and which could attract the students' interest. Educating the students to be critical thinkers is one of our educational goals. If we consider the citizens' understanding of political aspects which involve large numbers (e.g. the value of public investments in education, research or defense, or the number of civil servants in a territory such as doctors, policemen or teachers) we may encounter interpretational difficulties of certain issues. Sometimes it is the media that provide data which is hard to interpret, such as figures of water availability or the number of people who attended a demonstration. Estimation of large quantities is precisely the type of Fermi problems we used in our research project.

Inconceivable Magnitude Estimation Problems

Our classroom experience focuses on problems that are based on magnitudes we cannot perceptually estimate without some training, as well as magnitudes that can be imagined, but the value of which is difficult to interpret. If we think of magnitudes we are familiar with and that we have given meaning to (the size of a table, the amount of time passed during a film, or the number of people in a classroom), we can assert that they are familiar and conceivable. Some examples of magnitudes which are commonly considered inconceivable in this sense by most people may be the number of medical doctors in a state—which may not be enough to attend to the medical needs of the population—, the number of cars that pass by a certain point in a city—that may cause traffic problems or air pollution—or the number of persons in a demonstration. The specific case of demonstration attendance is especially interesting in regard to the present political situation in Catalonia, which is currently an autonomous region within the structure of the Spanish state. Actually, it was one of the territories that claimed this status of self-governance upon the arrival of democracy. The newspapers stated that more than a million people participated in a pacifist demonstration on the streets of Barcelona on September 11, 1977, in which different social classes took part and brought entire families to the streets to express their political discontent. In the year 1979 the claimed statute of autonomy was approved. This has not been the only large demonstration which has taken place in Catalonia in the last few decades. According to the newspapers, on February 15, 2003, around a million and a half citizens got together to ask the Spanish state not to send its troops to participate in the invasion of Iraq that year. In recent times political issues have triggered pacifist demonstrations, evidencing the population's discontent and claim for solutions. The Spanish constitutional court released the text of a new statute of autonomy for Catalonia and on July 10, 2010, Catalans again took to the streets to claim what they had previously approved in a referendum. Massive demonstrations have occurred lately on the national day of Catalonia (September 11) in both 2012 and 2013, asking for Catalonia to become a political nation state. In all these demonstrations, the most relevant aspect for evaluating their impact are participation figures. The press offers data without sharing the methods used to receive these numbers and controversy arises in order to maximise or minimise the political impact of demonstrations.

The set of conceivable magnitudes can be different for everyone, being relative to the experience and knowledge construction of each individual. For this reason, we find it appropriate to introduce this type of problem and classroom activities, which may allow all students to broaden the range of magnitude values that are significant to them. The examples dealt with involve large amounts which are hard to visualize but which are relevant to the public opinion. They represent values which are not usually approached and for which it is difficult to confer a meaning that would make sense to the students. Taking these ideas as a starting point, we define an inconceivable magnitude as a physical or abstract magnitude which is beyond our ability to interpret and for which we have not created any meaning (Albarracín 2011; Albarracín and Gorgorió 2014). Once we attempt to determine the value associated with an inconceivable magnitude, we must by definition work with approximate values, given the difficulty of obtaining accurate information for this kind of values. The most natural way of obtaining values for inconceivable magnitudes is to come to an estimation mainly through reasoning and including some calculations. The process by which these values are approximated can be included in activities oriented towards problem-solving and can be worked on in class (Ärlebäck 2009). In previous studies, we have proved that secondary-school students are able to create suitable resolution proposals for estimating these amounts

(Albarracín and Gorgorió 2014) and that they use relevant information from the context. This allows the students to model the situation (Albarracín and Gorgorió 2012).

Our aim is for the students to see the necessity of focusing on the essential components of the given situation while they solve inconceivable magnitude estimation problems and create meaning. One of the essential aspects is to know how to separate the key information from that which is not key and have procedures or methods at hand to extract the necessary data for solving the problem. In many occasions the necessary data is not within the students' reach. Some data is not made public due to private interests, at other times reliable records or counts are not available and therefore educated guesses are required. In order to make these estimations with a reasonable degree of reliability, a well-developed number sense is necessary. Giving students the chance to have this experience is one of the aims of the project.

Designing the Classroom Experience

We work with Fermi problems as a didactic sequence developed over several sessions. Since the problems are open and are not linked to any specific mathematical content, we consider it essential for the students to work in groups. We place the problems in a context that is familiar to the students, so that they can later on transfer the methods and models used to other problems. The Fermi problems which make up each of these sequences require estimating different quantities which are related to each other and are set in real-life contexts and therefore have a high degree of authenticity. These activities appear to give the students an integrated view on mathematics, allowing them to acknowledge its utility in solving problems in their proximity. One of the difficulties when dealing with large numbers is the fact that we find it hard to create a mental image of the amounts represented. For this reason, the first problem students deal with involves elements set in a context which is familiar to them. Such a setting could be the school itself, the environment or an element related to their family home. From the knowledge acquired in solving this first problem, the students will then be able to transfer the methods used and results obtained to other problems concerning less familiar situations. When learning problem-solving skills, it is crucial to approach a problem by developing an action plan. For this reason, the first exercise is the creation of an individual plan of action. After that, the students should discuss their action plans in teams and agree on one action plan, thus ensuring that each of the students has had enough time to reflect upon the problem by him/herself before getting into collective discussions. Teamwork and idea-sharing allows a larger amount of students to use valid action plans. Our experience has confirmed that not all students develop suitable action plans for the problems formulated, however, some do come up with proposals leading to a satisfactory resolution process (Albarracín and Gorgorió 2014). We assume that by working in teams of students with different learning styles they will be able to share ideas and thus develop more elaborate plans, yielding solutions suitable for the context of the problem.

The formulations of the employed problems are presented as follows. The first problem can be solved experimentally with data collection from the same educational centre.

A. How many people would there be room for in the high school playground if a concert was to be held there?

The rest of the problems we subsequently work on cannot be solved straightaway in the educational centre and require reconsidering and adapting the types of resolution used in problem A:

B1. How many trees are in New York's Central Park?
B2. How many people can stand in the Palau Sant Jordi event hall when attending a concert?
B3. How many people can there be in the city council square of Sabadell attending a demonstration?
B4. How many people can there be in Plaça Catalunya attending a demonstration?

Since problem A is set in an environment which is familiar to them, the students are allowed to carry out fieldwork in order to obtain all the information required. Deciding on the most relevant aspects of the resolution process and collecting the required data are exercises which connect mathematics with the students' reality. Once the students have applied their resolution proposal, gathered their data and reached a solution, they will be asked to write up a short report. Each of the teams should provide their own report, specifying the method used to solve the first problem.

Once accomplished the resolution of the first problem, we shared ideas in a feedback session in order to explain each of the teams' proposals, the difficulties encountered, the methods and models applied, their limitations and final results. In the discussion of methods, we follow Chapman (2006, p. 228) who suggests we should "allow students to use both formal and informal knowledge to make sense of the problem and the solution and communicate these understandings in class discussions". The discussion of methods and models brings ways of approaching the problem which are different to those used initially by the students and that could be used in future problems. After discussing problem A, we handed out the list of questions including four problems with mathematical formulations equivalent to that of the first problem but set in different contexts (B1–B4). In fact, we recommend posing problems set in contexts which are out of the students' reach. As they cannot access the context directly, students have to develop their own methods and use communication resources such as the internet to acquire information about the context. The students will then work on these problems in the same teams as before and a feedback session will take place again when finished. This time, however, we will take advantage of the internet to try and find answers to the questions posed and compare them critically to the students' solutions. We finally propose a problem C to the students in a different format. Instead of asking a question we give a set of contradictory data about participation at a demonstration of great political and social relevance. The objective of this is for the students to develop their own opinion of the number of assembled people:

C. On July 10, 2010, the streets of the centre of Barcelona were filled with people to participate in a mass demonstration. The organizational entity, Òmnium Cultural, stated that 1,500,000 people attended (La Vanguàrdia, July 11, 2010, p. 1), the Barcelona Traffic Police gave a figure of 1,100,000 and the firm Lynce, which specializes in people counting, gave a figure of 62,000.

In all cases, the students were asked to use their own methods and not to look up the information straightaway until they had obtained their own results, in agreement with all the members of their team. The results obtained could then be used to validate the information gathered from newspapers or websites, which possibly offer contradictory information without stating its origin.

The Classroom Experience

This activity was carried out with 21 students in a secondary school, lasting several sessions. Working on problem A—How many people would there be room for in the high school playground if a concert was to be held there?—a first model was developed by a group of students. It uses *population density* as a way to represent the distribution of people in the school playground. The model is based on the idea that the people who fill the playground will be distributed in a uniform way across its surface. The method employed by the students consists of experimentally determining the number of people who fit in a square metre, in order to take it as a reference point for all the accessible surface of the playground. Figure 1 shows a group of six students located within a bounded area of 1 m^2. Afterwards, the students took the necessary measurements to calculate the total surface of the rectangular playground. For obtaining a result, they finally multiplied the area of the playground with the density of people per square metre they determined as most appropriate.

The second type of model used was *iteration of a reference point* (Carter 1986). A reference point is a unit established as the basis of an estimation. The iteration of a reference point is a strategy which corresponds to measurement estimations, such as the use of the palm of your hand as a unit for the length of a table to approximate its value in centimetres. In the students' case, the model associated to the representation of the situation is based on considering that each of the people gathered in the playground will take up the room of one person standing up in a congregation. In order to obtain the area occupied by one person, the students marked the area and measured it. They calculated the total surface area of the playground and divided it by the total area filled by a standing person to obtain an estimate of the number of people that would fit in the playground.

The third estimation model proposed by the students is *grid distribution*. In this case, the model is based on the assumption that the amount of people fitting in the

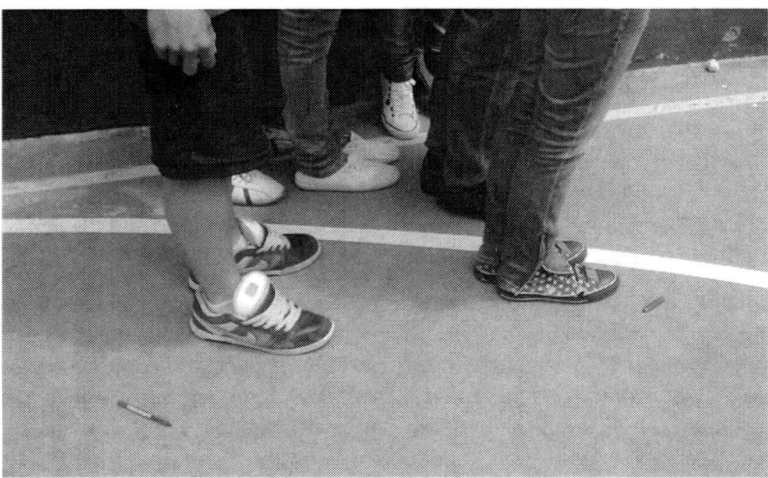

Fig. 1 A group of six students experimenting with population density

playground can be measured following a rectangular pattern where each person is placed in a certain location in an organised arrangement of rows and columns. With this method it is not necessary to be aware of the size of the playground, but it is required to determine the number of people that make up the rows and columns of the playground. The students who proposed this strategy arranged themselves one behind another in order to count the number of people in a row or column to cover the whole extent of the playground. In Fig. 2 a group of four students can be seen standing next to the wall and advancing while counting the number of people required to completely cover the side of the wall.

The results obtained by the students for a rectangular playground are shown in Table 1.

The next step of the pedagogic process was a discussion of these results in order to reach an agreement on the density of teenagers in a concert, which was five persons per square metre. Thus, the estimation of people that can fit in the playground leads to the result of 1,750 persons. This activity was useful to show that, although this number might not be exact, it provides a reasonable value for the magnitude. The class discussion was useful to question the result in relation to other contexts: What if the people we have to fit in are all adults? What if we would like to make sure that they can exit fast enough in case of an emergency? Would five people per square metre be comfortable long enough? The students proposed making some room in a corner of the classroom for all of them to stand together for some minutes. This allowed them to experience how it feels to be very close to other people, so they then reduced the number of people fitting comfortably in a certain space. In the case of adolescents, they established a density of three people per square metre. It is worth noting that during the process the students ended up exclusively using population density as the key concept to measure the amount of

Fig. 2 A group of four students experimenting with grid distribution

Table 1 Results for the number of people that fit in the playground

	Strategy	Area of playground (m^2)	Result
Group 1	Reference point	353	1,179
Group 2	Density	348	2,175
Group 3	Grid distribution	–	2,132
Group 4	Density	360	2,160
Group 5	Density	365	1,462
Group 6	Density	275	1,651

people that would fit in a given space, and therefore they gave this measure meaning and relevance themselves. This concept allowed them to unify the arguments that seemed valid in similar situations to those they could imagine.

The problems B1–B4 confronted the students with some added difficulties which forced them to rethink the methods used to solve problem A. In the concerts at Palau St. Jordi (a large indoor sports hall where concerts are often held), people can sit down in some areas and stand in others. Plaça Catalunya and the city council square do not have a rectangular shape and their area cannot be measured directly. Moreover, in both cases there is urban furniture which makes the calculation of the practicable surface difficult. In the case of the trees in Central Park, the students did not know their density yet and had to search for information on the type of trees in the park and their dimensions. Concerning the necessary data to solve the problems, the students observed that they could not make their own measurements directly. After a brainstorming, they suggested using Google Maps as a tool for obtaining the measurements.

In the particular case of Palau Sant Jordi, the students got estimations between 15,000 and 24,000 persons. After that, they looked for a given figure on the internet.

Some students found that the maximum capacity is 17,960 persons, limited by the council regulations. Others found that PSOE (Partido Socialista Obrero Español) had announced that more than 30,000 people attended a political meeting there in 2008. From their own results and their own mathematical productions, the students were able to assess the reliability of the statements made by the press, using their critical thinking to accept or refuse them. In this case, they were able to see that the data made available by the political party were excessive and that some of the media did not question them. When the students realised that there were deviations in the information given by the media, possibly biased due to political reasons—rendering the event more relevant and influential than it actually was—, they began to trust their own estimations that were backed up by reasoning and did not depend on other people's opinions.

As far as the number of trees at Central Park is concerned, Students found information of very different orders of magnitude. First, they were searching for references in Spanish to contrast their estimations. Finally, a group of students decided to look for the data at the official website of the park, and they were able to observe that the problem possibly lay in translation: some websites in Spanish had mistaken the number of trees for the number of shrubs and bushes. This is how the students were able to see how their reasoning became a tool to sensibly contradict certain information.

The last activity, problem C, focused on a case which triggered great controversy in the media. On July 10, 2010, there was a very crowded demonstration in Barcelona. The people protested against a resolution of the Spanish Constitutional Right's Court which limited and cut back on the new autonomous statute that had been democratically supported by 73.9 % of the Catalans. The different counts of demonstration participants were presented to the students and the question they had to solve was: Which of these numbers is the most reliable? The students were encouraged to use the knowledge acquired in the former activities to answer this question. In summary, they used two main strategies. The first one consisted of estimating the area occupied by participants (using Google Maps, again) and use the density to obtain an estimated number. The second strategy was to calculate the area needed to fit 1,500,000 people and try to imagine how long a street like Passeig de Gràcia would need to be. Due to the difficulty of the situation (people were moving, streets around the initial meeting point were occupied, street furniture) the students obtained a wide range of values (300,000–600,000 people). However, they were able to refute all the figures given by the different statements. At the end of the activity, the students had already touched upon methods and strategies of their own to carry out the estimation tasks. They still had not developed their own perceptive abilities to count large quantities but they had a way of operating which enabled them to estimate the results with enough precision to rule out some of the options presented initially. Therefore, the students developed their own resources to make their own decisions grounded on judgements supported by their own mathematical knowledge. Based on their experience, the students stated that different interests can influence the issue of information and concluded that it would not be very complicated to make an estimate, in the same way as those they had previously made, to scrutinize the information given by the media. It is worth noting that after solving several Inconceivable Magnitude Estimation Problems and validating their own solutions by means of different methods, the students developed confidence in their own results. In the case of the last problem, there is not a result validated by consensus of the community, since the official sources do not offer information on the methods used to make these estimates and the methodology of the company Lynce did not try to measure the amount of people when the street was most crowded.

Conclusions

In this article, we introduced Inconceivable Magnitude Estimation Problems as real context problems that can be solved through the use of partial estimates made by the students. We have seen that students produced many different strategies leading to successful solutions. From the perspective of learning mathematical concepts and procedures, the activity seems rich as it allows the students to exploit their own creativity and develop effective resolution processes based on modeling the studied realities. The students develop several mathematical models and adapt them to the characteristics of each situation. These models are based on identifying the essential parts of the situation and solving the subproblems separately, in the same way Fermi problems are solved. The working procedure allows the students to create and share concepts and methods, and thus all the groups can use suitable methods.

By using IMEP, teachers have access to open problems which can be discussed in an open manner due to the different approaches to their solutions and the connections to real life. The discussion of the methods employed and results obtained in a large group, with the aid of the teacher, allow for the comparison of the methods and the evaluation of the decisions made. Particularly, it is worth noting that the students in our example ended up choosing the method of *population density* as the most versatile and adaptable.

IMEP allow students to integrate school mathematics and real life tools to improve their skills for understanding their environment. In this process, the students gain confidence in their own methods and move on to give meaning to the obtained results. Specifically, this leads them to developing their own resolution methods and to obtain their own solutions. The potential political bias of published data can be appreciated, which allows for an interpretation from their own point of view and informed decision-making. In this experience, all the problems present a high degree of authenticity, which may lead the students to question some statements they find on the web or in the media. In this way, it is proven that students are able to develop their critical competence and obtain tools to evaluate some of the data made public in certain occasions. Especially, this possibility is relevant to information which involves non-public counting methods where the obtained data may influence the opinion of the citizens.

References

Albarracín, L. (2011). Sobre les estratègies de resolució de problemes d'estimació de magnituds no abastables. Unpublished PhD thesis, Universitat Autònoma de Barcelona.

Albarracín, L., & Gorgorió, N. (2012). Inconceivable magnitude estimation problems: An opportunity to introduce modelling in secondary school. *Journal of Mathematical Modelling and Application, 1*(7), 20–33.

Albarracín, L., & Gorgorió, N. (2014). Devising a plan to solve Fermi problems involving large numbers. *Educational Studies in Mathematics, 86*(1), 79–96.

Arcavi, A. (2002). The everyday and the academic in mathematics. In M. Brenner & J. Moschkovich (Eds.), *Everyday and academic mathematics in the classroom* (pp. 12–29). Reston: NCTM.

Ärlebäck, J. B. (2009). On the use of realistic Fermi problems for introducing mathematical modelling in school. *The Montana Mathematics Enthusiast, 6*(3), 331–364.

Ärlebäck, J. B. (2011). Exploring the solving process of groups solving realistic Fermi problem from the perspective of the anthropological theory of didactics. In M. Pytlak, E. Swoboda, & T. Rowland (Eds.), *Proceedings of the seventh congress of the European society for research in mathematics education* (pp. 1010–1019). Rzeszów: University of Rzeszów.

Camelo, F., Mancera, G., Romero, J., García, G., & Valero, P. (2010). The importance of the relation between the socio-political context, interdisciplinarity and the learning of the mathematics. In U. Gellert, E. Jablonka, & C. Morgan (Eds.), *Proceedings of the 6th international mathematics education and society conference* (pp. 199–208). Berlin: Freie Universität Berlin.

Carter, H. L. (1986). Linking estimation to psychological variables in the early years. In H. L. Schoen & M. J. Zweng (Eds.), *Estimation and mental computation* (pp. 74–81). Reston: NCTM.

Chapman, O. (2006). Classroom practices for context of mathematics word problems. *Educational Studies in Mathematics, 62*(2), 211–230.

Diamond, L. J., & Plattner, M. F. (2006). *Electoral systems and democracy*. Baltimore: Johns Hopkins University Press.

Efthimiou, C. J., & Llewellyn, R. A. (2007). Cinema, Fermi problems and general education. *Physics Education, 42*(3), 253–261.

English, L. D. (2006). Mathematical modeling in the primary school. *Educational Studies in Mathematics, 63*(3), 303–323.

Esteley, C. B., Villarreal, M. E., & Alagia, H. R. (2010). The overgeneralization of linear models among university students' mathematical productions: A long-term study. *Mathematical Thinking and Learning, 12*(1), 86–108.

Freudenthal, H. (1983). *Didactical phenomenology of mathematical structures*. Dordrecht: Kluwer.

Hogan, T. P., & Brezinski, K. L. (2003). Quantitative estimation: One, two, or three abilities? *Mathematical Thinking and Learning, 5*(4), 259–280.

Jurdak, M. E. (2006). Contrasting perspectives and performance of high school students on problem solving in real world situated, and school contexts. *Educational Studies in Mathematics, 63*(3), 283–301.

Jurdak, M., & Shahin, I. (1999). An ethnographic study of the computational strategies of a group of young street vendors in Beirut. *Educational Studies in Mathematics, 40*(2), 155–172.

Jurdak, M., & Shahin, I. (2001). Problem solving activity in the workplace and the school: The case of constructing solids. *Educational Studies in Mathematics, 47*(3), 297–315.

Lesh, R., & Harel, G. (2003). Problem solving, modeling, and local conceptual development. *Mathematical Thinking and Learning, 5*(2), 157–189.

Nunes, T., Schliemann, A. D., & Carraher, D. W. (1993). *Street mathematics and school mathematics*. Cambridge: Cambridge University Press.

Palm, T. (2006). Word problems as simulations of real-world situations: A proposed framework. *For the Learning of Mathematics, 26*(1), 42–47.

Palm, T. (2008). Impact of authenticity on sense making in word problem solving. *Educational Studies in Mathematics, 67*(1), 37–58.

Stocker, D. (2006). Re-thinking real-world mathematics. *For the Learning of Mathematics, 26*(2), 29–29.

Van den Heuvel-Panhuizen, M. (2005). The role of contexts in assessment problems in mathematics. *For the Learning of Mathematics, 25*(2), 2–10.

Verschaffel, L. (2002). Taking the modeling perspective seriously at the elementary level: Promises and pitfalls. In A. D. Cockburn & E. Nardi (Eds.), *Proceedings of the 26th PME international conference* (pp. 64–80). Norwich: PME.

Weinstein, L., & Adam, J. A. (2008). *Guesstimation: Solving the world's problems on the back of a cocktail napkin*. Princeton: Princeton University Press.

Winter, H. (1994). Modelle als Konstrukte zwischen lebensweltlichen Situationen und arithmetischen Begriffen. *Grundschule, 26*(3), 10–13.

Criticizing Public Discourse and Mathematics Education: A Commentary

Charoula Stathopoulou

Abstract The commentary on the chapters of Chassapis and Giannakopoulou and of Albarracín and Gorgorió points to the mutually constitutive relationship between in-school and out-of-school contexts. It reflects on the ways in which the two chapters integrate mathematics education with socio-political realities and social justice.

I would like to express my pleasure for the interesting and challenging opportunity to discuss the two chapters of this section, to share their common space, and to consider further perspectives. The two chapters might appear unconnected at first, concerned with research in different settings. Yet they have an underlying, connective tissue. Taken together, they provide a powerful analysis, grounding mathematics education in a sociocultural approach that emphasizes the political dimension of mathematics and mathematics education.

The chapter by Chassapis and Giannakopoulou, "Numbers on the Front Page: Mathematics in the News", explores media discourse, focusing on numerical discourse as it appears on the front pages of Greek newspapers. This research concerns mathematics outside of the classroom, but in a way that informs classroom practices. The other chapter, "Classroom Inconceivable Magnitude Estimation Problems to Improve Critical Thinking", by Lluís Albarracín and Núria Gorgorió, ostensibly analyzes field data on classroom practices; yet, their didactical approach strongly affects students' future reactions in out-of-school settings. So, though unintentionally, these two projects are in a dialogue, with both challenging the division of in-school and out-of-school mathematics by using mathematics/mathematical ideas consistently interlinking both fields. This dialogue is an interesting contribution to the discussion on the dichotomies of in- and out-of-school mathematics that typically begins with the assumption that they are separate entities. In addition, this dialogue contributes to the ongoing discussions regarding the role of mathematics education and of critical mathematics education for citizenship in a democracy.

C. Stathopoulou (✉)
Department of Special Education, University of Thessaly, Volos, Greece
e-mail: hastath@uth.gr

Numbers on the Front Page: Mathematics in the News

Chassapis and Giannakopoulou's research constitutes an example of criticizing public (media) discourse, and particularly a discourse articulated through and around numbers (numerical discourse) that appears in a characteristic place: Greek newspapers' front page headlines.

Numerical discourse, especially in the media, is a particular kind of discourse; the use of numbers usually appears as the objective measure of subjectivity, as the depiction of reality. Expressions like 'numbers speak for themselves', which accompany numerical data describing several social situations, dictate one and only one interpretation: that of the knowing writer. As the authors notice, quoting Alonso and Starr, acts of social quantification are "politicized" not in the sense that the numbers are somehow corrupted—although they may be—but because *political judgments are implicit in the choice of what to measure, how to measure it, how often to measure it and how to present and interpret the results.*

Furthermore, because readers often read and recall only the headlines, corresponding ideologies are passed along to the audience through these headlines. Through their textual function to frame the story, these paragraphs, which are typically but not necessarily one-sentence paragraphs and which stand alone at the beginning of the story, could also afford the development of a propagandistic discourse. As Chassapis and Giannakopoulou mention, most newspapers' headlines on front pages, as well as radio and TV news that announce or comment on governmental plans and decisions—and, as a consequence, most of the related political debates—are flooded with numbers and numerical indices, interpreted and commented upon according to the case, thus producing a specific numerical discourse.

The methodological framework the authors adopt in order to analyze the titles on front-pages is based on a version of Critical Discourse Analysis (CDA) introduced by Fairclough. This is a suitable interpretive framework since it helps to explain data through the broader context, in their case: Greece in the recent crisis and the main politics that are connected to that period. So, the messages on the front pages are analyzed in relation to the sociocultural and political situation. Chassapis and Giannakopoulou draw on work of Rogers et al. and Janks for clarifying three dimensions of discourse.

From a critical poststructuralist perspective, CDA focuses on the linguistic dimension of social power as a central terrain for struggle over other forms of social power. Chouliaraki (2010) sustains that, if the Saussurean view stresses the "referential" power of language, the critical and poststructuralist perspective stresses the performative power of language, "that is, the capacity of language to constitute the world in meaning at the moment that it claims to simply represent it". Every linguistic utterance arises from a position of social interest (be this race, gender, or class), and every linguistic utterance makes a claim to truth that seeks to reclaim these interests and reestablish their power through meaning. "In Foucault's terminology, linguistic relations appertain to particular systems of 'power/knowledge relations' specific to their historical juncture" (Chouliaraki 2008, p. 674). From a CDA perspective, media discourse "is a recontextualizing principle for appropriating other discourses and bringing them into a special relation with each other for the purposes of their dissemination and mass consumption" (Chouliaraki 1999, p. 39).

Chassapis and Giannakopoulou refer to the three levels of Fairclough's framework: the text, the discursive practice, and the sociocultural practice. In their textual analysis, they focus on the use of numbers in the newspapers' titles that appear in combination with sentences that *serve the main purpose of the text*. The text is presented in a multimodal way as pictures, tables and figures accompany and support it. Apart from the modality, another thing that is remarkable concerns the kind of numbers that are used in the titles of the front pages. As Chassapis and Giannakopoulou notice, although whole numbers, decimal numbers, and fractions appear, the majority are percentages. The dominance of percentages—as they rightly note—has to do with their property to be *suitable for numerical alchemies and appropriate for political manipulations projecting—among other ideas—a false sense of equality, hence equity*. For example, in the headlines the expression, "4 % reduction to salaries," *emphasizes the equal treatment of employees regarding the reduction of their salaries but conceals the inequality of the amounts corresponding to each particular salary*. This notice confirms the perception that on the one hand discourse is not only used in order to describe the reality but also to create this reality, and on the other hand that, very often, what is omitted is more important than what is presented.

Regarding the modes/genres that appear on the front pages of the newspapers, declaration and argument are identified as the most common; *each one [is] used … to serve a different purpose according to the political interests promoted*. Chassapis and Giannakopoulou notice that editors organize their reports in headlines around numbers in a way that suggests that these numbers present reality. The majority of the people who read the headlines—a different population from those who read the newspapers' articles—cannot decode the information given in a numerical discourse with which they are not familiar. The authors maintain that the use, here, of declarations and arguments with numerical data gives a sense of objectivity. Referring to Porter, *the language of mathematics is well suited to embody objective judgments and it is adopted when claims to knowledge need to gain trust and credibility beyond the bounds of locality and society*.

For the discourse practices, the primary concern is with the way texts are produced and consumed. It is through understanding of text production and consumption that the finding established at the micro level (textual analysis) can be properly interpreted. Fairclough calls for an understanding of at least three aspects of producing and consuming texts in order to interpret descriptions of texts: "the ways in which texts are produced by media workers in media institutions, how media texts are socially distributed, and the ways in which texts are received by audiences (readers, listeners, viewers)" (1995, p. 16). Chassapis and Giannakopoulou discuss how the texts are produced. They further look at their distribution, which depends greatly on the kind of media: Due to specific technological attributes, different kind of media may cause a different kind of impact on social behavior and social relations (Meyrowitz 1997; Sheyholislami 2008). Here, parallel to the analysis of the role of the headlines, we find another kind of discourse, one that differs from that of the actual articles for which the headlines have been written. A written text usually requires a particular familiarity with the language and the genre of the printed media; the headline is addressed to a broader audience than the people who read the whole newspaper article. We could say that the headline audience is closer to a TV audience. In their analysis of discursive practices, Chassapis and Giannakopoulou address differences across dimensions of textual consumption by the audience, yet they opt to discuss the texts they are examining as a potential situation. They do take the participants into consideration, mentioning that many readers, as participants in communicative events based on numerical discourses, are not in a symmetrical relationship with the other participant, the writer. They also indicate that the readers are *probably accustomed to the genre and the editor's strong voice in the text*, so that they do not contribute equally to the construction of a communicative event. They suppose that the use of numerical discourse reduces the full access to the text by everybody, that is, every potential audience.

The incorporation of audience research in the methodology of CDA research has been heavily criticized. In particular, it has been imputed that CDA analysts assume how audiences interpret texts without asking them. Although the value of audience research has been recognized, some researchers stress that this type of research often ignores issues of power associated with the text, thus undermines the political economy of media texts, the political and economic aspects of text production, and consumption (Madianou 2005). Regardless of whether one can incorporate other methods in a study or not, texts remain in the end as rich, intricate and appropriate sources of discursive and ideological analyses (Sheyholislami 2008).

In the third level of analysis—sociocultural practice—it is attempted to explain what social, cultural and political motives could be behind the ways in which texts are produced, distributed and consumed. The broader framework that explains the motives behind the text (newspapers' front page headlines) is neoliberalism in Greece in times of crisis. In a framework like this the aim of those in power—politicians and newspapers that support governmental policies and decisions—is to:

> persuade the public that the various consequences of the crisis that broke out in the country have to be borne equally by all citizens regardless of their individual income and its sources, since they have all been involved through their various actions, behaviors and stances in the creation of national fiscal deficits and thus in the causation of the crisis.

Again, as the authors point out, the use of a numerical discourse in front page headlines produces a particular rhetoric that is used to support politics. The examples of newspapers referred to mostly support the dominant politics through the use of numerical data. The fact that the numerical data is used in order to create an alibi for their policies becomes clearer after the example of a newspaper from the political opposition; here, the same data is used in a different way, leading to a different interpretation.

Chassapis and Giannakopoulou note that the "technicization of politics"—a term coined by Miller and Rose for describing situations that need to be presented by experts—is used in a way that excludes the majority of citizens, as non experts, from access to the information and moreover the interpretation of the information.

The use of numerical data on front pages and the way (mode, modality) these are presented, as they have pointed out, dictate one reading to the audience, an audience the majority of which has no tools for completely understanding the information, nor for distinguishing the information from its interpretation.

The demand the authors put forward at the end of their paper concerns the mathematics educators, and mostly mathematics educators ascribing to critical mathematics education: these educators *must teach the mathematical meanings of numerical concepts, e.g. percentage, and at the same time they have to select some of the referential meanings of these concepts and reject others, all of which are widely promoted and everywhere visible in public life, today.* It is a question asking how mathematical practices can contribute to the development of critical mathematics teaching; an issue that is also discussed in the next chapter of this section.

Solving Mathematics Problems in Real Life Contexts

Teaching mathematics as an abstract object for a long time has led students to perceive mathematics in a widely discussed dichotomy between in- and out-of-school mathematics. This creates difficulties in applying mathematics to real-life situations. As Van den Heuvel-Panhuizen (2001) comments, learning mathematics separate from students' experience leads students to quickly forget mathematics and to not be able to apply it. According to Freudenthal, *discussing problems with real-life contexts in the classroom may be very enriching for the students.* A widely discussed issue regarding real-life problems and their use in the classroom concerns their authenticity. As Albarracín and Gorgorió note, a *considerable feature of problems based on real-life is their level of authenticity*; authenticity of coursework according Palm is described as the extent to which a task can be moved to a real-life situation. He maintains that students who have extensive experience answering questions *with a higher level of authenticity, using knowledge from their everyday life, obtain more exact answers that are more consistent with reality than students who work mostly with problems with a lower level of authenticity.* The importance of solving problems in authentic, real-life contexts is also stressed by Gulikers et al. (2005, p. 511):

> An authentic learning environment provides a context that reflects the way knowledge and skills will be used in real life. This includes a physical or virtual environment that resembles the real world with real-world complexity and limitations, and provides options and possibilities that are also present in real life.

Authentic environments provide a realistic context to a (authentic) task.

The research project that Albarracín and Gorgorió present here concerns the estimation of large quantities. The wider framework informing their problem construction is the political situation regarding Catalonia's issue of autonomy, and particularly the numerous demonstrations in this geographical area in the last decade. Media reporting the number of people participating in these demonstrations make estimations and construct explanations without sharing the methods used to receive these numbers. The researchers begin with a problem in a context familiar to the students, then move to a similar context, and then focus on problems concerning estimation of populations participating in demonstrations. Apart from problem solving, students think about political issues and the ways media deal with them. In this way, students gain the opportunity to examine public discourse critically.

In the situations described, students are involved in solving authentic problems that demand creative responses for finding solutions. Students construct models for problem-solving that become more elaborate and richer from one problem to the next. Their involvement in these kinds of problems constitutes an exercise in critical thinking (Appelbaum 2000, 2008). In order to solve the problems, students need to take into consideration the framework of the problems, the real situations, and in this way they come closer to the quality of a conscious citizenship. Students not only have to solve problems in the framework of the mathematics classroom but also have to reflect on their solutions and examine the sustainability of their solutions. They need to move out of the classroom practically and symbolically, and incorporate this out-of-school experience into their work for mathematics learning.

Conclusion

Lluís Albarracín and Núria Gorgorió start their chapter quoting a statement of Diamond and Plattner: Democracy in its purest or most ideal form would be a society in which all adult citizens have an equal say in the decisions that affect their lives. In order for people to reach this point, it is necessary for them to have the capability of understanding and interpreting the situations that surround them. Albarracín and Gorgorió mention the need for interpretative tools. Their suggestion is that problem-solving—solving particular kinds of problems—could respond to this need. This sort of problem-solving demands not only that students process problem data, but that they also pursue additional data, and furthermore that they look for and take into consideration information from a broader context. It is in this sense that problem-solving alone does not fully capture the kinds of skills that democracy demands of its citizens. As Albarracín and Gorgorió show, problem-solving plus the inclination to pursue additional information might be more appropriate.

Dimitris Chassapis and Eleni Giannakopoulou demonstrate that the use and 'misuse' of numbers in public discourse can mislead people when they do not think critically, or do not have the training necessary to interpret, and perhaps question or challenge what appears to be the objective truth. As Harindranath (2009, p. 15) notes, "if public discourse is a constituent of democratic participation and knowledge and the interpretation of the media as the arena of public discourse are related to experience, then audience evaluations of what constitutes valid knowledge became crucial".

Taken together, the two chapters show the mutually constitutive relationship between in-school and out-of-school contexts: School mathematics promises to create citizens who will bring critical mathematics understandings and methods of interpretation into the public discourse. The use of public discourse examples in school can help students to become the kinds of citizens we might hope to have as fellow members of our democracies. At the same time, public discourse creates the need for education that prepares people for potentially misleading and manipulative news stories, which take their importance and apparently objective truth from the use of mathematics, and especially from the numbers in the news headlines on front pages, constituting the world while claiming to simply represent it.

Thus we can see the need for mathematical skills and concepts both in and out of school. We also see the need for mathematical skills and concepts to be integrated with critical thinking to raise citizens' ability to understand the use and 'misuse' of mathematics. We can furthermore understand from the interaction of these two chapters how the need for certain kinds of mathematical knowledge and the idea of what a mathematically literate citizen might be are mutually supportive of one another. In another historical context, Cline-Cohen (1999) explained how an increasing enthusiasm for numeracy in the early colonies of European powers led to both a uniquely colonial character defined by systems of census, mercantilism, power and control, and an accompanying set of skills expected of all members of that society. More recently, the rise of the social welfare state and its parallel interest in a progressive, student-centered education of mathematically literate citizens capable of making deliberate decisions in a social democracy, have been shown to create a collection of rational, logical thinkers who are in fact more easily governed through the use of reasoning and mathematical arguments (Walkerdine 1990). These are the preoccupations of critical mathematics education, which aims to integrate mathematics education with society in a common goal of social justice. Here we might take the papers in this section as examples of mathematics in action, observation and reflection; mathemacy as a critical mathematics education for citizenship, mixed together in the 'soup' of social responsibility (Skovsmose 2011).

References

Appelbaum, P. (2000). Eight critical points for mathematics. In D. Weil & H. Anderson (Eds.), *Perspectives in critical thinking* (pp. 41–56). New York: Peter Lang.
Appelbaum, P. (2008). *Embracing mathematics: On becoming a teacher and changing with mathematics*. New York: Routledge.
Chouliaraki, L. (1999). Media discourse and national identity: Death and myth in a news broadcast. In R. Wodak & C. Ludwig (Eds.), *Challenges in a changing world: Issues in critical discourse analysis* (pp. 37–62). Vienna: Passagen.
Chouliaraki, L. (2008). Discourse analysis. In T. Bennett & J. Frow (Eds.), *The SAGE handbook of cultural analysis* (pp. 674–696). London: Sage.
Chouliaraki, L. (2010). Discourse and mediation. In S. Allan (Ed.), *Rethinking communication: Keywords in communication research*. Cresskill: Hampton.

Cline-Cohen, P. (1999). *A calculating people: The spread of numeracy in early America*. New York: Diane.
Fairclough, N. (1995). *Critical discourse analysis: The critical study of language*. New York: Longman.
Gulikers, J. T. M., Bastiaens, T. J., & Martens, R. (2005). The surplus value of an authentic learning environment. *Computers in Human Behavior, 21*(3), 509–521.
Harindranath, R. (2009). *Audience-citizens: The media, public knowledge and interpretive practice*. New Delhi: Sage.
Madianou, M. (2005). *Mediating the nation: News, audiences and the politics of identity*. London: UCL Press.
Meyrowitz, J. (1997). Shifting worlds of strangers: Medium theory and changes in "them" versus "us". *Sociological Inquiry, 67*(1), 59–71.
Sheyholislami, J. (2008). *Identity, discourse and the media: The case of the Kurds*. Unpublished PhD thesis, Carleton University, Ottawa.
Skovsmose, O. (2011). *An invitation to critical mathematics education*. Rotterdam: Sense.
Van den Heuvel-Panhuizen, M. (2001). Realistic mathematics education in the Netherlands. In J. Anghileri (Ed.), *Principles and practices in arithmetic teaching* (pp. 49–63). Buckingham: Open University Press.
Walkerdine, V. (1990). *The mastery of reason: Cognitive development and the production of rationality*. London: Routledge.

Part VI
Organising Dialogue and Enquiry

Facilitating Deliberate Dialogue in Mathematics Classroom.................... 289
Ana Serradó
La Salle-Buen Consejo, Spain
Yuly Vanegas
Universitat Autònoma de Barcelona, Spain
Joaquim Giménez Rodríguez
Universitat de Barcelona, Spain

Inquiry-Based Mathematics Teaching: The Case of Célia 305
Luís Menezes
Escola Superior de Educação de Viseu, Portugal
Hélia Oliveira
Universidade de Lisboa, Portugal
Ana Paula Canavarro
Universidade de Évora, Portugal

Using Drama Techniques for Facilitating Democratic Access to Mathematical Ideas for All Learners.. 323
Panayota Kotarinou and Charoula Stathopoulou
University of Thessaly, Greece

Organising Dialogue and Enquiry: A Commentary 341
Lambrecht Spijkerboer
APS-International, Utrecht, The Netherlands
Leonor Santos
Universidade de Lisboa, Portugal

Facilitating Deliberate Dialogue in Mathematics Classroom

Ana Serradó, Yuly Vanegas, and Joaquim Giménez Rodríguez

Abstract This chapter develops the concept of 'deliberative dialogue' to explore crucial characteristics of mathematics classroom practice, in which the promotion of social participation is an explicit aim. It presents two case studies: one on future teachers facing deliberate dialogues, another on on-line deliberate dialogue in a secondary school mathematics classroom. It concludes that more pre-service and in-service teachers should be involved in design-based research activities that seek to facilitate participation and deliberate dialogue in mathematics classrooms.

Introduction

The use of internet and wireless access has brought a revolution in the world of communication, transforming the traditional communication of face-to-face interactions to new horizontal on-line interactions where actors are more independent of social institutions (Castells 2012). Considering that the school is a relatively weak public institution, we can ask if it is prepared for the challenge of this new kind of communication. The answer to this question reopens the analysis of the relationships between talk and debate, horizontal and vertical discourse, participation and dialogue, face-to-face and on-line interaction, autonomy and authority. Furthermore, it gives the possibility to answer some unanswered questions, such as: What kind of public debate is most likely to expand civic engagement and make it meaningful for

---

A. Serradó (✉)
Department of Science, Mathematics and Technology,
La Salle-Buen Consejo, Cádiz, Spain
e-mail: ana.serrado@gm.uca.es

Y. Vanegas
Departament de Didàctica de la Matemàtica i de les Ciències Experimentals,
Universitat Autònoma de Barcelona, Barcelona, Spain
e-mail: yulymarsela.vanegas@uab.cat

J. Giménez Rodríguez
Facultat de Formació de Professorat, Departament de Didàctica de les Ciències
Experimentals i la Matemàtica, Universitat de Barcelona, Barcelona, Spain
e-mail: quimgimenez@ub.edu

all people? Is there evidence of similar debates in mathematics classrooms? Is there a problem of defining mathematical communication when we talk about engagement in public discourse? Which kind of participation can be considered as deliberative dialogue in the mathematics classroom?

In this chapter, we explore theoretically the characteristics of mathematics classroom practices promoting participation that can be considered as deliberative dialogue. In fact, we know that two main variables influence mathematics participation and deliberation in rich tasks: the activity proposed, and the management of the activities (Gorgorió et al. 2000). Those characteristics form an initial theoretical framework that helps to overcome the difficulties of drawing the blurred lines between participation, deliberative dialogue and deliberative communication. The differences between them will be analysed on the basis of two case studies related to the analysis of the role of tasks and the role of the teacher when managing deliberate dialogue in mathematics classrooms.

Participation for a Deliberative Dialogue in Mathematics Classrooms

In this section, we explore theoretically the characteristics of participative practices in mathematics classrooms that can be considered promoters of deliberative dialogue. The analysis of the complexity of participation in mathematics classrooms has been elaborated in different ways. One of the simplest ways of analysing this complexity arises from the description of the structures of a dialogue when trying to clarify the differences between actively-talking and actively-listening participants (Hammond and Wiriyapinit 2005). From this analysis, three patterns of participation can be described: non-participation, quite-participation (reading), and communicative participation in an online context (Hammond 1999). We cannot consider these patterns as static terms of individual participation, since all the participants in a dialogue have moments of quite or communicative participation. The importance of this categorization for understanding the deliberative dialogue for a democratic mathematics education is that students, as communicative participants, have a commitment for their learning and feel the responsibility to promote the participation of others.

The belief of the importance of the responsibility in participation, as a democratic value, can be analysed through a theory of argumentation and of decomposition of the speaker's role (Krummheuer 2007), and a psychological perspective that characterises students' individual beliefs about their own role and others' roles when participating in mathematics classrooms (Cobb et al. 2001). The integrative and analytical theoretical framework, used by Cobb et al. (2001) for the analysis of participation in the mathematics classrooms, presents beliefs in relation to classroom social norms analysed through a socioconstructivist lens. Examples of those classroom social norms, which are jointly established by the teacher and the students, are: attempting to make sense of explanations given by others, indicating agreement or disagreement, and questioning alternatives when a conflict in interpretations becomes apparent. These examples of the participants' discursive interactions can be analysed from a linguistic perspective through four interactive properties: evaluative, informative, interpretative and negotiatory (Bairral and Powell 2013).

When teachers and students accept those social norms of participation and are engaged in interpretative and negotiatory discursive interactions, they can develop skills such as getting in contact, locating, identifying, advocating, thinking aloud, reformulating, challenging and evaluating—skills that teachers and students can bring into play in a deliberative dialogue (Alrø and Skovsmose 2002). However, the adoption of these social norms and skills in participation are not sufficient to assure a deliberative dialogue in the mathematics classroom. The deliberative dialogue can only be intentionally promoted by the teacher when s/he involves the students in a communicative process in which three points are considered attentively and carefully: (a) the reasons or lack of reasons for people's preliminary opinions and judgements before actually making a final statement, (b) the pros and cons of possible decisions before actually making them, and (c) the benefits and losses of possible courses of action, before engaging in them (Valero 1999).

A Critical Theory: Deliberate Dialogue and Deliberative Communication

One of the main differences between deliberate dialogue and deliberate communication is the teachers' intentionality to facilitate a democratic participation of all students. In the characterisation of the properties of discursive interactions for a deliberative dialogue, the meaning that is given to judgements is key. Students often use judgements as a neutral linguistic action while understanding and accepting, or not accepting, the conversational partner thinking aloud, expressing meaning. However, we consider that: "In making a judgment, people take into account the facts as they understand them and their personal goals and moral values and their sense of what is best for others as well as themselves" (Yankelovich 1999, p. 179).

The intentionality of the deliberate dialogue, when making explicit democratic mathematics classroom practices, can be understood, from the point of view of critical theory, as a more elaborate construct within the frame of deliberative communication. The role of the teacher is central in deliberate dialogue and crucial when it comes to management, power, responsibility, and judgement. In almost all cases, it is the teacher who has to make professional judgements about the possibility and suitability of initiating, authorising, and conducting (or continuing) deliberative communication, and if necessary about bringing it to an end if it seems to be unsuccessful or has been pursued as far as possible. The students are very important team workers, as both actively-talking and actively-listening participants, but it is the teacher who has the crucial role with regard to the direction, possible continuation, and conclusion of deliberative communication. It is considered that reciprocity, publicity, and accountability are the main principles for deliberative democracy (Gutmann and Thompson 1996).

That reciprocity should also grant the possibility to students as active-speakers and writers, as proposers of problems and questions (Hudson and Bruckman 2004), to overcome the deliberative contribution to meaning-creation and knowledge formation (Englund 2006), recognising the interactional suitability of the task (Giménez et al. 2013). Many authors assume that deliberative communication implies respect based upon differences and searching for consensus (Englund 2006; Habermas 1987) as fundamental criteria for democracy (Dewey 2008). Other characteristics mentioned are related to overcoming authorities or traditional views (represented, for example, by parents and tradition). They include the opportunity to challenge one's own tradition and the scope for students' deliberate activities without teacher control, i.e. for argumentative discussions between students with the aim of solving problems or decision-making through looking at problems from different points of view.

In a deliberative communication, as reported by Mark Gerzon, the "critical quality of dialogue lies in that participants come together in a safe space to understand each other's viewpoint in order to develop new options to address a commonly identified problem" (Pruitt and Thomas 2007, p. 20). Problems are open in their cultural, historical and political dimension, going beyond the mathematical contexts.

It is regularly assumed from the political perspective that deliberate dialogue has the following characteristics (McCoy and Scully 2002, pp. 120–128): (a) encourage multiple forms of speech and communication to ensure all people have a real voice, (b) make listening as important as speaking, (c) connect personal experience with public, (d) build trust and create a foundation for working relationship, (e) explore a range of views about the nature of the issue, (f) encourage analysis and reasoned argument, (g) provide a way for people to see themselves as actors and to be actors, (h) help people develop public judgement and create a common ground for action, (i) create ongoing processes, not isolated events. Deliberative democrats specifically accentuate the character of processes, with the starting point that different views have to be adjusted or confronted by means of argumentation in order to decide our common destiny on mutually acceptable terms (Englund 2006). That emphasizes responsibility and consequences, and implies that public socialisation introduces citizenship, giving meaning to schools as spaces for deliberative communication (Englund 2011).

From our perspective, deliberate communication in the global-technological world is an important topic, as our modern society requires citizens to contribute to political debates with democratic attitudes, by using dialogic behaviour (Valero 1999). Within such a framework, it is assumed that the role of the teacher is crucial, since s/he has both the real authority (in terms of the necessary knowledge and perspectives) to determine the discursive conditions for dealing with the problem in question and the formal authority to do so, which can always be misused. In addition, the student has the motivation and autonomy to accept the discursive norms established by the teacher to truly engage in discourse and establish common classroom social and mathematical norms to question alternatives when conflicts in solving problems or decision-making appear. When developing classroom social norms, the mathematical school and classroom practices become a privileged space in which it is possible to frame democracy, as an integral part of the public sphere.

We are convinced that deliberate dialogue in mathematics classrooms for democratic citizenship should enable different forms of communication that support face-to-face and on-line dialogue and negotiation through deliberative interaction as a form of participation, based on research attitudes that teachers incorporate in their practices for a full engagement of students. Furthermore, if this deliberate dialogue is intentionally initiated by the teacher for a democratic responsible engagement of the students in the mathematics classroom, we say that teachers engage students in a deliberate communication. In teacher education, it is also necessary to recognise capacity building and task analysis to identify social problems of democratic processes that are related to mathematics (Vanegas and Giménez 2012).

Within such a framework, our main questions here are related to both the role of the teacher and the role of the tasks: What is the teacher's role in deliberate dialogue in the mathematics classroom? Which conceptions do pre-service teachers hold about the teacher's role in a deliberate dialogue in the mathematics classroom? Which tasks facilitate deliberate dialogue in the mathematics classroom? Which conceptions do pre-service teachers hold about how tasks facilitate deliberate dialogue in the mathematics classroom?

In line with this background, two case studies are presented. In the first case study, we analyse the perspective of future teachers on deliberative communication and dialogue. The second case study shows experiences of how deliberative dialogue can be facilitated in secondary compulsory mathematics classrooms using a blend between face-to-face and on-line environments.

Case Study 1: Future Teachers Facing Deliberate Dialogues

In this study our aim is to identify the position of prospective teachers on the topic of deliberate dialogue, by asking their opinions about the teachers' role in a deliberate dialogue and about the kind of tasks to facilitate deliberate dialogue in mathematics classrooms. In order to identify these positions, we used two instruments: an initial task developed for prospective mathematics elementary teachers and a final professional task developed for future secondary mathematics teachers. We do not use any theoretical tool to analyse the questionnaire of the initial task. Meanwhile, for the final task we developed a didactical analysis. Furthermore, in order to identify the emergent characteristics about the deliberative dialogue, we classified future teachers' answers according to the following categories: (a) *compromise* in which something is explained about the importance of having judgements and decisions, and explanations in which ideas about collaborative reflection appear that include building questioning group processes, explaining the reasons or lack of reasons for people's preliminary opinions and judgements; (b) *openness to deliberative leadership* in which it is important to consider classroom dialogue having a priori reflections about possibilities, advantages and disadvantages of a situation, and (c) *transformation*, as argumentations in which ideas of using mathematical dialogues for transforming and mobilising social relations appear, explaining the benefits and losses of possible courses of action before engaging in them.

The initial instrument, consisting of a questionnaire developed in 2011 and improved in 2012 by Vanegas and Giménez (2012), asks two groups of prospective mathematics elementary teachers to analyze different school activities. The first activity was a quasi-deliberative debate in preschool, in which the teacher proposed observations about art and opened up a discussion about shapes. A second situation was started by explaining that water resources are limited. The main question proposed was to analyse the quantity of fresh water on Earth and to find the amount of river water, in order to explain why it is important to preserve these resources. When we, as experts, analyse the role of a situated context in the activity, we can see that the questions do not promote an immediate relation between proportions and the social problem. The teacher can decide about deliberative dialogue but it is not included in the text. A third situation introduced Hooke's law. By reading a table of data with missing values, students without any empirical experience were asked to establish the functional relation, to see the proportion and construct the equation of the corresponding straight line. No dialogue was presented in this case to see if the future teachers also considered the dialogue itself as important to describe the activity.

The didactical analysis of the final professional task for secondary mathematics teachers was done on the basis of data obtained from two groups of students (academic year 2011, 2012). The students' answers were classified, according to the emergent characteristics about the deliberative dialogue. The results in Table 1 show that a few statements relate to deliberative characteristics.

About the role of the tasks we found that some future teachers of the academic year 2011 told us that a certain type of tasks in specific contexts, such as ecological situations or environmental problems, promotes reflective compromises more than other situations. In these activities, the future teachers consider that the junior-school students can be introduced to social discussions. They also say that it is difficult to promote a deliberative dialogue without any question devoted to the historical context, and the possibility to introduce the idea of limited values in connection with Hooke's law.

In general, extra-mathematical context is considered as the main element to promote a questioning discussion. Thus, many statements focus on the impact of specific tasks, arguing that themes as "the need for water, offer more possibilities … giving responsibilities to the student" (Sergio). However, there is no evidence that the future teachers think of the use of the negotiation of mathematical meanings in order to

Table 1 Number of teachers referring to each deliberation category

	Future teachers (N)	Compromise & collaborative reflection	Openness for taking decisions	Transformation
2011	22	5	4	2
2012	34	5	0	0
Total	56	10 (17.8 %)	4 (7.14 %)	2 (3.57 %)

make deep judgements. Such a questioning participation seems not to be centred upon deliberate principles, and consensus seems to be related to extra-mathematical ethical issues (Vanegas 2013). Although the future teachers talk about social problems behind the tasks, they do not relate them to the mathematics needed for discussing the social problem, which is not included in the problem formulation.

About the role of the teacher, 2011-prospective teachers focus on consensus in participative classrooms in which "the teacher has a key role in driving such participation" (Lidia). Seeing sentences such as "the teacher introduces the reflection … and talks about planar shapes" we identify that some of the future teachers assumed a directive position. The idea of critical collaborative reflection interpreted as reasoning in order to promote personal opinions and learning from the others was scarce. When future teachers talk about the value of active participation, when building mathematical meanings to solve social problems, the statements relate only to teachers' intentions and attitudes. In some cases, their explanations state the need to evoke differences. We found explanations associated with the ideas of transformation and self-regulation. However, prospective teachers did not value how mathematical involvement might modify or improve the initial social conditions proposed by the task.

During the academic year 2012, a second version of the professional task was distributed to a new group of future secondary mathematics teachers. This new professional task was improved by including new activities similar to those described in the initial task with a description of a short class episode in a geometry situation using Geogebra. Furthermore, prospective teachers had been introduced to the analysis of the complexity of mathematical practices, focusing on participation and dialogue (Giménez et al. 2013). Also, teachers were asked to analyse their contribution as teachers to foster deliberation.

When involved in this activity, the future teachers revealed a more open questioning attitude than we observed during the initial questionnaire tasks. The value given to the communicative role was suddenly reduced to evaluation, and it was assumed that the main characteristic of a dialogue is to promote agreements among the students.

> I asked questions about what they are thinking, to assess their argumentation competencies. It was interesting to observe the disagreements and different opinions in some groups, and also how everyone tried to convince the others. Some groups had short dialogues, and I decided to stay with them to motivate the conversation, but not always with success. (Javier)

In order to reconstruct the effects of the training program in the two groups of future teachers (academic year 2011, 2012), we decided to observe the final-masters work, in which they analysed their pre-service school practice and gave proposals for improvement. Our aim was to find out which aspects they consider as facilitating participation, and which are their approaches to deliberative debates. Their final writings have been analyzed in terms of communicative arguments deeper than simple descriptions, questions and information. We categorized their statements, which express values of dialogic processes, according to the three main characteristics of deliberation.

An important result is that almost all of the future teachers' comments reveal that there is not enough participation in mathematics classrooms but just peripheral participation (Lave and Wenger 1991). 76 % of the future teachers claimed that a contextualized activity is necessary to achieve real participation. More than a half of the future teachers indicated that the type of questions and the role of the teacher must change to allow students to acquire autonomy and generate dialogue and communication. In some cases, the future teachers integrated this idea in an inclusive thinking.

> I consider that in the classroom dialogue and communication is possible, introducing/encouraging the interventions of the students with open questions like: Who thinks this or that? which create debates and different opinions among the students, through mathematics. (Susanna)

The future teachers identify the importance of collecting communicative processes as an instrument of the teacher's regulation. The main characteristic attributed to the students' dialogue is to give possibilities for solidarity *"identifying the needs of the colleagues and solving them"* (Roger) during the communication itself. Nevertheless, a few comments appear which recognise the potential of mathematics in contributing to making judgements or reaching consensus and to considering elements from theories.

Two future teachers mention explicitly the need of using digital communication and on-line tools (blog, email, Moodle etc.) to promote effective participation by giving responsibility to the students (according to Krummheuer 2007). This is an important change with respect to the use of dialogue as it refers to a change of the teacher's role.

> Promoting the exchange of group productions, generating discussions with the whole class in order to improve mathematical processes … as I could see when introducing Pythagorean rule … I think it was a key point to design a propaganda paper, because it offered the opportunity of discussing not only what is important, but also how to explain and present it to the others. (Roger)

Such a comment reflects a set of characteristics of the dialogue and the role of the teacher in favour of authentic participation. However, for the majority of future teachers, it still seems to be difficult to accept the rules of deliberative dialogue, even after the teaching training course. The process of overcoming these difficulties by a secondary school teacher is shown in the case study 2.

Case Study 2: On-Line Deliberate Dialogue in a Secondary School Mathematics Classroom

In order to answer the question of which tasks facilitate deliberate dialogue in the mathematics classroom, we carried out a design based research (Cobb et al. 2001) about the use of an asynchronous discussion forum of MOODLE as a tool for analysing deliberate dialogue in secondary mathematics classrooms. Two forums were created for grade-ten students (15–18 years old).

Forum 1 took place during September, October and November of 2008. The task requested students to participate in an on-line discussion forum to answer the question: How do scouts use trigonometry?

Forum 2 was initially structured in 2011 with the aim of analysing, comparing and interpreting two statistical graphs on the development of the economy and population during 1980 and 2005. In 2012 it was re-structured based on the five stages of Salmon (2004). The access and motivation stage, proposed by this author, was made in a face-to-face classroom environment, where the 19 students were asked to participate in the on-line forum that took place during January and February of 2013. They had to answer five questions:

(F21) Which data of the graphics about the variation of the economy and population led you to believe that there does, or does not, exist a relationship between them?
(F22) Can you affirm that when the economy grows there is a positive variation in the population?
(F23) Which are the social, political and economical factors that have caused a negative variation of the population?
(F24) In function of the data analysed, why do you think that the improvement of the Andalusians' life conditions are restricted by a work reform?
(F25) By looking at data on the internet from the period 2006–2012, explain how to understand the "possible relationship between the economic evolution and the decrease in the number of immigrants in the country?"

The students also had the instruction: "The first student writes up an idea, the second student explains if s/he agrees, then improves it and so on until you have created a corps of knowledge."

Different iterations for improving the instructional materials and teachers' experiments had been done together with retrospective analyses (Serradó 2009, 2012). The actual investigation reanalysed all the dialogues to identify the properties of the students' participation (NP non participatory, QP quite participatory, CP communicative participation). Every log has been coded to identify the main property of the interlocution: evaluative, informative, interpretative and negotiatory (Bairral and Powell 2013). In the case that the interlocution is mainly interpretative and negotiatory, the content analysis of the logs consisted in identifying the possible skills that students used in the dialogue: locating, identifying, advocating, thinking, reformulating, challenging and evaluating (Alrø and Skovsmose 2002). Finally, for those negotiatory processes, it has been analysed if the three conditions of deliberative dialogue established by Valero (1999) and the characteristics of deliberative communication described in the paper were accomplished. The third analysis of the students' logs led us to distinguish between participation, deliberative dialogue and deliberative communication.

In forum 1 ("How do scouts use trigonometry?") the on-line dialogue interplayed with the face-to-face learning of mathematics changing the structure of participation. When discussing the solution of the problem, we identified two different moments of dialogue: an on-line discussion before solving the problems of trigonometry with a communicative participation of eight students and the rest of the class silent, and a face-to-face discussion after learning how to solve trigonometry problems. The initial dialogue before learning how to solve a problem was:

Teacher: How do scouts use trigonometry?
José Luís: I think: Boy scouts [use it] to measure the height of a mountain knowing the hillside and the facts with the principles that sustain it.
Marian: They use it to measure the hillside of mountains, or what is the same the hypotenuse of a right triangle, for that they use Pythagoras's theorem, because it only works with this kind of triangles.
Teacher: Then you tell me that Pythagoras's theorem should not be used to calculate the height of the Leaning Tower of Pisa.
Valme: In my opinion, we can ... If we consider that the tower represents the hypotenuse of a right triangle, where sides b and c are the ground and the real height of the tower with respect to its base, we would be applying Pythagoras's theorem, no? And we could calculate the height using it.
Cristina: Ana (teacher), a question. With respect to what you have answered to Valme, asking her if she should apply Pythagoras's theorem to Pisa's tower, can we? I think, no, no? Because, it does not form a right-angled triangle with the ground.
Paco: I agree with Cristina, is it possible to calculate it with the Pythagoras? In my opinion, I think it is not.
Marian: Well I agree with Valme, because it is not the tower that has to form a right angle with the ground, if not an imaginary line k having b. The angle that the tower will form with the ground should be opposite to this imaginary side, angle B, no?

The discussion contains the four properties: evaluative, informative, interpretative and negotiatory. The initial informative utterances of José Luis and Marian answered the initial question proposed in the problem. After that initial moment, the contributions of Valme, Cristina and Marian were basically negotiatory. They began with the interpretation of the previous information, reinforced by the use of expressions such as "I think", "In my opinion", "I agree with". They complemented their explanations with the use of the expression "no?" to negotiate with the teacher and the other students about the coherence of their reasoning about how to solve the problem. In particular, Cristina asked directly for the participation of the teacher to confirm the veracity and authority of her argument.

In their discussion the students used skills characteristic of a deliberate dialogue such as: locating the situations in which scouts use trigonometry, identifying the properties of the triangles, thinking aloud, evaluating others' participation, supporting the reasoning presented by Cristina, challenging the correct process of resolution. The students analyse the pros and cons of how to draw that imaginary triangle in which they can apply Pythagoras's theorem.

We can conclude that the dialogue accomplishes the three properties for being considered deliberate in the sense of Valero (1999). However, we think that this deliberative dialogue cannot be considered an intentional deliberative communication, because the teacher and the eight students who participate do not show any responsibility to truly engage the other 21 classmates to promote a communicative participation in the sense of Hammond (1999), and to ensure that all students have a real voice (McCoy and Scully 2002).

This is visible in the participation in forum 2. The task used an initial face-to-face context in which students discussed with the teacher the on-line social norms of participation, which in the end facilitated communicative participation of 63 % (12 of 19) of the students. The teacher interviewed those four students who kept silent during the whole length of the discussion with the intention of facilitating a democratic access to all the students. The teacher offered the four students the possibility of participating in a face-to-face deliberate dialogue where they analysed the on-line discussions of their classmates and improved them. We include some excerpts in which the 12 students debate how to solve problem F35.

Alejandro, February 6th, 20:17. The fifth question was about looking up information on the internet, wasn't it?

Maria, February 6th, 21:05. Ana [teacher], I don't understand the fifth question can you explain it to us, please.

Elena, February 6th, 21:10. Ana, we need your help because I think that Sara, Maria and I don't understand what the question is referring to.

Rosa, February 6th, 22:26. Yes, Ana, if you could help us to understand what the fifth question means…

Alejandro, February 7th, 15:49. I imagine that we are supposed to look for information on the internet.

Sara, February 7th, 15:49. Ale, this is what we have to do, because it's written in the question, but what I don't understand is how we look for it.

Elena, February 7th, 16:03. What I don't understand is which data to look for.

Adrian, February 7th, 16:32. I looked on the internet and I have only found the evolution in Spain from 1996 to 2006. There is some interesting data but I don't know if this is what we are looking for. Here you have the page just in case http://www.compraverde.org/codesarrollo/documentos/File/Noticias/inmigrayecon.pdf

Gonzalo, February 7th, 20:25. And what will be the appropriate data for the question?

Nuria, February 7th, 21:01. I have been looking but I didn't find anything related.

Joel, February 7th, 21:21. I suppose that Ana is going to give us the webpage, because she said that she only participates to give us information about web pages.

Maria, February 7th, 21:24. Ana, we need your help, is the webpage that Adrian has shown correct.

Celia, February 7th, 22:09. I think that the webpage is correct, but which data is appropriate to answer the question?

Nuria, February 7th, 22:34. Ana, should we read everything and summarise the relationships?

Gonzalo, February 9th, 15:02. The data is correct but it talks about the bonus for exchange of risk in credits for the next 5 years in the European countries in the period 2009–2012. Is the data of the graphic useful in relation to the question or should we look for more web pages that have more optimal data?

The discussion contains the four properties of being evaluative, informative, interpretative and negotiatory. However, we can find differences in the characteristics of the informative and interpretative property of the interlocutions in the two forums. In forum 1 the informative logs presented basically facts; in forum 2, the students used expressions like "I don't understand" the facts introduced by others, confirming a previous evaluation, judgement and reflection of the information given. Furthermore, in forum 1, we have been categorizing as interpretative those interlocutions where it seems that the students were thinking and writing at the same time, using expressions like "I think". In this forum the students used expressions like "I imagine", "it is clear", that show a wider use of vocabulary.

The students use mainly the pronoun "I" in the evaluative, informative and interpretative interlocutions, using basically the plural pronoun "we" for the negotiating interlocutions. When the students use this plural pronoun, they show the intention and responsibility of opening the participation to all. An example of this responsibility is the interlocution of Elena: "Ana, we need your help because I think that Sara, Maria and I don't understand what the question is referring to". The fact that these students had not asked the teacher directly for help in the face-to-face or online context means that they have extended their dialogue to other contexts.

Without the help of the teacher, they negotiated the meaning of the task. The negotiations referred to two kinds of decision-making. The first concerned the place where they could find the data for answering the question. The negotiation about where to find it ended in the moment when Joel informed the group about the social norms of participation in the forum. The second was related to the nature of the data to solve the problem. The reflection about the accuracy of the data can be observed through analysing the last log of Gonzalo.

Along these interlocutions, students discussed the reasons for looking for information on the internet; they discussed the pros and cons of the data obtained from the internet. And, in particular when Nuria wrote: "should we read everything and summarise this relationships?" She reflected upon the benefits and losses that could occur from reading a mathematical text of more than 70 pages. The accomplishment of these three conditions led us to argue that the task has facilitated a deliberate dialogue between the students in the sense of Valero (1999).

In addition, three facts made us think that the task promotes deliberative communication. The first fact is that we have been talking about a participative communication, in which the students facilitated the learning and participation of their classmates. In the case of a quite participation, the teacher offered the students other forms of communication. Secondly, the students discussed about the nature of the data to solve the given statistical problem. This discussion surpassed the proposal of the task that had asked students to compare statistical distributions, and provided the students with informal knowledge about the importance of the data in the sampling process. Finally, when solving the given statistical problem, the students have to consider not only mathematical knowledge; they also have to reflect about the effects of social, cultural, political and economical variables.

Discussion

We have presented two case studies about deliberative communication in mathematics classrooms. The first case study analysed pre-service primary and secondary teachers' conceptions about the role of the teacher and the tasks in deliberative dialogue. The second case presented a design-based research study using on-line tasks in a secondary mathematics classroom (Cobb et al. 2001). In this section we discuss the resemblances to and dissimilarities between both case studies in relation to the role of the teacher and the characteristics of the tasks.

In both case studies, we can see the difficulties introducing authentic deliberative dialogues in mathematics classrooms. In fact, based on our observations in case study 1, we argue that a traditional teacher-centred background is still present in the future teachers' minds. Future teachers believe that deliberative mathematics dialogues are nearly impossible to be carried out in our regular classrooms of multicultural heterogeneous groups. Some future teachers considered that contextualisation is necessary for being critical and for promoting citizenship. Future teachers recognize that teachers have a crucial role in organising classroom dialogue, and the majority assumes that the role of the dialogue is to preserve and reveal students' content knowledge. The role of the dialogue is basically a modulation of opinions, sharing knowledge, and negotiation of mathematical consensus, but the need for deliberation in promoting democratic feelings and for accuracy in solving social problems through mathematics is not clarified.

The retrospective analysis of forum 1 in case study 2 led us to conclude the communicative participation of the students facilitated an unconscious deliberative dialogue around how to solve a trigonometry problem. We observed an obstacle for promoting on-line deliberate communication, when rapid answers from the side of the teacher constrained the autonomy of the secondary school students to facilitate their own decision-making during problem solving processes.

This obstacle was overcome in forum 2, where the teacher was conscious of the importance of engaging all students, managing a blend between a face-to-face and an on-line dialogue. This blend has given a fair opportunity to the students who lack technological opportunities. In the design and management of the forum, the teacher has considered as crucial three principles of deliberative democracy in the sense of Gutmann and Thompson (1996): accountability, and basic and fair opportunities.

In case study 1, the future teachers usually separated democratic attitudes from mathematical negotiation. The future teachers held the belief that the initial statement of a problem (even if open) is not enough to promote a critical dialogue, and the role of the teacher is crucial to provoke the pupils' reflection on how social, cultural, political and economical variables affect a situation.

In case study 2, the retrospective analysis of the tasks given to the students led us to the conclusion that not all tasks facilitate deliberative communication. In forum 2, the given tasks used previous social, cultural, historical knowledge of the students to provoke students' preliminary judgements. Through deliberative dialogue, the students created meaning of shared information and they negotiated with the aim of knowledge integration. In order to facilitate deliberate communication, we should create tasks related to wider contexts: to the situation given by the social, cultural and historical context, to the linguistic authenticity of the open question (Serradó 2012), and to the investigative role that it requires. Those tasks and the respective decision-making should be related to the role of mathematics in the world and to the nature of mathematics as a universe or a language (Kennedy and Kennedy 2011).

From the empirical perspective, we see that deliberation empowers people to engage in problem-posing and -solving and decision-making (Skovsmose and Valero 2001), at least in environments that blend face-to-face and on-line participation. Furthermore, the retrospective analysis of how to design instructional material and teaching experiments leads us to conclude that the teacher and the students should first of all delineate the on-line social norms of participation for a deliberate democratic communication.

We conclude that in order to surpass traditional models of instruction, which still dominate mathematics education practices, and construct investigative approaches, pre-service and in-service teachers should be involved in design-based research activities that seek to facilitate participation and deliberative dialogue in mathematics classrooms.

Acknowledgements The work presented was realized in the framework of the following Research Projects: (1) REDICE-10-1001-13 "A competencial perspective about the Master's Training for Secondary School Mathematics Teachers". (2) EDU2012-32644 "Development of a program by competencies in a initial training for Secondary School Mathematics". It was supported by the Agrupació de Recerca en Ciències de l'Educació in 2013 and the Comissionat per a Universitats I Recerca del DIUE from Generalitat de Catalunya (GREAV 2014 SGR 485).

References

Alrø, H., & Skovsmose, O. (2002). *Dialogue and learning in mathematics education: Intention, reflection and critique*. Dordrecht: Kluwer.
Bairral, M. A., & Powell, A. B. (2013). Interlocution among problem solvers collaborating online: A case study with prospective teachers. *Pro-Posições, 24*(1), 1–16.
Castells, M. (2012). *Redes de indignación y esperanza. Los movimientos sociales en la era de Internet*. Madrid: Alianza.
Cobb, P., Stephan, M., McClain, K., & Gravemeijer, K. (2001). Participating in classroom mathematical practices. *The Journal of the Learning Sciences, 10*(1&2), 113–163.
Dewey, J. (2008). *Democracy and education: An educational classic*. Radford: Wilder.
Englund, T. (2006). Deliberative communication: A pragmatist proposal. *Journal of Curriculum Studies, 38*(5), 503–520.
Englund, T. (2011). The potential of education for creating mutual trust: Schools as sites for deliberation. *Educational Philosophy and Theory, 43*(3), 236–249.
Giménez, J., Font, V., & Vanegas, Y. M. (2013). Designing professional tasks for didactical analysis as a research process. In C. Margolinas (Ed.), *Proceedings of ICMI Study 22 "Task design in mathematics education"* (pp. 581–590). Oxford: ICMI.
Gorgorió, N., Planas, N., & Vilella, X. (2000). The cultural conflict in the mathematics classroom: Overcoming its "invisibility". In A. Ahmed, J. M. Kraemer, & H. Wiliams (Eds.), *Cultural diversity in mathematics (education): CIEAEM 51* (pp. 179–185). Chichester: Horwood.

Gutmann, A., & Thompson, D. (1996). *Democracy and disagreement*. Cambridge, MA: Belknap.

Habermas, J. (1987). *The theory of communicative action* (Lifeworld and systems: A critique of functionalist reason, Vol. 2). Boston: Beacon.

Hammond, M. (1999). Issues associated with participation in on line forums: The case of the communicative learner. *Education and Information Technologies, 4*(4), 353–367.

Hammond, M., & Wiriyapinit, M. (2005). Learning through online discussion: A case of triangulation in research. *Australasian Journal of Educational Technology, 21*(3), 283–302.

Hudson, J. M., & Bruckman, A. S. (2004). The Bystander effect: A lens for understanding patterns of participation. *The Journal of Learning Sciences, 13*(2), 165–195.

Kennedy, N., & Kennedy, D. (2011). Community of philosophical inquiry as a discursive structure, and its role in school curriculum design. *Journal of Philosophy of Education, 45*(2), 265–283.

Krummheuer, G. (2007). Argumentation and participation in the primary mathematics classroom: Two episodes and related theoretical abductions. *Journal of Mathematical Behavior, 26*(1), 60–82.

Lave, J., & Wenger, E. (1991). *Situated learning: Legitimate peripheral participation*. New York: Cambridge University Press.

McCoy, M. L., & Scully, P. L. (2002). Deliberative dialogue to expand civic engagement: What kind of talk does democracy need? *National Civic Review, 91*(2), 120–128.

Pruitt, B., & Thomas, P. (2007). *Democratic dialogue: A handbook for practitioners*. Washington, DC: Trydells Tryckeri.

Salmon, G. (2004). *E-moderating: The key to teaching and learning online* (2nd ed.). London: Taylor & Francis.

Serradó, A. (2009). E-forum, a strategy for developing key competences of communication in, with and about mathematics. In L. Gómez Chova, D. Martí Belenguer, & I. Candel Torres (Eds.), *Proceedings of the international conference and new learning technologies (EDULEARN)* (pp. 1–12). Barcelona: IATED.

Serradó, A. (2012). How to question in an on-line forum to promote a democratic mathematical knowledge construction? *International Journal for Mathematics in Education, 4*, 369–374.

Skovsmose, O., & Valero, P. (2001). Breaking political neutrality: The critical engagement of mathematics education with democracy. In B. Atweh, H. Forgasz, & B. Nebres (Eds.), *Sociocultural aspects of mathematics education: An international research perspective* (pp. 37–56). Mahwah: Lawrence Erlbaum.

Valero, P. (1999). Deliberative mathematics education for social democratization in Latin America. *Zentralblatt für Didaktik der Mathematik, 31*(1), 20–26.

Vanegas, Y. M. (2013). Competencias ciudadanas y desarrollo profesional en matemáticas. Unpublished PhD thesis, Universitat de Barcelona.

Vanegas, Y. M., & Giménez, J. (2012). What future mathematics teachers understand as democratical values. *International Journal for Mathematics in Education, 4*, 457–462.

Yankelovich, D. (1999). *The magic of dialogue: Transforming conflict into cooperation*. New York: Simon & Schuster.

Inquiry-Based Mathematics Teaching: The Case of Célia

Luís Menezes, Hélia Oliveira, and Ana Paula Canavarro

Abstract This chapter discusses the instructional practice of a primary school teacher. It is based on a framework that we developed in the project "Professional Practices of Mathematics Teachers", which relates the teacher's intentions to her actions in an inquiry-based mathematics classroom. The framework covers the promotion of mathematics learning as well as the class management. It details the instructional actions of the teacher in terms of the launching of the mathematical task to the students, the support of the students' work, the orchestration of the discussion of the task and the systematization of the mathematical learning process.

Introduction

The curriculum transformations that occurred in several countries have inspired many teachers to seek for more-demanding learning objectives to integrate into their practice. Teachers aim for classrooms where students are encouraged to perform challenging tasks such as to communicate, to question, to reflect and to collaborate (Chapman and Heater 2010). Denominated as "inquiry-based teaching", this practice quite often poses significant and diverse challenges to teachers, as reported by the research (Cengiz et al. 2011; Franke et al. 2007; Oliveira 2009). However, there has been an evolution in how this practice is understood (Stein et al.

L. Menezes (✉)
Escola Superior de Educação de Viseu, Viseu, Portugal
e-mail: menezes@esev.ipv.pt

H. Oliveira
Instituto de Educação da, Universidade de Lisboa, Lisbon, Portugal
e-mail: hmoliveira@ie.ul.pt

A.P. Canavarro
Universidade de Évora, Évora, Portugal
e-mail: apc@uevora.pt

2008), starting by a phase in which attention was focused primarily on the characteristics of the proposed tasks and the teacher's role in encouraging students' interaction in their autonomous work, and in the effort of listening and understanding students' thinking (Stein et al. 2008). The emergence of dialogical perspectives on the learning of mathematics (Ruthven et al. 2011; Wells 2004) has led to a growing emphasis on the role of the teacher in the moments of collective discussion and synthesis of the mathematical ideas (Canavarro et al. 2012; Cengiz et al. 2011; Stein et al. 2008), without ignoring the importance and complexity of the remaining phases of the lesson for the teacher's practice.

This article discusses the practice of a primary school teacher, based on a framework that we developed in the project *Professional Practices of Mathematics Teachers*, which relates the teacher's intentions and actions in an inquiry-based mathematics classroom, with the aim of contributing to a better understanding of this kind of practice.

What Is Inquiry-Based Teaching in Mathematics?

Over the last decades many perspectives on mathematics teaching have emerged in opposition to the common vision and practice of knowledge transmission. Proposals for the transformation of the school mathematics curriculum are anchored in new approaches for students to learn mathematics meaningfully (Ministério da Educação 2007; NCTM 2000). The ideas of an inquiry-based approach to teaching, in opposition to the transmission model, echo Dewey's perspectives (Chapman and Heater 2010; Towers 2010; Wells 2004), where students' activity is central to the development of the learning process. In an inquiry-based approach, students' mathematics understanding is promoted as they get involved in rich mathematics tasks, such as investigations or problems, using their previous knowledge and experience, and are not directed by the teacher to a predetermined solution.

The perspective adopted in our work also follows Wells (2004) concerning the importance of the social dimension of learning. He considers that knowledge "is constructed and reconstructed between participants in specific situations, using the cultural resources at their disposal, as they work toward the collaborative achievement of goals that emerge in the course of their activity" (p. 105). For this author, the knowledge process is situated in cooperation with others, integrating action and reflection about what one has learned in the process.

This vision of teaching and learning has strong implications for classroom organization. There are, of course, different possibilities for structuring lessons in an inquiry-based approach. We adopted a model of four phases that resembles the model by Stein et al. (2008), although their model encompasses just three phases: the launching of the task; the exploration of the task by the students, and the discussion of the task and systematization (the latter considered as just one phase by those authors whereas we consider it as two phases).

In the first phase of the lesson (the launching of the task), the teacher's main role lies in presenting the mathematical task to the class to guarantee that all students understand the proposal and that they feel mathematically challenged by it. At this time, the teacher also needs to perceive if there is an appropriate social and physical environment for the students' work as well as if they have the required resources in order to succeed in the solution of the task (Anghileri 2006).

In the next phase (the exploration of the task), the teacher goes along with the students' autonomous work in solving the task, which may occur individually or in small groups. By monitoring the students' work, the teacher intends to guarantee that all of them get involved and are able to develop their work. However, her/his comments and answers to the students' doubts should neither reduce the cognitive demand of the task (Stein and Smith 1998), nor hamper the emergence of different strategies, in order not to impede the mathematics discussion that will follow this phase.

Important decisions have to be made in this phase, upon which the success of the subsequent phases depends. On one side, the teacher has to guarantee that students prepare their presentations and, on the other side, s/he must select and establish the sequence of their presentations for the whole-class discussion (Stein et al. 2008). The practice of selecting students' strategies is intended to support the discussion of important mathematical ideas that will be "illustrated, highlighted, and then generalized" (Stein et al. 2008, p. 328). In this phase, the teacher also has to think about the sequence of the presentation of the strategies. This action also contributes to having fruitful discussions, since it allows the teacher to highlight the connection between mathematical ideas, and to promote the development of students' mathematical thinking (Cengiz et al. 2011).

After the students' autonomous work, the class comes back to working collectively, in order to discuss their work while the teacher synthesizes the main mathematical ideas. The teacher's task in this phase is particularly complex and demanding. Taking as reference the script of the lesson and the observation of the students' autonomous work in the previous phase, there is a large set of possibilities for the teacher's intervention during the discussion. S/he has to coordinate the interaction among different students, orchestrating the discussion, promoting the mathematical quality of the presented explanations and argumentations. It is important to guarantee the comparison of distinct solutions and the discussion of their mathematical difference and efficacy (Ruthven et al. 2011; Yackel and Cobb 1996).

Still concerning the whole-class discussion, Cengiz et al. (2011) refer to important moments as 'extending episodes'. Extending episodes occur when the discussion focus moves to a different mathematical idea. These authors consider three different types of extending episodes: (i) the teacher encourages students' mathematical reflection, helping them to understand, compare, and generalize mathematical ideas, to consider and discuss relationships among ideas, to use multiple solutions and to consider the reasonableness of an argumentation; (ii) the teacher encourages students to go beyond initial-solution methods, looking for alternative solutions and trying more efficient strategies; and (iii) the teacher encourages students' mathematical reasoning, involving the justification of their own ideas and strategies and the engagement with the justifications given by others. There are different types of instructional actions that the teacher performs in each extending episode, namely, eliciting, supporting, and extending actions (Cengiz et al. 2011).

Finally comes the systematization of the mathematical learning, which is more teacher-centered than the previous phase. The teacher has an important role in orienting students to synthesize the main mathematical ideas that emerged from the whole-class discussion (Anghileri 2006). This moment should help students to recognize the concepts and procedures involved in the activity they have developed. It should also help them to establish connections with their previous learning and to strengthen central aspects of mathematical ability such as the use of appropriate mathematical representations, problem solving and mathematical reasoning.

Following these ideas and the analysis of some Portuguese teachers' practices (Canavarro et al. 2012), we developed a framework with the goal of describing practices of inquiry-based mathematics teaching. The framework synthesizes the teacher's instructional actions and the main intentions behind those actions in each of the four phases of the lesson (Table 1). The intentions are related to two different but connected goals: (i) the promotion of students' mathematics learning, and (ii) the management of students' work and of the class as a whole.

An Inquiry-Based Mathematics Classroom: Teacher Célia's Teaching Practices

Rich descriptions of inquiry-based mathematics classroom practice not only expand our knowledge about this complex practice but also constitute a useful resource for teacher education. With this conviction, we have developed in the Project *Professional Practices of Mathematics Teachers*, a study involving four teachers at different school levels (1–4, 5–6, 7–9 and 10–12 grades) who develop inquiry-based lessons on a regular basis. Our study has the following objectives: (i) to understand inquiry-based mathematics classroom practice, identifying the actions that the teacher performs and relating them to his/her intentions; and (ii) to construct multimedia cases that illustrate inquiry-based mathematics classroom practices for use in teacher education. These teachers have participated in different teacher-education programs over the last years, in which also some project members have been involved. All teachers were experienced, with more than 12 years in the profession, and declared themselves comfortable with having their lessons observed and videotaped.

Next, we present the teaching practices of one of these teachers (Célia), working with one fourth grade class. We begin by reporting the access to the teacher and the school context in which she conducts her work and then we focus on a lesson that unfolded around one mathematical task entitled "Cubes with stickers".

Table 1 Intentional actions of the teacher in an inquiry-based classroom practice

	Promotion of mathematics learning	Class management
Launching the task to the students	*Guarantee the comprehension of the task by the students:* Clarifying unfamiliar vocabulary Mobilizing and verifying prior knowledge Setting goals *Promote students' adhesion to the task:* Challenging them to work Requesting an expected result Establishing connections to students' prior experiences	*Organize students' work:* Establishing time for each phase of the class Setting forms of work organization (individual, pairs, small groups, whole-class) Organizing class materials
Supporting students' autonomous work on the task	*Guarantee the development of the task by the students:* Posing questions and giving clues Suggesting representations Focusing productive ideas Requesting clarifications and justifications *Keep the cognitive challenge* Promoting students' reasoning Trying not to validate the mathematical correctness of students' answers	*Promote the work of students/groups:* Setting interactions between students Providing materials *Guarantee the production of materials for students' presentation:* Requesting records Providing appropriate materials Providing specific time to prepare the presentation *Consider the selection and sequencing of students' presentations:* Identifying solutions; more or less comprehensive, complete or formal Identifying solutions with common errors Sequencing the selected solutions

(continued)

Table 1 (continued)

	Promotion of mathematics learning	Class management
Orchestrating the discussion of the task	*Promote the mathematical quality of the presentations:* Asking for clear explanations with mathematical evidence Asking for justifications of the outcomes and representation used Discussing the difference and the efficacy of the presented solutions. *Promote interactions among students in the discussion of mathematical ideas:* Encouraging questioning for the clarification of ideas Encouraging analysis, debate and comparison of ideas Identifying and making available to discuss questions or errors in the presentation	*Create favorable environment for presentation and discussion:* Putting an end to students' autonomous work Providing the reorganization of places to focus on a common resource (whiteboard, overhead…) Promoting an attitude of respect and attentiveness towards different presentations *Manage relationships among students:* Setting the order of presentations Justifying the reasons for not submitting some students' work (to avoid repetition…) and ensuring groups' rotation to the next task Promoting and managing students' participation in the discussion
Systematizing mathematical learning	*Institutionalize concepts or procedures on mathematical topics:* Identifying key mathematical concept(s) from the task, clarifying its definition and exploring their multiple representations Identifying key mathematical procedure(s) from the task, clarifying the conditions of its implementation and reviewing its use *Institutionalize ideas or procedures concerning the development of transversal capabilities:* Identifying and connecting the dimensions of transversal capabilities present Enhancing the key factors for its development *Establish connections with prior learning:* Highlighting links with mathematical concepts, procedures and transversal capabilities previously explored	*Create an appropriate environment for the systematization:* Focusing students on collective systematization Emphasizing the importance of this phase of the class for students' learning *Guarantee written record of the ideas that result from systematization:* Recording in computer or other physical resources (boards, interactive boards, transparencies, posters …) which may be done by students or the teacher Requesting students to record their work

The Context of the Research

The teacher[1] in this study is a primary school teacher (grades 1–4), who, after her initial teacher education, did other courses in Education and Mathematics Education. During the current school year, she developed a teaching experience for promoting students' algebraic thinking, which follows the new Portuguese mathematics curriculum (Ministério da Educação 2007),[2] in an urban school located in the periphery of Lisbon. At the time the research took place, the school had 11 teachers, eight primary school classes and two pre-school classes. It is attended by students who are mostly Portuguese, with a low and middle socioeconomic status, and who live near the school.

The fourth grade class in this study is composed of 19 students, 7 girls and 12 boys, aged between 9 and 10 years old. This group of students has largely remained together since the first grade of primary school, but it is the first year that Célia works with them. She considers the class very motivated and committed to classroom work, but at the beginning of the school year she felt they did not develop the mathematical understanding she had expected.

For Célia, as for all other teachers of the study, we observed two or three inquiry-based lessons (L) in order to get familiar with her teaching and the students, and to choose one of these lessons to study and construct a multimedia case. The data were collected in three moments—before, during and after the observation of the lessons—and with different purposes. The day before the lesson, we made an initial interview (I_1) focusing on the teacher's lesson plan and trying to understand her options for the development of the lesson. From this initial interview we knew the task selected by the teacher, the structure she had envisioned for the lesson, the mathematics exploration she had intended to promote in class (including some questions to pose to the students), the mathematical strategies and representations she had anticipated, and the resources she had planned to use and provide to the students.

After this interview, we observed and videotaped the lesson. The data collection involved the use of two video cameras to register the moments when the teacher was working with the whole class, as well as interaction episodes between the teacher and the students when they worked autonomously on the task.

One week after the lesson, we had a second interview with Célia (I_2) concerning her practice, with the purpose of obtaining her perspective on the development of the lesson and the explanations of her actions. In this post-lesson interview, watching the videos of some classroom episodes that we selected helped the teacher focus on particular events and realise how she had acted.

---

[1] Célia is not the regular teacher of this class. She developed the project with the class in cooperation with the school teacher and taught the lesson concerning the teaching experience on algebraic reasoning, as the teacher requested her. For this reason, we will refer to Célia as the teacher since she assumed this role in these lessons.

[2] The program just started to be implemented in 2009, and for this class in the previous year, when the students were in the third grade.

Data analysis incorporated elements from the classroom videos and interviews, and was complemented with the analysis of the lesson plan and of students' written work on the task. Célia's practice was the first one to be analysed in this project and contributed greatly to the developed framework. The majority of the intentional actions portrayed in Table 1 are present in this lesson. However, due to space limitations, we only describe and analyse the main ones here.

The Lesson "Cubes with Stickers"

In this lesson, Célia's general objective was to develop the algebraic reasoning of her students; in particular, to recognize an increasing "pictorial sequence" and the variables implicit in it, to identify the relations among the variables, and to express the general rule for the sequence in natural and symbolic languages. Célia proposed the task "Cubes with stickers", which followed other tasks with similar purposes but simpler relations (Fig. 1).

Célia developed the lesson in four phases for about 110 min (for each phase, she estimated the time): "introduction of the task" (10 min), "work in pairs" (45 min), "collective discussion" (45 min) and "systematization" (10 min). Her plan reveals a set of diverse detailed actions for each phase, some of them related to the mathematical learning (for example, "The teacher presents a cube and asks the students what they know about that solid") and others regarding the management of the class (for example, "then [the teacher] distributes two cubes with stickers to each pair of students"). These teacher's actions written in the plan and visible in the videotaped lesson have underlying intentions. Therefore, we will analyse some of Célia's actions and intentions in this lesson phase by phase, using the collected data.

Cubes with stickers
Joana builds a construction using cubes and stickers. She connects the cubes through one of their faces and forms a line of cubes. Then she glues a sticker on each of the cubes faces. The figure shows the construction that Joana has made with two cubes, where she used 10 stickers.
1. Find out how many stickers Joana used in a construction with:
1.1. Three cubes; 1.2. Four cubes; 1.3. Ten cubes; 1.4. Fifty-two cubes.
2. Can you find the rule that allows you to know how many stickers Joana used in a construction with any given number of cubes? Explain your thinking.

Fig. 1 Task "Cubes with stickers" (Adapted from Moss et al. 2005)

Inquiry-Based Mathematics Teaching: The Case of Célia 313

Introduction of the Task

The lesson begins with a moment of collective work in which the teacher introduces the task to the students. Together they read the text and interpret the situation, using a pair of big cubes to better visualize and analyze the situation:

Teacher:	Our task today is titled "cubes with stickers".
Daniel:	Sounds cool!
Teacher:	Sounds cool, says Daniel, "cubes with stickers". And I'll show you the text of the task. And ask … Matilde, can you read, from there? Aloud.
Matilde:	Joana builds a construction using cubes and stickers. (…)
Teacher:	Who is able to explain, in your own words, what this task says? João …?
João:	She is making constructions with cubes and she is putting stickers on each side that is visible, but she did not put stickers in the middle and they say she used 10 stickers and it is correct because she didn't put stickers in-between the cubes. They were together and it can't be … I mean, you can do it, but it doesn't make sense.
Teacher:	Why doesn't it make sense?
Several students:	Because you could not see them.
Teacher:	So, in that construction she made with two cubes, she used 10 stickers. I have here one construction with two cubes. Let's see in a very quick way, how she did this construction. Rita, do you want to help? (L)

Considering the nature of this mathematical task, Célia reports that one of her main intentions in this phase of the lesson is to guarantee the interpretation and understanding of the task by the students (see Table 1). For this purpose, she poses many questions to the class, and listens to the comments and questions posed by the students:

> I guess it should be a presentation of a challenge and it should be an interpretation of what the challenge requires… I think that interpreting and understanding what is needed [for the task's solution] are the two objectives. (…) I have the concern of dedicating enough time to them, to understand and to put their questions and doubts, too… Questioning them and waiting for their doubts also allows me to realize if they understood. (I_1)

In this phase of the lesson, besides guaranteeing the understanding of the task by the students, the teacher also wants to help them engage with the task and accept the challenge of solving it:

> (…) To predispose them to the task. It is not only to understand it, but also to assume the task as their own … Here emerges also the challenge; take the task as something that I want to solve. (I_1)

At this phase of the lesson, Célia reveals great concern about the availability of materials (cubes) to facilitate understanding of the task:

The first difficulty that I anticipate students to have is connected with the issue of visualization. I thought about the introduction of the task using the cubes and making the construction as Joana did... so, use two cubes, what the task says is that Joana joined the cubes by one of the faces and placed one sticker on each of the remaining faces. [I intend] to do this process with the group, so that they understand how Joana made this construction, and at the same time to help them see the faces that had stickers. Lastly, I want to give each pair of students one equal construction. (I_2)

Development of the Task

In this phase of the lesson, Célia invites the students to work in pairs. She provides each pair of students with a set of small cubes glue and stickers, and with one sheet with the text of the task. She lets the students start working on their own. After a few minutes, she starts monitoring the students' work, listening to their ideas and thoughts and putting some questions and comments. Célia recognizes that it is not easy for the teacher to do this work because, in some cases, the students' thinking is not plain and clear:

> When (...) I go from one pair of students to another, or from group to group, I try to understand how they are, what they see, how they are working—and I try to do this through questions but sometimes it is not easy. (...) I try to follow their reasoning but it is not easy because sometimes... we see that the student is right... but "how did he think"?! (I_1)

Knowing how her students are reasoning as they work on the task is seen by Célia as crucial for the teacher's monitoring process, because it is important to be aware of their solutions to be able to help them, at the right time, in case they need it. However, Célia explains her difficulties in monitoring the progress of the students' work, which requires, in her opinion, "a balance", not easy to achieve, between leaving them on their own, promoting their autonomy, and giving them some orientation as needed. Célia is also convinced that this strategy will support non-uniform solving strategies: "The aim is that they work on the task, that they work in different ways, as they are able to". (I_1).

In order to maintain the level of cognitive demand of the task and not constrain student's thinking—and consequently future collective discussion—, Célia's actions at this phase reveal her intention not to provide validation to the students (as we see in Table 1). Therefore, more than to say if an answer is right or wrong – a temptation for many teachers, even if only through their facial expression – Célia challenges her students to think about their answers, often through questioning them:

> If I notice that they are going into a completely wrong direction, I try not to say: "This is not correct." [Instead] I try to say: "How did you start? Go back to the beginning ...". For example, if they are mistaken for ten cubes, I ask them: "Is this way of thinking the same you used with the three cubes?" ... I try to make them reach a fruitful way [of solving the task] ... (I_1)

After 30 min of work on the task, Célia provided a transparency to each pair of students and asked them to record the solutions on it. She explained us that this is very important for the dynamics of the discussion, because she needs to ensure that all students are able to show and share their work in an efficient way.

One of the episodes discussed with the teacher has to do with a short period of reflection in action that we observed near the end of this phase of the lesson. In the post-lesson interview she explained that she had to decide which students' strategies she would choose to be discussed collectively in the classroom. In spite of having previewed some general ideas about different complementary aspects that these solutions should contemplate, she still considers it difficult to decide on the spot:

> To choose which of those solutions are important for the collective discussion—and this is very difficult because... it requires almost a detachment that is difficult to manage in the classroom with all those requests. (...) "Of these eight different solutions, which ones interest me to discuss in this particular lesson?". And that is one of the things that, in terms of planning, I also try to preview, that is "which will I choose [to discuss]?". (I_2)

Célia also defines the order for discussing the selected solutions, taking into account the potential that she recognizes in those for promoting the understanding of the task and for exemplifying productive forms of representation, as she explains concerning the ones she chose to be presented first:

> I have picked this one because of the visualization that it facilitates to the students... This [other] was because of the picture and the way they represented it, they were very clear here, 1 + 1 + 1... This [other] I thought that if there were any questions or if there were still difficulties, this was more an attempt to solve them because of its simplicity. (I_2)

Figure 2 shows the solution on a transparency chosen by Célia to start the presentation in the next phase of the lesson–the collective discussion. At the top of the transparency, students write the following rule: "The rule to find the number of stickers is to make the number of cubes (×4) sides of a cube and plus the two lateral faces as well".

Discussion of the Task

The teacher emphasizes this phase of the lesson because for her it is an important mathematics learning opportunity for all students in the classroom: "It is extremely important not only because they have been working in pairs, but also ... what it is in terms of the mathematics classroom, what it is that is discussed in the group." (I_1).

Célia asks those students whose solutions have been selected to expose their strategies and explain their reasoning to everyone, peers and teacher, as well as to answer each others' questions. The teacher tries not to dominate the discussion. Therefore she gives a bigger role to the students in commenting and asking questions: "the attempt is that the moment and the presentation are theirs, the questions are posed to the group in action (...) and they are constantly questioning their peers and are interested..." (I_1).

In this lesson, students seem to correspond to the teacher's intention when they address their peers to clarify ideas:

Boy:	This is not clear...
Girl:	Well, this is very confusing.
João:	I'll explain better ... there are eight cubes, right? And we represent the four sides, front, back, up and down.
Teacher:	If you want to exemplify, you have cubes to make [the construction]. Say what you mean by that. (L)

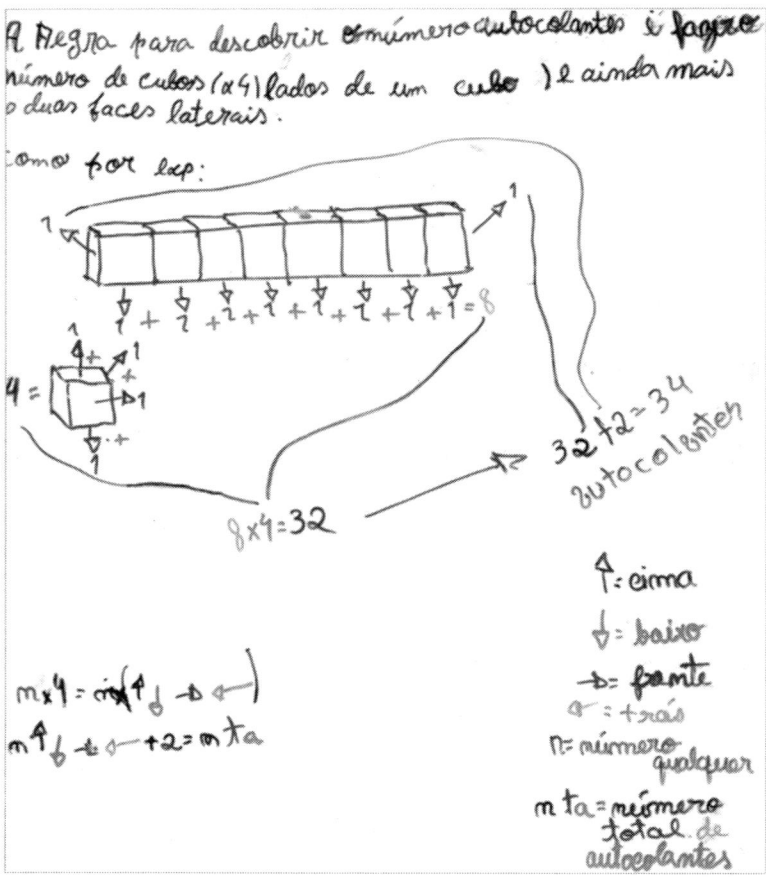

Fig. 2 First presentation by a pair of students

At this phase of the lesson, Célia's main intention is to focus on the orchestration of the students' interventions in order to promote the clarification of the emerging mathematical ideas (as we see in Table 1). For Célia, the discussion goes beyond the correction of the task's solution or simply making a presentation of different strategies and solutions. In her perspective, this moment should represent a mathematical enrichment for the students:

> I think if I present different [solutions], if they have the capacity to present different representations, and if then I can connect and establish connections between those ones, this is much richer than presenting just one (…). The idea is also to let emerge what comes from them. (I_1)

The next solution (Fig. 3), which Célia chose because the students use a symbolic representation and rely less on a visual representation (despite having used colors to make sense of the letters they use in the written expression), complies with

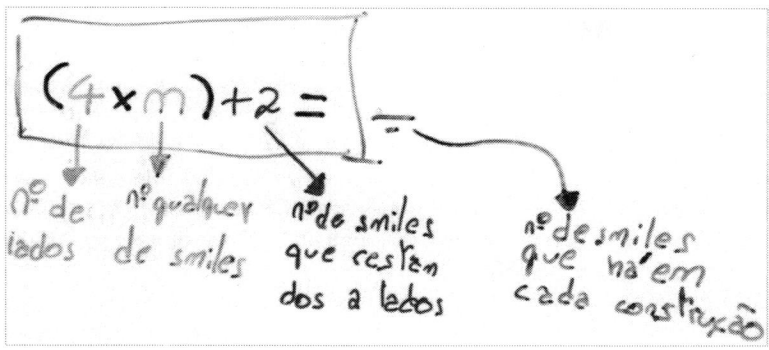

Fig. 3 Solution presented by another pair of students

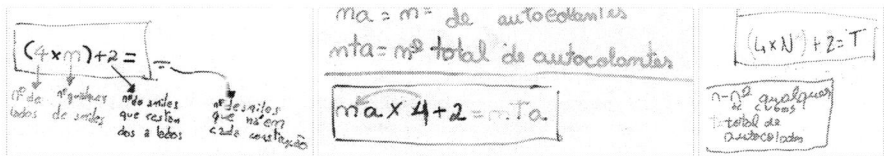

Fig. 4 Célia puts three solutions together on the overhead projector

her intention of introducing solutions to promote reasoning as a condition for the learning of mathematics. In Fig. 3, in red, the students write "number of sides"; in green, "number of smiles"; in blue, "number of smiles remaining on two sides"; and, finally, in brown, "number of smiles that are in each construction".

Although at this stage, the teacher is not the main protagonist in the classroom discourse, her role is not limited to managing the students' interventions or to listening to them. In this class, after several students' presentations, Célia recovers three of these and puts them together on the overhead projector to help the class to compare them (Fig. 4).

Her explanation for this action is based on her view of the role of the discussion of the task for the students' learning that particularly values an analysis and confrontation of ideas:

> There's one moment when I present three solutions. The idea is that there is a confrontation. [Discussion] is a learning moment, so ... it cannot be [only] a presentation, nor is it a correction, because it is not that ... The time is used for confrontation. It is used to think together about the different solutions, the different representations. This must emerge from the presentation ... it is a goal. ... To promote this critical sense: "In this situation, which is the representation, the solution ... what is the way of thinking that, in fact, helps me the most?" Because I can have many solutions, but, maybe, there is one that I can choose. (I_2)

Systematizing Mathematical Learning

Following up the discussion of the task, in which the class (the teacher and students) has agreed on a rule to calculate the number of stickers of any cube building, Célia plays a more directive role and aims for the systematization of the mathematical learning (see Table 1). For this, she presents an incomplete table which relates the number of cubes with the respective number of stickers on a transparency which she had already prepared, and challenges the students to think about another question that extends the previous task (Fig. 5).

Célia explains her intention for this phase of the lesson. She wants to build on the previous discussion of the task to emphasize the importance and the power of the rules with letters to explore situations when they work with a variable.

> After the work in pairs and the collective discussion, there is a part of systematization where I confronted... I started with this table, this number of cubes [52] [and I asked]: "How did you find the number of stickers?". So, we wrote the rule... (I_2)

Célia recalls also the inverse operation, claiming that the rule is read "the other way around", establishing connections with knowledge and procedures already discussed with students in previous situations:

> [I asked]: "And now, knowing the number of stickers, how do we know the number of cubes?" And then, there's the confrontation of the rules in order to see the inverse operations that they explored. And this was done in systematizing, at the moment of the systematization. (I_2)

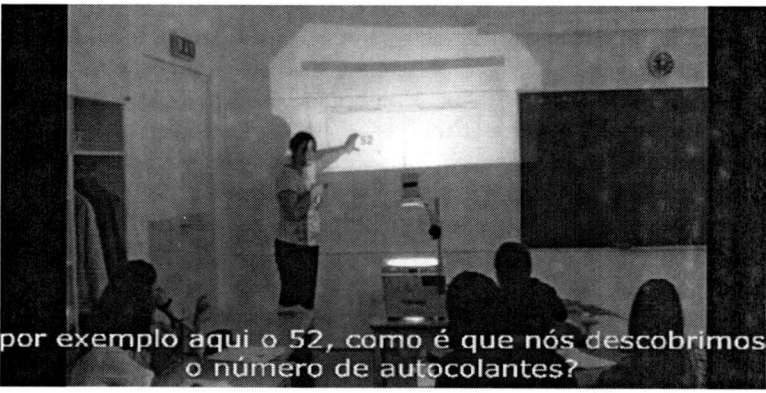

Fig. 5 Célia puts an incomplete table on the overhead projector to extend the initial task

Final Considerations

In the project *Professional Practices of Mathematics Teachers*, our aim is to understand mathematics inquiry-based classroom practice. We illustrated and analysed the practice of one primary school teacher supported by a framework that we developed (Table 1) and which connects the teacher's intentions and actions in an inquiry-based mathematics classroom. The lesson evolved through four phases with different purposes, and Célia performed set specific actions according to her goals that involve demanding learning objectives, as students solve a challenging task, collaborate, communicate and reflect altogether on their work.

The analysis of the teacher's practice, according to the two main purposes in the framework (the promotion of students' mathematics learning, and the management of the students' work and of the class as a whole) shows that these are closely interrelated and are present in all phases of the lesson, gaining specific contours in each one. Naturally, the actions performed by the teacher concerning class management are intended to create better conditions for students' mathematics learning. In fact, Célia performed the actions related to management with a great sense of their impact on the dynamic of the classroom and on the mathematics that students can learn. For instance, she realizes the importance of guaranteeing a good understanding of the task by the students, and therefore, she provides some materials for students (in this case, the cubes were important for the visualization of the sides with stickers). She also knows that if she does not pay attention to the order of the presentations of the solutions by the students, she may compromise the quality of the collective mathematical discussion and constrain the opportunities for student's learning. As several studies show, promoting whole-class discussion is a challenging undertaking for the teachers (Cengiz et al. 2011; Stein et al. 2008), which is also articulated by Célia. Therefore, she realizes the importance of making good preparation for this moment of the lesson, thinking on the mathematical ideas to be discussed and related questions that encourage students' mathematical reasoning.

Like Cengiz et al. (2011), we observe the teacher's effort in contributing to extending the students' thinking and the important contribution of the last phase of the lesson in this. There, Célia pushes the students to establish connections with prior learning, based on the work they have done in the lesson. This reinforces the pertinence of the proposed framework, which considers four phases for these lessons, the goal of each phase as well as their interrelatedness.

As it happened with other teachers who have collaborated in this research on mathematics inquiry-based classroom practice (Canavarro et al. 2012), we observe that it is globally a complex practice, which develops over time and arises from a strong intentionality of the teacher (Stein et al. 2008). The developed framework *Intentional actions of the teacher in an inquiry-based classroom practice*, although it was not intended to be normative for the teacher practice, may contribute to understanding that the demanding goals for students' learning require a high level of professional practice. This comprehensive framework is being used to analyse the teachers' practice in the context of multimedia cases of inquiry-based classroom

(Oliveira et al. 2012) in teacher education settings, where we expect (prospective) mathematics teachers to construct new visions and professional knowledge about how this demanding practice may develop.

Acknowledgements This paper is supported by national funds through FCT – Fundação para a Ciência e Tecnologia in the frame of the Project P3M – *Professional Practices of Mathematics Teachers* (contract PTDC/CPE-CED/098931/2008), coordinated by João Pedro da Ponte.

References

Anghileri, J. (2006). Scaffolding practices that enhance mathematics learning. *Journal of Mathematics Teacher Education, 9*(1), 33–52.

Canavarro, A. P., Oliveira, H., & Menezes, L. (2012). A framework for mathematics inquiry-based classroom practice: The case of Celia. In *Pre-proceedings of 12th international congress on mathematical education* (pp. 4137–4146). Seoul: International Commission on Mathematical Instruction (ICMI).

Cengiz, N., Kline, K., & Grant, T. J. (2011). Extending students' mathematical thinking during whole-group discussions. *Journal of Mathematics Teacher Education, 14*(5), 355–374.

Chapman, O., & Heater, B. (2010). Understanding change through a high school mathematics teacher's journey to inquiry-based teaching. *Journal of Mathematics Teacher Education, 13*(6), 445–458.

Franke, K. L., Kazemi, E., & Battey, D. (2007). Mathematics teaching and classroom practice. In F. K. Lester (Ed.), *Second handbook of research on mathematics teaching and learning* (pp. 225–356). Charlotte: IAP.

Ministério da Educação. (2007). *Programa de Matemática do Ensino Básico*. Lisbon: DGIDC.

Moss, J., Beaty, R., McNab, S. L., & Eisenband, J. (2005). *The potential of geometric sequences to foster young students' ability to generalize in Mathematics*, [PowerPoint file] presented at the algebraic reasoning: Developmental, cognitive and disciplinary foundations for instruction, September 14, 2005. Washington, DC: The Brookings Institution. Retrieved August 5, 2014, from http://www.brookings.edu/events/2005/09/14-algebraic-reasoning

National Council of Teachers of Mathematics [NCTM]. (2000). *Principles and standards for school mathematics*. Reston: NCTM.

Oliveira, H. (2009). Understanding the teacher's role in supporting students' generalization when investigating sequences. *Quaderni di Ricerca in Didattica (Matematica), 19*(Supplemento 4), 133–143.

Oliveira, H., Menezes, L., & Canavarro, A. P. (2012). The use of classroom videos as a context for research on teachers' practice and teacher education. In *Pre-proceedings of 12th international congress on mathematical education* (pp. 4280–4289). Seoul: International Commission on Mathematical Instruction.

Ruthven, K., Hofmann, R., & Mercer, N. (2011). A dialogic approach to plenary problem synthesis. In B. Ubuz (Ed.), *Proceedings of the 35th conference of the international group for the psychology of mathematics education* (Vol. 4, pp. 81–88). Ankara: PME.

Stein, M. K., & Smith, M. S. (1998). Mathematical tasks as a framework for reflection: From research to practice. *Mathematics Teaching in the Middle School, 3*, 268–275.

Stein, M. K., Engle, R. A., Smith, M. S., & Hughes, E. K. (2008). Orchestrating productive mathematical discussions: Helping teachers learn to better incorporate student thinking. *Mathematical Thinking and Learning, 10*(4), 313–340.

Towers, J. (2010). Learning to teach mathematics through inquiry: A focus on the relationship between describing and enacting inquiry-oriented teaching. *Journal of Mathematics Teacher Education, 13*(3), 243–263.

Wells, G. (2004). *Dialogic inquiry: Towards a sociocultural practice and theory of education.* Cambridge: Cambridge University Press.

Yackel, E., & Cobb, P. (1996). Sociomathematical norms, argumentation, and autonomy in mathematics. *Journal for Research in Mathematics Education, 27*(4), 458–477.

Using Drama Techniques for Facilitating Democratic Access to Mathematical Ideas for All Learners

Panayota Kotarinou and Charoula Stathopoulou

Abstract This chapter explores the dynamics of Geometry teaching in a classroom which uses "Drama in Education" techniques as a process that contributes to democratic access to mathematical ideas by all pupils. We describe a teaching experiment which aimed to motivate and actively engage through drama all 26 pupils of an 11th grade class and to encourage them to develop a critical attitude towards mathematical knowledge as being absolute, objective and irrefutable. The teaching experiment entitled "Is our world Euclidean?" was a drama-based teaching of the process of axiomatic definition of Euclidean and Non-Euclidean Geometries interrelated to the history of Euclid's 5th postulate. Our research reveals considerable evidence for the effectiveness of drama techniques as an alternative approach to creating appropriate learning conditions, activating all students as evidenced by their participation, and contributing to their development as critical citizens.

Introduction

A critical perspective promotes the need for a political vision of mathematics education and argues that one of its main objectives should be support for the creation of citizenship through experience and active participation in school mathematics education. According to Skovsmose (1996, p. 1267), mathematics education has the potential to contribute to the development of critical citizenship and to support democratic ideals as well as the development of citizens who participate actively and responsibly in discussions and processes in which the individual takes part in personal and public decisions. He stresses the "formatting power" of mathematics, an interaction between science and power, in which mathematics is not only a descriptive tool, but also a formatting tool, having the ability to affect, to produce or restrict social activities or to be a source of decision-making and the justification of actions (Skovsmose 1998; Skovsmose and Yasukawa 2000).

P. Kotarinou (✉) • C. Stathopoulou
Department of Special Education, University of Thessaly, Volos, Greece
e-mail: pkotarinou@uth.gr; hastath@uth.gr

The modern society requires, more than ever, mathematically competent citizens. A challenge for teachers, who are concerned about matters of democracy, is the provision of adequate mathematics education to the greatest number of students possible, as well as the cultivation of a critical citizenship through mathematical education. The notion of "democratic access" to mathematical ideas, which has been the focus of many mathematics educators, refers to the ability of providing mathematics education to all citizens, and, according to Skovsmose and Valero (2002), defines a mathematical education that contributes to the establishment of democratic social relations.

"Democratic access" is the creation of the appropriate learning conditions where all students have the chance to become critical thinkers and decision-makers while also developing the ability to solve and understand increasingly challenging problems in both classroom settings and in their life conditions. According to Skovsmose and Valero (2002), the concept of "democratic access" of mathematics education concerns three main levels: the class, the school, and the local and global society. Democratic relationships in a mathematics class, between teachers and students and amongst students, are considered those relationships that allow collaboration, discussion, dialogue, communication, contribution to and criticism of the mathematical content of the class.

According to Ernest (2008), mathematics can be a tool for democracy in the hands of every educated citizen, fostering citizens who are able to recognize, interpret, evaluate and criticize mathematics embedded in social and economic systems. Skovsmose (2004) argues that mathematics education has the potential to contribute to the cultivation of critical citizenship and that it can support democratic ideals. But, as he says in an earlier article titled "Mathematics and Democracy", we cannot expect any cultivation of such competences at school unless the teaching-learning process is based on dialogue and the curriculum is determined by the class itself. Bishop (1999) contends that for mathematics education to become more democratic—i.e. to address as many people as possible—the values transmitted through mathematics education should be scrutinised.

One of the main questions emerging in this case concerns the kind of learning environments and classroom practices that are needed in order to promote democratic access to mathematical ideas for all students.

Gerofsky (2009) claims that a classroom, where students' active participation and ideas are clearly valued, might prepare the youth to live in a more participatory political system where their voices and ideas are taken seriously. She suggests to reconfigure our conception of mathematics classrooms, learning to include participatory performance both as a model and a means to stimulate democratic participation.

We claim that Drama in Education (DiE) is a teaching technique which promotes participatory performance and hence democratic access to mathematical ideas for all pupils. At the same time, it can create a framework that offers innovative ways for the students to develop critical thinking about 'mathematics in society'. It also cultivates those skills and abilities necessary for a responsible citizen in a democratic society—main features of the teaching in the context of critical mathematics education.

Drama in Education and Critical Mathematics Education

DiE, according to O'Neil and Lambert (1990, p. 11), is a mode of learning in which pupils can learn to explore issues, events and relationships through their active identification with imagined roles and situations. It is an art form with pedagogical character that has as the basic aim the understanding of ourselves and of the world (O'Neil and Lambert 1990). It is also a dynamic and creative methodological tool for the various contents of the curriculum (Somers 1994). According to Heathcote (1984, pp. 203–204), "drama can be a learning tool because it demands co-operation, makes factual experience come to an active employment, uses fiction and fantasy but makes people more aware of reality, stresses agreeing to all trying to sustain mutual support while allowing people a chance to work differently, makes people find precision in communication, stresses the use of reflection." Drama is essentially a social art and includes by definition contact, communication and negotiation of meaning. Students are partners contributing to the process new perspectives, opinions and ideas and a number of previous experiences and therefore they are able to teach and learn from each other (Catterall 2007). "Within the safe framework of the make-believe, individuals can see their ideas and suggestions accepted and used by the group, they can learn how to influence others; how to marshal effective arguments and present them appropriately" (O'Neil and Lambert 1990, p. 13). At the cognitive and metacognitive domain, DiE may foster students' divergent thinking and creativity, and liberate imagination and originality (Annarella 1992). Further, it may encourage reflective thinking (Neelands 1984), the development of problem-solving abilities (Bolton 1985; De La Roche 1993) and decision-making skills (De La Roche 1993). In addition, DiE may stimulate students' self-confidence and foster their self-esteem, flexibility and tolerance (Yassa 1999). It may also enhance communication and conflict resolution skills (Catterall 2007). The use of drama-related techniques in education has been proposed, among others, as a context that offers innovative ways for teaching varied disciplinary curriculum areas and as a creative way for students to reflect on broader—social and political—learning issues (Kotarinou et al. 2010). DiE is a very important pedagogical tool of critical pedagogy for increasing the awareness of social justice (Teoh 2012).

One of the arguments for the educational value of drama is that it involves critical thinking, and that critical thinking can be promoted by drama (Bailin 1998). According to Bailin, critical thinking is not a generic skill but highly contextual. This involves thinking about what to believe or how to act in problematic situations, demanding of the thinker to make reasoned judgements. The thinker must draw on a range of intellectual resources in order to respond to those challenges, including knowledge, strategies and habits of mind. Critical thinking can best be supported by infusing it into every curricular practice in which our students are involved (Bailin et al. 1999).

According to Appelbaum (2004, p. 310), "for the last century, teachers of mathematics have been figuring out how to drop the teaching of critical thinking in favor of establishing environments that allow for the critical thinking that is possible through discussion and interaction." Such a learning environment fosters students' cooperation, enables students to analyse and evaluate the mathematical thinking and strategies of others in their class, and allows students to feel free and safe to express their ideas as in a real democracy (Appelbaum 2004).

We claim that the use of drama in teaching contributes to the creation of a democratic community of inquiry, favouring the development of critical thinking. Through working in teams, the students get opportunities, encouragement, and support for speaking, writing, reading, and listening mathematics while cultivating critical thinking skills.

The Experiment: Design and Implementation

This chapter, drawing on our qualitative research that examined the potential of DiE with regards to the learning and teaching of mathematics, focuses on the possibilities that DiE entails as a mean that promotes "democratic access" to mathematical ideas for all learners.

Setting: Participants, Framework and Methods

Data for the reflections presented in this chapter arose from our explorations of the contribution of drama in the formation of a teaching that fosters students' participation and critical thinking in geometry in higher secondary schooling. The participants included a group of 26 11th grade students and their secondary school teachers from varied curricular areas (mathematics, physics, drama, language etc.) who all came together to work for this particular project. The setting of our study was an urban school located in the greater area of Athens. It took place in the academic year 2010–2011. Its duration was 4 months and it was based on weekly and daily meetings.

In terms of the methods used, we designed and implemented an interdisciplinary didactical intervention based on a teaching experiment methodology (Cobb et al. 1991, cf. Chronaki 2008 for an overview of its potential). The teaching experiment focused on a detailed design of the teaching of the axiomatization of Euclidean and non-Euclidean geometries as well as the history of Euclid's fifth postulate through using DiE techniques. The initial theme—the axiomatization of geometry by Euclid—did not presuppose much prior knowledge, so that all students were able to get involved in the activities. Utilising ethnographic techniques as part of the teaching experiment implementation helped us to gather data:

- participant observation of students' mathematics classes in their ordinary lesson and participatory observation during the teaching experiment for exploring the way DiE techniques affect the instructional classroom norms and practices,

- questionnaires concerning students' beliefs about geometry before the teaching experiment,
- student interviews (2 months later) exploring the mathematics achievement and retention of knowledge of each student, their image of geometry as the result of the teaching experiment as well as the reasons for students' motivation and active participation through specific acts within the teaching experiment.

All presentations were videotaped and recorded in order to analyse the proper use of mathematical notions in their dialogues, while some episodes of students' group work were analysed regarding the role of drama as a mediating tool for the negotiation of meaning and the development of understanding.

Designing a Geometry Project: 'Is Our World Euclidean?'

The teaching experiment entitled: "*Is our world Euclidean?*" was carried out by the first author, as researcher in teaching role as part of an engagement in participant-observation in 25 teaching periods during 7 weeks in geometry, history, language, literature and Ancient Greek language classes. Our teaching experiment was designed to introduce students to the discovery of non-Euclidean geometries, an epistemological rupture in the history of mathematics, assuming that this would have the potential to challenge students' beliefs about geometry and generally their image of mathematics as a science of absolute truth.

Through the design of the teaching experiment we aim to motivate and actively engage all students of a mathematics class through drama; to develop students' understanding of the axiomatization of Euclidean, hyperbolic and elliptic geometry and the general concept of the axiomatic foundation of a science; and to challenge students' stereotypical beliefs about geometry, encouraging the students to develop a critical stance towards mathematical knowledge as absolute, objective and infallible. Specifically, the following stages were encountered as entries to the teaching intervention:

- a lecture enhanced with digital projection was provided as an introduction to the topic
- the students were asked to work in teams, using appropriate bibliographical recourses such as digital or haptic material and books, for them to acquire suitable knowledge regarding their presentations
- a summing-up activity ensued, where there was ample chance to discuss ideas in public
- the teams prepared their presentations under drama conventions
- after rehearsing students performed their presentations
- a concluding and reflective session followed, while at the end of the teaching experiment an entire class period was dedicated to the same purpose.

The project consisted of the following units:

1. **Euclid's Elements and the axiomatisation of Euclidean geometry** (6 hours)

   The main teaching aims of this unit were the students' understanding of Euclidean geometry as an axiomatic system and a clarification of different concepts such as: definition, 'common notions', postulate, and proof. In the Ancient Greek class, the relation between Euclid's axiomatic foundation of geometry and Aristotle's *Logic* was presented by the second author. Students were encouraged to read and translate from ancient to new Greek excerpts from Euclid's Book I concerning the definitions, the postulates and common notions. During the geometry class, students, divided into six larger groups, were asked to answer questions concerning the mathematics of the text after having studied appropriate literature, e.g., some relevant excerpts of the book *The Historical Roots of Elementary Mathematics* (Bunt et al. 1981). Then students prepared their presentations on the postulates, the common notions and some definitions of Euclid's axiomatic foundation of geometry and made their performances using drama techniques such as 'role-playing', 'reportage', 'alter-ego', and 'interview'. As we realized, listening to students' answers during the reflection, some students had not yet understood the axiomatic foundation of geometry by Euclid, we used the drama technique 'teacher in role' for a short recapitulation of the topic. The researcher (first author) in the role of Euclid and the maths teacher of the class in the role of Hilbert talked about their respective foundation of geometry. Then the researcher 'out of role', addressed the students with open questions in order to identify how the majority of students had understood the five postulates set by Euclid. Through the intervention, the students were able to develop a broader understanding concerning Euclid's role within the foundations of geometry.

2. **Euclid and the historical, cultural and political frame of his era** (4 hours)

   The main aim was for the students to understand mathematics as a cultural and human creation whose development is influenced by the political, economic and cultural environment. In order for the students to know the historical context in which Euclid's *Elements* were written, a presentation was held by the first author in the history class, concerning Alexandria in the Hellenistic period, while excerpts from Denis Guedj's (2000) book *The Parrot's Theorem* were read concerning the history of this era, as well as the reasons for the blossoming of mathematics in this historic period and area. Knowing that dramatisation is an important tool in the repertoire of a teacher for humanizing and contextualizing the development of mathematical concepts (Hitchcock 1996; Ponza 2000), the chapter 'Euclid's conceit' from J. P. Luminet's (2003) book *Le Bâton d'Euclide*, presenting Euclid and his era, was read expressively by some students in the literature class, while some scenes of the same chapter concerning differences in thought between Pythagorians and Euclid as well as historical anecdotes about Euclid were dramatised.

3. **'History in shadow': The controversy of Euclid's fifth postulate until the eighteenth century** (5 hours)

   The purpose for all the activities of the unit 'History in shadow' was to get the students to understand, through the history of contestation of the fifth postulate, how an intellectual problem—in our case a mathematical problem—can be a challenge for mathematicians for almost twentieth centuries. Through the students' acquaintance with the problem, especially with the ongoing and unsuccessful attempts to solve it, we also aimed to challenge the student's stereotypical images of mathematics as an absolute and certain body of knowledge. At the same time, on a declarative cognitive level, the students have been expected to learn various propositions equivalent to the fifth postulate characterizing Euclidean geometry and to learn about the logical error of Petitio Principii (circular argument), an error that the students themselves make quite often in proofs. In the mathematics class, the history of the fifth postulate as one of the five set by Euclid and the challenging of it as an independent of the other postulates by mathematicians of classical Islam and by Western mathematicians until the nineteenth century were presented. More specifically, the unsuccessful attempts of the Arab mathematicians Thabit ibn Qurrah, Al-Haytham, Omar Khayyam and Nasir al-Din al-Tusi as well as of Saccheri and Lambert to prove the fifth postulate were presented, while their errors in all these efforts—using equivalents to the fifth postulate proposition to prove it—were interpreted. During the history class, the development and the reasons of the development of mathematics in the Islamic world, from the eighth to the thirteenth century, were discussed. A short extract from Denis Guedj's book *The Parrot's Theorem* (pp. 269–272), which refers to the "House of Wisdom" in Baghdad and its role in the collection and translation of the work of the ancient Greeks, was read. A combination of 'shadow theatre' and 'role playing' was used for presenting all these unsuccessful efforts as well as the errors made, and the students prepared their performances after having studied relevant excerpts from the books *A history of mathematics* (Boyer and Merzbach 1997) and *Great moments in mathematics* (Eves 1983).

4. **János Bolyai, Lobachevski, Riemann: The creators of non-Euclidean geometries** (3 hours)

As the biographical allusions serve the purpose of humanising concepts (Ponza 2000), the students studied the biographies of the two latter mathematicians in their Greek language class. The aim of students' study of the biographies of the founders of non-Euclidean geometries was to change stereotypical images of mathematicians and see them as people who are not cut off from their social environment but who participate in it facing personal problems. They are persons with a special interest in mathematics who spend a lot of time studying the works of other mathematicians of the past. For Lobachevski's biography, the book Men of Mathematics was used, a book on the history of mathematics written by the mathematician Bell (1993). For presenting Lobachevski, the drama convention 'role-on-the-wall' was used. The students were asked to write within the outline of the figure ideas and feelings that, according to the various elements of the biography, Lobachevski himself might have had, and outside the outline their own feelings and thoughts about him. During the second class period, students were asked to read the biography of Riemann, from Boyer's and Merzbach's book, and to underline the fundamental points of it as well as the points that impressed them. To present the second biography, the drama technique 'portrait' was utilised, in which one student took the role of Riemann and was asked to come alive to answer questions and queries of the students. In the third teaching period, the technique, 'letters' was chosen, in which three students in the roles of Gauss, Farcas Bolyai (father) and Janos Bolyai (son) read three letters from the correspondence between them concerning their work with the fifth postulate, taken from Mankiewicz's (2002) book *The History of Mathematics*. Two more DiE techniques, 'conscience alley' and 'conflicting advice', followed. The 'conscience alley' concerned Janos Bolyai's struggle to continue or to stop working on the fifth postulate. At a crucial moment, when he had to take a decision, Bolyai walked between two rows of students who gave him different advices on the decision he had to take.

Bolyai's conflict either to prove the fifth postulate or to replace it with another non-equivalent postulate was manifested by the technique 'conflicting advice', i.e. by 'voices in his head' which gave him conflicting advice. The students took the role of 'voices' arguing in favour of one or the other mathematical approach he had to follow. The scene ended with Gauss's reading of his letter to Farcas Bolyai, in which he refers in a very typical way to the work of Janos, a letter that embittered and discouraged Janos regarding his mathematical work.

5. **Hyperbolic geometry and the Poincaré model** (6 hours):

    The unit's aims were to enable the students: (a) to perceive the axiomatic foundation of hyperbolic geometry, (b) to perceive the role of the postulates in every axiomatic system, (c) to redefine Euclidean geometry, (d) to perceive the role of a model in mathematics, (e) to gain a deeper understanding of geometry by comparing the similarities and differences of the hyperbolic geometry with the Euclidean. In the history class, a presentation with historical data, key concepts and theorems of hyperbolic geometry, elements of the Poincaré model as well as of the works of Escher was conducted by the first author. A discussion with students followed about the notion of an axiomatic system, about its consistency, the independence of the axioms, and the meaning of a model of an axiomatic system. In the Modern Greek language class (1 hour), the students studied in groups excerpts from the chapter 'Platterland' from Ian Stewart's book *Flatterland* concerning Poincaré's model of hyperbolic geometry, in order to prepare a radio show with the same name. In the geometry class (1 hour), the students used the interactive Java software 'NonEuclid' by J. Castellanos, Joe Austin, Ervan Darnell, and Maria Estrada for visualizing the Poincaré model, the axioms and basic concepts of this non-Euclidean geometry. The students worked on computers using worksheets in groups of two or three. They explored the model by drawing points, lines, segments, angles and perpendiculars to a given straight line. They also measured segments and angles and they wrote their comments about the construction of equal circles, line segments of equal length. Finally, the axiomatic foundation of hyperbolic geometry was taken from the Poincaré model. Students prepared their own texts for the 'radio broadcasts' on hyperbolic geometry and its Poincaré model and presented them in two teaching periods in the Modern Greek language class. The radio broadcasts were presented from behind a screen so that the students were not seen by the audience.

6. **Spherical geometry and the axiomatization of elliptic geometry** (1 hour)
   The students studied the axioms and the basic notions of spherical and elliptic geometry through the 'Lénárt sphere' and other haptic tools, like table tennis balls (Lénárt 1996). A role-playing activity was used for the evaluation of their developing knowledge.

7. **The film** (during the class breaks)
   A documentary film entitled 'Our lives with Euclid' was created around these drama-activities with the students in the role of narrators who wrote their own texts of the narration after having studied the relevant literature. A student cameraman filmed all the narrations and two others wrote and performed the music for the film. Students were taught the program 'Movie Maker' for making their own montage of the film, but there was not enough available time and the montage was finally made by the first author.

Observations

In this section, the outcomes of implementing the above teaching experiment will be discussed in relation to its impact on students' experiences in geometry and drama's contribution to critical mathematics education.

Drama-Based Teaching and Students' Learning of Geometry Notions

Drama techniques helped new practices to be formed in a class of active students who cooperated towards the development of mathematical knowledge. The social interaction and communication between the students created opportunities for the negotiation of meaning and the development of understanding. During the students' cooperation with each other, each student's personal solutions were explained to the members of the team and every member also tried to understand the explanations of others.

Stefanos: … and you would see the satisfaction on the other guy's face, the moment that he perceived what you had explained to him. I understood what he said and he was happy.

As students said, they understood better the mathematical concepts through explaining them to classmates.

Stefanos: Some of the concepts were too difficult for me to explain to the others, but trying to help the others you teach yourself …, trying to explain it to the other, you actually see how it's done.

Sofia: A girl told me: I got it better from you than from the teacher in the class.

Regarding students' understanding of the axiomatisation of Euclidean and non-Euclidean geometries, the analysis of the dialogues in students' drama performances suggests that students conceived the mathematical notions that they had to present, integrating them correctly in their performance dialogues.

Students' retention of knowledge of these concepts was evaluated via relevant questions in specifically organised interviews 2 months after the end of the teaching experiment. Students' responses indicate that drama based teaching had positively affected the students' learning and retention of the axiomatisation of Euclidean and hyperbolic geometry. They stated that drama had made learning geometry notions easier, more meaningful and permanent.

Ioanna: In the way we did all this, I think that one learns more easily. What we learned (through this procedure) wasn't designed for learning by heart or to show to the teacher that we really learned it. I think it was easier to learn in this way and to keep this knowledge in our mind.

Drama in Education and Critical Mathematics Education

Through the observation and the interviews we realized that drama was a significant factor for all students' active participation, it provided the framework for the development of critical thinking and also cultivated in students abilities and skills necessary for a responsible citizen.

1. All students' active participation

From observation and interviews it appears that the innovative drama-based teaching motivated all students' active participation in the teaching experiment.

Tzina: In this particular project we all worked, irrespective of the direction of our majors.

Christos: In the project, we worked in teams, we had a good time, we all participated.

Often when students work in groups, some leave all the responsibility and initiatives to the so-called 'good' students. Through DiE conventions all students participated in the preparation of the dialogues for the scene, with more or less contribution to the presentation.

Stefanos: The thing that most impressed me was that in all other cases when we have to do teamwork, in the end, only one person does the writing, while through the methods we used, all students had to work to reach the final result.

The main motivation for the participation of all students in the teaching experiment, according to them, was the drama activities (cf. Fig. 1).

Fig. 1 Dramatization

Yolena: Personally, I was yearning for the next class with Ms Kotarinou. Every time the door knocked I wished we had another class with her rather than another boring session.

This was due to the following reasons:

- to the innovative nature of drama based teaching of geometry.

    Virianna: It was something new, it was so very different for everyone, for me personally, without having any particular relation with theater and everything around it, it was very original and it was so very different for everyone, because something similar has not been tried before.
- to the experiential nature of drama.

    Antony: We felt as if we were discussing it with others for them to learn as we were the teachers Euclid, Gauss and Lobachevski. We were more involved in profound knowledge.
- to drama creative character, which gave them the opportunity to express themselves in their own way and for everyone to bring out his/her special abilities and talents. The students felt that the sketches were their own creation and due to this they wanted to have the best result in their presentations.

    Antony: Without drama this would not have been a part of us. Through drama we prepared it, we presented it, we felt it as our thing.

    Virianna: … it was very creative and everyone could show his abilities, the different ones that everybody has.

Fig. 2 Radio broadcast

- to the atmosphere of game, joy and fun that drama created (cf. Fig. 2).

   Petros: All this dramatic, theatrical thing is like a game ...

   Zoe: The way we did it, co-operating many kids together, was hilarious and fun and made time fly quickly and enjoyably, leaving us wanting more.
- to the impact that drama had on students' identities as mathematics learners, stimulating their confidence as individuals who can express their opinion, participate in the writing of the texts and in the presentation of the different scenes.

   Charis: Yes, surely (I felt more capable) because before I hardly participated, while here I said my opinion, I wrote and I took part in the presentation. I wouldn't have imagined this before.
- to the cooperation of students and the sense of security that this cooperation offered to them.

   Gina: It was the first time we cooperated with the whole class ... Now with this project we all came closer.

2. **Drama, critical thinking and students' epistemological beliefs**

The conception of critical thinking according to Bailin (1998) and Bailin et al. (1999) involves three dimensions: (a) *critical challenges*—tasks that require reasoned judgement, like solving problems, resolving dilemmas, evaluating theories, creating and interpreting works of art, and so provide the impetus and context for critical thinking, (b) *intellectual resources*—the background knowledge and critical attributes such as knowledge of the principles of quality thinking and critical concepts, repertoire of strategies or heuristics, attitudes or habits of mind, and (c) *critically thoughtful responses*—evaluating the quality of the response to the challenge, e.g. discussions students have while they are working, evaluation of the dramatic performance.

Drama as a teaching method promoted the opportunities for critical thinking. The question '*Which is the geometry which best describes our physical world?*' was the critical challenge involving drama, posing an intriguing dilemma for critical thinking. For responding to this critical challenge, knowledge related to the issue was crucial. So the students would need to draw upon many intellectual resources to deal with the challenge. Students then studied the other geometries and evaluated them. The empathetic and affective understanding in drama was related to students' attitudes, such as open-mindedness and a willingness to listen to and consider the views of others. These were central to the making of reasoned judgements and ultimately to critical thinking. Drama offered a suitable environment for the development of critical thinking, fostering a climate of critical inquiry where, during the preparation of scenes and acting the roles, questioning, debating, expressing, explaining, discussing, criticizing and justifying ideas in drama activities were encouraged in an atmosphere of mutual respect and communal inquiry.

The response to the challenge took the form of drama presentations. These scenes demonstrated an understanding of the context, and sound reasoning of the issue and also an understanding of a skilful use of dramatic principles (Figs. 3 and 4).

This critical stance to the problem challenged the students' images of geometry and enabled them to perceive mathematics as a creation under constant negotiation, modifying thus their epistemological beliefs about mathematics and scrutinizing the dominant belief that Euclidean geometry is the only model which interprets and represents our real world, shaking thereby another certainty: mathematics as a science of absolute truth.

Fig. 3 Alter ego

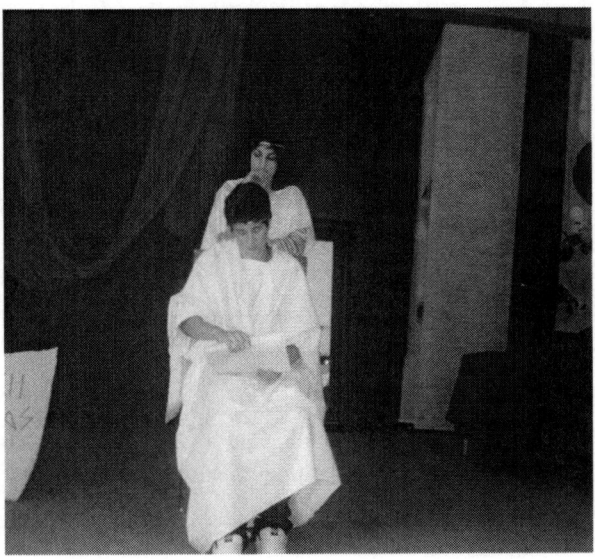

Fig. 4 Shadow theatre

Stefanos: Certainly the plasticity of mathematics emerged and how mathematics is created and is changed depending on the needs of mathematicians, of scientists and of human beings generally. It is clear that mathematics is a complex creation which is not restricted to one way of understanding reality.

Angeliki: Finally there are other views and we cannot say which is absolutely right and which is not.

3. Enhancing citizenship through drama based teaching

Through drama work, the students learned to cooperate, to work together, to take collective decisions and to help each other. Drama cultivated among students cooperation,

Thodoris: Even the guys who are not interested in the lesson, they cooperate and try to do their best.

collectivity,

Virianna: This feeling that we are all together and we must do our best in order for both us and the audience to be satisfied as well.

and solidarity,

Christos: It was hard for someone to be indifferent. Not because he would not like to participate, but somehow as an obligation to other classmates

i.e., abilities and skills necessary for a responsible citizen in a democratic society, skills that are not usually met in a traditional mathematics classroom.

Collectivity among students: Drama as a social art encourages the development of social skills, cultivating interpersonal relations among students and creating a climate of trust and confidence. For preparing their drama presentations, the students had to learn to listen carefully to others, exchange their ideas, criticize others' ideas, make decisions concerning the drama convention they should use, negotiate their roles and presentation and decide what elements of the issue are necessary to be contained in their texts. The students thus learned to work together, to take collective decisions, to appreciate the abilities of the other, to have responsibility as members of a team and to contribute to the team according to their capabilities and particular inclinations.

Christos: Several ideas were dropped and several times we combined them and so we came to a result.

Marietta: We were separated in groups and everybody tried to give his knowledge in order for a nice result to come out. We had ambitions for something nice to come out.

In our teaching experiment, the students emphasised the sense of team-spirit that developed between the members of the team.

Thodoris: Especially with all these I believe that the sense of a team was cultivated among us, because usually in school the grades are the main goal.

Tatiana: ... no one took initiatives to say: I'll do that and the rest hold your tongue. We felt we were a team.

Cooperation in groups also helped team members to get to know each other, it blunted the controversies and created friendly relations.

Thodoris: With this project, I contacted people with whom I don't even talk and this was very pleasant. When you don't talk enough to a person, it's nice to work with him even for that moment.

Solidarity among members of the team: During the preparation of their drama presentations, the students learned to help each other.

Christos: Basically everyone had a role in the team. And if he couldn't fulfil the requirements of his role, another member of the team would help him.

Students of different directions of studies cooperated and everyone brought to the team knowledge of his/her own field of studies.

Virianna: And everybody could show his skills, the different abilities that one has. Through the project not only the qualities of those more involved with mathematics could be seen, but there were some moments when the knowledge of students of humanities majors were needed and so everyone helped each other and the cooperation had a very good result.

Sofia: We all combined our knowledge in a remarkable way i.e. the girl who knew math better than me helped me in this lesson, while I, who knew literature better, helped her in that lesson.

Solidarity among team members helped the students feel more secure and confident about themselves, knowing that the other team members would give them appropriate assistance.

Tatiana: Definitely, a team works better than a person alone and has far better results, so everyone felt more sure of himself, had no anxiety and knew that he had some help, because a team works together and this gives more courage.

Concluding Remarks

Our experiment offered considerable evidence for the effectiveness of the use of drama techniques as an alternative approach in the creation of the appropriate learning conditions where all students participate in teaching/learning processes and become critical thinkers and decision-makers in the classroom.

The analysis of the texts prepared by the student groups for their presentations indicates that the students understood the mathematical notions that they had been assigned to present. Meanwhile, the analysis of the interviews, nearly 2 months after the presentations, also indicates that drama-based teaching had positively affected the students' retention of knowledge. This finding on students' achievement is coherent with previous studies (Duatepe and Ubuz 2004; Kotarinou and Stathopoulou 2008; Omniewski 1999; Saab 1987) which provided evidence for the efficiency of drama-based teaching in facilitating understanding of mathematics concepts at primary and high-school level. Drama techniques are an alternative teaching approach in fostering students' creativity by providing them opportunities to think critically. Through role playing, dramatisation and other drama techniques, we were able to discuss concepts like non-Euclidean geometries and historical events of mathematics that are not often discussed in the class context, broadening in this way students' perceptions of the nature of mathematics. This modification of students' epistemological conceptions about mathematics supports the findings of a previous study (Kotarinou et al. 2010) concerning the establishment of the 'meter' as a unit of length measurement.

Drama-based teaching gave students lessons about democracy, meaning that the students learned to cooperate with each other and also encouraged them to take collective decisions. Through DiE they learned to face differences and to reach agreements with the team members through compromise and negotiation. Students learned to pay serious attention to others, to help each other, to appreciate the abilities of the other, to have responsibility as members of a team and to contribute to the team work according to their capabilities and particular inclinations. From the students' comments, describing aspects of drama based teaching, it seemed that the use of DiE conventions has acted as an effective mediating tool for the creation of collectivity bonds and solidarity amongst them; qualities necessary for the development of citizenship. All previously stated conclusions are consistent with

Yassa's (1999) and Catterall's (2007) research findings about the development of these skills through drama.

We believe that by the aforementioned activities with DiE techniques all students experienced mathematics not only mentally but emotionally and physically. It seems that such projects can help towards the participation of all students in courses promoting critical thinking and decision making, both major goals of critical mathematics education.

References

Annarella, L. A. (1992). *Creative drama in the classroom* (ERIC Document Reproduction Service ED391206).
Appelbaum, P. (2004). Mathematics education. In J. Kincheloe & D. Weil (Eds.), *Critical thinking and learning: An encyclopedia for parents and teachers* (pp. 307–312). Westport: Greenwood.
Bailin, S. (1998). Critical thinking and drama education. *Research in Drama Education, 3*(2), 145–153.
Bailin, S., Case, R., Coombs, J., & Daniels, L. (1999). Conceptualizing critical thinking. *Journal of Curriculum Studies, 31*(3), 285–302.
Bell, E. T. (1993). *Men of mathematics* (Vol. 2). Heraklion: Crete University Press.
Bishop, A. J. (1999). Mathematics teaching and values education: An intersection in need of research. *Zentralblatt für Didaktik der Mathematik, 31*(1), 1–4.
Bolton, G. (1985). Changes in thinking about drama in education. *Theory Into Practice, 24*(3), 151–157.
Boyer, C. B., & Merzbach, U. C. (1997). *History of mathematics*. Athens: Pnevmatikos.
Bunt, L., Jones, P., & Bedient, J. (1981). *The historical roots of elementary mathematics*. Englewood Cliffs: Prentice-Hall.
Catterall, J. (2007). Enhancing peer conflict resolution skills through drama: An experimental study. *Research in Drama Education: The Journal of Applied Theatre and Performance, 12*(2), 163–178.
Chronaki, A. (2008). The teaching experiment: Studying the process of teaching and learning. In B. Svolopoulos (Ed.), *Connections of educational research and practice*. Athens: Atrapos.
Cobb, P., Wood, T., & Yackel, E. (1991). A constructivist approach to second grade mathematics. In E. von Glasersfeld (Ed.), *Radical constructivism in mathematics education* (pp. 157–176). Dordrecht: Kluwer.
De La Roche, E. (1993). *Drama, critical thinking and social issues* (ERIC Document Reproduction Service ED379172).
Duatepe, A., & Ubuz, B. (2004, July 4–17). Drama based instruction and geometry. In M. Niss (Ed.), *Proceedings of 10th international congress on mathematical education*, Copenhagen, Denmark.
Ernest, P. (2008, November 21–23). What does the new philosophy of mathematics mean for mathematics education. In *Proceedings of the 25th Panhellenic Congress in mathematics education of Hellenic Mathematical Society*, Volos, Greece.
Eves, H. (1983). *Great moments in mathematics (after 1650)*. Athens: Trohalia.
Gerofsky, S. (2009). Performance mathematics and democracy. *Educational Insights, 13*(1).
Guedj, D. (2000). *The parrot's theorem*. Athens: Polis.
Heathcote, D. (1984). Dorothy Heathcote's notes. In L. Johnson & C. O'Neil (Eds.), *Heathcote, Dorothy: Collected writings on education and drama*. London: Hutchinson.

Hitchcock, G. (1996). Dramatizing the birth and adventures of mathematical concepts: Two dialogues. In R. Calinger (Ed.), *Vita mathematica: Historical research and integration with teaching* (pp. 27–41). Washington, DC: MAA.

Kotarinou, P., Chronaki, A., & Stathopoulou, C. (2010). Debating for 'one measure for the world': Sensitive pendulum or heavy earth? In U. Gellert, E. Jablonka, & C. Morgan (Eds.), *Proceedings of mathematics education and society 6th international conference* (pp. 322–329). Berlin: Freie Universität Berlin.

Kotarinou, P., & Stathopoulou CH. (2008, November). Role – Playing in mathematics instruction. In *Proceedings of the 13th international congress of drama/theatre in education* (pp. 108–118). Ankara, 21–23.

Lénárt, I. (1996). *Non Euclidean adventures on the Lenart sphere: Activities comparing planar and spherical geometry*. Emeryville: Key Curriculum Press.

Luminet, J.-P. (2003). *Euclid's bar*. Athens: Livanis.

Mankiewicz, R. (2002). *The story of mathematics*. Athens: Alexandria.

Neelands, J. (1984). *Making sense of drama: A guide to classroom practice*. London: Heinemann.

O'Neil, C., & Lambert, A. (1990). *Drama structures: A practical handbook for teachers*. Kingston: Stanley Thornes.

Omniewski, R. (1999). *The effects of an arts infusion approach on the mathematics achievement of second-grade students*. Unpublished PhD thesis, Kent State University, USA.

Ponza, M. V. (2000). Mathematical dramatisation. In J. Fauvel & J. A. van Maanen (Eds.), *History in mathematics education: The ICMI study* (pp. 335–342). Dordrecht: Kluwer.

Saab, J. (1987). *The effects of creative drama methods on mathematics achievement, attitudes and creativity*. Unpublished PhD thesis, West Virginia University.

Skovsmose, O. (1996). Critical mathematics education. In A. J. Bishop, K. Clements, C. Keitel, J. Kilpatrick, & C. Laborde (Eds.), *International handbook of mathematics education* (pp. 1257–1288). Dordrecht: Kluwer.

Skovsmose, O. (1998). Linking mathematics education and democracy: Citizenship, mathematical archaeology, mathemacy and deliberative interaction. *Zentralblatt für Didaktik der Mathematik, 98*(6), 195–203.

Skovsmose, O. (2004, July 4–11). *Critical mathematics education for the future*. Paper presented at ICME 10, Copenhagen, Denmark.

Skovsmose, O., & Valero, P. (2002). Democratic access to powerful mathematical ideas. In L. English (Ed.), *International research in mathematics education* (pp. 383–408). New York: Routledge.

Skovsmose, O., & Yasukawa, K. (2000, March 26–31). Formatting power of mathematics: A case study and questions for mathematics education. In *Proceedings of the 2nd MES conference*, Montechoro, Algarve, Portugal.

Somers, J. (1994). *Drama in the curriculum*. London: Cassell.

Teoh, J. (2012). Drama as a form of critical pedagogy: Empowerment of justice. *The Pedagogy and Theatre of the Oppressed International Journal, 1*(1), 4–26.

Yassa, N. (1999). High school students' involvement in creative drama: The effects on social interaction. *Research in Drama and Theatre in Education, 4*(1), 37–51.

Organising Dialogue and Enquiry: A Commentary

Lambrecht Spijkerboer and Leonor Santos

Abstract The commentary on the chapters of Menezes, Oliveira and Canavarro, Kotarinou and Stathopoulou and Serradó, Vanegas and Giménez introduces a model called 'OBIT' to differentiate between 'deep' and 'surface' approaches to learning. It discusses the three chapters in the light of the model.

These days, the learning of mathematics is very challenging. It is no longer sufficient to know something, it is required that learning is developed by understanding (NCTM 2000). There are several conditions that may be favourable for students' learning of mathematics. Organising dialogue and enquiry is one of these approaches. The approaches are recognised as facilitating students' development of mathematical content knowledge in mathematical processes, such as mathematical reasoning and communicating mathematical ideas, thinking skills and collaborative skills (Chapman 2013). This is the reason why it is obvious to dedicate one section of the volume to this theme. In the next pages, a commentary on the three contributions of the section will be presented. Quotes from the three chapters will be used to underline the different aspects of the analysis. The quotes are written *in italics* and the tag numbers (1), (2) or (3) refer to the chapters by: Menezes, Oliveira and Canavarro (1), Kotarinou and Stathopoulou (2) and Serradó, Vanegas and Giménez (3).

Reading the different contributions in this part of the book, we notice that there are several similarities in how researchers describe classroom practice. First of all, they all focus on learning mathematics by the active participation of the learners. The situation in the mathematics classroom is organised in different ways where students are invited to communicate, to argue, to reflect together, to propose ideas, to negotiate and to build consensus. Working together in this sense is called

---

L. Spijkerboer (✉)
APS-International, Utrecht, The Netherlands
e-mail: l.spijkerboer@aps.nl

L. Santos
Instituto de Educação, Universidade de Lisboa, Lisbon, Portugal
e-mail: mlsantos@ie.ulisboa.pt

'cooperative learning', 'participatory performance' or 'deliberate dialogue'. Inviting students to build their abilities in mathematics this way, there should be appropriate tasks and a special demand on the role of the teacher in order to provide the suitable classroom environment for the learning to take place in the way it was intended.

Before going into detail about the tasks and role of the teacher in the described classroom practice, we shortly describe a learning-theory called 'OBIT-model' about a surface and a deep approach of learning (Biggs 1987; Marton and Säljö 1984; Smith and Colby 2007). The OBIT-model, which is concerned with different learning activities, is helpful for understanding the different demands of (future) teachers, students and researchers. In the kind of mathematic lessons where students are actively involved, they perform with learning activities of the higher stages of the OBIT-model.

OBIT-Model*

This model distinguishes two types of learning called the surface approach and the deep approach (Smith and Colby 2007). In the Netherlands, the two approaches are modelled by the so called 'OBIT-model'. OBIT contains learning activities called **O**nthouden = Remembering (reproduction), **B**egrijpen = Understanding (explain in own words), **I**ntegreren = Integrate (make combinations with former knowledge), **T**oepassen = creative application (in new situations).

A surface approach contains remembering and understanding. With these learning activities, knowledge is mostly built up in the short-term memory. This is called a surface approach because after some days or weeks, the knowledge can vanish when it is not connected to other experiences. Knowledge in the short-term memory should be repeated continuously to get it into the long-term memory for later use.

---

*OBIT-Model Lower Stages

Onthouden = Remembering

For doing remembering activities it is not necessary to know what you are learning exactly, it is only a copy-paste activity. You learn by heart the Pythagoras theorem as a formula $a^2 + b^2 = c^2$. Remembering is based on reproduction of the knowledge you know, not necessarily understand.

Begrijpen = Understanding

The learning activity understanding is shown when a student is able to explain what was learned during the lessons in his/her own words. S/he is able to copy the way of working, the way of solving problems in the same way as demonstrated by the teacher or with help of the explanation in the book.

(continued)

> Understanding is a straightforward activity, and can be done by students who do their homework and are mentally present in class. The thinking steps are given, not to be made by the students themselves. Understanding questions are: solve this linear equation, draw a graph related to this formula, compute the angle in this right-angled triangle, ... All activities were done by teachers before, so it is a rehearsal.

If learning takes place by using learning activities like integration and creative application, the knowledge is stored in the long-term memory. The knowledge is connected to other knowledge and memories like emotions and information, given by other senses. For that reason, the results of this learning activity are not easily forgotten, which is why this type of knowledge building is called a 'deep' approach. The OBIT-model is based upon Bloom's taxonomy but it uses different words, especially the word 'application' has a different meaning in the OBIT-model. Another difference is that Bloom's is a taxonomy while OBIT is not.

> ***OBIT-Model Higher Stages**
> *Integreren = Integration*
> The learning activity integration is focused on the relation between different parts of knowledge. Integration means you use your insight in the situation. You make use of your knowledge and connect it to the new information achieved. Compared to understanding, during integration activities there are more thinking steps, and the learners add something themselves, it is a productive activity, not only a reproductive one.
>
> Examples of integration questions are: try to figure out the relation between, what do you know, what don't you know. Explain why a parabola cannot have more than two intersection points with any straight line, ...
>
> Especially for this learning activity you need to argue, explain, try, search, etc. and it can best be done in communication with classmates.
>
> *Toepassen = Creative Application*
> In the learning activity creative application, the thinking steps are not given anymore, like in integration, but the students have to engage in a sometimes creative thinking process to solve the problem; What do I know? What can I use? How to connect the different know-how I have or I want to gain to make sure the correct solution of the problem? In this learning activity, the students really make use of their own skills, as well as their (subject) knowledge. There is a design process going on, which is stimulated by a new situation, never faced before in context.
>
> Research assignments are mostly seen in experiments inside or outside the classroom. Students have to find out certain phenomena and try to explain

(continued)

> why they exist. Because students can learn so much from each other, especially application activities can very well be carried out in groups. Cooperative work gives space for different approaches, different tasks and different learning styles. Connection between real life and mathematics has to be made. The teacher has to take care of the complexity of these kinds of assignments, because the learning activity application can confuse the students easily. It is important that they approach the tasks open minded and with self-confidence, without being prepared to make a lot of calculations. The focus is really to find out, apply their knowledge and explore.

OBIT in Education

The OBIT-model was introduced by Ebbens and Ettekoven (2007) based on the research of Boekaerts and Simons (2003) and is used for observing teaching and learning in class. This model is also often used to analyse tests and to build reasonable test designs (Spijkerboer et al. 2007).

We can also recognise these different learning activities in the mathematics classroom. As described in (1), the students have to figure out the relation between the number of cubes and the number of stickers. This topic can be taught by a teacher showing the mathematical relationship by putting numbers on a table, by making a drawing of the connected graph or even by constructing a formula. Usually students learn by recognition after remembering and understanding, the two lower stages of the learning activities in the OBIT-model. However, in the case of the enquiry-based mathematics teaching, other learning activities like integration are addressed. In the task in drama in education (2), the students are highly motivated to perform and study the topic of mathematics in a way called integration and application, the two higher stages of learning activities in the OBIT-model.

Findings from a study examining teaching practices and student learning outcomes of 46 teachers in 17 different states of the U.S.A. indicate that most of the learning in these classrooms is characterized by reproduction, categorization of information or replication of a simple procedure (Smith et al. 2005). Daily practice in mathematics classrooms in other countries is also much more focused on the lower stages of the OBIT-model. The three research papers in this section refer to other ways of teaching and learning as mainstream. *Over the last decades many perspectives on mathematics teaching have emerged in opposition to the common vision and practice of knowledge transmission* (1). Knowledge transmission is focused on learning activities like remembering and understanding. However, the three research papers we read, stated that students learn mathematics in a way that is focused on the higher stages of the OBIT-model. For these learning activities students' active participation is essential. The learning for the long-term memory takes place when students are invited to communicate with others, to reflect and to argue, to negotiate and to build consensus. Different solutions, strategies, ideas, or proposals are helpful for the learning of every participant in classroom discussions. Where the learning of mathematics is based on dialogue and reflection, also *ongoing and unsuccessful attempts* (2) contribute to the learning.

Particularly interesting is the recognition of the new generation; future teachers *recognise the potential of mathematics in contributing to making judgements or reaching consensus* (3), but the authors complain that there is not enough participation in mathematics classrooms. The majority of future teachers in the research group believe that *deliberative mathematics dialogues are nearly impossible to be carried out in our regular classrooms of multicultural heterogeneous groups* (3). They also claim that *the initial statement of a problem (...) is not enough to promote a critical dialogue!* (3)

Enquiry-based education, as well as drama in education and deliberate dialogue, are obviously answers to the question of how to motivate students to cooperate in the learning of mathematics. *A contextualized activity is necessary to achieve real participation* (3). As a student reports: *a team works better than a person alone and has far better results.* (2)

It is helpful to read about front research facilitating the learning activities integration and creative application in very different types of education: in basic education (1), secondary education (2) and higher educational environments (3). Learning mathematics is not for the sake of recognition and understanding, but proves its value with integration and application. Students' motivation is found in doing this kind of learning activities in class: The students of Celia (1) are very motivated and committed to classroom work, they address their peers to clarify ideas.

Tasks

Obviously, the mathematics teacher should provide appropriate tasks in order to invite students to do integration and application. A good task is mathematically demanding and challenging for the students. *The curriculum transformations that occurred in several countries have inspired many teachers to seek for more demanding learning objectives to integrate into their practice* (1). Because of the curriculum transformations, some teachers have been inspired to design challenging tasks, although the need for good tasks in classical curricula is also evident.

Very well-presented tasks that make the students assume the tasks as their own are suitable for challenging mathematics lessons. Like in drama in education, students want to solve their problem and avoid being hopelessly lost in front of the audience (2).

Not all tasks facilitate deliberative communication. For that purpose, *we should create tasks with relation to the wider context* (3)—tasks with an open character and with many different ways to solve, with possibilities for different approaches and with more than one answer which are more suitable for communication and explanation in the mathematics lessons focussing on integration and application (the higher stages of the OBIT model). Tasks that invite students to communicate, to question and to reflect are suitable for reaching the goals for democratic citizenship as well as understanding mathematical reasoning (3). To recognise the value of learning mathematics, an extra-mathematical context such as ecological situations and environmental problems is considered the main element to promote a questioning discussion. In such a case mathematics is needed for discussing the social problem (3) and shows the value of mathematical thinking skills for both students and teachers.

With all tasks that invite students to achieve mathematical knowledge and skills, the most important activity to facilitate the learning process is an ongoing reflection. We recognised this phase in the lessons described: phase 4 in the theoretical framework of (1)—systematizing mathematical learning; drama in education is always followed by a reflection afterwards; and also deliberate dialogue gains value in the evaluation. *The role of the teacher is crucial to provoke the pupils' reflection* (3).

Role of the Teacher

In all three contributions the important role of the teacher to facilitate learning is often stressed. The role of the teacher is to encourage students' interaction and to invite them to explain to each other (1). The teacher must feel the responsibility to promote participation with others. The teacher decides how to design the learning, not only for mathematics sake but also with the *notion of "democratic access"* (i.e. to be addressing as many people as possible) to mathematical ideas (2). It is the teacher who has the crucial role with regard to direction, possible continuation and conclusion of deliberative communication (3). This means the teacher is not always giving subject-based input, like explaining, showing or performing mathematically. It can be hard but helpful for teachers to restrict themselves and to not validate fast students' ideas but to keep questioning and not giving clues too early. In enquiry-based education as well as in drama in education and deliberative dialogue, the teacher has to approach the students with open questions. The teacher has to observe how the students are reasoning in order to intervene at the right moment with the right action and to decide on the spot.

The way students act and react can be helpful information for the teacher to identify if and how the students understand. To observe students' thinking it is also helpful for the teacher to know what the solutions are to bring in for *collective discussion and synthesis of the mathematical ideas* (1) and to choose the sequence of the presentations of students' solutions. In the collective reflection phase of the lessons the teachers illustrates, highlights and generalizes the ideas and synthesises the main mathematical ideas. It is important to guarantee the comparison of distinct solutions (1). The teacher needs to perceive if there is an appropriate social and physical environment. The risk is that teachers—because of their feeling of responsibility—take the lead in pre-task and post-task moments in class too quickly to give input to mathematical content needed. In that case the value given to the communicative role of the teacher is suddenly reduced to the evaluation of the ideas to promote agreement among students. The teachers' reaction on students' doubts should not hamper the emergence of different strategies (1). A balance between leaving them on their own and giving them some orientation is not easy to achieve. That is the teacher's challenge.

Preparing lessons this way also means for the teacher to think about questions to pose to students in case these are needed. Level-raising questions are to be prepared in order to facilitate differences in the mathematics classroom. Dealing with differences in the classroom is the ability to save time for students to help each other. Teachers sometimes evoke differences to facilitate the learning. Working in pairs can be very helpful, but it still gives a lot of different solutions to be collected in a full-size group. If pairs present the solutions to each other, the same interaction can be organised in less time.

The teacher has the role of inviting all learners to participate. In drama in education there is no chance to hitchhike with others, because the task is demanding for all. There is a group-responsibility and the drive is not to fail in front of the audience. A student reports: *We had ambitions for something nice to come out* (2).

Final Considerations

Because the three contributions refer to modern tendencies in mathematics education, there can be some questions about the implementation of these ideas into practice.

1. The teacher's role in enquiry-based mathematics lessons is seen as a complex demanding practice (1), both by experienced teachers (Célia) in Portugal and by pre-service teacher students in Spain. What does teacher education, in-service and pre-service, have to offer to prepare (future) teachers for this role?
2. Is it necessary that enquiry-based mathematics lessons are set in between lessons with traditional set-ups with knowledge transmission or is it possible to use this approach all through the weeks? Is it necessary to have projects like drama in education in between lectures or should the students keep seeing those non-regular and challenging tasks as another way of thinking (and learning)? What will confuse them more?
3. Would teachers agree on mathematics lessons based on the same learning theories (like the OBIT-model) in order to achieve that mathematical education in school is more coherent and gives a clear presentation of the goals to be reached in the mathematics lessons: reasoning, arguing, decision-making strategies, etc.?
4. Do we need, as a consequence of enquiry-based education, drama in education and deliberative dialogue, to put more emphasis on cooperative learning, dealing with differences in the classroom, and democratic citizenship in schools?

References

Biggs, J. (1987). *Student approaches to learning and studying*. Hawthorn: Australian Council for Educational Research.

Boekaerts, M., & Simons, P. R. J. (2003). *Leren en instructie, psychologie van de leerling en het leerproces*. Assen: Van Gorcum.

Chapman, O. (2013). High school mathematics teachers' inquiry-oriented approaches to teaching algebra. *Quadrante, 22*(2), 6–28.

Ebbens, S., & Ettekoven, S. (2007). *Actief leren*. Groningen: Wolters Noordhoff.

Marton, F., & Säljö, R. (1984). Approaches to learning. In F. Marton, D. Hounsell, & N. Entwistle (Eds.), *The experience of learning* (pp. 39–58). Edinburgh: Scottish Academic Press.

NCTM. (2000). *Principles and standards for school mathematics*. Reston: NCTM.

Smith, T. W., & Colby, S. A. (2007). *Teaching for deep learning*. Washington, DC: Heldref.

Smith, T. W., Gordon, B., Colby, S. A., & Wang, J. (2005). *An examination of the relationship between depth of student learning and national board certification status*. Boone: Office for Research on Teaching, Appalachian State University.

Spijkerboer, L., Bootsma, G., & Denijs, W. (2007). *Determinatie en toetsen*. Utrecht: APS.

Part VII
Providing Information Technology

**Educational Laptop Computers Integrated into
Mathematics Classrooms**.. 351
Maria Elisabette Brisola Brito Prado and Nielce Meneguelo Lobo da Costa
Anhanguera University of São Paulo, Brazil

**Technology and Education: Frameworks to Think
Mathematics Education in the Twenty-First Century**............................... 365
Gilles Aldon
Ecole Normale Supérieure de Lyon, France

**Technology in the Teaching and Learning of Mathematics
in the Twenty-First Century: What Aspects Must Be Considered?—
A Commentary**.. 383
Fernando Hitt
Université du Québec à Montréal, Canada

Educational Laptop Computers Integrated into Mathematics Classrooms

Maria Elisabette Brisola Brito Prado and Nielce Meneguelo Lobo da Costa

Abstract This chapter discusses a national program, i.e. an initiative of the Brazilian Federal Government and the Ministry of Education, which was developed between 2007 and 2012. The initiative aimed at the insertion of computers into public schools, which is a conflictive issue as some educators fear that computers could replace teachers in the classroom. In this chapter, we are reflecting on experiments and surveys involving mathematics teachers' use of educational laptop computers within classroom practices.

Introduction

The insertion of computers into public schools in Brazil began in the 1980s and caused a conflict in the educational context, as many educators feared that computers could replace teachers in the classroom. There were, among them, serious differences of opinion. A few idealistic and adventurous teachers argued in favor of installing computers in basic schools, while others, the conservative and cautious ones, were against it. In the light of this discussion, Paulo Freire's words were able to cause reflections about this new reality and also to raise awareness in order to cautiously examine the question, especially with an investigative spirit that could lead to understanding implications of this technology in teaching and learning.

M.E.B.B. Prado (✉)
Post-graduation Program of Mathematics Education,
Anhanguera University of São Paulo, UNIAN, São Paulo, Brazil

Applied Information Technology Center of Education,
State University of Campinas, UNICAMP, Campinas, Brazil
e-mail: bette.prado@gmail.com

N.M. Lobo da Costa
Post-graduation Program of Mathematics Education,
Anhanguera University of São Paulo, UNIAN, São Paulo, Brazil
e-mail: nielce.lobo@gmail.com

Paulo Freire's phrase was as follows[1]: "Preventing or hindering access of this technology to young students in public schools is an attitude that only reinforces a different kind of education to the oppressed social class." Freire focused on the youth, who most depend on public policies for access to information, as a fundamental step in the process of overcoming social stratification.

According to Freire (1981) and to D'Ambrosio (1986), information access is the starting point for learners that can increase their power of creation and also their ability to reconstruct knowledge, through interaction with others (including the teacher). The digital technologies such as the internet, cell phones and mobile computers interfere with the ways people understand each other regarding relations and communication as well as the scope and speed of information dissemination. Technology is able to change the expression and production capacity, using different languages (artistic, mathematical, scientific), and also the representations of knowledge.

Research that has been done around the theme of integrating technology into the school curriculum, for instance by Valente (1999) and Prado (2008), shows improvement, i.e. progress towards teachers integrating technology into their teaching practices. However, these advances are still only occasional experiments and they point to the importance of continuous investments in training school staff. Likewise, research has also shown that computer labs are still working as an appendage for the classroom; this means that systematic use of computer resources integrated into the curriculum seems to be an exceptional situation. We know that, in fact, this integration does not happen so easily and immediately with the installation of the laboratory, because it requires teachers to reconstruct their pedagogical, mathematical and didactical knowledge, which goes beyond the mere application of computational resources (Almeida and Prado 2009a). In fact, the integration of digital technologies into teaching practices is not limited to a juxtaposition of resources in teaching methodology. This process—as experienced by the teacher—involves a reconstruction of the structure of thought which now includes the "thinking with" and "thinking about" technology (Papert 1985), and later on it will critically include technology in the teaching process. This is the moment when the teacher learns to work the 'new' and is able to understand this 'new' in order to reinvent and recreate her practice.

Many teacher-education initiatives have taken place in Brazil. Regarding the mathematics teacher education projects, research has indicated the need of reflective and critical approaches, which should occur concomitantly with the pedagogical actions, considering specific characteristics of each school's reality (Lobo da Costa 2010). Furthermore, they should incorporate more global elements through a technological learning network, which promotes reflection and confrontation of ideas among teachers from various realities of schools across the country by means of exchanging experiences. The process of teacher education promoting the use of

---

[1] Pronouncement of Paulo Freire at the beginning of his function as Municipal Secretary of Education of São Paulo, in 1988, during the implementation of the Genesis Project: Educational computing in public schools.

technology is still under development; however, especially after 2007, a new educational scenario began to evolve in the country, with the arrival of educational laptop computers inside school classrooms. This framework has been set up in several countries, where similar projects are under development, showing that democratizing the access to information for all students is a global concern. A new reality is taking shape and must be studied in order to understand the implications involved when laptops are in students' hands inside and outside classrooms, surrounding the school environment. This is a different situation from the one when computers are used at the school computer lab, in which teachers can exert substantial control over the computer use. The student with a laptop computer has free access to computing resources (applications, software, games, wireless internet) at any time, which sets new demands for schools and especially for teachers. This is an unusual educational environment, although the idea of each student with a computer was established a long time ago. According to Valente (2011), when Alan Kay met Seymour Papert at the Massachusetts Institute of Technology (MIT), in 1968, and saw Papert's work with Logo programming language—even before the advent of microcomputers—he was deeply impressed, especially because he saw children solving complex mathematical problems using computers. In that moment, Kay and Papert envisioned the possibility of every child having its own laptop computer. Papert (1985, p. 16) said: "Computer presence could contribute to mental processes, not only as an instrument, but conceptually, influencing people's thinking even when they are physically distant."

The innovative dream of that time is now becoming reality in Brazil with the implementation of the *Um Computador por Aluno* program (UCA)—one laptop computer per student—in public elementary schools (children from 7 to 15 years old). The development of new technology for laptop computers allowed for lower production costs of this equipment and made it possible to spread its use at schools and educational activities. Several countries such as the United States and India, and also African countries such as Rwanda, and South American countries such as Uruguay began to implement technological inclusion projects and professional development projects using these laptop computers since 2006 (Bebell and O'Dwyer 2010; Ceibal 2010). Taking into account this development, the aim of this chapter is to discuss a national program, UCA, an initiative of the Brazilian Federal Government and the Ministry of Education, which was developed between 2007 and 2012. We are going to analyze some experiments and surveys which involved the use of educational laptop computers in classroom practices of mathematics teachers.

The UCA Program: The Beginning

Educational laptop computers started to be introduced to Brazilian public schools by the UCA program. The program was inspired by experiments developed at MIT's Media Lab in the United States, under Seymour Papert and Nicholas Negroponte at the beginning of the year 2000, which were made possible through the One Laptop per Child project. The main goal of providing one laptop computer per student (young children and teenagers) was to create educational opportunities so that everyone could have access to laptop technology. This goal encompassed a strong social desire to provide digital inclusion for those students who otherwise could not afford to have access to such technology to connect them to the world. Likewise, the UCA program began with the aim of promoting the digital and social inclusion of public school students. It was developed in two phases, an initial phase (2007–2009) and a pilot phase (2010–2012). In the initial phase of the program (Phase 1), five educational experiments conducted in public schools in different states (São Paulo, Rio de Janeiro, Brasilia, Rio Grande do Sul and Tocantins) served to substantiate the pilot phase (Phase 2), in which the project was expanded to 300 schools.

The UCA program was developed with the purpose of promoting the use of laptop computers in education to enhance the quality of education, the digital inclusion, and the Brazilian insertion into the manufacturing process and maintenance of this kind of computers (UCA Princípios 2007). The innovative points of the proposal were the use of laptop computers by all public school students and educators in an environment that allows for their immersion in the digital culture; the mobility of the equipment in environments both inside and outside the school; connectivity, through which the process of using the laptop computer and the interaction between students and teachers is done through a wireless network connected to the internet; and the use of different educational media, available for the educational laptop computer. Such innovative aspects, which focused on the daily classroom practices, required UCA program developers to think of management strategies to ensure the program implementation at schools while providing the university researchers with a basis to analyze the pedagogical implications of using laptop computers in classroom practices. Four types of interconnected actions were projected and developed to achieve this goal:

1. Infrastructure: Actions related to the purchase of laptop computers, the adaptation of the physical space and the electrical grid of the participating schools in this phase of the UCA program, wireless network installation and setting up local teams of technical support.
2. Development: Creation and implementation of a teacher training called "Brazil's Teacher Development" (Formação Brasil, in Portuguese) for both teachers and school managers, and also the creation of logistics for an online development course which included universities, education departments and schools.
3. Evaluation: An assessment approach was developed to support both infrastructure and development actions by means of diagnostic evaluation, in a process that implied systematic and impacting monitoring.
4. Research: Projects were presented by a number of universities to the Federal Agency For Research Advancement (CNPq, in Portuguese). All these on-going research projects aim to investigate themes related to the teaching-learning processes in which laptop resources are integrated into the curriculum.

It is important to underscore the inception of the course "Brazil's Teacher Development", which included the production of educational material and activities to meet the shared needs of the country's different regions, while, at the same time, it established consortiums with different local universities so that the latter would adapt the course to fulfil the regional needs by including extra materials and activities which were meaningful for the reality of each of the participating schools in the UCA program (UCA Formação Brasil 2009). The flexibility of a teacher development course is especially important as Brazil has continental dimensions with significant differences among its regions. The south and southeast areas are highly developed, but they have a good deal of problems regarding social inequality and internal migration, while the north and northeast regions have large territories, most of which have rivers that go across forests with restricted accessibility, which require alternative actions to leverage their development. Brazil has a wide economic and cultural diversity which demands specially-designed logistics to enable the implementation of a national program. The "Brazil's Teacher Development" course offers this possibility as it can be adapted individually by local universities in each region while preserving its educational principles. In this course, the focus lies on the school. Managers and teachers of all areas learn how to use laptop resources under the learning-by-doing perspective as proposed by Dewey (1979). In this learning process, the pedagogical mediation is conducted by university professors, who were working in partnership with the IT and education professionals of the local education departments. The course takes place on-site at the schools and online through a virtual learning environment. The course's approach emphasizes the principles of reflective practice, the articulation of theory and practice, the interaction and sharing of ideas, reflections and experiences between peers and the developers drawing on the group's experiences regarding the pedagogical use of laptop computers with school children.

Research Development

Some surveys were developed during Phase 1 of the program. Among them Mendes (2008), which was conducted in one school of Tocantins, showed that during the laptop-computer implementation process, there were signs of changes in the classroom management and in space and time reorganization following decisions proposed by the school teachers, managers and technicians from state educational agencies. The group was engaged in a process of curriculum adaptation based on the inclusion of laptop computers in the classroom. The research also found that laptop computers brought new forms of interaction and communication among students and between students and teachers. According to Mendes and Almeida (2011, p. 52): "The association of educational laptops and connectivity can help the teacher to expand not only learning spaces, but also the participation in learning networks, the collaborative work and to enable contact with other cultures." Another research in the context of the São Paulo state was developed by Saldanha (2009). He recognized that indicators of a flexible curriculum are key for the feasibility of using laptop computers in school. Besides, Almeida and Prado (2009b) showed that laptops in students' hands pose a new challenge, which requires a new management of teacher's pedagogical practices. This can cause a curriculum-revision movement, which implies various new decision-making styles within the educational system.

These initial experiments supported Phase 2 of the UCA program, which was then implemented in order to include schools of all states. From 2010 to 2012 the Ministry of Education purchased 150,000 Classmate[2] laptop computers. The UCA program expanded to cover 300 schools (in urban and rural areas) throughout 27 states, in which all students have been given a laptop computer. In addition, six small municipalities, characterized as economically disadvantaged, were chosen to be "digital cities", which implied that they were given priorities in the sense that all schools were equipped with educational laptop computers.

In the second phase of the implementation of these digital cities, the aim was to understand the impact of the widespread distribution of laptop computers on the daily routines of a community, such as how the use of laptop computers affects routines, relations, businesses, services, families and the students themselves. On the one hand, this experience showed that aspects which can enable a more global digital inclusion require local governments to take a stand and develop political as well as management actions to make the necessary infrastructure and teacher development viable at schools. On the other hand, such digital cities were reported to have a positive effect on the digital inclusion of the local population. However, we know that this is not enough to enhance the pedagogical use of laptop computers. For the latter, it is necessary to interrelate the school's community and the teacher development courses and to have a continuous investment in learning to increasingly explore the potentialities of technological resources under a perspective that integrates syllabi. According to Almeida and Valente (2011, p. 34), based on Freire's liberating and problematizing concept, it should be underscored that

> ... in the process of integrating technologies into the curriculum, it is essential to seek the development of a human being who is dialogic, inquisitive, reflective, critical, capable of changing himself and the world. The use of Digital Technologies of Information and Communication (DTIC) allows for identifying the students starting point, i.e., his way to interpret the world ... and to create the conditions to write his own history, his understanding of himself as a person of his time, a member of a community with whom he socially and historically shares and builds knowledge, values and experiences. (p. 12)

As an example of the research developed in Phase 2 of the UCA program, we report our study that originated from the course "Brazil's Teacher Development". Here, our role was to prepare schoolteachers to work with the laptop resources under a pedagogical approach based on action-research in mathematics teaching. Our qualitative methodology followed the principles of action research, which is a

---

[2] The laptop computers used by the UCA program have devices such as webcams with photo and video capabilities, microphones and speakers, and ports for robotics devices. They also come equipped with programs to produce animation, games and narratives using a number of digital media resources.

type of social research based on observation of phenomena associated with the action and resolution of problems, and where the researcher participates in an active and cooperative process (Thiollent 1988).

Classroom Action-Research at the Mathematics Class

It is undisputable that providing laptop computers to students represents a highly motivational innovation that can be used to build knowledge. However, it is still common practice inside classrooms that students passively watch their teacher's explanations of certain contents, and afterward do a list of exercises which have to be handed in. The notion that students learn through the teacher's explanation and emulation (Skovsmose 2008) still holds true for many. Learners' critical thinking and investigative, creative and authorship skills are often forgotten.

The common pedagogical model does not offer the students an investigative environment in which their curiosity could be aroused for exploration, hypotheses formation and analysis so that they become active, inquisitive and autonomous participants in their learning process. In order to develop the kind of teaching which allows the learner to take hold of his own learning process through elaboration, exploration and disclosure, the teacher needs operational and conceptual understanding of the mathematical structures which form the basis of a rich learning environment for students. Regarding the discussion of working with curriculum contents, it should be added that it is still a great challenge for teachers to integrate computers and other available resources into the school context (Lobo da Costa et al. 2013; Prado et al. 2013). Hence, laptop use in mathematics classes might lead to a destabilization of teachers. To manage learning actions requires pedagogical and technological proficiency as well as openness towards educational change.

An investigative approach to the learning of mathematics refers to activities that preserve the fundamental features of a science (Ponte et al. 2003). This means that, using a syllabus content-related theme, students should form questions and hypotheses with clear goals aiming at gathering information from trusted sources, and develop practical actions to achieve results. It also implies data analysis and interpretation, the production of means to represent and record achieved results, and publication and sharing of the findings. In other words, teaching through an investigative approach means to create conditions for students to commit to scientific research fundamentals and to learn in an active way through investigation projects (Valente 2008). The term 'learning based on investigations' has been used in international contexts. For authors such as Ramos et al. (2009), who research about the development of educational environments mediated by DTIC, learning based on investigations is an interesting path to overcome traditional teaching and learning models, by placing the student in action so that he 'learns by doing' as proposed by Dewey (1979). The starting point for learning is in the students' inquisitiveness that compels them to investigate. Only transmission of information through the teacher is not enough to develop students' knowledge about facts, concepts and studied phenomena. Students must be led to engage in "sustained inquiry, i.e. activities that include formulating authentic, meaningful questions, planning tasks, gathering resources and information, predicting outcomes, debating the value of information, evaluating information, collaborating with others and reporting findings" (Krajcik 2002, p. 411).

Unfortunately, in Brazilian school contexts, developing 'research work' has frequently been understood as a procedure that can be summed up as the creation of student groups who gather information about an issue using technology, and then produce a text report to be handed to the teacher. The proposal for teacher development based on an investigative approach in the teaching of mathematics using UCA laptop computers in Phase 2 included the discussion of contexts and dynamics, depicting scenarios that represented possible paths for students to become investigators. Regarding the possibilities offered by the use of laptops in the teaching and learning of mathematics, we underscore that the ability to animate objects on the screen can become an important tool to support, or even replace many of the activities designed for pencil and paper. In mathematics, animation can also have an important role in designing dynamic graphs, allowing for the variation of a few parameters to produce immediate effects on the graph (Bairral 2007).

The software to perform simulations of physical, chemical, biological and environmental phenomena or to explore a number of mathematics themes can be found on the internet, for example on the PhET website, the Interactive Simulation Project developed by the University of Colorado (PhET 2010). Other dynamic graphs software programs, such as WinPlot or Graphmatica, can be used by both mathematics teachers and students, as reported in several studies (Castro 2011; Lobo da Costa 1997; Maia 2007). There is also a Brazilian government website called "Portal do Professor" ("Teacher's Portal"; http://portaldoprofessor.mec.gov.br/index.html), which offers digital resources and activities for educational use, giving teachers the opportunity to select situations that can be dealt with under the perspective of an investigative learning of mathematics.

Figure 1 shows the graph plotter for quadratic equations in the Interactive Simulation Project (PhET 2010). Activities in this application can be far more interesting and meaningful than simply finding the roots, which correspond with the points where the curve crosses the x-axis, as done with pencil and paper. The graphs are dynamic, and by changing the parameters of these functions, new graphs are immediately plotted. The point where they intercept the x-axis can be viewed and identified in the graph, but an extremely interesting aspect is to understand the role of parameters and how they influence the forms of the different resulting curves. Figure 1 shows the draft of a curve for a quadratic equation according to the selected parameters.

The student activity can follow the investigative approach with the goal to understand the role of each parameter for the curve's shape, such as 'aperture' and 'concavity'. In an investigative activity, the learner should form hypotheses, test them and record results to gradually develop an explanation for each of the parameters. The same investigative spirit should be used in other activities, such as to find

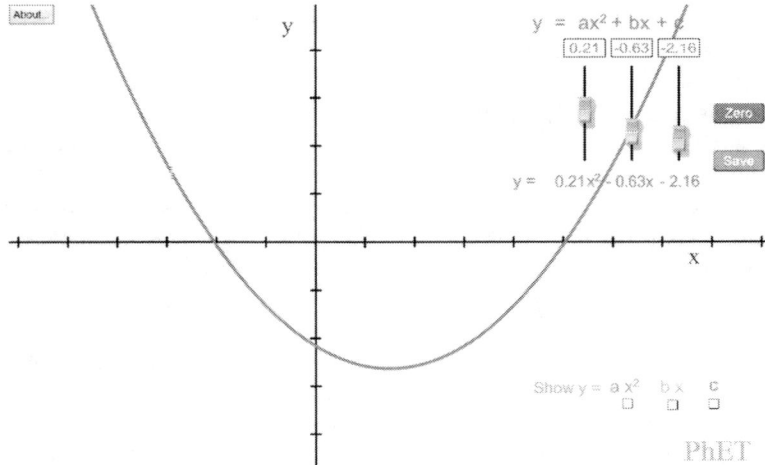

Fig. 1 Quadratic function graph and selected parameters

the minimum and maximum value points in the parabola, the representation of quadratic functions and the use of graphs as a strategy for problem-solving.

For the development of the investigative pedagogical approach using laptop resources in the classroom it is essential to discuss among the teachers the management of an investigative lesson using technological resources. It is important for the teachers to realize that the use of technological tools is not sufficient per se for the learning of mathematics. Even if some teachers can imagine interesting mathematical activities, the management of such activities with the students is complex.

Another main aspect of the implementation of the investigative pedagogical approach using laptop resources in the classroom is the possibility for the student to abandon the role of a mere consumer of teacher-based information, and to become a content-creator. The investigation performed by students, together with the knowledge built, can be described and recorded, and further shared at the school and with other schools, both on-site and using web-based resources. The idea is that DTIC users let go of the consumer's role and become active agents in the "production and dissemination of information and knowledge, transmuting consumer-users into citizen-users" (Mota and Tome 2005, p. 64). This knowledge—once it is properly documented and described—can be shared in discussions with peers, or through lectures, through presentations on boards or other media. These experiences produced in an organized format can become a bank of 'knowledge' generated by the school. Contextualized experiences in the school's reality that can be shared and used as reflection-objects for the school itself or for other people and educational institutions will create the possibility to leap onto higher levels of building knowledge through practice (Prado 2003; Prado and Valente 2002). Besides, the dissemination of this kind of knowledge and the DTIC can be useful in establishing a network of schools that share experiences and reflections, as well as management strategies used in this new educational approach (Vallin 2004).

Furthermore, our research showed that some mathematics teachers were not comfortable to use educational software, such as Geogebra, WinPlot, Ruler and Compass, in their lessons. When they did use them, the application was restricted to visualization in order to illustrate the lesson's content, without exploring the investigative advantages made possible by the technological resources. The possibilities for students to put their hypotheses to active testing, while comparing and altering their effect on the laptop screen, were hardly explored. The realization of this situation showed that the integration of DTIC into the curricular content is more complex as it requires the mathematics teachers to resort to a deeper and more consistent content-knowledge to be able to re-contextualize and represent it in a new format. Besides this, the necessary teaching skills are different, because in a classroom dynamic where investigation takes place, the educational focus is on the students while the teacher is the one who carefully performs pedagogical mediation, including providing the students with the formalization of the concepts involved. This is, therefore, a challenge for the teacher development. This realization led us to seek other paths that could prepare mathematics teachers for developing investigative activities with their students, using laptop resources. Bearing this in mind, some workshops were developed for teachers at the participating schools of the UCA program. One of these workshops took place in São João da Ponta, one of the six "digital cities", which is 230 km away from the capital city of Para, in northern Brazil. There, the UCA program was developed in all 14 schools (urban and rural), involving 83 educators (teachers and administrators) and approximately 1,600 students. While guiding the educators, as researchers, we collected data through semi-structured interviews, questionnaires and, especially, through the teachers' narratives registered in the virtual learning environment as well as by observation of their actions at the schools. The workshops held at São João da Ponta emphasized practices to teach students how to select a problem in their reality and seek information to propose solutions. One of the activities used the Webquest methodology aiming to provide students with the development of a guided investigation with meaningful tasks involving mathematical concepts to be experienced in the activity using laptop resources. This form of learning by experiencing situations, which instigates searching, interpretation and data selection that help understand the reality, enhances knowledge-building in a critical and reflexive thinking process.

Another investigative activity was project-based work. In this situation, the students identified a problem and organized an information and knowledge search, while developing the ability to critically analyze a given situation and contextualize mathematical contexts and other knowledge areas to solve problems. The project-based methodology encourages students to work in groups, debate ideas, develop arguments and produce something in which they are involved both cognitively and emotionally. The learning process through projects is as rich as its product, which expresses the learner's creative and authoring potential. Many of the resources in the laptop computer can be used in integration with the project and even serve as a means to publicize on the web, demanding accountability and ethical attitudes from the learners.

Discussion

Further data reveals that there is a significant commitment of teachers to create opportunities for students to interact with each other and by means of the educational laptop computer. Teachers realize that this technology can improve students' learning processes. The development of investigative activities, such as project-based methodology, showed that mathematical knowledge can be acquired in a more encompassing way than in the traditional setup of mathematics lessons. In articulation with contents from other areas of knowledge, it may lend an interdisciplinary character to students' learning. The technology allows this interdisciplinary character and the teachers could create lessons to link sciences and mathematics, for example, by using spreadsheets to create graphs to represent and interpret data in order to comprehend a phenomenon. Regarding the use of educational software such as Geogebra, Winplot, Ruler and Compass, among others, teachers reported the importance of using these resources as they allow for a better display of geometrical forms, their characteristics, specific values and properties, which can facilitate students' learning. However, in practical terms, using this software was mostly restricted to its display mode only.

Mathematics teachers who took part in this research are in the process of integrating the resources into their teaching practice. So far, technology has been used to illustrate classes and not to its full potential of exploring in order to develop mathematical knowledge by performing tasks, for example, in asserting hypotheses, building relationships, and supporting the construction and validation of arguments that could be experienced by the student in interaction with the software in a reflective and investigative classroom environment.

Another aspect highlighted in this research is related to teachers' perceptions regarding the development of learner autonomy in terms of handling the laptop computer to navigate the web. It has become evident to teachers that a student with a laptop computer has much more autonomy to explore new possibilities, enabled by this technology. Hence, it is important to include debates on issues related to ethics, authorship and values in pedagogical practices that enable students to use the internet and social networks securely and responsibly.

Concerning curriculum-related content, it is still a great challenge for teachers to work with both computational and other available resources within the school context. The prevailing view still is that students learn when teachers explain, and that the student has to practice (Skovsmose 2008). Students' critical thinking, their investigative capacity and creative authorship are very often left aside. Indeed, in our research we found a mismatch between the possibilities of the new learning environment and the formal view of education, which is unsuitable for today's society. Moreover, there is another worrying factor for mathematics teachers which is related to their mathematical knowledge. During the program of mathematics-teacher education to develop the educational use of the laptop computers, they mastered the operation of computing resources. However, the difficulty in mastering the mathematical content is noticeable, which makes it hard to reconstruct their practices to encourage investigative approaches and student authorship.

These results allow us to go back to the initial concern of many educators faced with technology at school. Technology really could 'replace' teachers, but only those who simply retransmit information, as computers can do this in a much more effective and attractive way. Now, for the teacher who understands mathematics education—with or without the use of technology—as critical thinking, focused on the development of values, realizes that his role is broader, acting as an advisor and mediator in the process of student learning, not only taking into account their intellectual abilities, but also taking them in their wholeness, represented in their sensitivity, history, background and culture.

Acknowledgements We would like to thank the grant received through the UCA program, funded by the Ministry of Education, which made this study viable.

References

Almeida, M. E., & Prado, M. E. (2009a). *Formação de Educadores para o uso dos Computadores Portáteis: Indicadores de Mudança na Prática e no Currículo*. Proceedings of the VI Conferência Internacional de Tecnologias de Informação e Comunicação na Educação, Braga, Portugal.

Almeida, M. E., & Prado, M. E. (2009b). *Formação de Educadores e o Laptop Educacional: Uma experiência vivenciada no Projeto-UCA de Tocantins*. http://www.uca.gov.br/institucional/downloads/estudoDeCasoTO_1.pdf. Accessed 15 July 2013.

Almeida, M. E., & Valente, J. A. (2011). *Tecnologias e currículo: trajetórias convergentes ou divergentes?* São Paulo: Paulis.

Bairral, M. A. (2007). *Discurso, interação e aprendizagem matemática em ambientes virtuais de aprendizagem*. Seropédica: Editora Universidade Rural.

Bebell, D., & O'Dwyer, L. M. (2010). Educational outcomes and research from 1:1 computing settings. *The Journal of Technology, Learning, and Assessment, 9*(1), 5–16.

Castro, A. L. (2011). *Tecnologias Digitais da Informação e Comunicação no Ensino de funções Quadráticas: Contribuições das diferentes representações*. Unpublished master thesis, Universidade Bandeirante de São Paulo, São Paulo. Available in: http://www.matematicaepraticadocente.net.br/pdf/teses_dissertacoes/dissertacao_AnnaLuisadeCastro_2011_Bette.pdf. Accessed 24 Feb 2015.

Ceibal. (2010). *Plan Ceibal*. http://www.ceibal.edu.uy/. Accessed 15 July 2013.

D'Ambrosio, U. (1986). *Da realidade à ação – reflexões sobre a Educação Matemática*. São Paulo: Sammus.

Dewey, J. (1979). *Experiência e Educação* (3rd ed.). São Paulo: Morata.

Freire, P. (1981). *Educação como prática de liberdade*. Rio de Janeiro: Paz e Terra.

Krajcik, J. S. (2002). The value and challenges of using learning technologies to support students in learning sciences. *Research in Science Education, 32*(1), 411–414.

Lobo da Costa, N. M. (1997). *Funções Seno e Cossenno: Uma sequencia de ensino a partir dos contextos do "Mundo Experimental" e do computador*. Unpublished master thesis, Pontifícia Universidade Católica de São Paulo, São Paulo. Available in: http://www.matematicaepraticadocente.net.br/pdf/teses_dissertacoes/dissertacao_nielce_meneguelo_lobo_da_costa.pdf. Accessed 24 Feb 2015.

Lobo da Costa, N. M. (2010). Reflexões sobre tecnologia e mediação pedagógica na formação do professor de matemática. In W. Beline & N. M. Lobo da Costa (Eds.), *Educação matemática, tecnologia e formação de professores: Algumas reflexões* (pp. 85–116). Campo Mourão: FECILCAM.

Lobo da Costa, N. M., Prado, M. E., & Pietropaolo, R. C. (2013). *Mathematics teachers continuing education and technology: A necessary practice in a globalized context.* Proceedings of CIEAEM 65, Torino, Italy.

Maia, D. (2007). *Função quadrática: um estudo didático de uma abordagem computacional.* Unpublished master thesis, Pontifícia Universidade Católica de São Paulo, São Paulo.

Mendes, M. (2008). *Introdução do laptop educacional em sala de aula: Indícios de mudanças na organização e gestão de aula.* Unpublished master thesis, Pontifícia Universidade Católica de São Paulo, São Paulo.

Mendes, M., & Almeida, M. E. (2011). Utilização do Laptop em sala de aula. In M. E. Almeida & M. E. Prado (Eds.), *O computador portátil na escola: Mudanças e desafios nos processos de ensino e aprendizagem* (pp. 49–59). São Paulo: Avercamp.

Mota, R. E., & Tome, T. (2005). Uma nova onda no ar. In A. Barbosa Filho, C. Castro, & T. Tome (Eds.), *Mídias digitais: Convergência tecnológica e inclusão digital* (pp. 51–84). São Paulo: Paulinas.

Papert, S. (1985). *Logo: Computadores e Educação.* São Paulo: Brasiliense.

PhET. (2010). *Interactive simulation project.* http://phet.colorado.edu/. Accessed 15 July 2013.

Ponte, J. P., Brocado, J., & Oliveira, H. (2003). *Investigações matemáticas na sala de aula.* Belo Horizonte: Autêntica.

Prado, M. E. (2003). *Educação a distância e formação do professor: Redimensionando concepções de aprendizagem.* Unpublished PhD thesis, Pontifícia Universidade Católica de São Paulo, São Paulo. Available in: http://www.matematicaepraticadocente.net.br/pdf/teses_dissertacoes/tese_Bette_Prado.pdf. Accessed 24 Feb 2015.

Prado, M. E. (2008). Os princípios da informática na educação e o papel do professor: Uma abordagem inclusiva. In D. Raiça (Ed.), *Tecnologias para a educação inclusiva* (pp. 55–66). São Paulo: Avercamp.

Prado, M. E., & Valente, J. A. (2002). A educação a distância possibilitando a formação do professor com base no ciclo da prática pedagógica. In M. C. Moraes (Ed.), *Educação a distância: fundamentos e práticas* (pp. 27–50). Campinas: UNICAMP/NIED.

Prado, M. E., Lobo da Costa, N. M., & Galvão, M. E. (2013). *Teacher development for the integrated use of technologies in Mathematics teaching.* Proceedings of CIEAEM 65, Torino, Italy.

Ramos, P., Giannella, T. R., & Struchiner, M. (2009). *A pesquisa baseada em design em artigos científicos sobre o uso de ambientes de aprendizagem mediados pelas tecnologias da informação e da comunicação no ensino de ciências: Uma análise preliminar.* VII Encontro Nacional de Pesquisa em Educação em Ciências, Florianópolis.

Saldanha, R. P. T. (2009). *Indicadores de um currículo flexível no uso de computadores portáteis.* Unpublished master thesis, Pontifícia Universidade Católica de São Paulo, São Paulo.

Skovsmose, O. (2008). *Desafios da reflexão em educação matemática crítica.* Campinas: Papirus.

Thiollent, M. (1988). *Metodologia da pesquisa-ação.* São Paulo: Cortez.

UCA Formação Brasi. (2009). *Projeto, planejamento das ações, cursos.* Brasilia: Ministério de Educação.

UCA Princípios. (2007). *Orientadores para o uso pedagógico do laptop na educação escolar.* Brasilia: Ministério de Educação.

Valente, J. A. (1999). *O computador na sociedade do conhecimento.* Campinas: UNICAMP/NIED.

Valente, J. A. (2008). A escola como geradora e gestora do conhecimento: O papel das tecnologias de informação e comunicação. In A. J. H. Guevara & A. M. Rosini (Eds.), *Tecnologias emergentes: Organizações e educação* (pp. 21–40). São Paulo: Cengage Learning.

Valente, J. A. (2011). Um laptop para cada aluno: Promessas e resultados educativos efetivos. In M. E. Almeida & M. E. Prado (Eds.), *O computador portátil na escola: Mudanças e desafios nos processos de ensino e aprendizagem* (pp. 20–33). São Paulo: Avercamp.

Vallin, C. (2004). *Projeto CER: Comunidade Escolar de Estudo, Trabalho e Reflexão.* Unpublished PhD thesis, Pontifícia Universidade Católica de São Paulo, São Paulo.

Technology and Education: Frameworks to Think Mathematics Education in the Twenty-First Century

Gilles Aldon

Abstract Improving the quality of teaching and learning by effective use of technology is a common goal that brings together teachers, researchers, students and, more widely, other citizens. However, the roads leading to this goal are often quite different. What are the main changes, for teachers, for students and in the interactions between students and teachers? Different theoretical frameworks provide tools to analyse and understand what happens in the classroom: multirepresentation and multimodality; instrumental and documentational genesis; role of technology in an experimental part of mathematics, didactical incidents. Starting from experiments, this chapter shows how these frameworks can be combined to analyse the role of technology, the difficulties and some success in mathematics education.

A Positioning of the Problem

From the Babylonian clay tablets, which have been used in the scribes' schools both by teachers and students (Proust 2012), up to the mathematical machines (Maschietto and Trouche 2010), not to mention the geometrical tools, mathematics and mathematics education have always had to do with technologies. Tools and instruments, a ruler or a compass as well as a mathematical result, are part of the toolbox and of the documentary set of mathematicians and students. In the twenty-first century, to understand and to use digital technologies are both part of the mathematical act and thus part of the documentary set and of the toolbox of mathematicians and students. To be interested in mathematics education nowadays leads naturally to studying the available resources and tools from both the point of view of the content and of the medium, or container.

Following Serres (2012), three main revolutions occurred in the human history. These three revolutions have something to do with the communication between

G. Aldon (✉)
Ecole Normale Supérieure de Lyon, Lyon, France
e-mail: gilles.aldon@ens-lyon.fr

people and more precisely with a modification of the relationship between media and contents.

The first revolution is the appearance of writing about 5,000 years ago. The world changed dramatically from an oral to a written civilization. Written documents had not only a value of communication but more deeply appeared to have a role of witness, a value of evidence and a value of information. The code of Hammurabi (about 1770 BC) is a good example of the change brought by the existence of a written law: "He was the first monarch to make a united Babylonia" (Prince 1904, p. 602), justice and right should supplant the oral traditions. There was a common document, universal reference for life. This new age is the beginning of the modern civilizations due to the possible, incontestable and durable organization of the world through a new mode of relationship between people. The consequences on social relationships, economy, religion (the birth of the "Religions of the Book") and education were fundamental. In this time of Païdeia,[1] education became an intentional act in the city in coherence with the changes that life in the city brought to the development of children.

The second revolution has also to do with the dialectics of media and contents. It is the dissemination of writing, the advent of printing in the fifteenth century. It is the time of the Reformation when each person can read directly in the Book: Luther's translation of the Bible into the vernacular instead of Latin made the Book accessible for everybody: "In what concerns the word of God and the faith, every Christian is as good a judge for himself as the pope can be for him" (Luther 1904, p. 78). It is the time of the huge development of trade, the development of experimental sciences: the scientific work is no more the learning of the existing science, which is externalized and disseminated in books, but the experimentation of this knowledge through experience. In the sixteenth century, Montaigne wrote that it should be better for education to build "une tête bien faite, plutôt qu'une tête bien pleine" (1854, p. 64; "a well-made head rather than a well-filled head"). The externalization of knowledge enabled people to develop other skills and to experiment the written knowledge. In terms of education the humanist school was born and would be developed during the next centuries. It is interesting to notice about the first two revolutions that these were not revolutions regarding knowledge itself but revolutions in the relationship between knowledge and media, which deeply modified the way knowledge was shared among people. In turn, these societal transformations allowed the emergence of new knowledge: the dialectics of media and content implied an evolution of techniques of writing and of development of knowledge.

Adding 'digital' to the word documents and the resources is about to trigger a new revolution. We are living in the time of the beginning of the third revolution in which the dialectics of media and contents are fundamentally renewed by the possibilities offered by digital technologies in terms of information and communication. It opens a time of social networks, optimizing learner success through learning

---

[1] Παιδεια from παιδοσ: child and αγειν: to lead; literally to lead children through the city. The word pedagogy has the same root.

analytics (Morency et al. 2013) where techniques are at the service of personalization of education: "Theoretically, LA [Learning Analytics] has potential to dramatically impact the existing models of education and to generate new insights into what works and what does not work in teaching and learning" (Siemens 2012). Economic, religious, and social consequences will surely be as important as they were in the two previous revolutions, as well as changes in education. By bringing out new paradigms and new values, often one revolution suppresses the previous one. The digital revolution highlights communication and cooperation values. Is it then possible to learn from history and to keep the humanist and civic values in this new age? This is surely a great challenge to our new world where knowledge is widely available but not necessarily easy to understand: as an example, googleling "Wiles-Fermat theorem proof" brings the text of Wiles's proof in less than half a second. In this sense, knowledge is available, or more precisely the text of knowledge is available, but still hard work is required in order to integrate and to understand all the actual concepts and notions at stake in this proof.

In these dialectics of media and contents, we are confronted with a paradox because digital media are instantaneous, whereas knowledge construction is a long process which requires time and effort. It is possible to find information very quickly about any subject, it is possible to download the text of knowledge, but having available information is not yet synonymous with understanding this knowledge and using the knowledge in a particular context. The process of the transformation of available data into usable knowledge is a challenge of education. At the same time, digital technologies offer new tools to allow new approaches to mathematical notions. The example of dynamic geometry software is symptomatic of a new approach to Euclidian geometry and to its teaching and learning. In addition, multi-representations, which contribute to the understanding of a specific notion, are facilitated by the use of software that offers a connection of different applications (typically Geogebra or TI-*n*spire with geometry, spreadsheet, algebra, and CAS windows). For example:

> When presented with the TI-*n*spire, we assumed that these developments could offer new possibilities for students' learning as well as teachers' actions. They could foster increased interactions between mathematical areas and/or semiotic representations. They could also enrich the experimentation and simulation methods, and enable storage of far more usable records of pupils' mathematics activity. However, we also hypothesized that the profoundly new [sic] nature of this calculator and its complexity would raise significant and partially new instrumentation problems both for students and teachers and that making use of the new potentials on offer would require specific constructions, and not simply an adaptation of the strategies which have been successful with other calculators. (Artigue and Bardini 2010, p. 1,172)

What are the main changes for teachers, for students and in the interaction of students and teachers? What tools can research bring to teachers to enhance mathematical education in this beginning digital era? Present work of the education community brings elements of answers to these complex questions.

Towards a Framework That Takes the Complexity of the Topic into Account

Pedauque (2006) places the modernization of documentary activity in the center of the intertwining of media and contents. The main four functions of documentary production are described as two cognitive functions—mnemonics and organization of ideas—seen as the fundamental basis for a documentary production, the function of creativity and the function of transmission. These four functions can be crossed by three levels of mediation: the individual level (the documentation of the personal library), the group or collective level (the documentation shared by a specific community) and the public level where documents are made available for everybody. Even if the boundaries between the three levels as well as between the four properties are not clear-cut, this classification allows the functions of a document to be related to the different spaces of mediation. The modernization due to digital technologies moves the boundaries between the levels of mediation: individual documents appear in the collective or the public level which makes quality tracking difficult, especially in the world of education. However, this model is insufficient to make understandable the relationships between actors in a teaching/learning ratio. The notions of "document" and "resources" need to be clarified and explained in the context of education. For this purpose, I subsequently explain and follow a documentational approach to didactics (Gueudet and Trouche 2009, 2010).

Another, and linked, important aspect is the modification of tasks that digital technologies bring into the classroom. Because most of the answers to classical mathematics exercises are directly available on the web, the students' tasks must evolve in order to develop new skills. The mathematical tasks evolve with technology through new possibilities of manipulating some of the representations of mathematical objects. The experimental part of mathematics, which is not born with technology, changes with the use of technology. These "augmented mathematics" and the role of experience in the learning of mathematics also need to be clarified.

Documentational Genesis

Coming back to a basic question: what are resources? it is interesting to have a look at the etymology of the word: it comes from the Latin *resurgere* which means "to rise again" and that the resource was somewhere hidden and is brought into light in relation to a specific goal. Thus, resources for teachers, and students as well, could have a wide range of origins: a discussion with colleagues or mates, data kept in a computer or in the "cloud", books and encyclopedic knowledge, tools and software, meetings, talks and course. When a goal has to be reached, when a task has to be completed, available resources are "risen again" to produce something operational, a *document* which will be the result of a process called by Gueudet and Trouche (2009) the documentational genesis. This model extends the model of instrumental genesis (Rabardel 1995), which has been used in mathematics education (Artigue 2002; Lagrange et al. 2003; Trouche 2004) to explain the process of appropriation of artifacts and their transformation into instruments when associated with schemes. As in instrumental genesis, the process of transformation (the instrumental or the documentational genesis) is the result of a double movement from the subject to the artifact (or the resource) and from the resource (or the artifact) to the subject. The first is called instrumentalization and the second instrumentation. In the process of instrumentation, the subject's behavior is modified by the artifact or the resource, whereas instrumentalization describes how the subject fashions the resource or the artifact for his/her own use. The document, which is the result of the documentational genesis, becomes part of the set of resources of the subject. Students as well as teachers build their documents in such a documentational genesis and lean on their own set of resources to build their own documents related to their learning intentions. The set of resources may be individual, collective or public and the document itself becomes individual before perhaps entering a collective level where it is shared or distributed.

A good example of this process is the content of students' calculators. It is now possible in most calculators to store data and students may organize the memory with different contents. It is also possible to communicate between calculators and to download files. But a specific content that has been downloaded or copied is not yet usable. Before it becomes a document, students have to modify it, to organize the data for their own use. However, knowing that data is reachable in the calculator modifies the way they learn.

Resources become a document related to a certain goal and students have learning intentions whilst a teacher has teaching intentions. Both include their conceptions, beliefs and knowledge. Therefore, the documentational genesis of students and teachers are developed simultaneously and the confrontation of the two geneses may provoke *incidents* (Aldon 2011, 2014). These incidents can be called didactic incidents because they have to do with the knowledge at stake. Analyzing these incidents gives clues for understanding the misunderstanding between teachers and students in relation to the management of the resources. Coming back to the example of calculators' contents, what can be seen by students as an organization of ideas and a good use of the mnemonics property of technology may appear from the point of view of the teacher as cheating or refusing to memorize. This misunderstanding may induce a withdrawal at the individual level or at a collective level excluding the teacher. The consequence may be a deep perturbation in the integration of the calculator in the mathematics course as expressed by a teacher in an interview: "Sometimes it's difficult because they [the students] do not know their lessons. They do not learn because they believe that they have everything inside their calculators." Looking at calculators with their different potentialities, we can consider them as artifacts with possibilities of calculation and representation (properties of creation) and as digital resources with possibilities of data processing and data sharing (properties of memorization, organization of ideas, and communication).

Didactic Incidents

Didactic incidents are phenomena of the didactic relationship and concern the interaction in the classroom. Some of them are linked to the interaction between teacher and students, others are related to the mathematical content or the pedagogical environment. I elsewhere (2010, 2011) distinguish five types of incidents, each of them causing perturbations in the dynamic of the classroom:

1. An outside incident corresponds to an event not directly linked to the learning situation but often important in the classroom. This type of incident can strengthen a previously caused perturbation.
2. A syntactic incident is linked to the conversion between semiotic registers of representations. In a technological environment, these incidents mainly come from the feedback of the machine.
3. A friction incident corresponds to the confrontation of two situations in the interaction between students and teachers.
4. A contract incident occurs when an event breaks or modifies significantly the didactical contract. This modification is strongly correlated to the appearance of didactical bifurcations where students invest a situation differently from the situation intended by the teacher.
5. A mathematical incident when a mathematical question is asked and not answered.

An example of a contract incident is the place where students access technological tools. The following dialogue took place when students had to look for the 1000th term of a sequence. The environment of the computer laboratory seemed to favor a contract authorizing any available tool. In this brief excerpt, however, we see G1 offended when she saw other students using a spreadsheet (G indicates a girl, B a boy):

B1: Oh là là! We must calculate the thousandth!
G1: Wait! They are cheating!
B1: Who?
G1: They cheat, they use Excel!

Even if the spreadsheet is an element of her set of resources, it is not visible in the particular contract sensed by G1. What is highlighted in this example is also very present when it comes to using (or not using) a particular feature of the technology whose status is not clearly defined. A typical friction incident occurred when the teacher took part in a discussion with students previously working alone:

T: Do you understand what I said?
B1: How do we calculate?
T: Then how to calculate … can you move this point, it's always positive, okay?
B1: Yes but …

The teacher, when continuing her discourse, did not take into account the actual work of the student who asked a technical question and, instead, she answered conceptually.

When working with software, feedback given by the program is often at the origin of a syntactic incident. There is a necessary interpretation and a transposition from a register of representation to another that may provoke a misinterpretation of the feedback. It is interesting to set this dialogue against the following observation in an observed lesson at the beginning of the school year; the teacher spoke to the whole class whilst students were working with their calculators:

T: Then you open the catalogue and type the first letter of the command, well for the moment, R and you just have to go down, okay, you see Randint, it's here. Well. [*T is demonstrating on the white board whilst speaking.*]
T: Well. I have simulated the throw of a dice. The question now is: How are you going to simulate the throw of two dice and how will you obtain the value of the difference of the greater minus the smaller?

At the same time, S1 and S2 have been working with their calculators:

S1: We have to type a blank.
S2: Do you think so?
S1: It's six.
S2: Yes, randint one six minus randint one six?
S1: And, how do you type the absolute value? ... It doesn't work.
S2: [*Watching the screen of E1's calculator.*] Missing?
S1: And now it gives six, ahhh!
S2: Ahhhh!
S1: It doesn't work!
S2: Too many arguments?
S1: I can't do that!

The gap between the talk of the teacher and the students' difficulties originates in instrumentation problems. The syntactic incident is caused by incomprehension of the feedback of the machine. At first, instead of typing *randint(1,6)*, S1 typed *randint 1 6*; the feedback of the machine was *Missing)*, but the bracket was not read by the students.

This didactic incident serves to highlight key elements of the students' trajectories in the classroom's dynamic. It gives elements of explanation of divergent or amplified dynamics. It is particularly important when interactions are mediated by technology to understand why the teacher's intentions do not meet the students' will of learning. Through the documentational approach and the analysis of didactic incidents, a general approach of knowledge construction takes into account the main properties of digital documents, that is to say organization of ideas and mnemonics, communication and creativity.

Coming back to the dialectics of media and contents and simultaneously taking up the dialectics of representation and object, I would like to emphasize two particular properties of technology: the possibility of multi-representation and the link to the role of experiences in creating and learning mathematics.

Multi-representation and Experiences in Mathematics

Mathematical objects that mathematicians handle have various representations and the mathematicians' work is about some of the representations: "The semiotic representations are productions made of signs belonging to a system of representation which has its own constraints of significance and operating" (Duval 1991, p. 234). Mathematical objects can be considered as the equivalence class of their representations modulo the equivalence relation defined by: two representations are equivalent if they represent the same object. This observation has two important consequences:

- A mathematical object can be mastered in a particular context and is difficult or unknown in another context.
- Converting a register of representation into another is essential for the understanding of a mathematical object. This conversion requires a translation in which some elements of meaning are lost and others are added. Changing the significant, that is, the way to designate the object, on the one hand modifies and enriches, on the other hand impoverishes the signified, that is, the designated object. The thesis of indeterminacy of translation (Quine 1960) tends to explain that the translation between two languages cannot be complete. More precisely, Quine argues that it is always possible to build different interpretations, semantically coherent, of a given text. His famous "gavagai", word (or sentence?) uttered by a primitive watching rabbits running into the forest, is an example of holophrastic indeterminacy, that is, the indeterminacy of sentence translation: Does it mean 'rabbit', or 'stages of rabbits' or 'rabbithood'? "In each case the situations that prompt assent to 'gavagai' would be the same as for 'rabbit'. Or perhaps the objects to which 'gavagai' applies are all and sundry undetachable parts of rabbits, again the same stimulus meaning would register no difference" (Quine 1960, p. 47).

The issues raised by Quine are not restricted to translation between different languages but can also be present within a language, through the interpretation of a word or a sentence. They can also occur when different significants refer to a sole signified. That is the case of semiotic registers of representation of mathematical objects. Consider, for example, the Julia set known as the Douady's rabbit; it can be defined in a topological perspective as the closure of the set of repelling periodic points of $f(z) = z^2 - 0.123 + 0.745i$. Or in an analytical (or algorithmic) perspective: For all but at most two points $z \in X$, the Julia set is the set of limit points of the full backwards orbit:

Or in a graphical perspective as shown in Fig. 1., and so on. All of these representations give information about the structure and the properties of this Julia set, but also lose information or properties. Thus, the graphical representation, even if it is not calculable, gives precious information about the dynamic of this set and it is not surprising that the work of Julia remained mainly unknown until computers allowed for this kind of graphical representations.

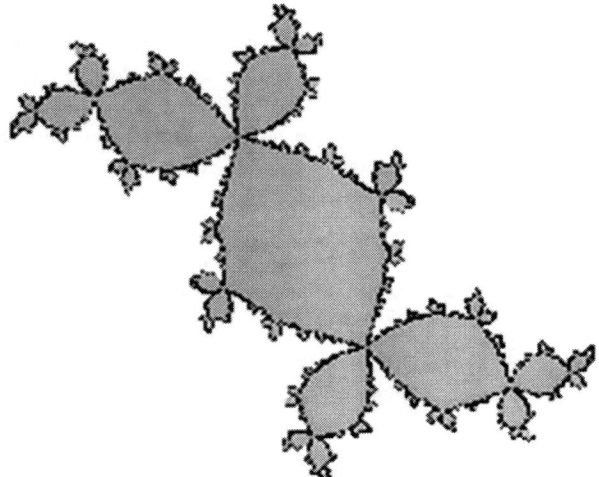

Fig. 1 Douady's rabbit

The assumptions that underpin my work are then that technology offers opportunities for multiple representations that facilitate the understanding of mathematical objects. Following the work of Arzarello and Robutti (2010), technology can play within an internal representation with multiple software representing the same object (spreadsheet, DGS, CAS) but also externally by providing through communication different approaches from different points of view. The notion of multimodality emphasizes the many ways people experience and develop understandings and the two aspects of multirepresentation and multimodality can be seen as the two faces of the same coin: multirepresentation being the technological way and multimodality the cognitive way of understanding mathematics. "Instrumental activity in technological settings is multimodal, because action is not only directed towards objects, but also towards people" (Arzarello and Robutti 2010, p. 718).

In order to study the multiple representations offered by technology, it seems important to me not to stop at the graphing and calculation properties, but also to consider the properties of organization of ideas, creativity and communication involved in the implementation of external representations of the studied mathematical objects, that is to say, to include documentational properties within technology.

Scientific phenomena or observable occurrence can be included within an experimental device only if they can be considered as objects. Mathematics is not an exception and the relationship between mathematical objects and reality has to be clarified. In a Kantian perspective, mathematics is a human construction which builds and defines its objects *a priori*. The different levels of reality enable considering the mathematical objects as elements of this model: the perceptible reality which is perceived by the five senses, the empirical reality which can be experimented and the objective reality on which it will be possible to build mathematical experiments. The reality *per se*, also called "unattainable" reality, remains inaccessible. Experience takes into account the perceptible, empirical and objective reality, and what can be called an experience in mathematics is a work on naturalized representations of mathematical objects defined in a system of signs. The word "naturalized" is understood as the mastery of internal transformations within a register of representation or conversions from one register to another. A mathematical experience allows us to define and explore the properties of a particular object in relation to a theory. Thus mathematical concepts, even if they are created in mind, are fully realized in the relationships with empirical phenomena:

> Gedanken ohne Inhalt sind leer, Anschauungen ohne Begriffe, sind blind. Daher ist es eben so nothwendig, seine Begriffe sinnlich zu machen, (d. i. ihnen den Gegenstand in der Anschauung beyzufügen), als seine Anschauungen sich verständlich zu machen, (d. i. sie unter Begriffe zu bringen). (Kant 1781, p. 51)[2]

These philosophical considerations on the nature of the objects are of great importance for education; indeed, especially in a technological environment, the role of experience on mathematical objects seems to be a widely shared assumption. However, experiments are built not on objects fundamentally synthetic in nature but on representations of these objects that allow extending the studied concept to make it perceivable.

The dialectics of media and contents and of resource and document discussed above have then to be put in relation with objects and representations. As a resource becomes a document in the documentational genesis, the understanding of a mathematical object is built through and by the experiments on some of its representations. As an example, we can consider the clay tablet shown on Fig. 2. The mathematical content is synthesized on the tablet with different representations of the same object: symbols of the sexagesimal Babylonian writing of numbers, a

Fig. 2 Clay tablet YBC 7,289, Yale University (http://nelc.yale.edu/babylonian-collection)

---

[2] "Thoughts without content are void; intuitions without conceptions, blind. Hence it is as necessary for the mind to make its conceptions sensuous (that is, to join to them the object in intuition), as to make its intuitions intelligible (that is, to bring them under conceptions)." (Translation by J.M.D. Meiklejohn; http://www.gutenberg.org/files/4280/4280-h/4280-h.htm)

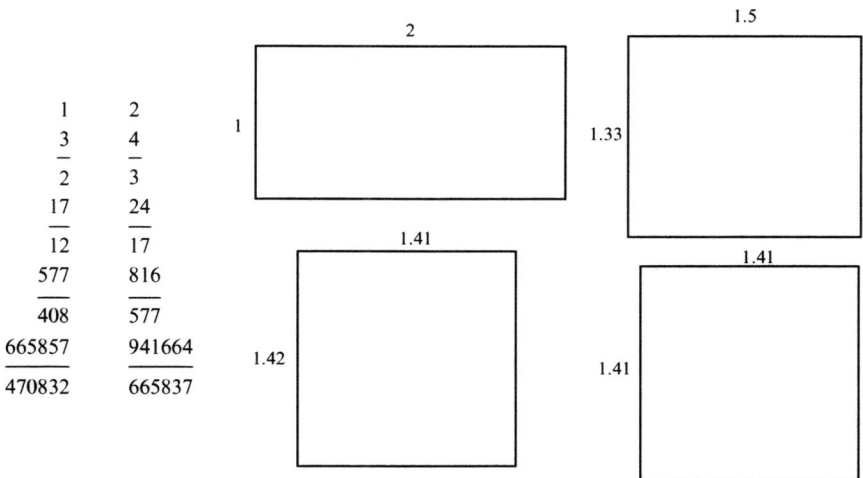

Fig. 3 Babylonian algorithm in two registers of representation

square and its diagonals. The Babylonian algorithm of determination of the square root of two, even if not present on the tablet, is present through the result of the calculation: the side is 30 (≪≪) and the diagonal is 30 times square root of 2 (I ≪ IIII ≪<≪ I <: $1 + 24/50 + 51/3{,}600 + 10/60^3 \approx 1.41421296$) and the two approximate values, result of the algorithm, are written on the diagonal.

In this example, links between registers of representations bring into light the necessary experiment that combined methods and concepts to create new knowledge. Even if no document describes the calculation done by the scribe to reach this precision, we can imagine that the so-called "Babylonian algorithm" came from the combination of drawings and calculation; starting with a rectangle 1×2 of area 2, then replacing the side of length 2 by the mean of 1 and 2 and the side of length 1 by a side such that the area is still 2, and so on. When the rectangle becomes (almost) a square, the calculation gives the result (almost) written on the tablet as illustrated in Fig. 3. Extending to a register of analysis, it is interesting to notice that this algorithm can be described by the algorithm of Newton with the function f: $x \rightarrow x^2 - 2$: the sequence $u_0 = 2$, $u_n + 1 = u_n - f(u_n)/f'(u_n)$ gives as first terms: 2, 3/2, 17/12, 577/408, 665,857/470,832.

Summarizing the Framework

Examining mathematics education in the twenty-first century brings us to consider, within the relationships between students, teachers and knowledge, the subtle games played in the documentary process between resource and document as well as

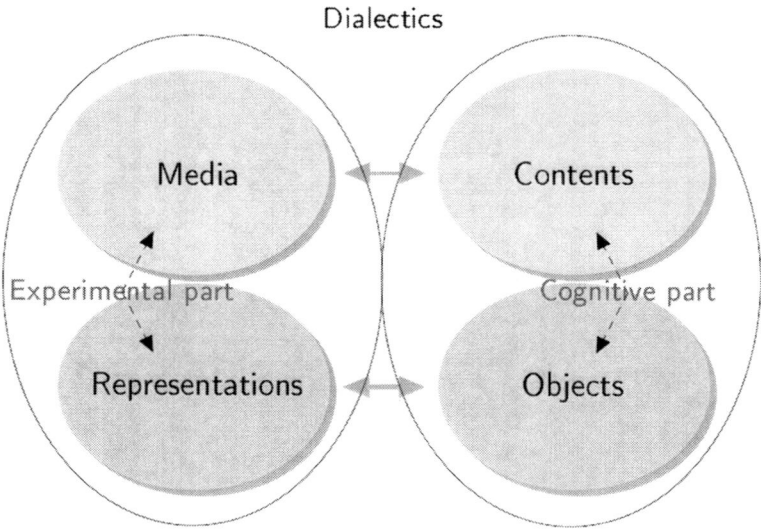

Fig. 4 The experimental part of mathematics

between representations and objects. The experimental part of mathematics can be considered as all that can be done with representations and media when the cognitive part joins the associated contents and the mathematical objects as illustrated in Fig. 4. In this context, didactic incidents are events which occur, modify and flow into the dynamics of knowledge construction.

Using the Framework for Classroom Research

By the back-and-forth from the experimental to the cognitive part of mathematics, mathematical objects are built progressively through experience of some of their representations. In this section, I illustrate the working of the model by showing examples taken from classroom experiments.

Context

Experiments took place in different research settings where the main question was to discuss the experimental part of mathematics in a technological context (Aldon 2011; Aldon et al. 2010). These investigations were situated in the perspective of a collaboration between researchers and teachers, each of them bringing in their competencies. Different classes were observed, with teachers experts (or not) in the use of technology, accustomed to use problem solving in their teaching (or not).

The common denominator of the experiments was that classrooms were 'ordinary' classrooms in the sense that it was the teacher who was responsible for the content of the lesson and for the choice of problems. Researchers were external observers trying to disappear from the interactions as much as possible.[3] In all these settings, some students and the teacher were interviewed after the lesson. The examples illustrate the experimental part of mathematics in act, leaning on incident analysis.

Examples

The first example took place in a class of 17–18 year-old students working on a probabilistic problem. The question was to determine the difference of two dice that comes up with highest probability. Students, after having simulated the throw of dice with a spreadsheet, try to conjecture the result. In the dialogue, T is the teacher, S a student.

T: Something important. Remember! Ctrl R to refresh … as if you made a new experiment. Abs for absolute value! Remember!
S: Oh! Five fives in a row!
T: Finally, better than reality, huh?

This very short excerpt illustrates the possible back-and-forth from practical experience to theory and, unfortunately, a mathematics incident where the teacher does not take into account the mathematical question posed by the result of the experience. Is it exceptional to obtain five 5 in a row in a probabilistic experience of throwing dice 30 times? The back and forth between experience and theory was broken and there is a lack in the construction of mathematical thinking.

The second example is from an experiment where a whole class was equipped with calculators TI-Nspire. The IT environment is one of the central elements of this work. The choice of the particular TI-Nspire technology is justified by the innovative nature of the technology that makes it a paradigmatic example of other existing or developing technologies. I assume that the results obtained through experiments with this technology are possibly generalizable to other technologies. The main characteristics of this technology are firstly the possibility of multi-representation through different software (CAS, spreadsheet, DGS, notepad), secondly the possibilities of storing and of organizing data, and finally the possibility to work on a computer or with a handheld calculator with possibilities of communication between the two.

Often not very much taken into account, the case of statistics is however very interesting to study because the description of data often requires different kinds of representation, each of them showing and hiding properties; a complete understanding of the data requires exploration of these different representations and conversion

---

[3] Despite this intention, an observer in a class observation is always a perturbation which needs to be taken into account. The "outside incidents" described above are very often related to the observer's presence or to the sound or video recording material.

from one to the other. Starting from this hypothesis, we[4] elaborated a class situation whose goal was to measure the reaction time of students. Students were invited to measure their reaction time to a visual stimulus, the apparition of a red disk on the screen. The question asked by the teacher was: "What is your reaction time?" In order to answer this question, students had to treat their data statistically. The questions of the description, of the comparison and of the communication of data led to define the main characteristics of the statistics, but also to explore inferential statistics. The observed group is a group of four 16 year-old boys (B1, B2, B3, B4) in a science class. After having experimented their reaction time, they analyzed the data beginning by some commentaries about their performance:

B2: Too bad ... uh, zero forty-nine.
B3: Oh no, in fact I'm fine. Oh no, I'm zero thirty-nine.
B2: The second is better. No, not even.
B1: Yes, me there, oh no, the first is better.
B2: Where is your time?

The students then began to process data and they decided to calculate the arithmetic average and the median of the whole series; however, the calculator gave a set of answers (mean, median, standard deviation) that confused the students. The syntactic incident, due to a misunderstanding of the feedback of the calculator, postponed the interpretation of the experience until the teacher's intervention provoked a beginning of reflection:

B3: What the hell, one variable statistics?
B2: Yeah.
B4: Well, you have your average or not?
B2: No, no, no, yeah, okay, and then you put the b, no, b, nb put nb.
B3: Why nb?
B2: Go! Yes! And then you put there.
B4: But do you have the mean? Where do you see the average?
B2: Ben I dunno, I understand nothing, sum x, x squared.
B4: Sir! Sir!
B2: There's the median, but there's no average.
B4: Uh! It has not! Average, we dunno how to calculate!
T: Ah! It is x.
B4: x.
B2: Zero point two eight!
B3: How?
B4: But it is not yours?
B2: Yeah, it is B1!
B4: It is zero point three one.

---

[4]This situation has been developed in the EdUmatics project, 50,324-UK-2009-COMENIUS-CMP; European Development for the Use of Mathematics Technology in Classrooms, http://www.edumatics.eu

In that excerpt, the interpretation of the experience is blocked by the disturbance provoked by a syntactic incident. It is possible to see the fundamental role of the teacher who is able to suppress the blockage even though the students were not able to understand the feedback of the machine. Syntactic incidents, mainly caused by problems of conversion between semiotic registers of representation, appear as generators of disturbances that can prevent students' from entering the process of reflection about the results of an experience.

In other circumstances, incidents come from a different documentational genesis and the position of the technology is in the level of mediation. In the next example, the teacher in a class observation disconnected the experiment on the calculator and the mathematical knowledge which provoked a didactic incident:

S: Do we save our work?
T: You save if you want, but tomorrow we will do the theoretical part.

By this answer, the teacher notified that the calculator is part of the student's set of resources and did not promote a shared documentational genesis. As a consequence, the student investigated the memorization property of the technology privately which, later, provoked a deep perturbation in the integration of the calculator in the mathematics course. The same resource became a different document for the student and the teacher. The calculator could not be used on a collective level of mediation although the knowledge construction could have been reinforced if the collective level of mediation had been taken into account:

> Sometimes, I don't remember how to do it, it happened once, I didn't remember the names, you know, it was, a x plus b plus c, or something like that, I was not able to go on; then I typed on my calculator. I didn't explain the result, but I had one and I was able to answer the next questions. Otherwise. I would have had nothing right in the exercise. (Interview, student)

This example shows that the documentary activity of students in relation with the properties of memorization, organization of ideas, creativity and communication can contribute to the dynamics of knowledge construction. When the instrumental approach focuses on the transformation of an artifact into an instrument, the documentational approach considers the documentary properties of calculators. The significance given to the calculator in the private domain by students and teachers is not necessarily shared. Therefore, documentary geneses become distinct, unrelated and sometimes divergent, which causes didactic incidents and generates disturbances. It is also important to notice that incidents reinforce the barrier between the usual digital tools that students use in their everyday life on a collective level of mediation and the school tools that have to be used in the specific school community or on a private level of mediation.

Conclusion

"Generation Y, millennial generation, digital natives" are expressions used to designate teenagers born at the time when the development of communication and information technologies began. There are many voices claiming that these changes will disrupt the world. However, the question of how to change teaching and learning in order to adapt schools and society is an open question. The paradox of learning in an instantaneous information world needs to be considered, if education still ascribes to the values of humanism and citizenship while adapting to the digital world.

Looking at theoretical frameworks to understand and to describe the new situation is a reasonable approach that researchers can produce in connection with teachers who have a vivid perception of the changes. Joint work of teachers and researchers is surely a way to face the challenge of a changing education. Documentational genesis crossed with incident analysis, leading to considering computing environments as part of the documentary system of teachers and students, is one of the proposals that allows us to understand and to follow the dynamics of teaching and learning in a technological context.

By analyzing didactic incidents we can find reasons for breaks in the dynamics of the classroom. Didactic incidents arise out of various causes: the mathematical knowledge, the relationship between students and teachers, a misunderstanding of the teacher's intention, and so on. However, incidents can also come from the different documentational geneses of students and teachers. In order to teach, teachers use a wide range of resources: discussions with peers, their teaching experience, external resources in the "cloud", academic courses, readings, and more. These resources often relate to the documentational properties of artifacts like calculators with properties of communication and storage of data. The process of transformation of these resources into a document for a special use in a particular context is the result of a double movement where actors modify and combine resources whilst resources modify the actors' behavior. At the same time, the students transform and combine their own resources to build their own documents in an individual or collective way. The junction of the two geneses may provoke incidents, particularly when the documentational properties of tools are used in different ways, as shown in the example of communication properties or storage of information. The framework of didactic incidents may increase a teacher's sensitivity for students' work, particularly when observing and facilitating the institutionalisation of knowledge based on the actual activities of the students.

Analyzing contributions of technology to the construction of mathematical knowledge, the development of an experimental part of mathematics can be considered as linking representations and object as well as media and contents. The possibility of multi-representations inherent in the technology gives numerous occasions for a better grasping of the "reality" of mathematical objects and thus for new ways to teach and learn mathematics.

References

Aldon, G. (2010). Handheld calculators between instrument and document. *ZDM – The International Journal on Mathematics Education, 42*(7), 733–745.

Aldon, G. (2011). *Interactions didactiques dans la classe de mathématiques en environnement numérique: construction et mise à l'épreuve d'un cadre d'analyse exploitant la notion d'incident*. Unpublished PhD thesis, Université de Lyon, Lyon. https://tel.archives-ouvertes.fr/tel-00679121.

Aldon, G. (2014). Didactic incidents: A way to improve the professional development of mathematics teachers. In A. Clark-Wilson, O. Robutti, & N. Sinclair (Eds.), *The mathematics teacher in the digital era: An international perspective on technology focused professional development* (pp. 319–343). Dordrecht: Springer.

Aldon, G., Cahuet, P.-Y., Durand-Guerrier, V., Front, M., Krieger, D., Mizony, M., & Tardy, C. (2010). *Expérimenter des problèmes de recherche innovants en mathématiques à l'école (CD-Rom)*. Lyon: INRP.

Artigue, M. (2002). Learning mathematics in a CAS environment: The genesis of a reflection about instrumentation and the dialectics between technical and conceptual work. *International Journal of Computers for Mathematical Learning, 7*, 245–274.

Artigue, M., & Bardini, C. (2010). New didactical phenomena prompted by TI-NSPIRE specificities: The mathematical component of the instrumentation process. In V. Durand-Guerrier, S. Soury-Lavergne, & F. Arzarello (Eds.), *Proceedings of the sixth congress of the european society for research in mathematics education* (pp. 1171–1180). Lyon: Institut National de Recherche Pédagogique.

Arzarello, F., & Robutti, O. (2010). Multimodality in multi-representational environments. *ZDM – The International Journal on Mathematics Education, 42*(7), 715–731.

Duval, R. (1991). Structure du raisonnement déductif et apprentissage de la démonstration. *Educational Studies in Mathematics, 22*(3), 233–261.

Gueudet, G., & Trouche, L. (2009). Towards new documentation systems for mathematics teachers? *Educational Studies in Mathematics, 71*(3), 199–218.

Gueudet, G., & Trouche, L. (2010). *Ressources vives. Le travail documentaire des professeurs en mathématiques*. Rennes: PUR & Lyon/INRP.

Kant, I. (1781). *Critik der reinen Vernunft*. http://www.deutschestextarchiv.de. Accessed 7 Sept 2013.

Lagrange, J. B., Artigue, M., Laborde, C., & Trouche, L. (2003). Technology and mathematics education: A multidimensional study of the evolution of research and innovation. In A. J. Bishop, M. A. Clements, C. Keitel, J. Kilpatrick, & F. K. S. Leung (Eds.), *Second international handbook of mathematics education* (pp. 237–269). Dordrecht: Kluwer.

Luther, M. (1904). *The life of Luther, written by himself, collected and arranged by M. Michelet translated by William Hazlitt*. London: George Bell & Sons.

Maschietto, M., & Trouche, L. (2010). Mathematics learning and tools from theoretical, historical and practical points of view: The productive notion of mathematics laboratories. *ZDM – The International Journal on Mathematics Education, 42*(1), 33–47.

Montaigne, M. (1854). *Essais*. Paris: Firmin-Didot.

Morency, L.-P., Oviatt, S., Scherer, S., Weibel, N., & Worsley, M. (2013). ICMI 2013 grand challenge workshop on multimodal learning analytics. In *Proceedings of the 15th ACM international conference on multimodal interaction* (pp. 373–378). Sydney: ACM.

Pedauque, R. T. (2006). *Le document à la lumière du numérique*. Caen: C&F.

Prince, J. D. (1904). The code of Hammurabi. *The American Journal of Theology, 8*(3), 601–609.

Proust, C. (2012). Masters' writings and students' writings: School material in Mesopotamia. In G. Gueudet, B. Pepin, & L. Trouche (Eds.), *From text to 'lived' resources: Mathematics curriculum materials and teacher development* (pp. 161–180). Dordrecht: Springer.

Quine, W. O. (1960). *Word and object*. Cambridge: MIT Press.

Rabardel, P. (1995). *L'homme et les outils contemporains*. Paris: A. Colin.

Serres, M. (2012). *Petite poucette*. Paris: Le Pommier.

Siemens, G. (2012). *Learning analytics: envisioning a research discipline and a domain of practice*. Paper presented at the 2nd International Conference on Learning Analytics & Knowledge, 29 April–2 May, Vancouver, Canada. http://learninganalytics.net/LAK_12_keynote_Siemens.pdf. Accessed 4 Sept 2013.

Trouche, L. (2004). Managing the complexity of human/machine interactions in computerized learning environments: Guiding students' command process through instrumental orchestrations. *International Journal of Computers for Mathematical Learning, 9*, 281–307.

Technology in the Teaching and Learning of Mathematics in the Twenty-First Century: What Aspects Must Be Considered?— A Commentary

Fernando Hitt

Abstract The commentary on the chapters of Prado and Lobo da Costa and of Aldon starts with casting a sceptical glance on the impact of instruction in computational environments on mathematics classroom practice. It discusses the difficulties large-scale projects on computers in mathematics education face. As a key variable it identifies the role of representations provided by technologies and their relation to the spontaneous representations that students develop when engaged in non-routine mathematical tasks.

The two chapters this commentary is based on consider two topics quite different in nature. The first shows peculiarities in implementing a large-scale program in Brazil, on the use of the portable computer in the mathematics classroom. In the second, the author proposes a theoretical framework for thinking about mathematics education.

In relation to the first chapter, the Brazilian project has been named "Um Computador por Aluno" (UCA; one computer per student), henceforth I will refer to the "UCA project". This project began with the following objective: *UCA was developed with the purpose of promoting the use of laptop computers in education to enhance the quality of education*. UCA is an extremely ambitious project. The UCA project has been developed over two periods, one from 2007 to 2009, and the second from 2009 to 2012. It would be interesting to know what has happened with the project more recently, i.e. in the period 2012–2014. The characteristics of the two stages are:

- Phase 1 (2007–2009): Five experiments conducted in public schools in five Brazilian cities.
- Phase 2 (2009–2012): The UCA project was expanded to 300 schools (the government purchasing 150,000 laptop computers).

F. Hitt (✉)
Département de Mathématiques, Université du Québec à Montréal, Montréal, QC, Canada
e-mail: ferhitt@yahoo.com

First, I would like to introduce Artigue's (2000) work, as this will permit us to analyse the UCA project, and at the same time, give us a general vision of the use of technology in the mathematics classroom. Artigue states that over the 20 years of instruction in computational environments since 1980, there had not been any real impact in the mathematics classroom, and she points to four reasons:

1. The poor educational legitimacy of computer technologies as opposed to their social and scientific legitimacy.
2. The underestimation of issues linked to the computerisation of mathematical knowledge.
3. The dominant opposition between the technical and conceptual dimensions of mathematical activity.
4. The underestimation of the complexity of instrumentation processes (Artigue 2000, pp. 8–9).

The UCA project included all subjects that teachers teach in the school, however, the authors restrict their document only to mathematics and we are doing the same. On the first page, the authors, Brito Prado and Lobo de Costa, touch on Artigue's first point about how technology in everyday social use can change the way we communicate. This is really important. Indeed, technological development has promoted the use of iconic representations. Nowadays, visual images are more frequent in mathematical textbooks, including e-books that incorporate dynamic representations. Technology has permitted a shift from static to dynamic mathematics (Moreno-Armella et al. 2008). However, who uses this kind of mathematics? As Artigue asks: How come that technology plays such a minor role in schools, while people use and highly appreciate it in their every day lives? As the authors point out, the literature shows that incorporation of technology in school goes at a slower pace than outside school. We can ask: Where does this problem come from?

Let us look at the critique made by Brito Prado and Lobo de Costa. They say that in Brazil the teachers only use the software they studied in their training courses, and only use it to show things. They state that:

> Unfortunately, in Brazilian school contexts, developing 'research work' has frequently been understood as a procedure that can be summed up as the creation of student groups who gather information about an issue using technology, and then produce a text report to be handed to the teacher.

We can say that, as in every large project, difficulties emerge in the process of its implementation and the question is, how the academic staff in charge of the project manages to solve them. The authors give one example about the quadratic function that the student can find on Internet, and mention that the Brazilian government has developed a site "Portal do Professor".

In relation to this mathematical content, the quadratic function, we can find at that address an example showing the students how to find a mathematical model to 'lose calories when doing sport'. If all students have a laptop in the classroom, teachers must have access to activities that can be implemented every day. From my point of view, what is proposed to teachers is not sufficient. It seems that teachers do not have enough examples to use in the classroom. The authorities give some examples dealing with a large project that the students can be asked to develop. They are also asking the teacher to think about activities that s/he can use in the mathematics classroom every day! Prado and Lobo are aware of this and consider it a real problem (Lobo da Costa et al. 2013; Prado et al. 2013).

The teacher must follow a syllabus rooted in the institutional framework of Brazil's education system. To do that, s/he must implement activities every day in the classroom. That is the problem! In the past, a similar problem arose in the 1980s when the LOGO project dazzled the academic community and LOGO was implemented in primary schools, and BASIC in secondary schools. Academic authorities thought that training teachers to manage technically to program with LOGO and BASIC "would do the trick" Authorities thought that once the teachers had managed to learn those languages they would elaborate wonderful activities in the classroom. This was not the case, and in some countries education authorities were confronted with this problem, confirming Artigue's second point.

It is true that nowadays we have access to a lot of information on the internet. The authors claim that: *The student with a laptop computer has free access to computing resources (applications, software, games, wireless internet) at any time, which sets new demands for schools and especially for teachers.* Yes, I agree with that, but what we find on the net is a "sort of mathematics already done". What I mean is that on the internet we find a lot of things, and now we must have the ability to discard what is not relevant. And this work is not easy. Here, I can introduce one of the elements highlighted in the second chapter by Aldon, about the problems related to documentational genesis (Gueudet and Trouche 2009).

Let us look at an example using the internet: If I write "Pythagorean theorem", 1,760,000 results appear on the screen in 0.32 s. If I choose one, let's say http://www.purplemath.com/modules/pythagthm.htm, I get the following:

> Back when you first studied square roots and how to solve equations, you were probably introduced to something called "the Pythagorean Theorem". This Theorem relates the lengths of the three sides of any right triangle ... The legs of a right triangle (the two sides of the triangle that meet at the right angle) are customarily labelled as having lengths "a" and "b", and the hypotenuse (the long side of the triangle, opposite the right angle) is labelled as having length "c". The lengths are related by the following equation: $a^2+b^2=c^2$.

With this approach, students are losing the important part of how this theorem was discovered. The visual part of this theorem has been postponed as if the learning of the formula is more important than the visual approach related to areas of the squares constructed on the sides of the triangle along the right angle and on its diagonal. Analysing the history of mathematics related to this issue, we can find Plato's dialogue about Socrates: The Menon. Here, a problem is discussed: Given a square, how to construct another square with double its area?, and the geometric solution of this problem, using an isosceles right triangle, seems to offer ideas to formulate a general theorem. This example illustrates Artigue's third point, namely, that the operational approach has been selected instead of a conceptual approach and, also, it shows the complexity of documentational genesis.

The same problem arises if you are interested to learn about the "Moons" of Hippocrates de Chios (http://fr.wikipedia.org/wiki/Théorème_des_deux_lunules); the visual images they give are the general case of the "lunules", with a general right-angled triangle, that is not related to the isosceles right-angled triangle that was used by Hippocrates. With this approach, the opportunity to discover the mathematical result is hidden. With the net, on the one hand, you can find fragmented issues from the history of mathematics and, on the other hand, educational authorities are asking teachers and students to navigate through the net to do a documentational process to reconstruct mathematical results from "mathematics already done". Indeed, I agree with Aldon's remark that *having available information is not yet synonymous with understanding this knowledge and using knowledge in a particular context.*

The authors of the first chapter claim that teachers are using technology in a pedestrian approach to the problem. I think that researchers and authors of textbooks and e-books do not have enough applications to provide teachers and students with a wide range of examples to permit them to approach the teaching and learning of mathematics in a better way. Indeed, another variable highlighted in the first chapter is that not only do we need to provide teachers with good mathematical examples about the use of technology, but we need to take care of their mathematical knowledge as well.

Aldon's introduction in the second chapter gives us a global overview of three main revolutions in human history. The first is related to the appearance of writing 3000 years B.C., the second is about the dissemination of writing using printing machinery (1500 A.D.), and the third is related to digital documents and communication in this century.

As I said at the beginning of this commentary, the main part of the second chapter deals with four interesting topics that we must take into account in mathematics education: (a) documentational genesis, (b) didactic incidents, (c) multi-representation and experiences in mathematics, and (d) the role of technology in an experimental approach to mathematics. Aldon is proposing a framework dealing with these issues to analyse and understand better what happens in the mathematics classroom.

For several generations, Vygotskian reflection about transforming an artefact into a tool in an active process has been at the heart of some research. Some important post-Vygotskian thinking came to surface with the work of Rabardel (1995) concerning "instrumental genesis" that deals with processes of instrumentation and instrumentalisation related to technology. An important issue that has been worked on nowadays expands this process with a new theoretical approach named "documentational genesis" (Gueudet and Trouche 2009). Instrumental genesis centred mostly on the student while documental genesis also includes the teacher. From these authors' point of view, a document is an operational product that emerges with the action taken when dealing with available resources. As I pointed out in my comments related to the first chapter of this section, this process is a complex one, and it requires a lot of work for the teacher, or for the students, to have a document that can serve their purposes.

Aldon introduces us to the notion of *didactic incident*. In general, we find that the word "incident" is related to an occurrence or event that interrupts a normal procedure or precipitates a crisis. Aldon (2011) quotes Roditi (2001, p. 350) providing a definition that fits our purposes in the classroom:

> Nous proposons donc de compter comme un incident toute manifestation publique (au sens où elle s'intègre dans la dynamique de la classe) d'un élève ou d'un groupe d'élèves en relation directe avec l'enseignement en jeu, en décalage par rapport à l'objectif visé de cet enseignement. (Ibid, p. 350). (p. 24)
>
> We therefore propose to count as an incident any public exhibition (in the sense that it fits into the dynamics of the class) of a pupil or group of pupils in direct relationship with the teaching, disruption in relation to the objective of this teaching. (Ibid, p. 350). [Translation FH]

In the mathematics classroom incidents can be related to an obstacle to learning or, on the contrary, they can provide assistance to the students in the learning process. Aldon presents a classification of five types of incidents that can be found in the mathematics classroom: outside incident, syntactic incident, friction incident, contract incident, and mathematical incident. He provides us with several examples of incidents that occurred in the mathematics classroom. However, the name "mathematical incident" (according to the author, when a mathematical question is asked and not answered) seems not to be a good name; it is so general that it does not give a direct idea about the incident, unlike the others. Maybe we could use: "Not-answered incident".

I agree with Aldon about the importance of analysing what happens in a mathematics classroom through the lens of incidents, but something that bothered me is the fact that the examples are isolated. The reader needs to know more about the mathematical content; that is, what happened with the performances of the pupils when solving a mathematical task. This is also the case in some articles about pupils' gestures when solving mathematical tasks; the authors show photos of some gestures, but what is missing is an articulation of those gestures with the pupils' performance when solving the mathematical task. For the reader, it is important to know how the incident affected the acquisition of knowledge.

Concerning the syntactic incident, Aldon makes an association with the syntax of computers and calculators that pose an obstacle in the acquisition of knowledge. Also, as stated by Artigue (2000, 2002), this is due to the complex task related to the process of instrumentation and instrumentalisation.

The third component of the author's framework is about multi-representation and experiences in mathematics. In the middle of the 1980s, theoretical approaches to representations began to arise among researchers (e.g., Janvier 1987) and in the early 1990s, Duval (1993, 1995) presented the notion of a register of representations, consolidating theoretical ideas about representations. In the past, the notion of mental representation was a priority, and semiotic representations on paper and on screens were not really taken seriously in theories of learning. Indeed, Duval's theoretical approach shows the other side of the coin. Duval created a definition of a register of representations restricting a sign system to three cognitive functions: recognition and production of a representation in a sign system, transformation of a representation inside a register, and conversion between representations. The main cognitive activity is the third one; conversion among representations. This is fundamental in the construction of mathematical concepts, based on the fact that a representation of a mathematical object is always partial to what it represents. In the learning of mathematics it is essential that the construction of an articulation in the pupil's mind (also in teacher's mind, see, e.g., Hitt 1998) is based in processes of conversions among representations.

Digital resources have constantly influenced the mathematics curriculum; for example, when technology allowed the possibility of presenting graphics on computer screens and calculators the Triple Representation Model (TRM) emerged in the curriculum (see Schwarz and Dreyfus 1995) where tabular, graphical, and algebraic settings were integrated. We could think that it was the perfect marriage, a theory of representations that could give theoretical support to the TRM curriculum. But, in reality, the teaching of mathematics using technology did not work as expected (Artigue 2002). Indeed, there are two problems that new research must clarify: one that is well known today from a theoretical point of view about the process of instrumentation and instrumentalisation when dealing with an artefact (Hoyles et al. 2004; Rabardel 1995; Trouche 2004), and the other, the cognitive aspects required in a conversion process; that is, the awareness of the significant units for each representation that one must take into account in order to achieve the conversion process (Duval 1995).

Yes! Technology does allow the conversion between representations to be made in an efficient way, but, following Duval, it is the student who needs to make the conversions between representations that permit him/her to construct an internal articulation among the representations. In the past, Tall et al. (1991) constructed the software "Graphing calculus," including a section related to "Guess my function". I do not understand why producers do not take into consideration the important didactical variable that was useful in the process of the construction of an articulation among representations. Indeed, the views of authors of both chapters agree that technology provides the user with a dynamic approach that is needed to understand some mathematical concepts.

Aldon states that:

> A mathematical experience allows [us] to define and explore the properties of a particular object in relation to a theory. Thus mathematical concepts, even if they are created in [the] mind, are fully realized in the relationships with [the] empirical phenomena.

In fact, following this quotation, there is another variable that needs to be taken into account: the diversity of spontaneous representations that pupils and students produce when solving a non-routine mathematical task (in a modelling process). Duval (1993, 1995) showed us the importance of institutional representations in the construction of mathematical objects, and technology goes well with these kinds of representation (if considering, e.g., Tall et al.'s (1991) ideas), but spontaneous representations are also important and we need a new theoretical approach to take them into account in the construction of mathematical concepts (see, e.g., diSessa et al. 1991; Hitt and González-Martín 2015; Hitt and Kieran 2009; Hitt et al. 2013, 2015).

What I claim is that the productions of non-institutional representations in future could be incorporated into documents, regularly, in a process of documentational genesis (in the sense of Gueudet and Trouche 2009) using touch screens.

Aldon stresses the role of technology in an experimental part of mathematics. In this way, technology has made great progress unifying the experimental part of mathematics related to physics. Nowadays there are some free computer programs like "Tracker" that allow students to analyse videos and to capture data, and then to copy and paste this data onto other software as "GeoGebra"; in this way, we can process data and construct a mathematical model. For example, a student could take a video of a team-mate running; then put it into Tracker 2015 to capture the data, using time as the independent variable and distance as the dependent variable. The pupil could copy and paste this into GeoGebra 2015 and plot a discrete graphic representation, then ask the program for a regression equation related to a curve s/he thinks is related to the form of the discrete graphic, and then construct a continuous model of the phenomenon. This way we are dealing with a rich approach to the experimental part not only of mathematics, but also sciences, and approaching a solution to the fourth problem signalled by Artigue (2000). Today, teacher training and students' problems of learning in a technological environment must be seen, as Aldon suggests, taking into account the complex process of the documentational genesis introduced by Gueudet and Trouche (2009), but centred in *ad hoc* designing activities.

Finally, Aldon stresses the idea that in thinking of mathematics education in the twenty-first century two main branches will be worth considering: the process between resource and document, and representations and objects. I think we must consider the important role of mathematical modelling (Blum et al. 2007; Hitt and González-Martín 2015) that, from my point of view, includes the problematic about representations and objects, and, as Hoyles et al. (2004) stress, that integrating technology into classroom practice needs more empirical research and assimilating this into our theoretical approaches.

References

Aldon, G. (2011). *Interactions didactiques dans la classe de mathématiques en environnement numérique: construction et mise à l'épreuve d'un cadre d'analyse exploitant la notion d'incident*. Unpublished PhD thesis, Université de Lyon, Lyon.

Artigue, M. (2000). *Instrumentation issues and the integration of computer technologies into secondary mathematics teaching*. Proceedings of the annual meeting of GDM, Potsdam. http://webdoc.sub.gwdg.de/ebook/e/gdm/2000. Accessed 3 June 2014.

Artigue, M. (2002). Learning mathematics in a CAS environment: The genesis of a reflection about instrumentation and the dialectics between technical and conceptual work. *International Journal of Computers for Mathematical Learning, 7*, 245–274.

Blum, W., Galbraith, P., Henn, H., & Niss, M. (Eds.). (2007). *Modelling and applications in mathematics education: The 14th ICMI study*. New York: Springer.

diSessa, A., Hammer, D., Sherin, B., & Kolpakowski, T. (1991). Inventing graphing: Meta-representational expertise in children. *Journal of Mathematical Behavior, 10*, 117–160.

Duval, R. (1993). Registres de représentation sémiotique et fonctionnement cognitif de la pensée. *Annales de Didactique et de Sciences Cognitives, 5,* 37–65.

Duval, R. (1995). *Sémiosis et pensée humaine: Registres sémiotiques et apprentissage intellectuels.* Bern: Peter Lang.

GeoGebra. (2015). *Dynamic mathematics & science for learning and teaching (version 5).* http://www.geogebra.org/cms/. Accessed 22 Feb 2015.

Gueudet, G., & Trouche, L. (2009). Towards new documentation systems for mathematics teachers? *Educational Studies in Mathematics, 71*(3), 199–218.

Hitt, F. (1998). Difficulties in the articulation of different representations linked to the concept of function. *Journal of Mathematical Behavior, 17*(1), 123–134.

Hitt, F., & González-Martín, A. (2015). Covariation between variables in a modelling process: The ACODESA (collaborative learning, scientific debate and self-reflexion) method. *Educational Studies in Mathematics, 88*(2), 201–219.

Hitt, F., & Kieran, C. (2009). Constructing knowledge via a peer interaction in a CAS environment with tasks designed from a task-technique-theory perspective. *International Journal of Computers for Mathematical Learning, 14,* 121–152.

Hitt, F., Saboya, M. & Cortés, C. (2013). *Structure cognitive de contrôle et compétences mathématiques de l'arithmétique à l'algèbre au secondaire: Les nombres polygonaux.* Actes du congrès CIEAEM65, Turin, Italie, pp. 134–146. http://math.unipa.it/~grim/quaderno23_suppl_1.htm

Hitt, F., Saboya M., et Cortés C. (2015). *La pensée arithémtico-algébrique dans la transition primaire-secondaire et le rôle des représentations spontanées et institutionnelles.* Actes du congrès CIEAEM66, Lyon, France, pp. 252–257. http://math.unipa.it/~grim/CIEAEM%2066_Pproceedings_QRDM_Issue%2024,%20Suppl.1.pdf

Hoyles, C., Noss, R., & Kent, P. (2004). On the integration of digital technologies into mathematics classrooms. *International Journal of Computers for Mathematical Learning, 9,* 309–326.

Janvier, C. (Ed.). (1987). *Problems of representation in the teaching and learning of mathematics.* London: Lawrence Erlbaum.

Lobo da Costa, N. M., Prado, M. E., & Pietropaolo, R. C. (2013). *Mathematics teachers continuing education and technology: A necessary practice in a globalized context.* Proceedings of CIEAEM 65, Torino, Italy.

Moreno-Armella, L., Hegedus, S., & Kaput, J. (2008). From static to dynamic mathematics: Historical and representational perspectives. *Educational Studies in Mathematics, 68*(2), 99–111.

Prado, M. E., Lobo da Costa, N. M., & Galvão, M. E. (2013). *Teacher development for the integrated use of technologies in mathematics teaching.* Proceedings of CIEAEM 65, Torino, Italy.

Rabardel, P. (1995). *Les hommes et les technologies, approche cognitive des instruments contemporains.* Paris: Armand Colin.

Roditi, E. (2001). *L'enseignement de la multiplication des déximaux en sixième, étude de pratiques ordinaires.* Unpublished PhD thesis, Université Paris 7, Paris.

Schwarz, B., & Dreyfus, T. (1995). New actions upon old objects: A new ontological perspective of functions. *Educational Studies in Mathematics, 29*(3), 259–291.

Tall, D., van Blokland, P., & Kok, D. (1991). *A graphic approach to the calculus.* Warwicks: Rivendell Software.

Tracker. (2015). *Video analysis and modeling tool (version 4.87).* http://www.cabrillo.edu/~dbrown/tracker/. Accessed 22 Feb 2015.

Trouche, L. (2004). Managing the complexity of human/machine interactions in computerized learning environments: Guiding students' command process through instrumental orchestrations. *International Journal of Computers for Mathematical Learning, 9*(3), 281–307.

Part VIII
Transcending Boundaries

Family Math: Doing Mathematics to Increase the Democratic Participation in the Learning Process 393
Javier Díez-Palomar
University of Barcelona, Spain

Service-Learning as Teacher Education 409
Peter Appelbaum
Arcadia University, Philadelphia, USA

The Learning and Teaching of Mathematics as an Emergent Property Through Interacting Systems and Interchanging Roles: A Commentary 425
Fragkiskos Kalavasis
University of the Aegean, Greece
Corneille Kazadi
Université du Québec à Trois-Rivières, Canada

Family Math: Doing Mathematics to Increase the Democratic Participation in the Learning Process

Javier Díez-Palomar

Abstract This chapter discusses the participation of families in the process of learning mathematics. It introduces evidence of different types of participation and relates these to positive and negative effects on students' learning of mathematics. The special case of the research project FAMA is used to explore how the involvement of families in mathematical practices can foster democratic participation in the learning process. The chapter ends with recommendations for further research.

Efforts to improve students' mathematics learning have focused on improved teacher education, modified curriculum, and school-wide programs (Ball 1993; Cuevas and Driscoll 1993; Knapp 1997; Smith and Hausafus 1998). In this chapter I will focus on another component, which also has been analysed by many authors: the role of families (and members of the community) to improve students' mathematics performance. In fact, parental involvement is recognized as a crucial outside-school aspect in children's mathematics achievement (Dias et al. 2011; Díez-Palomar and Kanes 2012).

According to Hoover-Dempsey and Sandler (1995) and Hoover-Dempsey and Bassler (1997), parents become involved in their children's educational process mainly due to three different reasons: psychological motivations, perception of others' invitations to get involved, and perception that life context variables allow and enable them to become involved.

Family engagement has a strong impact on students' mathematics achievement. Researchers have provided much evidence on the positive effects that family engagement has on students' achievement in mathematics. Benefits of that impact have been well known to the scientific community for many decades (Catsambis 2001; Epstein 1991, 2005; Henderson and Berla 1994; Ho and Willms 1996; Keith et al. 1993; Simon 2004). In 1968, Mildred Smith published the results of the project *School and Home*, involving over 1,000 students and their families. Children were asked to take books home. On the books there were some notes like "Please, read me," and other similar messages addressed to parents to engage them in an

J. Díez-Palomar (✉)
Department of Mathematics and Science Education,
University of Barcelona, Barcelona, Spain
e-mail: jdiezpalomar@ub.edu

active support of their children's learning. Results obtained by those children in reading and writing tests were very positive, and higher than the ones achieved before doing this type of practices. Analogous results were provided by another study, the *HighScope Perry Project*. In this case the research team followed 123 children, between 1962 and 1967. The children came from poor families. For 5 years, teachers offered them high-quality curriculum activities, with the families' participation. Years later, when those children were in their forties, the researchers investigated what happened with their lives. They were able to contact 97 % of the original sample. They found that adults at age 40 who had passed the preschool program had higher earnings, were more likely to hold a job, had committed fewer crimes, and were more likely to have graduated from high school than adults who did not attend preschool.

Drawing on this and other research, we know that developing partnerships between teachers and parents leads students to obtain higher achievements in mathematics, better attendance, more course credits, and more responsible preparation for class (Epstein 2005). The crucial element here is the participation of families or community members.[1] Data available allow us even to question old explicative models of school success (Baudelot and Establet 1971; Bowles and Gintis 1976). Those models used to present the socio-economical status and the social class as the main independent variables to explain why some children get better scores than others.

More recent research refutes this kind of explanation. Epstein (2005) points to Catsambis and Beveridge's (2001) contributions. Using hierarchical linear modelling analysis, these authors concluded, "students in neighborhoods with high concentrations of poverty had lower math achievement test scores, but this effect was ameliorated by on-going parental involvement in high school" (Epstein 2005, p. 2). Similar conclusions are reported by Flecha (2012) or Díez-Palomar et al. (2011) who used qualitative approaches.

In this chapter I examine the participation of families in the process of learning mathematics. First I have a retrospective look on the literature of research about family involvement. I introduce evidence of different types of participation, because research-based evidence suggests that not all kinds of family and community involvement produce the same positive effects on students' mathematical learning. The type of participation with better results is the one grounded on a democratic participative approach. I discuss evidence emerging from *FAMA – Family Math for Adult Learners* to illustrate such an approach. I conclude this chapter by opening up lines for further research to extend our knowledge about the impact that families may have on the improvement of the students' mathematics learning.

---

[1] I am using here the term 'families', but research points out that members of the community being volunteers in the school may also have a positive effect on students' mathematical performance.

Recent Research About Family Engagement Learning Processes

Research about the role played by families in the process of learning led to the creation of many research networks, such as the *European Research Network About Parents in Education* (ERNAPE), founded in 1993 to exchange knowledge of research developments in Europe and to stimulate research about parents in education at all levels. In the USA, there are many centres focusing on the study of the binomial family-education. This is the case with the *Center on School, Family, and Community Partnerships* or the *National Network of Partnership Schools* at the Johns Hopkins University, directed by Joyce L. Epstein; the *Harvard Family Research Project* directed by Heather Weiss; or the *Center for the Mathematics Education of Latinos/as* directed by Marta Civil; among others.

As we mentioned at the beginning of this chapter, maybe one of the most significant contributions of families supporting their children's learning processes is the fact that this involvement is effective in overcoming inequalities among children coming from vulnerable groups in risk of exclusion. According to classic studies in education (e.g., Bernstein 1993; Bourdieu 1984; Bourdieu and Passeron 1970), academic achievement is somehow connected to the socio-economic status. This is because our society is stratified in social classes; hence each class has its own cultural and social capital. In some way these kinds of capitals explain that individuals tend to reproduce their class position from one generation to the next. However, nowadays we know that this is not fully true. Success or failure in school depend on many different aspects, and we know that a teacher using a high-quality curriculum can make a huge difference. International studies such as PISA have attributed a significant impact to the variable "level of studies achieved by the mother" of the student interviewed. There are data showing that students whose mothers have achieved a university level of education tend to perform better than students whose mothers barely had the opportunity to go to school. This statement indicates that mothers are the ones in charge of supporting their children at home when doing homework and other academic activities and, apparently, the higher the educative level of a mother, the better they help their children to solve academic questions.

Studies conducted during the last two decades contributed to deepening this analysis. According to them, families have a strong impact in terms of academic success, but also in terms of motivation and behaviour (Dias et al. 2011). Children's home environments affect their attitudes toward mathematics (Balli 1998; Parsons et al. 1982). Parents' beliefs and expectations for their children in mathematics predict student achievement in elementary and middle school mathematics (Entwisle and Alexander 1996; Gill and Reynolds 1999; Halle et al. 1997; Holloway 1986). Learning activities conducted at home with the families also predict students' achievement (Cai et al. 1997; Ho and Willms 1996; Keith et al. 1993).

In fact, students demonstrate a more positive attitude towards school and learning as well as higher achievement and improved school attendance when teachers and parents work together (Christenson and Sheridan 2002; Côté et al. 2011; Epstein

2001; Henderson and Mapp 2002). Even in difficult situations with high risk for drop-out and school failure, such as the transitions between elementary and middle school, the tremendous impact of motivation over academic success has been demonstrated. Studies on student transitions have shown that declines in students' achievement motivation beliefs (e.g., self-competence, self-image, value of the school) accompany declines in achievement (Eccles et al. 1993; Jacobs et al. 2002; Wigfield et al. 1991). Moreover, those changes have been associated with school characteristics and practices (Roeser et al. 2000). The positive effects of family engagement on students' education may suggest that fostering these types of interactions has the potential to lessen the negative impact that dramatic changes may provoke on students' performance (Sheldon and Epstein 2005).

These studies suggest that there is a clear relationship between students' academic achievement and students' behavior. However, the most relevant fact is that such connexion is not related to the socio-economic status. One possible explanation lies in the assumption that usually parents deposit their hope and their expectations on their children's activities. We rarely find parents who do not want their children to succeed in school and to have a successful future. This happens in all social groups, including privileged ones, working class, ethnic minorities; (it always happens). Moreover, we also have data suggesting that practices based on Successful Educational Actions (SEAs) make possible that overcoming inequalities does not depend on the cultural or social capital, nor on the habitus, but on the implementation of such SEAs (Flecha and Soler 2013; INCLUD-ED 2012). This has been an educational revolution because it demonstrates that we do not need to wait for the next generation to raise the academic results of the children of families living in poor conditions (as we can read behind PISA's theoretical approach when inquiring the level of studies of the mother) (García 2011). The INCLUD-ED project has contributed a list of SEAs that have proved their effectiveness in many European countries (Aubert 2011; Flecha et al. 2009).

In fact, families are envisioned as resources. Civil has described families as keepers of what Luis Moll called *funds of knowledge* (González et al. 2005; Moll et al. 1992). González et al. (2001) portray the mathematical practices done by Latino families in Arizona. Families have large repositories of non-formal and informal resources, mobilizing those resources to help their children at home to do the mathematics tasks requested by teachers (Civil 1999; Díez-Palomar et al. 2011; Krüger and Michalek 2011). However, not all types of family engagement produce the same positive effects on students' learning and development. When a father or a mother approaches the school to argue with the teacher about the methods used, or to question the curriculum, or to generate conflict, then the result is not improving children's academic achievement (Flecha 2012; INCLUD-ED 2012). Research suggests that the more democratic the participation of the families, the better the children's academic results (Gatt et al. 2011).

There have been many efforts to categorize the participation of families. According to Hoover-Dempsey and Sandler (1995), and Hoover-Dempsey (2005), after families decide to become involved in their children's learning they choose among different forms of parental involvement (level 2 in their model).

The mechanisms highlighted by those authors through which parental involvement influences students' outcomes include:

1. Encouragement
2. Modelling
3. Reinforcement
4. Instruction: (a) closed-ended and (b) open-ended.

Joyce L. Epstein talks about six different types of participation, including the following five (Sheldon and Epstein 2005, p. 197):

Type 1 Parenting: Helping all families establish supportive home environments for children
Type 2 Communicating: Establishing two-way exchanges about school programs and children's progress
Type 3 Volunteering: Recruiting and organizing parent help at school, home, or other locations
Type 4 Learning at home: Providing information and ideas to families about how to help students with homework and other curriculum-related materials
Type 5 Decision-making: Having parents from all backgrounds serve as representatives and leaders on school committees

In Europe, the Centre of Research in Theories and Practices that Overcome Inequalities (CREA) analysed the participation of thousands of families across the continent as part of the INCLUD-ED project. Drawing on such empirical evidence, CREA elaborates another typology of family engagement. The categories of this typology include(cf. INCLUD-ED Consortium 2009, p. 54):

(1) Informative. Not participating in decision-making. Families are informed about school activities, school functioning and decisions. They attend meetings.
(2) Consultative. Being consulted in decision-making processes. Participation in official bodies.
(3) Decisive. Participation in decision-making processes through their representation in official bodies. Monitoring of the school progress (accountability).
(4) Evaluative. Participation in both student and school evaluation.
(5) Educative. Participation in students' learning during school hours and after school. Participation in family education.

There is a gradation in the possibilities of participation for the families, from (1) "informative," where families barely participate in the school activities, to (5) "educative," where they may participate in the learning process. The most effective type of participation in terms of children's academic achievement is (5), which is the most democratic type of participation since it involves the voices of all members of the educative community within the learning process.[2]

---

[2] Other researchers have also pointed out the importance of 'democratic family participation' to improve children's scores (Sheldon and Epstein 2005).

One of the greatest difficulties to implement democratic forms of family engagement is to eliminate the gap between the work done by the research community and the practice in schools. We already know that the more democratic the participation the better is the children's mathematics achievement. However, what are the specific actions that produce such success? To move towards a democratization of the practices done in a particular school, a condition *sine qua non* is knowledge about SEAs. Neither teachers nor families are going to implement forms of democratic participation closer to type (5) ("educative") rather than type (1) ("informative") without clear information about what works to engage all educational actors in the learning process, in terms of SEAs.

Another action for success is the introduction of 'critical friends'. With this label Allexsaht-Snider and Buxton (2011) designate the type of people who are invited to collaborate in a series of sessions conducted with teachers and researchers. The aim of such sessions is to improve teachers' practices drawing on critical discussions about examples or real practices. The persons invited are scholars with expert knowledge to share with the audience their expertise in an egalitarian dialogue aiming to improve teachers' practices based on rigorous discussion. These critical friends may also be persons in positions of responsibility in the education system. This approach has been replicated in a similar vein in teacher training programs. Data suggest that this kind of 'critical discussions' with experts may induce improvement among both, in-service teachers as well as pre-service teachers (Vanegas et al. 2013).

There are more cases of successful practices. Knopf and Swick (2008) and Goldman (2006) propose *home visits* to learn from families and support children's learning processes. The effectiveness of these home visits depends on the quality of practice (Gomby et al. 1999; Knopf and Swick 2008). Teachers and families have to develop a positive affective relationship to be successful at the home visiting (Sweet and Appelbaum 2004). Most of the studies related to home-visit report cases in which the teacher comes to the student's home just to meet the parents and to know a bit about the environment (Kahraman and Derman 2011). Other studies report cases in which teachers really come to the homes to see the kind of mathematics that children do at home, including both formal mathematics (such as school homework) and non-formal and informal practices (such as going to do shopping, playing games, etc.) (Goldman 2006; Goldman et al. 2010).

Workshops with parents have been proven successful many different times, improving both family engagement and children performance (Balzano 2011; Díez-Palomar and Molina 2009). This type of practice mainly explains to families how teachers teach mathematics nowadays. The pedagogic differences among strategies followed by the teachers to promote critical thinking among students, as well as to provide parents with a plethora of resources to better support their children at home, are discussed during the sessions (Dias et al. 2011). The lack of knowledge that some families show about strategies currently being used by teachers in schools has been identified as a difficulty (for families to become involved). However, according to recent research, one of the main obstacles for family engagement is teachers' institutional resistance to let families access schools (and classrooms) to decide on how to teach, or some teachers' misunderstanding of such 'pedagogical innovation'. It is not strange to hear teachers complaining about families discussing with them about the curriculum in mathematics and how they (the teachers) implement such curriculum. The category number (5) identified within INCLUD-ED ("educative participation") is very difficult to carry out if a context of conflict and resistance appears between teachers and families. The type of participation which may be successful is the one that manages to create partnerships between these two actors (school and community), grounded on the idea that the collaboration is to improve students' mathematics achievement, not to let parents and teachers fight for the academic authority. Questioning the model of academic authority does not produce any kind of success in terms of students' achievement. On the contrary, to improve students' academic performance, a partnership has to emerge from the collaboration of all actors involved in the educational process. In Catalonia i.e., during 2013, a project called *Clau* [*Key*] has been implemented by the Catalonian Federation of Parents Associations. This project arose from parents' claim to know what the SEAs are in order to have this information when looking for a school for their children. This is a bottom-up process of democratic participation. García (2011) reports the case of *La Paz* school, in Albacete, Spain. This was a very conflictive school in one of the poorest neighbourhoods in Europe. In this case, the movement to change the school practices came from the inspection. They built a partnership with teachers, families, NGOs involved in the school activities, the Church, volunteers, etc., to include all educative actors' voices in the process of transformation. They agreed to sign a contract to implement only SEAs in the school. Two years later, the school stands out because of its positive results.

Families Doing Mathematics for a Democratic Participation in the Learning Process: The Case of *FAMA* – *Family Math for Adult Learners*

The teaching of mathematics has greatly improved in the last decades, thanks to the research that has been done in this area. Research results (Carpenter et al. 1989; Erlwanger 1973; Good and Grouws 1979; Lampert 1990) have been the basis for many innovations and new viewpoints that have reformed the way we teach mathematics in our current classrooms. The curricular innovations implemented during the last decades in mathematics (NCTM 1989, 2000) sometimes have meant a loss of sense for many families, who do not understand the new teaching methods (Civil and Bernier 2006; Meyer et al. 1996; Peressini 1998; Remillard and Jackson 2006).

One of the main disadvantages when parents want to help their children is their lack of knowledge in mathematics (Civil 2001; Jackson and Remillard 2005; Rockliffe 2011). In some cases, however, it is not lack of knowledge, but low self-confidence (Civil and Bernier 2006; Díez-Palomar and Molina 2009). Teaching and learning of mathematics looks unfamiliar to most parents (Civil and Bernier 2006;

Remillard and Jackson 2006). The 'lack of knowledge' uses to appear related to different ways and strategies to solve mathematical problems. Nowadays teachers use different tools and strategies to teach mathematics, drawing on the US-reform of mathematics (Hiebert 1999; Kilpatrick et al. 2003; Kilpatrick et al. 2001; NCTM 1989, 2000). That makes it hard for parents to help their children. In addition, school mathematics is an old body of knowledge that some parents have simply forgotten. They remember the main topics, but not the specific procedures.

Some studies argue that one part of the difficulties experienced by students when learning mathematics could be explained by the type of support they receive in their home environments (Sheldon and Epstein 2005). Research in the field of family involvement tries to solve this type of difficulties. As we have shown in the last section, family engagement may adopt many different forms. In Table 1 we introduce in a succinct way the main types of family engagement in mathematical practices found within FAMA. FAMA was a research project analysing how parents become involved in their children's mathematical learning. In the frame of that project, we conducted a very detailed literature review. As a result, we found five types of family engagement in the realm of mathematics education.

During the project several parents were interviewed. Some of these parents had sent their children to more than one school, thus they had experienced different types of participation. Most of them highlighted the importance of feeling welcomed by the teachers when attending the school looking for information regarding their children's learning. Drawing on their experiences a key conclusion emerges: the teachers' openness encourages families' involvement. Many times families feel reluctant (and even resistant) to attend school meetings, teachers' appointments, and other forms of informative participation because there is a lack of communication between teachers and families. This usually happens either when teachers claim that teaching mathematics is their exclusive responsibility (and nobody else's) or when families question teachers' professional work. FAMA found cases of teachers

Table 1 Types of support of families for learning mathematics

Type	Definition
At home	Family members try to support their children with mathematics at home (or out of school, such as shopping) by helping them to do their homework, encouraging them to do mathematical activities, reading, games, etc.
Going back to school	Family members attend mathematics courses, such as *parent maths nights*, *maths workshops*.
Private teaching	Family members send their children to private lessons, or they pay for a personal teacher at home, thus extending their children's learning time.
Appointments with the teacher	Family members ask for appointments with their children's mathematics teacher in order to allay concerns and doubts, clarify misunderstandings, etc., and thus become able to better help their children.
Social networking	Family members look for support among their community of peers, neighbours, relatives, etc.

claiming that the only thing they request from families is to make sure that their children go to school. On the opposite, FAMA also found cases of parents complaining about the methods used by teachers because they felt that their children were not learning mathematics properly. When there is a clear will to establish a dialog between teachers, parents, volunteers, and other members of the community, then resistance (as well as the conflict deriving from it) disappear.

Drawing on INCLUD-ED's typology of family types of involvement, data collected in different schools all over Europe confirm that opening participation in the educational activities provided in the school to the broadest range of people possible is a way to reduce resistance and conflict in the school. Children gain from an environment focused on high expectations in terms of learning. Using SEAs makes a difference between different schools. However, it is not enough to open the school to family involvement. It is also crucial that this participation is democratic, based on an egalitarian dialogue (in terms of Flecha 2000). When parents, teachers, and students participate with a clear will to discuss using arguments based on valid claims to share their respective knowledge, then results are more rich and positive for all of them. Authority here comes from evidence, not from the position that somebody has in terms of hierarchy or status.

Communication is a key feature in terms of promoting participation. Communication between parents and teachers brings benefits to both children and teachers (Epstein 2001). For this reason, successful experiences are those that provide the participants with ways or spaces for communication, including 'family corners' on the school website, meeting points, appointments with teachers/tutors that are clearly defined from the beginning of the semester, flexibility in terms of schedules to meet with parents and other family members, etc.

Knowing that family involvement is a dynamic process (Hoover-Dempsey and Sandler 1995), successful actions are those that build structures to manage this communication. If there is not a space for this communication to happen, then it is unlikely to happen (Díez-Palomar and Molina 2009). We need to create egalitarian spaces for promoting communication and dialog.

Students whose parents attended training and information workshops and obtained materials to help their children at home made greater gains in mathematics than students did whose parents did not attend the workshops (Shaver and Walls 1998; Starkey and Klein 2000). The case of Montse, a mother attending one of the workshops provided in a middle and high school in Barcelona, is instructive, here. Montse had two children: a daughter and a son. When she started attending the workshop, her daughter was in the second year of the middle school. Montse was very worried because she noticed the negative impact on her daughter of moving from elementary to middle school. It was a "shock" for her. As a consequence her grades in mathematics went down dramatically. Montse tried to talk with the teacher, but somehow was not productive, because, as she reported in an interview, the teacher felt questioned by Montse. Thus the relationship was not cordial at all. Her daughter had to bear the consequences. There was no room for talking and a just dialogue in that situation. Then, the possibility of participating in a workshop appeared. Montse was enthusiastic from the very beginning. Sometimes she even came to the workshop with her daughter to share questions, uncertainties, to request additional information, resources, etc. After one semester, her daughter improved her grades in mathematics significantly. Two years later her daughter participated in a conference at the university. On her way home she asked for information regarding university degrees in engineering. Her mother started to bring also her son (who was younger) to the workshop. Democratic participation really opens this type of opportunities for all.

Further Research: New Horizons for the Coming Years

Some studies have put parents in the sideline, leaving it up to educators and other professionals to decide how mathematics learning should take place. However, research clearly demonstrates the importance of families within the process of learning, although not all types of participation have the same effect. The form of participation producing best results is the one that involves more people in education. The more democratic the practices are, the better the results. Those patterns reinforce the need for educators to exert extra efforts to revise the mathematics curriculum, instructional approaches, quality of teaching, and family and community partnerships to improve students' skills and test scores (Sheldon and Epstein 2005). It is necessary to see how we can promote this type of guidance in our teacher training programs. There is some experience that includes the work with families within teacher-training programs (Díez-Palomar 2013; Vanegas et al. 2013). This is a promising line of work for the immediate future. We need to know more about the real impact of including families in teacher-training programs. Does it improve the practices of future teachers?

Finally, another challenge that we face in this field is to find more effective ways to transfer research results into practice, in order to improve students' mathematics learning. Desforges and Abouchnar (2003) argue that we have a plethora of studies concerning achievement through parental involvement. However, little evidence exists on how to apply this knowledge. Lusse (2011) proposes the use of design science research to reduce the gap between schools and families in terms of democratic participation, but we do not have many more examples with a similar goal. This opens the possibility (and necessity) for further research on implementing the results emerging from research into the practice—to make sure that there is a social impact of research.

Acknowledgement This chapter is partly based on work supported by the European Commission under FAMA – Family Math for Adult Learners, project number 504135-2009-LLP-ES-GRUNDTVIG-GMP, and the subprogram Ramon y Cajal funded by the Ministerio de Ciencia e Innovación of the Spanish Government. Some contributions also belong to my involvement in CREA (University of Barcelona), as well as my participation in the Center for the Mathematics Education of Latinos/as at the University of Arizona, as a Fulbright Visitor Scholar. This publication only reflects the views of the author, and the funding institutions cannot be held responsible for any use, which may be made of the information contained therein. I would like also to mention the Project EDU2012-32644 Development of a program by competencies in an initial training for secondary school for providing me the time to write this chapter.

References

Allexsaht-Snider, M., & Buxton, C. (2011). *Engaging middle school students and families together in bilingual science learning: Can we challenge rising anti-immigrant discourses and open pathways to postsecondary learning?* Paper presented at the European Network about Parents in Education (ERNAPE) conference, Milano, Italy.

Aubert, A. (2011). Moving beyond social exclusion through dialogue. *International Studies in Sociology of Education, 21*(1), 63–75.

Ball, D. L. (1993). With an eye on the mathematical horizon: Dilemmas of teaching elementary school mathematics. *The Elementary School Journal, 93*(4), 373–397.

Balli, S. J. (1998). When mom and dad help: Student reflections on parent involvement with homework. *Journal of Research and Development in Education, 31*(3), 142–146.

Balzano, E. (2011). Science laboratory activities for kids and parents in Naples. In M. Pieri, A. Pepe, & L. Addimando (Eds.), *Home, school and community: A partnership for a happy life?* (pp. 24–25). Milano: I Libri di Emil.

Baudelot, C., & Establet, R. (1971). *L'école capitaliste en France (cahier 213)*. Paris: Maspero.

Bernstein, B. (1993). *Clases, códigos y control. La estructura del discurso pedagógico*. Madrid: Morata.

Bourdieu, P. (1984). *La distinction: Critique social du jugement*. Paris: Minuit.

Bourdieu, P., & Passeron, J. C. (1970). *La reproduction: éléments pour une théorie du système d'enseignement*. Paris: Minuit.

Bowles, S., & Gintis, H. (1976). *Schooling in capitalist America: Educational reform and the contradictions of economic life*. New York: Basic Books.

Cai, J., Moyer, J. C., & Wang, N. (1997). *Parental roles in students' learning of mathematics: An exploratory study*. Paper presented at the annual meeting of the American Educational Research Association (AERA), Chicago, IL.

Carpenter, T. P., Fennema, E., Peterson, P. L., Chiang, C. P., & Loef, M. (1989). Using knowledge of children's mathematics thinking in classroom teaching: An experimental study. *American Educational Research Journal, 26*(4), 499–531.

Catsambis, S. (2001). Expanding knowledge of parental involvement in children's secondary education: Connections with high school seniors' academic success. *Social Psychology of Education, 5*(2), 149–177.

Catsambis, S., & Beveridge, A. A. (2001). Does neighbourhood matter? Family, neighborhood, and school influences on eighth grade mathematics achievement. *Sociological Focus, 34*(4), 435–457.

Christenson, S. L., & Sheridan, S. M. (2002). *Schools and families: Creating essential connection for learning*. New York: Guilford.

Civil, M. (1999). Parents as resources for mathematical instruction. In M. van Groenestijn & D. Coben (Eds.), *Mathematics as part of lifelong learning (Proceedings of the fifth international conference of adults learning mathematics – A research forum)* (pp. 216–222). London: Goldsmiths College.

Civil, M. (2001). Mathematics for parents: Issues of pedagogy and content. In L. Johansen & T. Wedege (Eds.), *Proceedings of the eighth international conference of adults learning mathematics – A research forum* (pp. 60–67). Roskilde: Centre for Research in Learning Mathematics.

Civil, M., & Bernier, E. (2006). Exploring images of parental participation in mathematics education: Challenges and possibilities. *Mathematical Thinking and Learning, 8*(3), 309–330.

Côté, P., Dumoulin, C., & Tremblay, N. (2011). Family-school collaboration: Activities implemented by elementary school teachers involved with high-risk students from underprivileged class. In M. Pieri, A. Pepe, & L. Addimando (Eds.), *Home, school and community: A partnership for a happy life?* (pp. 38–40). Milano: I Libri di Emil.

Cuevas, G., & Driscoll, M. (1993). *Reaching all students with mathematics*. Reston: NCTM.

Desforges, C., & Abouchnar, A. (2003). *The impact of parental involvement, parental support and family education on pupil achievements and adjustment: A literature review*. London: Department for Educational Skills.

Dias, M. C., Tomás, C., & Gama, A. (2011). School achievement in risk situations: The impact of government and non-government programmes in the fight against social exclusion. In M. Pieri, A. Pepe, & L. Addimando (Eds.), *Home, school and community: A partnership for a happy life?* (pp. 35–36). Milano: I Libri di Emil.

Díez-Palomar, J. (2013). *Be a teacher in a global society: Challenges and opportunities drawing on a sociological approach to mathematics teacher training*. Paper presented at CIEAEM 65, Torino, Italy.

Díez-Palomar, J., & Kanes, C. (Eds.). (2012). *Family and community in and out of the classroom: Ways to improve mathematics' achievement*. Bellaterra: UAB Press.

Díez-Palomar, J., & Molina, S. (2009). Contribuciones de la educación matemática de las familias a la formación del profesorado. In M. J. González, M. T. González Astudillo, & J. Murillo (Eds.), *Investigación en Educación Matemática XIII* (pp. 211–225). Santander: SEIEM.

Díez-Palomar, J., Gatt, S., & Racionero, S. (2011). Placing immigrant and minority family and community members at the school's centre: The role of community participation. *European Journal of Education, 46*(2), 184–196.

Eccles, J. S., Midgley, C., Wigfield, A., Buchanan, C. M., Reuman, D., Flanagan, C., et al. (1993). Development during adolescence: The impact of stage environment fit on young adolescents' experiences in schools and families. *American Psychologist, 48*(2), 90–101.

Entwisle, D. R., & Alexander, K. L. (1996). Family type and children's growth in reading and math over the primary grades. *Journal of Marriage and Family, 58*, 341–355.

Epstein, J. L. (1991). Effects on student achievement of teacher practices of parent involvement. In S. Silvern (Ed.), *Literacy through family, community, and school interaction* (pp. 261–276). Greenwich: JAI.

Epstein, J. L. (2001). *School, family, and community partnerships: Preparing educators and improving schools*. Boulder: Westview.

Epstein, J. L. (2005). *Developing and sustaining research-based programs of school, family, and community partnerships: Summary of five years of NNPS research*. http://www.csos.jhu.edu/p2000/pdf/Research%20Summary.pdf

Erlwanger, S. H. (1973). Benny's conception of rules and answers in IPI mathematics. *Journal of Children's Mathematical Behavior, 1*(2), 7–26.

Flecha, R. (2000). *Sharing words: Theory and practice of dialogic learning*. Lanham: Rowman & Littlefield.

Flecha, A. (2012). Family education improves student's academic performance: Contributions from European research. *Multidisciplinary Journal of Educational Research, 3*(2), 301–321.

Flecha, R., & Soler, M. (2013). Turning difficulties into possibilities: Engaging Roma families and students in school through dialogic learning. *Cambridge Journal of Education, 43*(4), 451–465.

Flecha, A., García, R., Gómez, A., & Latorre, A. (2009). Participation in successful schools: A communicative research study from the INCLUD-ED project. *Cultura y Educación, 21*, 183–196.

García, R. (2011). *Contrato de inclusión dialógica en el colegio La Paz: Una propuesta educativa de éxito para superar la exclusión social*. Unpublished PhD thesis, University of Barcelona, Spain.

Gatt, S., Ojala, M., & Soler, M. (2011). Promoting social inclusion counting with everyone: Learning Communities and INCLUD-ED. *International Studies in Sociology of Education, 21*(1), 33–47.

Gill, S., & Reynolds, A. J. (1999). Educational expectations and school achievement of urban African American children. *Journal of School Psychology, 37*(4), 403–424.

Goldman, S. (2006). A new angle on families: Connecting the mathematics in daily life with school mathematics. In Z. Bekerman, N. Bubules, & D. Silberman-Keller (Eds.), *Learning in places: The informal education reader* (pp. 55–76). Bern: Peter Lang.

Goldman, S., Pea, R., Blair, K. P., Jimenez, O., Booker, A., Martin, L., & Esmonde, I. (2010). Math engaged problem solving in families. In K. Gómez, L. Lyons, & J. Radinsky (Eds.), *Learning in the disciplines: Proceedings of the 9th International Conference on the Learning Sciences (ICLS 2010)* (pp. 380–388). Chicago: ICLS.

Gomby, D. S., Culross, P. L., & Behrman, R. E. (1999). Home visiting: Recent program evaluations: Analysis and recommendations. *The Future of Children, 9*(1), 4–26.

González, N., Andrade, R., Civil, M., & Moll, L. (2001). Bridging funds of distributed knowledge: Creating zones of practices in mathematics. *Journal of Education for Students Placed at Risk, 6*(1–2), 115–132.

González, N., Moll, L. C., & Amanti, C. (Eds.). (2005). *Funds of knowledge: Theorizing practices in households, communities, and classrooms.* Mahwah: Lawrence Erlbaum.

Good, T. L., & Grouws, D. A. (1979). The Missouri mathematics effectiveness project: An experimental study in fourth-grade classrooms. *Journal of Educational Psychology, 71*(3), 355–362.

Halle, T. G., Kurtz-Costes, B., & Mahoney, J. L. (1997). Family influences on school achievement in low-income, African-American children. *Journal of Educational Psychology, 89*(3), 527–537.

Henderson, A. T., & Berla, N. (Eds.). (1994). *A new generation of evidence: The family is critical to student achievement.* Columbia: National Committee for Citizens in Education.

Henderson, A. T., & Mapp, K. L. (2002). *A new wave of evidence: The impact of school, family, and community connections on student achievement* (Annual synthesis). Austin: National Center for Family Community Connections with Schools, Southwest Educational Development Laboratory.

Hiebert, J. (1999). Relationships between research and the NCTM standards. *Journal for Research in Mathematics Education, 30*(1), 3–19.

Ho, E. S., & Willms, J. D. (1996). Effects of parental involvement on eight-grade achievement. *Sociology of Education, 69*(2), 126–141.

Holloway, S. (1986). The relationship of mothers' beliefs to children's mathematics achievement: Some effects of sex differences. *Merrill-Palmer Quarterly, 32*(3), 231–250.

Hoover-Dempsey, K. V. (2005). *Research and evaluation of family involvement in education: What lies ahead?* Paper presented at the annual meeting of the American Educational Research Association (AERA). Montreal, Canada.

Hoover-Dempsey, K. V., & Bassler, R. B. (1997). Parents reported involvement in students' homework: Strategies and practices. *The Elementary School Journal, 95*(5), 435–450.

Hoover-Dempsey, K. V., & Sandler, H. M. (1995). Parent involvement in children's education: Why does it make a difference? *Teachers College Record, 97*(2), 310–331.

INCLUD-ED. (2012). *Final INCLUD-ED report: Strategies for inclusion and social cohesion in Europe from education* (CIT4-CT-2006-028603). Brussels: European Commission.

INCLUD-ED Consortium. (2009). *Actions for success in schools in Europe.* Brussels: European Commission.

Jackson, K., & Remillard, J. T. (2005). Rethinking parent involvement: African American mothers construct their roles in the mathematics education of their children. *The School Community Journal, 15*(1), 51–73.

Jacobs, J. E., Lanza, S., Osgood, D. W., Eccles, J. S., & Wigfield, A. (2002). Changes in children's self-competence and values: Gender and domain differences across grades one through twelve. *Child Development, 73*(2), 509–527.

Kahraman, P. B., & Derman, M. T. (2011). The views of primary school and preschool teachers about home visiting: A study in Turkey. In M. Pieri, A. Pepe, & L. Addimando (Eds.), *Home, school and community: A partnership for a happy life?* (pp. 22–24). Milano: I Libri di Emil.

Keith, T. Z., Keith, P. B., Troutman, G. C., Bickley, P. G., Trivette, P. S., & Singh, K. (1993). Does parental involvement affect eight-grade student achievement? Structural analysis of national data. *School Psychology Review, 22*(3), 474–496.

Kilpatrick, J., Swafford, J., & Findell, B. (Eds.). (2001). *Adding in up: Helping children learn mathematics.* Washington, DC: The National Academies Press.

Kilpatrick, J., Martin, G. W., & Schifter, D. (2003). *A research companion to principles and standards for school mathematics*. Reston: NCTM.

Knapp, M. S. (1997). Between systemic reform and the mathematics and science classroom: The dynamics of innovation, implementation and professional learning. *Review of Educational Research, 67*(2), 227–266.

Knopf, H. T., & Swick, K. J. (2008). Using our understanding of families to strengthen family involvement. *Early Childhood Education Journal, 35*(5), 419–427.

Krüger, J., & Michalek, R. (2011). Parents' teachers' cooperation: Mutual expectations and attributions from a parents' point of view. In M. Pieri, A. Pepe, & L. Addimando (Eds.), *Home, school and community: A partnership for a happy life?* (pp. 66–68). Milano: I Libri di Emil.

Lampert, M. (1990). When the problem is not the question and the solution is not the answer: Mathematical knowing and teaching. *American Educational Research Journal, 27*(1), 29–63.

Lusse, M. (2011). How to break through the knowledge paradox in home-school partnership? In M. Pieri, A. Pepe, & L. Addimando (Eds.), *Home, school and community: A partnership for a happy life?* (pp. 68–70). Milano: I Libri di Emil.

Meyer, M. R., Delagardelle, M. L., & Middleton, J. A. (1996). Addressing parents' concerns over curriculum reform. *Educational Leadership, 53*(7), 54–57.

Moll, L., Amanti, C., Neff, D., & González, N. (1992). Funds of knowledge for teaching: A qualitative approach to developing strategic connections between homes and classrooms. *Theory Into Practice, 31*(2), 132–141.

NCTM. (1989). *Curriculum and evaluation standards for school mathematics*. Reston: NCTM.

NCTM. (2000). *Principles and standards for school mathematics*. Reston: NCTM.

Parsons, J. E., Adler, T. F., & Kaczala, C. M. (1982). Socialization of achievement attitudes and beliefs: Parental influences. *Child Development, 53*(2), 310–321.

Peressini, D. (1998). The portrayal of parents in the school mathematics reform literature: Locating the context for parent involvement. *Journal for Research in Mathematics Education, 29*(5), 555–582.

Remillard, J. T., & Jackson, K. (2006). Old math, new math: Parents' experiences with *Standards-based* reform. *Mathematical Thinking and Learning, 8*(3), 231–259.

Rockliffe, F. (2011). The perspectives of parents as students, newly qualified or early career teachers on strategies for developing effective parental partnership in mathematics education. In M. Pieri, A. Pepe, & L. Addimando (Eds.), *Home, school and community: A partnership for a happy life?* (pp. 95–96). Milano: I Libri di Emil.

Roeser, R. W., Eccles, J. S., & Sameroff, A. J. (2000). School as a context of early adolescents' academic and social-emotional development: A summary of research findings. *The Elementary School Journal, 100*(5), 443–471.

Shaver, A. V., & Walls, R. T. (1998). Effect of parent involvement on student reading and mathematics achievement. *Journal of Research and Development in Education, 31*(2), 90–97.

Sheldon, S. B., & Epstein, J. L. (2005). Involvement counts: Family and community partnerships and mathematics achievement. *The Journal of Educational Research, 98*(4), 196–207.

Simon, B. S. (2004). High school outreach and family involvement. *Social Psychology of Education, 7*(2), 185–209.

Smith, F. M., & Hausafus, C. O. (1998). Relationship of family support and ethnic minority students' achievement in science and mathematics. *Science Education, 82*(1), 111–125.

Starkey, P., & Klein, A. (2000). Fostering parental support for children's mathematical development: An intervention with head start families. *Early Education and Development, 11*(5), 659–680.

Sweet, M. A., & Appelbaum, M. I. (2004). Is home visiting as effective strategy? A meta-analytic review of home visiting programs for families with young children. *Child Development, 75*(5), 1435–1456.

Vanegas, Y., Díez-Palomar, J., Font, V., & Giménez, J. (2013). *Considering extrinsic aspects in the analysis of mathematics teacher-training programs*. Paper presented at the 37th conference of the international group for the Psychology of Mathematics Education (PME), Kiel, Germany.

Wigfield, A., Eccles, J. S., Mac Iver, D., Reuman, D. A., & Midgley, C. (1991). Transitions during early adolescence: Changes in children's domain specific self-perceptions and general self-esteem across the transition to junior high school. *Developmental Psychology, 27*(4), 552–565.

Service-Learning as Teacher Education

Peter Appelbaum

Abstract This research is based on four different service learning projects bringing together secondary students with future and current teachers for a once-weekly four-month after-school "intergenerational math circle". Each group took on its own unique character with its own particular types of activities and goals, self-defined by the group participants. The chapter discusses service-learning as teacher education focusing on the changing role of assessment, on the ethics of service learning and on the horizon it may provide for future teachers.

Teacher education often leaps over the issues that emerge when educators seek to work with and through youth culture, the enabling and disabling characteristics of classroom culture, the contrasts among playful and non-routine approaches to teaching and learning, the uses and misuses of mathematical models, and so on. Future teachers jump from being a student constructed in school activities as a passive consumer of mathematics uncritical of the implicit mathematization of society into training that reproduces the practices that maintain such an atheoretical and acritical stance. They tend to leave preparation programs unable to bring mathematical practices (implicit or explicit) outside of the classroom into the foreground of students' learning, lacking the intellectual tools and practical skills that might address issues of social injustice and differentiated access to mathematics (education) practices, unable to invent for themselves instructional strategies to facilitate access and participation, dissatisfied with their ability to facilitate students' discovery of the power of mathematics, and disenfranchised when it comes to fostering mathematical activities that bridge education and life-work boundaries. Future and current teachers appear to have a deficit in terms of knowledge and skills in these areas, and find themselves relying on pre-packaged scripts or formulaic curriculum materials, instead of facilitating a robust and life-enhancing experience of mathematics.

P. Appelbaum (✉)
Department of Curriculum, Cultures and Child/Youth Studies,
School of Education, Arcadia University, Philadelphia, PA, USA
e-mail: appelbap@arcadia.edu

Perhaps what teachers and future teachers 'need' are opportunities to discover for themselves the funds of knowledge that they might bring to the educational encounter. My research suggests this is indeed the case: given time to work together with youth, interrogating the personal, social, political, etc., power people may gain by acting mathematically, and reflecting on uses of mathematics, inside and outside of the classroom, educators, future educators, and youth can explore how different forms of mathematical knowledge and different mathematical pedagogies can enable different cultural and social groups to articulate their relationships with mathematics. In community service learning projects that bring together teachers, future teachers and youth in 'underserved' or 'at risk' communities, most of the youth might be characterized as 'disposable,' marginalized and living in 'modern ghettos' that 'can be considered a dumping ground for people who have no role to play in the informational society,' while most of the future and current teachers could be described as 'consumers of mathematics,' people who are reading or listening to a range of offers, opinions, statements and reports, confronted with justifications of decisions based on mathematics (Castells 1998; Skovsmose 2007). Few of the participants' relationships with mathematics would be appropriately interpreted as sharing the attributes of 'operators' or 'constructors.' 'Operators' employ practices in which they have to make decisions on the uses and decisions based on mathematical practices. They experience their life-world as "rich in implicit mathematics" (Skovsmose 2005, p. 142), and are not only prepared for their tasks in terms of the content of their mathematical training, but also accustomed to the 'habit' of following rules as a consequence of the hidden curriculum of school mathematics. 'Constructors' specifically invent and create mathematical practices, exercising power over operators and consumers.

When disposable and consumer mathematicians come together outside of school to pursue projects that they invent together based on community needs, it is easier to document evidence that these people are not adequately described with such terms. They act as operators and constructors, and interweave in their constant redefinition of their projects an explicit critique of common school-based positions of differentiated mathematical power relations, belying commonly held assumptions that others hold for these participants, and reconceptualizing their own roles in perpetuating or transforming mathematics education in and out of school, with arguments that are insightful and ethically and politically responsible. The experience is surprisingly 'efficient' in transforming participants' relationships with mathematics in such a short period of time and with so little direct attention to the details, or curriculum, of the experience. The youth are merely treated as 'apprentices' rather than 'students' (Appelbaum 2009; Ladson-Billings 1995), and because of this, come to see themselves as constructors and operators. The future and current teachers are simply given time to work with youth mathematically without a prescribed curriculum or specific objectives or learning outcomes; their experience changes their perspective on teacher-student differences, and compels them to 'need' to know the youth and their future students personally. They desperately seek deep comprehension of the everyday life experiences of young people as the foundation for conceptual and skill-based learning.

Research Context

This research is based on four different service learning projects bringing together secondary students with future and current teachers for a once-weekly 4-month after-school 'intergenerational math circle.' Group 1 included 5 pre-service teachers, 1 early-childhood teacher, and 1 secondary mathematics teacher, and 16 grade 10–11 youth who were selected by their teachers as 'needing extra help in mathematics;' Group 2 involved 5 current teachers (1 primary teacher, 2 middle-school—ages 12–15—teachers, and 2 secondary teachers) and 4 grade 10–11 students who were self-identified as interested in mathematics. Groups 1 and 2 were transported by their regular school teacher to the nearby university to meet with the teachers as part of a program that also gave them opportunities to learn about university life and the university application process; the secondary school from which students in Groups 1 and 2 came is a highly racially, economically and ethnically diverse suburban school. Groups 3 and 4 met together as part of an after-school program run by a local church across the street from the neighborhood secondary school, located in a high-poverty, African-American, urban community. Group 1 included 8 pre-service teachers, 2 current secondary teachers, 5 current middle-school teachers, and 17 grade 9–12 youth; Group 4 included 3 pre-service teachers, 1 current secondary teacher, and 20 grade 9–12 youth. Groups 1 and 2 were given no constraints regarding activities, topics, projects, etc. Groups 3 and 4 were given the explicit requirement that they address personal finance and entrepreneurship education, since materials and resources were supported by a service-learning grant with this orientation.

What Happens When Intergenerational Groups Come Together Mathematically?

Each 'math circle' followed a five-part curricular structure to define their purposes and to organize their work, forming sub-groups that pursued their own investigations (Appelbaum 2008). The first part of the work together was an 'opening,' during which participants got to know each other, and during which they explored 'seed' activities that could provide the basis of investigations, if and only if they provoked further questions. Part 2 involved trying out a possible project and designing an exploration. Part 3 involved pursuit of a specific investigation/project, along with mini-lessons facilitated by the current and future teachers on particular mathematical skills and concepts that would propel the groups' projects further, or otherwise assist the group in carrying out its work. Part 4 included 'taking action,' during which participants found a way to interact with an audience outside of the program in order to make an impact on the world. And Part 5 involved an 'archaeology' encounter, designed by the current and future teachers, through which the youth were assisted in identifying what they had learned and applying new skills and concepts to other contexts and situations.

Outcomes

Each group took on its own unique character with its own particular types of activities and goals, self-defined by the group participants.

Self-Identification as a Mathematical Actor

Group 1, which included the youths identified by their mathematics teacher as 'needing help' with mathematics, brought this 'need' to the experience. Confronted with learners who wanted help with school mathematics, the university students returned on the second day with open-ended, problem-based activities that would help the youth practise skills they were developing in school, having spent the first day collecting the kinds of topics that they had recently studied in their mathematics classes. They hoped that these initial problem situations would lead to student-posed questions that the youths found themselves wanting to work with after the initial set of 'seed' experiences. It quickly became clear that the youths seemed incapable of seeing mathematics in the problem contexts presented, that is, of interpreting the statement of the problem contexts using mathematical concepts. There was no better way to help the future and current teachers see the futility of common school approaches, the result of which is young people who are at best able to carry out scripts for determining 'solutions' to formulaic word problem statements. Together, the teachers and youths worked through their frustrations in a terrible, challenging hour-and-a-half of trying to reach a common understanding of their contrasting orientations to mathematics. The result was a set of compromises through which the entire group took on a joint project of working on the creation of mathematical questions. Both teachers and students would bring to each meeting either ambiguous problem situations or formulaic word problems, and together they would compose new questions out of those they brought to the meeting. The youths reported a fundamental transformation in how they spent their time in their in-school classes thanks to the math circle out-of-school experiences: They now routinely *asked questions* in their own mind, silently, as their teacher or other students worked with mathematics. For example, they would think to themselves: What questions could be asked with this new mathematical idea? What questions could be answered with this new tool? How is this idea connected to other mathematics we have been learning this year? How is this question similar to or different from others we have asked this week? Typically silent in their school mathematics classes, they reported routinely asking for clarifications of things they did not understand or with which they disagreed, having practiced this in the math circle.

The Potential of Almost Any Mathematics to Be Meaningful

Group 2 took a very different path from Group 1, possibly because of the different relationships with mathematics that the youths brought to the encounter. Left to decide the topics and nature of the work, the group began with a collection of open-ended mathematical questions, contributed by each teacher or secondary-student member, which they explored in order to discuss what they would like to pursue as an exploration during their time together. The questions ranged from a comparison of racial injustices through data on housing costs and incomes to the best angle of one's foot for kicking a soccer ball, to trying to understand Fermat's Last Theorem to several mathematical logic puzzles. Serendipitously, the entire group got 'sidetracked' by a seemingly silly school word problem that most members had absolutely no interest in and which none of the teachers thought was pedagogically useful: 'Seven girls each bring seven cats onto a bus, and each cat has seven kittens. How many cats and kittens are on the bus?' They created four different ways to answer the question, one algebraically, one arithmetically, one using small objects to model the situation, and one phone call to the friend of the participant who had voiced the problem to her earlier in the day when she had suddenly remembered she was supposed to bring a math question to the circle that afternoon. As the organizer of the math circle I followed my standard line, 'OK, having answered this question, do you have new questions that you want to ask?' assuming this would lead nowhere. On the contrary: each participant had several interesting questions! Is this possible? Could such a number of girls and cats *fit* on a bus? What if we cared about the health of the cats? Do they need a minimum amount of space? The list went on and on. The ensuing investigation involved detailed research on buses, carrying boxes for animals, humane treatment of animals, conversations with veterinarians and bus drivers and cat owners; the youths invited mathematics professors to visit the circle to help them think through Pascal's triangle and the mathematical generalization of the problem situation. Reluctant at first to even offer an idea for how to 'solve the problem,' 4 weeks later each participant, teacher and high-school student had initiated a carefully designed telephone inquiry with an expert, grilled the mathematicians with mathematical questions, and carried out a series of generalized mathematical explorations that placed the initial question in a broader mathematical context. Each participant had analyzed data on animal populations and irresponsible cat care and breeding in the greater metropolitan region. By the end of the program, each participant had impressed both the animal rights experts and the mathematicians with which they worked with the intentionality that they brought to their conversations.

Entrepreneurs Who Can Make a Difference

Group 3 seeded the math circle encounter with a collection of finance and entrepreneur contexts, hoping that each secondary student would find in one of them the spark of a potential investigation: creating a business plan that could be pursued in the next math circle; planning a fund-raiser by buying ingredients, baking cookies, and selling them for a profit; analyzing mathematical strategy games in order to design one's own board or video game; analyzing basketball statistics in order to design a 'dream team,' incorporating a budget for salaries; and 'street math' of choosing a mobile phone plan and buying a car. The secondary students refused to select one area to pursue, insisting that each of these topics was fascinating and important, so the group developed a plan for continuing to work on every one of them together, rather than forming small groups around specific projects. Activities included video interviews of local business leaders who offered concrete feedback on the business plans and specific advice on how to seriously identify investors for their projects. When the after-school program, which collaborated with the math circle by providing the meeting space in the church, lost its funding, the youths proposed a presentation to the church board on the value of the program and demanded to be taught about the current budget. They were successful with their presentation in convincing the board to find the funds to maintain the program.

Philanthropy Is Also Entrepreneurship

A majority of the secondary students in Group 4 were continuing their work from Group 3, and were ready to enact their business plans. However, a petite grade 9 student came in on the second week suggesting that maybe some people should go back to their community needs ideas from the last semester, and use their business sense to create a 'business' that could help the neighborhood rather than to just make some money to spend themselves at MacDonald's. This raised a number of interesting perspectives on the use of mathematics to model business practices, and in the process, made it clear to younger and older participants that mathematical assumptions format the experience of non-profit as well as for-profit business plans. The only way to make sense of the similarities and differences was to enter into conversations about everyday life in the high-poverty neighborhood in comparison with the everyday life of most of the teachers and future teachers who did not live in this neighborhood; the juxtaposition of assumptions, dreams and expectations led to a provocative change in everyone's understanding of available data and available forms of expertise in this community around the school and church, which required adjustments on the part of every member.

Service Learning as the Context for Research

A note on the methodology of this research is in order. Because these math circles were embedded in courses about mathematics teaching and assessment practices, a lot of qualitative data was available for analysis, both as a component of the enacted curriculum and as a component of the research. University students collected assessment data in the form of secondary-student work samples, anecdotal records from observations, and informal and impromptu interviews. They analyzed this assessment data as part of their ongoing inquiry into the lives of their students and as a component of their ongoing planning and organization of the math circle. The assessment data as well as the analysis of this data on the part of the current and future teachers were available for analysis by the older participants and the university professor/researcher and comprised the main body of research data that this report is based on. The dual nature of the research as teacher training/development and as empirical research warrants ethical and logistical reflection. It was noted between Groups 3 and 4 that the secondary students were the only members of the math circle who were not part of the ongoing research about the math circle. Because of this, the adolescents participating in Group 4 were invited to join in the post-circle meetings that involved reflection on the day's experiences and planning for the following week's meeting. Their perspective on the math circle and on the research was interesting because they felt that their participation took on a political aspect: they were now representatives of high school students in urban neighborhood schools, and not merely children looking for an interesting way to spend the afternoon before they went to the church gym to play basketball. Their use of handheld video cameras to interview local business leaders and entrepreneurs now expanded to their own narrative comments on the experience itself, and this became a simultaneous research analysis and ongoing action research project that occurred simultaneously with the experience.

Service Learning as Teacher Education: Implications

Typically in the U.S., teachers and future teachers practise what they are learning about mathematics education in regular school classrooms, under the presumption that classroom contexts are the most like where they will do their work. A significant implication of this research is that much learning can take place in contexts other than school classrooms. A more significant implication is that some types of learning—about the nature of mathematics, about ways to learn about students' lives, and about one's own relationship with mathematics, are possibly best learned outside of school contexts. A recommendation is to include non-school environments in any teacher preparation program or professional development experience. Because there is no prescribed curriculum with specified learning outcomes, the teachers and future teachers can more easily focus on teacher skills that help them learn about students' lives and the mathematics of those lives, and practice understanding mathematics from the learner's perspective. They can practice skills for fostering curiosity and self-confidence in mathematics, and learn how to support young people making the transition from disposables into operators and constructors, so that later work on precise mathematical skills and concepts can be layered on top of these critical teacher practices. More significantly, these future teachers consistently recognize students themselves and families as resources; the families, always present in the work of the group in terms of the experiences that the secondary students bring with them to the activities, are repositories of formal and informal resources, rather than the barriers that families are commonly imagined to be by teachers who have never met them, physically or metaphorically (see Díez-Palomar, chapter "Family Math: Doing Mathematics to Increase the Democratic Participation in the Learning Process", this volume). Working together with young people on mathematical investigations and projects is a uniquely valuable opportunity for all of this to take place.

The Changing Role of Assessment

One might wonder, if the future teachers are not provided with a clear set of learning outcomes, what would they possibly be looking for in terms of their assessment practices? While the 'teachers' in these math circles certainly had ideas about mathematics skills and concepts that they wanted their 'students' to develop and master, we began our work with more generalized ideas about a successful mathematics classroom. Assessment for us (in the service learning projects) was the collection and analysis of information that can be used to make decisions about what should or could be happening in the short-term and long-term. Assessment in this sense is very different from 'evaluation,' in which a teacher judges the quality of student work, scores or ranks it, compared with other students or compared with a standard or norm. While there may be a system of grading or scoring involved in assessment, the primary goal is not to evaluate, but to make professional choices. One set of assessment questions that we used was suggested by Susan Ohanian (1992):

1. Do my students see themselves as mathematicians?
2. So my students see mathematics as covering a wide range of topics?
3. Are my students developing a flexible repertoire of problem-solving strategies?
4. Are my students able to communicate their problem-solving strategies to others? Can they talk and write about how they solve math problems?
5. Are my students able to assess themselves? Are they able to develop and use criteria to evaluate their performance?
6. Do my students engage in mathematical thinking without a specific assignment? For example, if they have 'free time' do they choose math?
7. Are my students developing the attitudes of independent and self-motivated thinkers and problem solvers?

8. Do my students welcome challenges in math? Are they able to focus on math problems of increasing complexity for longer periods of time?
9. Do my students recognize the importance of math in the real world outside school?
10. Do my students use mathematics to solve problems outside math class?

In our experiences, we simply used whatever method of information collection we could think of to try to answer these questions on an ongoing basis, including interviewing individuals or small groups of students, observing students at work and taking notes on what we observed, asking the students to interview each other, video recording of group work, analysis of student work samples on paper or in video presentations, and short, informal surveys. The 'teachers' reflected on what they knew about their students, and invented activities that would have the potential to enable students to demonstrate that they were developing the attitudes and habits implicit in Ohanian's questions.

Another set of assessment questions grew out of our reading of Boaler and Humphrey's (2005) *Connecting Mathematical Ideas*. Before we began our math circles, we spent several weeks with this book, which includes a CD of video cases from Humphrey's classroom. Through our viewing of the videos and reading Boaler's and Humphrey's analysis of Humphrey's teaching, we identified as a collaborative group exemplary mathematics teaching/learning practices; for example, in the Group 4 mentioned in this chapter, the following list guided planning and assessment during the math circle:

1. The development of a 'questioning disposition' over time. This needs to take two forms:

    (a) As a form of 'academic literacy,' where students on their own eventually approach each mathematical item, problem, situation, application, etc., as something to ask questions about. Students would then use methods of working as a mathematician to turn their questions into investigations.
    (b) As a tool of critical thinking, where students would be able to evaluate a question as more or less 'powerful.'

2. A culture or environment – sometimes referred to as a 'community of mathematicians,' where one finds mathematical talk in an atmosphere that values participation, and a culture of trust where 'wrong answers' are as useful and helpful as anything correct. Reasoning and thinking are thus more important than memorizing correct procedures.
3. The specific pedagogical strategy and later learning tool of convincing yourself, convincing a friend, convincing a skeptic.
4. The specific mathematical approach of seeing every mathematical item, problem, situation application, etc., as a 'special case.' Each mathematical entity might be a special case of more general collections of such things in more than one way. Using entities as special cases leads to the mathematical activity of generalizing from these special cases to develop conjectures that can be discussed, tested with artfully chosen further special cases, modified

through the identification of counter-examples that lead to newly stated conjectures, etc.
5. The teacher working to ask questions that promote higher order thinking. A variation on this is labeled questions of 'type 3' in the Boaler and Humphrey's (2005) text and videos.

The service-learning context enabled a separation of content and mathematical development, so that the future teachers, 'teachers' in the math circle, could focus on mathematical ways of being, thinking and collaborating, as distinct from school-mathematics curriculum learning goals. Because of this, the 'teachers' were able to understand the difference between 'living and working as a mathematician' and 'practicing and mastering skills.' This also enabled the 'teachers' to understand their students' growth over time in ways independent of traditional models of school mathematics progress, to recognize talents, dispositions, and funds of knowledge that could be capitalized on in group activity, and to support a respectful environment grounded in the present encounters as differentiated from the participants' school histories (which often included unhappy stories of persistent failure). The students responded in powerful ways, engaging in activities, requesting more, sharing skills and knowledge they ordinarily would not be able to share in school classrooms, and connecting the mathematics to concrete life situations. This form of assessment was a tool of community building that fostered richer comprehension of the connections between mathematics and the community, helping the 'teachers' to learn more about their students rather than to judge their progress.

Most future teachers have never seen their mathematics teachers carrying out such assessment in their classrooms, and very few observe this in traditional school-based field work or student-teaching. Of course, when one is a student, one is not necessarily noticing what the teacher is doing, busy as one is with one's own work as a student. For example, if a teacher were taking notes on what was happening in small group work, a student is not likely to realize it is happening – unless the teacher were to refer to his/her note-taking directly in classroom conversations. In my own experience with teacher education, it has been a challenge to convince future teachers that such work is part of the job of teaching, or that it is worth doing. Those future teachers who have participated in community math circles have reported that they could not imagine teaching without it, after having had this experience.

Radical Transformation and a Horizon

The future teachers also report another significant transformation in their understanding of teaching, thanks to their participation in the service learning projects. They describe the context of service learning as creating a long-term sense of purpose and direction for their work with these youth, shifting the focus away from practicing an incremental skill added onto one from a previous lesson toward establishing a change in the community. The university students described realizations in later reflections that this is usually missing from school mathematics classrooms. While they had come to the course expecting to be trained in techniques for motivating, explaining, and testing learners, they now saw their future jobs in terms of what David Kirshner (2000, 2002) calls enculturation and acculturation. The service-learning context, independent of the classroom culture and separable from school mathematics learning goals, helped them understand school mathematics culture and goals in two new ways: (a) as important, yet easily accomplished in the context of large-scale problem solving projects, introduced in mini-lessons as needed and as serving the needs of the project (i.e., what the students were doing and cared about), rather the needs of the curriculum; and (b) as only important if recognized as influencing the ongoing, emerging sense of self and mathematics for each individual in the classroom. Both in the classroom and in the service-learning projects, mathematical enculturation and acculturation are happening at the same time, unfolding and intermingling in innumerable and complex ways. Enculturation is the transition of students into a somewhat alien, or 'second,' mathematical culture, with its own unique set of languages and practices; students are treated as foreigners immigrating to a new land. Acculturation refers to the changes in attitudes, habits, customs and social institutions when 'home' or everyday mathematical culture interacts with the new mathematical cultures spawned by the study of new mathematical topics.

Because the university students, the 'teachers,' had spent a sustained period of time describing for themselves how their students were changing and how their mathematical understandings were evolving over time, they now explained 'learning' as a process of radical transformation of self, rather than a constant accretion of new information and techniques. They now believed that such transformations only occur when one is invested in a project that has a 'horizon,' a long-term sense of purpose and direction. While they understood the value of specific mathematics content goals, and understood that a school curriculum expected mastery of specific concepts and skills within a certain period of time, they also believed that such goals were best met when they supported this more significant, radical self-transformation. Without the ongoing self-transformation, they suggested, there was no development of community, understanding, meaning, or mathematical dispositions and ways of being.

An Ethics of Service-Learning

In the U.S. context, field experiences are plentiful in teacher education. Over the course of a 4-year university degree, education students spend increasing amounts of time in schools, taking on more and more responsibilities of the regular classroom teacher. Commonsense suggests that one learns how to be a teacher through carefully sequenced apprenticeships, and this tradition buttresses such commonsense assumptions. The dilemma, alluded to above in the case of mathematics teachers who probably do not carry out alternative assessment practices, is that the mentor teachers are not necessarily exemplary teachers; while future teachers learn the skills of 'fitting in' to a school culture, they do not necessarily practice innovative instructional strategies or even research-supported practices. In some cases, the pre-service teacher is not given the freedom to experiment. In other situations, the pre-service teacher has not yet mastered minimal teaching skills and should not be given such freedom, much as it would be valuable for the teacher-in-training. When inexperienced not-yet-teachers take on leadership roles in non-school settings, are there different ethical questions? In the research reported here, the 'teachers' explained their background and education to the participants, and routinely reminded their 'students' with examples of how the experience was helping them to become better teachers in the future. Participation in the group was voluntary and could end at any time. The programs were supervised by university faculty who conferenced with the 'teachers' throughout the program. In some sense, then, precautions were taken to make sure that no overt physical or emotional harm would be wielded upon any participant. In a broader sense, service-learning does more than meet the immediate needs of the participants. Service-learning further aims to meet the concerns, needs and hopes of *communities*. Service-learning also requires analysis of the experience in social, political, and cultural contexts in order for it to be more than mere volunteer work.

> If school students collect trash out of an urban streambed, they are providing a valued service to the community as volunteers. If school students collect trash from an urban streambed, analyze their findings to determine the possible sources of pollution, and share the results with residents of the neighborhood, they are engaging in service-learning.
>
> In the service-learning example, in addition to providing an important service to the community, students are learning about water quality and laboratory analysis, developing an understanding of pollution issues, and practicing communications skills. They may also reflect on their personal and career interests in science, the environment, public policy or other related areas. Both the students and the community have been involved in a transformative experience. (CNCS 2013)

The analogous distinction between volunteering at a community program as a tutor or group facilitator and an academic service-learning experience requires that the future teachers carefully analyze their findings about how to best facilitate the learning of mathematics in the context of community projects in the specific neighborhood in which they are working. The future teachers in this study report that participation of this kind made a significant impact on them. A majority in each group noted in final reflections on the program and their own learning from the program that they understood their work as part of a broader community project greater than their own training in mathematics education and richer than helping the youth in the program to learn mathematics. They described this inclusion in a project with longer–term impact, building relationships among university students and faculty and community schools and agencies, in ways that echoed the notion of an horizon discussed above. What we found is that the future teachers understood their own changes over time as intimately intertwined with social actions and as embedded in a broader project of social justice; they understood their own perspective on youth, mathematics, learning and group activity as continually reinvented in every moment of participation. In other words, they left the experience framing teaching as an ethical stance upon the world (Block 2003), and their ongoing, emerging sense of themselves as teachers as an ethical relation developed in face-to-face encounters with the youths that became their co-mathematicians, co-designing group projects in their community. As teachers participate in collaborative curriculum design, they are no longer subjects of power but are active constructors of knowledge, working within and across regimes of truth and power. Such active design may be taken as an articulation of the ethical stance that teaching embodies (Appelbaum and Dávila 2007).

Service-learning encourages reflection on commitment to the common good, counteracting dominant educational structures grounded in competitive individualism (Karlberg 2005). More important, however, is the need for analysis that pushes participants to consider the stories that the youths in these projects tell about their own experience. The 'teachers' do not have to agree with these stories, but they do need to dialogue with them. Tensions often emerge between comprehending the experiences from the perspectives present in a marginalized community and the need to sometimes critique harmful discourses that are part of these communities. Similar tensions arise when the 'teachers' confront the distance between school-mathematics expectations and the needs and goals of the youth with which they work in community contexts. As these future teachers reflect on their changing relationship with mathematics, with teaching and learning, and with service learning as a context for learning, they speak of the ethical obligations to live with their future students as co-inquirers, fostering not just the development of conceptual understanding and mastery of skills, but more importantly to enable 'perspectival growth' (Hoops 2011), that is, transformation in one's comprehension of what it means to think and act mathematically for the common good.

Service-Learning as Teacher Education: Discussion

Because service-learning contexts may take place outside of the constraints of traditional classrooms, they make it easier for future teachers to engage with issues that are not incorporated in commonly-used pedagogical practices. However, there is no guarantee that service-learning will automatically make this happen, nor should we assume that such opportunities are impossible in traditional classroom settings. For example, a community project outside of school constraints offers many invitations to work with and through youth culture. However, it is certainly possible for a future teacher to remain unaware of these invitations, not recognize them, or refuse them. Similarly, some student teachers in typical school fieldwork settings find ways to engage with youth culture. Nevertheless, it is commonly the case that such engagement is not readily supported or even understood in traditional settings, and that the outside-of-school context makes it readily apparent to these university students that their collaborative work across generations will be more successful if they learn more about the lives and interests of their collaborators. More to the point, we established the project with the assumption that we would build activities out of the concerns and needs of the community in which we were working; such a platform demands engagement with the everyday lives of those in the community. This need to learn about the community in order to support their interests then leads to the realization on the part of the future teachers that such understanding is powerful in the support of mathematics teaching and learning. They report thinking that it would now seem impossible to be a successful teacher without such understanding of the everyday lives of their students, and without legitimizing it in their future classrooms.

In this way, it becomes similarly apparent through service-learning encounters that traditional classrooms are both enabling and disabling in specific ways. Often, the university 'teachers' would wistfully dream of how easy it would be to organize a straightforward lecture on a topic if they were only in a school classroom. Even more often, though, they would note in their assessment analyses how much more rewarding and efficient it was to be working in the community context outside of the classroom. This was especially the case with helping their students to consider the uses and misuses of mathematical models, given that the models were always considered and modified in light of concrete projects. But this was not only limited to applications of the mathematical models. The context of practical applications made it possible to shift in and out of that mode of thinking, and to set aside time and place for playful explorations not tied to routine applications. In later reflections after the service-learning project, the university students noted that it felt ironic at first that they were able to engage with their collaborators in imaginative fantasies that extended beyond the specific problem being solved, into analogous and fanciful extensions of the original problem.

Extensions of problems into discussions and imaginative explorations were also possible in the realm of uncovering hidden mathematizations of everyday life, such as in the technologies of supermarkets and banking systems. Critiques and complaints about the way things worked led in several cases to the reimagining of alternatives, through side-inquiries about how these systems worked and why they have become so pervasive in contemporary society. The university students thought that these discussions would not have been possible in a typical school classroom, since there would be no space for them. They then strategized what they will do in their own classrooms to find the time and space for such discussions, pointing to documented evidence for how these discussions and explorations came to help their participants think mathematically and learn more routine mathematical concepts and skills more readily later on.

Because we worked in non-traditional locations and in non-traditional ways for mathematics teachers, the future teachers needed to invent their own ways of facilitating the groups. In later reflections they began to articulate how this did not feel like 'teaching' yet nevertheless facilitated learning. Once they began to feel comfortable with a redefinition of 'teaching' specifically as enabling learning to take place, they were able to see that they had invented instructional strategies of their own, simply by paying attention to the needs and interests of their 'students,' who were constructed as co-collaborators in a community project. These experiences in redefining teaching and learning, teacher and student, established a field of knowledge about inventing pedagogical practices, rather than being a passive implementer of prepackaged curriculum. For teachers who have had the privilege of facilitative groups in service-learning projects, teaching is first and foremost about the power of mathematics to be a resource for people ordinarily disenfranchised to find skills and knowledge in themselves and their community, and to use mathematics as a tool for joy, change, and the pleasures of a can-do attitude.

Acknowledgements This material is partly based on work supported by the Corporation for National Service under Learn and Serve America: Higher Education Grant 09LHAPA001. The dissemination of this material is supported by the Corporation for National Service under Learn and Serve America: Higher Education Grant 09LHAPA001. Opinions or points of view expressed in this document are those of the author and do not necessarily reflect the official position of the Corporation for National Service or the Learn and Serve America Program.

References

Appelbaum, P. (2008). *Embracing mathematics: On becoming a teacher and changing with mathematics*. New York: Routledge.
Appelbaum, P. (2009). Taking action: Curricular organization for effective teaching and learning in mathematics. *For the Learning of Mathematics, 29*(2), 38–43.
Appelbaum, P., & Dávila, E. (2007). Math education and social justice: Gatekeepers, politics, and teacher agency. *Philosophy of Mathematics Education Journal, 22*.
Block, A. (2003). They sound the alarm immediately: Anti-intellectualism in teacher education. *Journal of Curriculum Theorizing, 19*(1), 33–46.
Boaler, J., & Humphreys, C. (2005). *Connecting mathematical ideas: Middle school video cases to support teaching and learning*. Portsmouth: Heinemann.
Castells, M. (1998). *The information age: Economy, society and culture. Vol. III, end of millennium* (Vol. III). Oxford: Blackwell.
CNCS (Corporation for National and Community Service). (2013). *National service learning clearinghouse*. http://www.servicelearning.org/what-is-service-learning. Accessed 4 July 2013.
Hoops, J. (2011). Developing an ethic of tension: Negotiating service learning and critical pedagogy ethical tensions. *Journal for Civic Commitment, 16*(Jan), 1–15.
Karlberg, M. (2005). Elevating the service in service-learning. *Journal of the Northwestern Communication Association, 34*(Spring), 16–36.
Kirshner, D. (2000). Exercises, probes, puzzles: A crossdisciplinary typology of school mathematics problems. *Journal of Curriculum Theorizing, 16*(2), 9–36.
Kirshner, D. (2002). Untangling teachers' diverse aspirations for student learning: A crossdisciplinary strategy for relating psychological theory to pedagogical practice. *Journal for Research in Mathematics Education, 33*(1), 46–58.
Ladson-Billings, G. (1995). Making mathematics meaningful in multicultural contexts. In W. G. Secada, E. Fennema, & L. B. Adajian (Eds.), *New directions for equity in mathematics education* (pp. 126–145). Cambridge: Cambridge University Press.
Ohanian, S. (1992). *Garbage pizza, patchwork quilts, and math magic: Stories about teachers who love to teach and children who love to learn*. New York: Freeman.

Skovsmose, O. (2005). *Travelling through education: Uncertainty, mathematics, responsibility*. Rotterdam: Sense.
Skovsmose, O. (2007). Mathematical literacy and globalization. In B. Atweh, A. Calabrese Barton, M. Borba, N. Gough, C. Keitel, C. Vistro-Yu, & R. Vithal (Eds.), *Internationalisation and globalisation in mathematics and science education* (pp. 3–18). Dordrecht: Springer.

The Learning and Teaching of Mathematics as an Emergent Property Through Interacting Systems and Interchanging Roles: A Commentary

Fragkiskos Kalavasis and Corneille Kazadi

Abstract The commentary on the chapters of Díez-Palomar and of Appelbaum introduces a complex model of structures in education. It argues for an interdisciplinary complexity-theory-approach to understand and develop educational designs for mathematics education practices.

It seems difficult to include the scheme of interactions and the management of the boundaries between explicit, formal, implicit and informal phenomena, structures or objectives into an operative syllogism. In school mathematics education, there is not only the technical difficulty of this interactive scheme and management, it is rather the difficulty of the global perception of all the components of the learning process, which involve and reflect the interior procedures with the exterior factors and their representations.

There is a particular epistemological difficulty to perceive the importance of the retroaction of some external systems, such as the family or the teachers' community in the progress of individual learning. The core of this difficulty consists of the impression that the only learning subject is the pupil and all the other components have to confine themselves to teaching, helping at home and encouraging the learning subject. In this frame, we have observed all kinds of useful training programs that might improve the learning situation, but have not changed the trend of massive failure of school mathematics education. It is a fact that a great variety of training activities have not traversed the frontiers of this educational paradigm based on discrete rules.

F. Kalavasis (✉)
Faculty of Humanities, Department of Sciences for Preschool Education and Educational Design, University of the Aegean, Rhodes, Greece
e-mail: kalabas@aegean.gr

C. Kazadi
Université du Québec à Trois-Rivières, Trois-Rivières, Canada
e-mail: corneille.kazadi@uqtr.ca

If we try to perceive the school system as a learning organization, in which the knowledge and its communication and transformation give life, we can understand that the construction of mathematical thinking is not confined to the students' efforts but emerges from the interactions among students, teachers, administration, families and societal environment. The coordination of this complex net of interactions needs the development of learning and teaching abilities for all the partners, but with respect to the limits that are connected to the different kind of partners. The complexity of this net is due to the interchangeable roles between the partners. Teachers take on the role of the parents, administrators or social actors and vice versa. In their chapters, Javier Díez-Palomar and Peter Appelbaum introduce two models for the development of mathematical learning and teaching abilities of the families and of the teachers respectively. In these models, we can observe how their involvement in the learning of mathematics conduces to the transformation of the paradigm and the functional variation of stereotypes about mathematics and mathematics education, which was the core of an epistemological difficulty. To overcome this kind of difficulty or obstacle, we need more dynamic and flexible theories which can include the phenomenon of learning and teaching of mathematics in a genetic perspective, considering both the historical evolution of mathematics and the learning subject as well as the evolution of the interaction of each individual with the broader environment.

In the case of mathematics and of mathematics education, this hard work of paradigm change has been started by the Didactics of Mathematics, which has produced a phenomenology of the learning and teaching of mathematics for more than 45 years. Among the genitors of this interdisciplinary field are mathematicians, psychologists and educators, such as Gustave Choquet, Jean Piaget, and Caleb Gattegno who met after World War II, founding the *Commission Internationale pour l'Etude et l'Amélioration de l'Enseignement des Mathématiques* (CIEAEM). In the evolution of the Didactics of Mathematics, we can observe a continuous enlargement of frameworks outside of the classroom, however interacting with what happens inside of the classroom. As some of the difficulties in the learning of mathematics are linked to epistemological obstacles that we can find in the evolution of mathematics, other learning or teaching problems are connected with professional habits and social stereotypes or with family beliefs and socio-cultural practices in a similar way. Some of the difficulties are related to the confusion between the phenomenon and the symptom or between what is the happening and what is its meaning. This kind of confusion is developed by the *observer's paradox,* by the fact that the observer is participating in the phenomenon as well as by his/her beliefs and expectations. Thus, Javier Díez-Palomar's chapter assumes that families, who usually like to play the role of the evaluator of the school system's quality, interact with this quality and we thus have to design a pathway of how to incorporate their involvement. Furthermore, it should be noted that the involvement of the family in order to help the pupils may have limitations with respect to age and mathematics content, but the explicit recognition of these limits can reinforce the bridge of the coordination between partners.

An important difficulty of the management of the boundaries is related to the variability of the borders due to the continuous change of the status among implicit and explicit knowledge. Just as on the border of the beach, the movement of the waves continuously changes the line and the identity, one moment there is sea and in the next moment there is sand. In the same way we can observe that the cognitive tools or some mathematical techniques and expressions of the teachers change from being implicit to being explicit. This unconscious confusion is a fact that interacts with the diversity of the students' paths to mathematics and could increase the divergence between those who understand and those who feel excluded from the secret mechanisms of mathematics education.

We perceive a large set of relations, which contain the relationships that the school system constructs with the system of the family and of the society with the use of mathematics in everyday life, with the teachers' beliefs or their representations about mathematics and about mathematics education. We can imagine this large dynamic set as a variable borderline. This porous borderline can describe the dynamic of the classroom of mathematics while the learner constructs the semantic status and the expression model of an emerged mathematical concept. This construction emerges from the interactions, which occur between himself/herself and the mathematical knowledge, the teacher, the classmates, the school system, the family and the societal environment.

However, this involvement of all partners should not ignore the deep belief that confuses the primary or pilot role of the teacher in the school system with the impression of an exclusive influence to the learning process. In various places, parents feel responsible for the discipline dimension of education (politeness, good manners, civility, etc.) but they leave the exclusive authority in terms of academic education and instruction (learning abilities, science and humanities, pedagogical skills) to the school. The model of relations, which the school unit constructs or avoids constructing with the other systems, has an important influence on the model of the respective relations of the student which influence the learning process. In the self-similarity of the pentagon below we can see a representation of this complex network of interactions (see Fig. 1).

In this scheme the porous borderline is the perimeter of the inside pentagon. This pentagon is interactive and auto-similar with the outside one which describes the implicit and explicit relationships that the school unit constructs with mathematics knowledge, with other educative units, with the school system, with the family and with society. Peter Appelbaum's chapter suggests a project in which most of these interactions are included in a learning community.

The description becomes more complex if we imagine the same pentagonal development for the construction of school mathematical knowledge as the center of a network of interactive relationships of school discipline with academic knowledge (didactic transposition), with other disciplines (interdisciplinary), with the school system, with the family system and with society. The interdisciplinary and complex approach of the learning and teaching of mathematics in a school unit conduces to a perception of school as a learning organization in which the systems of the

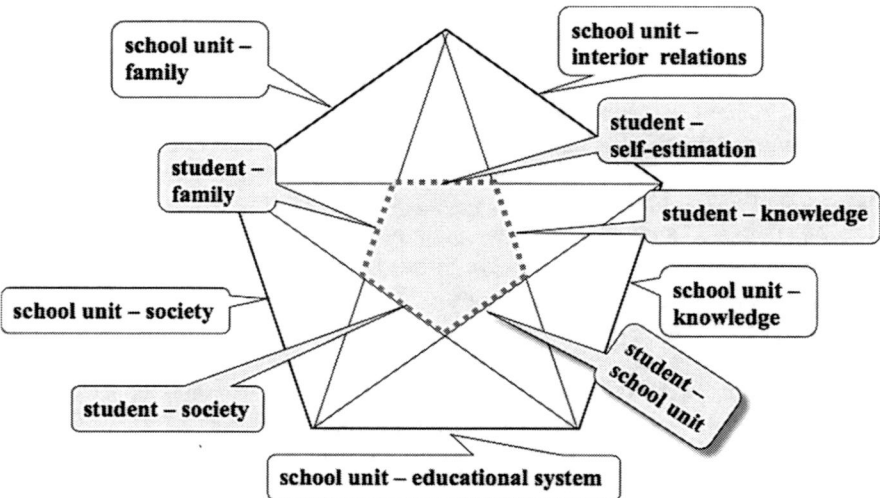

Fig. 1 A model of the complex structures in education (Adapted from Kalavasis 2012). The pentagon is self-similar

environment, principally of the family, the larger educational system and the society system are included. By this perception, the phenomenon of learning and teaching is connected with the involvement of the learning and teaching activities of all partners, as well as it tries to challenge some of the social stereotypes of discrimination or exclusion practices.

There is a necessary transition to a new epistemology for the understanding of this description, which valorizes the particular role and the involvement of all partners in mathematics education. We have to reject the stereotypical concept of learning as an individual hypothesis and adopt the ambivalent concept of learning as an emerging property of an interactive system with interior processes and exterior procedures. A same kind of epistemological transition leads to the construction of intermediate terms such as "glocalization" (emerging from the simultaneous procedure of globalization and localization) or "coopetition" (emerging from the simultaneous procedure of competition and cooperation). With these kinds of concepts, the difficulties concern the identification of the coexistence of formal procedures and informal processes as well as the exchange of roles between them.

Once adopted and connected into an interdisciplinary complexity theory, this new kind of complex concepts can help us to describe and to understand the possibility of an educational design for mathematics instruction. It has to be capable of identifying and synchronizing the informal and the formal activities inside and outside of the classroom, transcending the reflective and porous framework of the boundaries between roles, partners and their involvement in learning or teaching activities. In this orientation, the adequate and dynamic organization of the teachers and of the families is crucial, as proposed by Díez-Palomar and Appelbaum. The interdisciplinary approach has to promote opportunities for collaboration among various fields of knowledge and practice, without regarding it as a juxtaposition, but rather as an interchange.

Reference

Kalavasis, F. (2012). Modélisation des structures complexes en éducation, en vue de la gouvernance des unités scolaires. *Les Cahiers de Recherche de LAREQUOI, 1*, 46–71. Université de Versailles St Quentin, France.

Appendices

Appendix A: Themes and Places of the CIEAEM Conferences

CIEAEM 1 – 1950 Debden (UK)	Relations between the curricular of mathematics in the secondary schools and the intellectual capacities development of the adolescent /
	Relations entre les programmes mathématiques des écoles secondaires et le développement des capacités intellectuelles de l'adolescent
CIEAEM 2 – 1951 Keerbergen (Belgium)	The teaching of geometry in the first years of the secondary schools /
	L'enseignement de la géométrie dans les premières classes des écoles secondaires
CIEAEM 3 – 1951 Herzberg (Switzerland)	The functional curriculum from the maternal school to the university /
	Le programme fonctionnel: de l'école maternelle à l'université
CIEAEM 4 – 1952 La Rochette par Melun (France)	Mathematical and mental structures /
	Structures mathématiques et structures mentales
CIEAEM 5 – 1953 Weilerbach (Luxemburg)	Relations between the teaching of mathematics and modern science and technical requirements /
	Les relations entre l'enseignement des mathématiques et les besoins de la science et la technique moderne
CIEAEM 6 – 1953 Calw (Germany)	Connections between the pupil's thinking and the teaching of mathematics /
	Les rapports entre la pensée des élèves et l'enseignement des mathématiques
CIEAEM 7 – 1954 Oosterbeek (The Netherlands)	Modern mathematics at school /
	Les mathématiques modernes à l'école
CIEAEM 8 – 1955 Bellano (Italy)	The pupil coping with mathematics – A releasing pedagogy
	L'élève face aux mathématiques – Une pédagogie qui libère

(continued)

CIEAEM 9 – 1955 Ramsau am Dachstein (Austria)	Probability and statistics teaching at the university and the school / L'enseignement des probabilités et des statistiques à l'université et à l'école
CIEAEM 10 – 1956 Novi Sad (Yugoslavia)	Primary school teacher training / La formation mathématique des instituteurs
CIEAEM 11 – 1957 Madrid (Spain)	Teaching materials / Matériel d'enseignement
CIEAEM 12 – 1958 Saint Andrews (Scotland)	The question of the problems in mathematics teaching / La question des problèmes dans l'enseignement des mathématiques
CIEAEM 13 – 1959 Nyborg & Aarhus (Denmark)	The universities and the schools coping with their mutual responsibilities / Les universités et les écoles devant leurs responsabilités mutuelles
CIEAEM 14 – 1960 Krakow (Poland)	Basic mathematics / Mathématiques de base
CIEAEM 15 – 1961 Lac Léman (Switzerland)	Languages of mathematics / Langages de la mathématique
CIEAEM 16 – 1962 Morlanwelz (Belgium)	Experimental and axiomatic attitudes in the teaching of mathematics / Attitudes expérimentales et axiomatiques dans l'enseignement de la mathématique
CIEAEM 17 – 1963 Digne (France)	A reconstruction of the mathematics for the teaching of 10 to 18 years olds / Reconstruction de la mathématique dans l'enseignement de 10 à 18 ans
CIEAEM 18 – 1964 Oberwolfach (Germany)	The contribution of psychology to a modern mathematical teaching / Enseignement mathématique moderne et apports de la psychologie
CIEAEM 19 – 1965 Milano Marittima (Italy)	The place of geometry in modern mathematical teaching / Place de la géométrie dans un enseignement moderne de la mathématique
CIEAEM 20 – 1966 Dublin (Ireland)	First steps in calculus in the secondary school / Les débuts de l'analyse dans l'enseignement secondaire
CIEAEM 21 – 1968 Gandia (Spain)	The teaching of mathematics for those between 6 and 12 / L'enseignement de la mathématique au premier niveau (de 6 à 12 ans)
CIEAEM 22 – 1970 Nice (France)	Progresses in mathematics after 1945. The study of new concepts / Progrès en mathématique depuis 1945. Etude de concepts nouveaux
CIEAEM 23 – 1971 Krakow (Poland)	The mathematical logic at school / La logique mathématique dans l'enseignement
CIEAEM 24 – 1972 Morlanwelz (Belgium)	Algorithmic thinking in the school / Pensée algorithmique et enseignement
CIEAEM 25 – 1973 Québec (Canada)	Development of mathematical activity in education / Développement de l'activité mathématique dans l'enseignement
CIEAEM 26 – 1974 Bordeaux (France)	Probability and statistics in primary and secondary education / Probabilités et statistique dans l'enseignements primaire et secondaire
CIEAEM 27 – 1975 Tunis (Tunisia)	Mathematics, why? / Pourquoi la mathématique?
CIEAEM 28 – 1976 Louvain-la-Neuve (Belgium)	Some questions related to the use of problems in the teaching of mathematics / Problématique et enseignement de la mathématique

(continued)

CIEAEM 29 – 1977 Lausanne (Switzerland)	Evaluation in the teaching of mathematics / Evaluation et enseignement mathématique
CIEAEM 30 – 1978 Santiago de Compostela (Spain)	Connections between the teaching of mathematics and the subjects which it serves and from which it is induced / Relations entre l'enseignement de la mathématique, la réalité et les autres branches qu'elle sert et qui l'inspirent
CIEAEM 31 – 1979 Veszprem (Hungary)	Mathematics for all and for everyone (6 to 16 years) / Mathématiques à la portée de tous et adaptées a chacun (6 à 16 ans)
CIEAEM 32 – 1980 Oaxtepec (Mexico)	The processes of mathematisation and applying mathematics: mathematical and pedagogical aspects / Processus de mathématisation et d'application de la mathématique: aspects mathématiques et pédagogiques
CIEAEM 33 – 1981 Pallanza (Italy)	Processes of geometrisation and visualization / Processus de géométrisation et de visualisation
CIEAEM 34 –1982 Orléans (France)	Means and materials for the teaching of mathematics: present state and future perspectives / Moyens et medias dans l'enseignement des mathématiques: bilans et perspectives
CIEAEM 35 – 1983 Lisbon (Portugal)	Mathematics education in relation to the reality of schools and society / Didactique de la mathématique et réalité scolaire et sociale
CIEAEM 36 – 1984 Frascati (Italy)	Restricted meeting: Aims, priorities and future modes of action of CIEAEM / Réunion restreinte: Buts, priorités et modalités de l'action future de la CIEAEM
CIEAEM 37 – 1985 Leiden (The Netherlands)	Mathematics for all … in the computer age / Mathématiques pour tous … à l'âge de l'ordinateur
CIEAEM 38 – 1986 Southampton (UK)	Mathematics for those between 14 and 17: Is it really necessary? / Mathématique pour les élèves de 14 à 17 ans: Est-ce qu'ils en ont vraiment besoin?
CIEAEM 39 – 1987 Sherbrooke (Canada)	The role errors play in the learning and teaching of mathematics / Rôle de l'erreur dans l'apprentissage et l'enseignement de la mathématique
CIEAEM 40 – 1988 Budapest (Hungary)	Restricted meeting: Structure and politics of the commission / Réunion restreinte: Structure et politique de la commission
CIEAEM 41 – 1989 Brussels (Belgium)	Role and conception of mathematics curricula / Rôle et conception des programmes de mathématique
CIEAEM 42 – 1990 Szcyrk (Poland)	The teacher of mathematics in the changing world / Le professeur de mathématiques dans un monde qui change
CIEAEM 43 – 1991 Locarno (Switzerland)	Restricted meeting: Preparation of the meetings of 1992 and 1993 / Réunion restreinte: Préparation des rencontres en 1992 et 1993
CIEAEM 44 – 1992 Chicago (USA)	The student confronted with mathematics / L'élève face aux mathématiques
CIEAEM 45 – 1993 Cagliari (Italy)	Assessment focused on the student / L'évaluation centrée sur l'élève

(continued)

CIEAEM 46 – 1994 Toulouse (France)	Graphical and symbolic representations from primary school to university / Représentations graphiques et symboliques de la maternelle à l'université
CIEAEM 47 – 1995 Berlin (Germany)	Mathematics and common sense / Mathématiques et sens commun
CIEAEM 48 – 1996 Huelva (Spain)	Restricted meeting: The present and the future of CIEAEM / Réunion restreinte: Le présent et le futur de la CIEAEM
CIEAEM 49 – 1997 Setubal (Portugal)	Interactions in the mathematics classroom / Les interactions dans la classe de mathématiques
CIEAEM 50 – 1998 Neuchâtel (Switzerland)	Relationships between classroom practice and research on didactics of mathematics / Les liens entre la pratique de la classe et la recherche en didactique des mathématiques
CIEAEM 51 – 1999 Chichester (UK)	Cultural diversity in mathematics (education) / La diversité culturelle dans l'enseignement des mathématiques
CIEAEM 52 – 2000 Amsterdam (The Netherlands)	Restricted meeting: Preparation of future meetings / Réunion restreinte: Préparation des futures rencontres
CIEAEM 53 – 2001 Verbania (Italy)	Mathematical literacy in the digital era / Littéracie mathématique à l'ère digitale
CIEAEM 54 – 2002 Vilanova i la Geltrú (Spain)	A challenge for mathematics education: to reconcile commonalities and differences / Un défi pour l'éducation mathématique : le commun et les différences
CIEAEM 55 – 2003 Plock (Poland)	The use of didactic materials for developing pupils' mathematical activities / L'utilisation de matériels didactiques pour développer des activités mathématiques des élèves
CIEAEM 56 – 2004 Paris (France)	Restricted meeting: CIEAEM in today's world / Réunion restreinte: La CIEAEM dans la monde d'aujourd'hui
CIEAEM 57 – 2005 Palermo (Italy)	Changes in the society: a challenge for mathematics education (I) / Changements dans la société: un défi pour l'enseignement des mathématiques (I)
CIEAEM 58 – 2006 Srni (Czech Republic)	Changes in the society: a challenge for mathematics education (II) / Changements dans la société: un défi pour l'enseignement des mathématiques (II)
CIEAEM 59 – 2007 Dobogókő (Hungary)	Mathematical activity in classroom practice and as research object in didactics: two complementary perspectives (I) / L'activité mathématique dans la pratique de la classe et comme objet de recherche en didactique: deux perspectives complémentaires (I)
CIEAEM 60 –2008 Paris (France)	Restricted meeting: Complexity and mathematics education Réunion restreinte: Complexité et éducation mathématique
CIEAEM 61 – 2009 Montreal (Canada)	Mathematical activity in classroom practice and as research object in didactics: two complementary perspectives (II) / L'activité mathématique dans la pratique de la classe et comme objet de recherche en didactique: deux perspectives complémentaires (II)

(continued)

CIEAEM 62 – 2010 London (UK)	Restricted meeting: Mathematics as a living, growing discipline: CIEAEM's contribution to making this explicit /
	Réunion restreinte: Les mathématiques: une discipline en pleine croissance: Le rôle de la CIEAM pour exposer cette situation
CIEAEM 63 – 2011 Barcelona (Spain)	Facilitating access and participation: mathematical practices inside and outside the classroom /
	Faciliter l'accès et la participation: pratiques mathématiques à l'intérieur et à l'extérieur de la classe
CIEAEM 64 – 2012 Rhodes (Greece)	Mathematics education and democracy: learning and teaching practices /
	Education en mathématiques et démocratie: les pratiques d'enseignement et d'apprentissage
CIEAEM 65 – 2013 Torino (Italy)	Mathematics education in a globalized environment /
	L'enseignement des mathématiques dans un environnement globalisé
CIEAEM 66 – 2014 Lyon (France)	Mathematics and realities /
	Mathématiques et réalités
CIEAEM 67 – 2015 Valle d'Aosta (Italy)	Teaching and learning mathematics: resources and obstacles /
	Enseigner et apprendre les mathématiques: ressources et obstacles

Appendix B: Presidents of the Commission Internationale pour l'Etude et l'Amélioration de l'Enseignement des Mathématiques

The following persons acted as Presidents of CIEAEM:

Caleb Gattegno (UK)	1950–1952 (Directeur de la rencontre)
Gustave Choquet (France)	1952–1963
Georges Papy (Belgium)	1963–1970
Anna Zofia Krygowska (Poland)	1970–1975
Claude Gaulin (Canada)	1975–1979
Emma Castelnuovo (Italy)	1979–1981
Stefan Turnau (Poland)	1981–1982
Dieter Lunkenbein (Canada)	1982–1984
Hans Freudenthal (The Netherlands)	1984–1985
Michele Pellerey (Italy)	1985–1988
Izzie Weinzweig (USA)	1988–1993
Lucia Grugnetti (Italy)	1993–1997
Christine Keitel (Germany)	1997–2003
Juliana Szendrei (Hungary)	2003–2007
Corinne Hahn (France)	2007–2014
Uwe Gellert (Germany)	2014–

References

Abbott, A. (2012). It's your money they're after! Using advertising flyers to teach percentages. In A. Hector-Mason & D. Coben (Eds.), *Proceedings of the 19th international conference on adults learning mathematics* (pp. 175–179). Hamilton: The National Centre of Literacy & Numeracy for Adults, University of Waikato.

Albarracín, L. (2011). Sobre les estratègies de resolució de problemes d'estimació de magnituds no abastables. Unpublished PhD thesis, Universitat Autònoma de Barcelona.

Albarracín, L., & Gorgorió, N. (2012). Inconceivable magnitude estimation problems: An opportunity to introduce modelling in secondary school. *Journal of Mathematical Modelling and Application, 1*(7), 20–33.

Albarracín, L., & Gorgorió, N. (2014). Devising a plan to solve Fermi problems involving large numbers. *Educational Studies in Mathematics, 86*(1), 79–96.

Aldon, G. (2010). Handheld calculators between instrument and document. *ZDM – The International Journal on Mathematics Education, 42*(7), 733–745.

Aldon, G. (2011). *Interactions didactiques dans la classe de mathématiques en environnement numérique: construction et mise à l'épreuve d'un cadre d'analyse exploitant la notion d'incident*. Unpublished PhD thesis, Université de Lyon, Lyon. https://tel.archives-ouvertes.fr/tel-00679121

Aldon, G. (2014). Didactic incidents: A way to improve the professional development of mathematics teachers. In A. Clark-Wilson, O. Robutti, & N. Sinclair (Eds.), *The mathematics teacher in the digital era: An international perspective on technology focused professional development* (pp. 319–343). Dordrecht: Springer.

Aldon, G., Cahuet, P.-Y., Durand-Guerrier, V., Front, M., Krieger, D., Mizony, M., & Tardy, C. (2010). *Expérimenter des problèmes de recherche innovants en mathématiques à l'école (CD-Rom)*. Lyon: INRP.

Allexsaht-Snider, M., & Buxton, C. (2011). *Engaging middle school students and families together in bilingual science learning: Can we challenge rising anti-immigrant discourses and open pathways to postsecondary learning?* Paper presented at the European Network about Parents in Education (ERNAPE) conference, Milano, Italy.

Almeida, M. E., & Prado, M. E. (2009a). *Formação de Educadores para o uso dos Computadores Portáteis: Indicadores de Mudança na Prática e no Currículo. Proceedings of the VI Conferência Internacional de Tecnologias de Informação e Comunicação na Educação*. Portugal: Braga.

Almeida, M. E., & Prado, M. E. (2009b). *Formação de Educadores e o Laptop Educacional: Uma experiência vivenciada no Projeto-UCA de Tocantins*. http://www.uca.gov.br/institucional/downloads/estudoDeCasoTO_1.pdf. Accessed 15 July 2013.

Almeida, M. E., & Valente, J. A. (2011). *Tecnologias e currículo: trajetórias convergentes ou divergentes?* São Paulo: Paulis.
Alonso, W., & Starr, P. (Eds.). (1987). *The politics of numbers*. New York: Russell Sage.
Alrø, H., & Skovsmose, O. (2002). *Dialogue and learning in mathematics education: Intention, reflection and critique*. Dordrecht: Kluwer.
Alrø, H., Skovsmose, O., & Valero, P. (2008). Inter-viewing foregrounds: Students' motives for learning in a multicultural setting. In M. César & K. Kumpulainen (Eds.), *Social interactions in multicultural settings* (pp. 13–37). Rotterdam: Sense.
Anghileri, J. (2006). Scaffolding practices that enhance mathematics learning. *Journal of Mathematics Teacher Education, 9*(1), 33–52.
Annarella, L. A. (1992). *Creative drama in the classroom* (ERIC Document Reproduction Service ED391206).
Appelbaum, P. (1995). *Popular culture, educational discourse, and mathematics*. New York: SUNY Press.
Appelbaum, P. (2000). Eight critical points for mathematics. In D. Weil & H. Anderson (Eds.), *Perspectives in critical thinking* (pp. 41–56). New York: Peter Lang.
Appelbaum, P. (2004). Mathematics education. In J. Kincheloe & D. Weil (Eds.), *Critical thinking and learning: An encyclopedia for parents and teachers* (pp. 307–312). Westport: Greenwood.
Appelbaum, P. (2008a). *Embracing mathematics*. London: RoutledgeFalmer.
Appelbaum, P. (2008b). *Embracing mathematics: On becoming a teacher and changing with mathematics*. London: Taylor & Francis.
Appelbaum, P. (2008c). *Embracing mathematics: On becoming a teacher and changing with mathematics*. New York: Routledge.
Appelbaum, P. (2009). Taking action: Curricular organization for effective teaching and learning in mathematics. *For the Learning of Mathematics, 29*(2), 38–43.
Appelbaum, P. (2011). Carnival of the uncanny. In E. Malewski & N. Jaramillo (Eds.), *Epistemologies of ignorance and studies of limits in education* (pp. 221–239). New York: IAP.
Appelbaum, P. (2012). *The shape and feel of "participation": Flipping public spaces, performing democracy and playing with the exteriority of thought*. Vancouver: American Association for the Advancement of Curriculum Studies. April 10–13.
Appelbaum, P., & Dávila, E. (2007). Math education and social justice: Gatekeepers, politics, and teacher agency. *Philosophy of Mathematics Education Journal, 22*.
Apperly, I. A., Samson, D., & Humphreys, G. W. (2005). Domain-specificity and theory of mind: Evaluating evidence from neuropsychology. *Trends in Cognitive Science, 9*, 572–577.
Apperly, I. A., Samson, D., Chiavarino, C., Bickerton, W., & Humphreys, G. W. (2007). Testing the domain-specificity of a theory of mind deficit in brain-injured patients: Evidence for consistent performance on non-verbal, "reality unknown" false belief and false photograph tasks. *Cognition, 103*, 300–321.
Apple, M. (2004). *Ideology and curriculum*. New York: Routledge.
Appleton, K. (2003). How do beginning primary school teachers cope with science? Toward an understanding of science teaching practice. *Research in Science Education, 33*(1), 1–25.
Araújo, J. L. (2007). Relação entre matemática e realidade em algumas perspectivas de modelagem matemática na educação matemática. In J. C. Barbosa, A. D. Caldeira, & J. L. Araújo (Eds.), *Modelagem matemática na educação matemática Brasileira: pesquisas e práticas educacionais* (pp. 17–32). Recife: SBEM.
Arcavi, A. (2002). The everyday and the academic in mathematics. In M. Brenner & J. Moschkovich (Eds.), *Everyday and academic mathematics in the classroom* (pp. 12–29). Reston: NCTM.
Ärlebäck, J. B. (2009). On the use of realistic Fermi problems for introducing mathematical modelling in school. *The Montana Mathematics Enthusiast, 6*(3), 331–364.
Ärlebäck, J. B. (2011). Exploring the solving process of groups solving realistic Fermi problem from the perspective of the anthropological theory of didactics. In M. Pytlak, E. Swoboda, & T. Rowland (Eds.), *Proceedings of the seventh congress of the European society for research in mathematics education* (pp. 1010–1019). Rzeszów: University of Rzeszów.

References

Armstrong, B. E., & Bezuk, N. S. (1995). Multiplication and division of fractions: The search for meaning. In J. T. Sowder & B. P. Schappelle (Eds.), *Providing a foundation for teaching mathematics in the middle grades* (pp. 85–119). Albany: SUNY Press.

Aronowitz, S., & De Fazio, W. (1997). The new knowledge work. In A. H. Halsey, H. Lauder, P. Brown, & A. S. Wells (Eds.), *Education: Culture, economy, and society* (pp. 193–206). Oxford: Oxford University Press.

Aronowitz, S., & Giroux, H. (1991). *Postmodern education: Politics, culture and social criticism*. Minneapolis: University of Minnesota Press.

Artigue, M. (2000). *Instrumentation issues and the integration of computer technologies into secondary mathematics teaching*. Proceedings of the annual meeting of GDM, Potsdam. http://webdoc.sub.gwdg.de/ebook/e/gdm/2000. Accessed 3 June 2014.

Artigue, M. (2002). Learning mathematics in a CAS environment: The genesis of a reflection about instrumentation and the dialectics between technical and conceptual work. *International Journal of Computers for Mathematical Learning, 7*, 245–274.

Artigue, M., & Bardini, C. (2010). New didactical phenomena prompted by TI-NSPIRE specificities: The mathematical component of the instrumentation process. In V. Durand-Guerrier, S. Soury-Lavergne, & F. Arzarello (Eds.), *Proceedings of the sixth congress of the European society for research in mathematics education* (pp. 1171–1180). Lyon: Institut National de Recherche Pédagogique.

Arzarello, F., & Robutti, O. (2010). Multimodality in multi-representational environments. *ZDM – The International Journal on Mathematics Education, 42*(7), 715–731.

Arzarello, F., Paola, D., Robutti, O., & Sabena, C. (2009). Gestures as semiotic resources in the mathematics classroom. *Educational Studies in Mathematics, 70*(2), 97–109.

Arzarello, F., Bikner, A., & Sabena, C. (2010). Complementary networking: Enriching understanding. In V. Durand-Guerrier, S. Soury-Lavergne, & F. Arzarello (Eds.), *Proceedings of the 6th congress of the European society for research in mathematics education* (pp. 1545–1554). Lyon: INRP.

Ashton, D. N., & Sung, J. (1997). Education, skill formation and economic development: The Singaporean approach. In A. H. Halsey, H. Lauder, P. Brown, & A. S. Wells (Eds.), *Education: Culture, economy, and society* (pp. 207–218). Oxford: Oxford University Press.

Astington, J. (1991). Intention in the child's theory of mind. In D. Frye & C. Moore (Eds.), *Children's theories of mind: Mental states and social understanding*. Hillsdale: Lawrence Erlbaum.

Atweh, B., Bleicher, R. E., & Cooper, T. J. (1998). The construction of the social context of mathematics classrooms: A sociolinguistic analysis. *Journal for Research in Mathematics Education, 29*(1), 63–82.

Atweh, B., Graven, M., Secada, W., & Valero, P. (Eds.). (2011). *Mapping equity and quality in mathematics education*. New York: Springer.

Aubert, A. (2011). Moving beyond social exclusion through dialogue. *International Studies in Sociology of Education, 21*(1), 63–75.

Auslander, P. (1999). *Liveness: Performance in a mediatized culture*. New York: Psychology Press.

Austin, J. L. (1971). *Palabras y acciones: cómo hacer cosas con palabras*. Barcelona: Paidós.

Bailin, S. (1998). Critical thinking and drama education. *Research in Drama Education, 3*(2), 145–153.

Bailin, S., Case, R., Coombs, J., & Daniels, L. (1999). Conceptualizing critical thinking. *Journal of Curriculum Studies, 31*(3), 285–302.

Bairral, M. A. (2007). *Discurso, interação e aprendizagem matemática em ambientes virtuais de aprendizagem*. Seropédica: Editora Universidade Rural.

Bairral, M. A., & Powell, A. B. (2013). Interlocution among problem solvers collaborating online: A case study with prospective teachers. *Pro-Posições, 24*(1), 1–16.

Bakker, A. (2014). Special issue: Characterising and developing vocational mathematical knowledge. *Educational Studies in Mathematics, 86*(2), 151–156.

Ball, D. L. (1993). With an eye on the mathematical horizon: Dilemmas of teaching elementary school mathematics. *The Elementary School Journal, 93*(4), 373–397.

Balli, S. J. (1998). When mom and dad help: Student reflections on parent involvement with homework. *Journal of Research and Development in Education, 31*(3), 142–146.

Balzano, E. (2011). Science laboratory activities for kids and parents in Naples. In M. Pieri, A. Pepe, & L. Addimando (Eds.), *Home, school and community: A partnership for a happy life?* (pp. 24–25). Milano: I Libri di Emil.

Barbosa, J. C. (2001). *Modelagem Matemática: concepções e experiências de futuros professores.* Unpublished PhD thesis, Universidade Estadual Paulista (UNESP Rio Claro), Instituto de Geociências e Ciências Naturais, Rio Claro.

Barbosa, J. C. (2006a). Mathematical modelling in classroom: A socio-critical and discursive perspective. *Zentralblatt für Didaktik der Mathematik, 38*(3), 293–301.

Barbosa, J. C. (2006b). Mathematical modelling in classroom: A critical and discursive perspective. *ZDM – The International Journal on Mathematics Education, 38*(3), 293–301.

Barnett, J. H., Lodder, J., & Pengelley, D. (2014). The pedagogy of primary historical sources in mathematics: Classroom practice meets theoretical frameworks. *Science & Education, 23*(1), 7–27.

Baron-Cohen, S., Leslie, A. M., & Frith, U. (1985). Does the autistic child have a "theory of mind"? *Cognition, 21*(1), 37–46.

Baroody, A. J., & Hume, J. (1991). Meaningful mathematics instruction: The case of fractions. *Remedial and Special Education, 12*, 54–68.

Barry, B. (1995). *Justice as impartiality.* Oxford: Clarendon.

Bartolini Bussi, M., & Boni, M. (2009). The early construction of mathematical meanings: Learning positional representation of numbers. In O. A. Barbarin & B. H. Wasik (Eds.), *The handbook of child developmental and early education: Research to practice* (pp. 455–477). New York: Guilford.

Bateson, G. (1975). Some components of socialization for trance. *Ethos – Journal of the Society for Psychological Anthropology, 3*(2), 143–155.

Bauchspies, W. (2009). Potentials, actuals and residues: Entanglements of culture and subjectivity. *Subjectivity, 28*, 229–245.

Baudelot, C., & Establet, R. (1971). *L'école capitaliste en France (cahier 213).* Paris: Maspero.

Bauman, Z. (1998). *Globalization: The human consequences.* New York: Columbia.

Bazerman, C. (1997). The life of genre, the life in the classroom. In W. Bishop & H. Ostrum (Eds.), *Genre and writing* (pp. 19–26). Portsmouth: Boynton/Cook.

Bazzini, L. (2007). The mutual influence of theory and practice in mathematics education: Implications for research and teaching. *ZDM – The International Journal on Mathematics Education, 39*, 119–125.

Bazzini, L., Sabena, C., & Villa, B. (2009). Meaningful context in mathematical problem solving: A case study. In *Proceedings of the 3rd international conference on science and mathematics education* (pp. 343–351). Penang: CoSMEd.

Beales, A. C. F. (1956). El desarrollo histórico de los "métodos activos" en educación. *Revista Española de Pedagogía, 55.*

Bebell, D., & O'Dwyer, L. M. (2010). Educational outcomes and research from 1:1 computing settings. *The Journal of Technology, Learning, and Assessment, 9*(1), 5–16.

Beeli-Zimmermann, S., & Hollenstein, A. (2011). Financial literacy of microcredit clients: Results of a qualitative exploratory study and its implications for educational schemes. In T. Maguire, J. J. Keogh, & J. O'Donoghue (Eds.), *Proceedings of the 18th international conference on adults learning mathematics* (pp. 26–36). Adults Learning Mathematics – A Research Forum.

Bell, E. T. (1993). *Men of mathematics (Vol. 2).* Heraklion: Crete University Press.

Berg, P., & Dasmann, R. (1990). Reinhabiting California. In C. Van Andruss & J. Plant (Eds.), *Home! A bioregional reader.* Philadelphia: New Society.

Berliner, P. (1994). *Thinking in jazz.* Chicago: University of Chicago Press.

Bernstein, B. (1970). Education cannot compensate for society. *New Society, 15*(387), 344–347.

Bernstein, B. (1971). *Class, codes and control. Volume 1: Theoretical studies towards a sociology of language.* London: Routledge.

Bernstein, B. (1993). *Clases, códigos y control. La estructura del discurso pedagógico*. Madrid: Morata.

Bernstein, B. (1996). *Pedagogy, symbolic control and identity: Theory, research, critique*. London: Taylor & Francis.

Bernstein, B. (1999). Vertical and horizontal discourse: An essay. *British Journal of Sociology of Education, 20*(2), 157–173.

Bernstein, B. (2000). *Pedagogy, symbolic control and identity: Theory, research, critique* (Revth ed.). Lanham: Rowman & Littlefield.

Berry, W. (1977). *The unsettling of America: Culture & agriculture*. San Francisco: Sierra Club.

Bertau, M. C. (2007). On the notion of voice: An exploration from a psycholinguistic perspective with developmental implications. *International Journal of Dialogical Science, 2*(1), 133–161.

Best, S., & Kellner, D. (2003). Contemporary youth and the postmodern adventure. *Review of Education, Pedagogy, and Cultural Studies, 25*(2), 75–93.

Bezemer, J. (2008). Displaying orientation in the classroom: Students' multimodal responses to teacher instructions. *Linguistics and Education, 19*(2), 166–178.

Bezuk, N. S., & Bieck, M. (1993). Current research on rational numbers and common fractions: Summary and implications for teachers. In D. T. Owens (Ed.), *Research ideas for the classroom: Middle grades mathematics* (pp. 118–136). New York: Macmillan.

Biesta, G. (1998). Say you want a revolution … Suggestions for the impossible future of critical pedagogy. *Educational Theory, 48*(4), 499–510.

Biesta, G. (2005). Against learning: Reclaiming a language for education in an age of learning. *Nordisk Pædagogik, 25*(1), 54–55.

Biesta, G., & Egéa-Kuehne, D. (2001). *Derrida & education*. London: Routledge.

Biggs, J. (1987). *Student approaches to learning and studying*. Hawthorn: Australian Council for Educational Research.

Bishop, A. J. (1988). *Mathematical enculturation: A cultural perspective on mathematics education*. Dordrecht: Kluwer.

Bishop, A. J. (1999). Mathematics teaching and values education: An intersection in need of research. *Zentralblatt für Didaktik der Mathematik, 31*(1), 1–4.

Bishop, A. J., Clements, K., Keitel, C., Kilpatrick, J., & Laborde, C. (Eds.). (1996). *International handbook of mathematics education*. Dordrecht: Kluwer.

Bishop, A. J., Clements, M. A., Keitel, C., Kilpatrick, J., & Leung, F. K. S. (Eds.). (2003). *Second international handbook of mathematics education*. Dordrecht: Kluwer.

Bissel, J. (2002). Teachers' construction of space and place: A study of school architectural design as a context of secondary school teachers' work. *Dissertation Abstracts International, 64*(02), 311A. (UMI No. 3082110).

Black, L., Mendick, H., & Solomon, Y. (2009). *Mathematical relationships in education: Identities and participation*. New York: Routledge.

Blackman, S. (2005). Youth subcultural theory: A critical engagement with the concept, its origins and politics, from the Chicago school to postmodernism. *Journal of Youth Studies, 8*(1), 1–20.

Blair, C., & Razza, R. P. (2007). Relating effortful control, executive function and false belief understanding to emerging math and literacy ability in kindergarten. *Child Development, 78*, 647–663.

Blake, P. R., & Rand, D. G. (2010). Currency value moderates equity preference among young children. *Evolution and Human Behavior, 31*, 210–218.

Block, A. (2003). They sound the alarm immediately: Anti-intellectualism in teacher education. *Journal of Curriculum Theorizing, 19*(1), 33–46.

Blömeke, S., Hsieh, F.-J., & Schmidt, W. H. (2013). Introduction to this special issue. *International Journal of Science and Mathematics Education, 11*, 789–793.

Blommaert, J., & Huang, A. (2010). Semiotic and spatial scope: Towards a materialist semiotics (Working papers in urban languages and linguistics, 62). London: King's College.

Blum, W., Galbraith, P., Henn, H., & Niss, M. (Eds.). (2007). *Modelling and applications in mathematics education: The 14th ICMI study*. New York: Springer.

Boaler, J. (1998). Nineties girls challenge eighties stereotypes: Updating gender perspectives. In C. Keitel (Ed.), *Social justice and mathematics education: Gender, class, ethnicity and the politics of schooling* (pp. 278–293). Berlin: Freie Universität Berlin.

Boaler, J. (2002). The development of disciplinary relationships: Knowledge, practice and identity in mathematics classrooms. *For the Learning of Mathematics, 22*(1), 42–47.

Boaler, J., & Humphreys, C. (2005). *Connecting mathematical ideas: Middle school video cases to support teaching and learning*. Portsmouth: Heinemann.

Boekaerts, M., & Simons, P. R. J. (2003). *Leren en instructie, psychologie van de leerling en het leerproces*. Assen: Van Gorcum.

Bolton, G. (1985). Changes in thinking about drama in education. *Theory Into Practice, 24*(3), 151–157.

Bonotto, C. (2013). Artifacts as sources for problem-posing activities. *Educational Studies in Mathematics, 83*(1), 37–55.

Borasi, R. (1984). Some reflections on and criticisms of the principle of learning concepts by abstraction. *For the Learning of Mathematics, 4*(3), 14–18.

Boreham, N. (2004). Orienting the work-based curriculum towards work process knowledge: A rationale and a German case study. *Studies in Continuing Education, 26*(2), 209–227.

Bourdieu, P. (1984). *La distinction: Critique social du jugement*. Paris: Minuit.

Bourdieu, P., & Passeron, J. C. (1970). *La reproduction: éléments pour une théorie du système d'enseignement*. Paris: Minuit.

Bourne, J. (1992). *Inside a multilingual primary classroom. A teacher, children and theories at work*. Unpublished PhD thesis, University of Southampton.

Bowles, S., & Gintis, H. (1976). *Schooling in capitalist America: Educational reform and the contradictions of economic life*. New York: Basic Books.

Boyd, R. (2006). The puzzle of human sociality. *Science, 314*, 1553.

Boyd, R., & Richerson, P. J. (2009). Culture and the evolution of human cooperation. *Philosophical Transactions of the Royal Society B, 364*, 3281–3288.

Boyer, C. B., & Merzbach, U. C. (1997). *History of mathematics*. Athens: Pnevmatikos.

Brigham, F. J., Wilson, R., Jones, E., & Moisio, M. (1996). Best practices: Teaching decimals, fractions, and percents to students with learning disabilities. *LD Forum, 21*, 10–15.

Brosnan, S. F., Schiff, H. C., & de Waal, F. B. M. (2005). Chimpanzees' (Pan troglodytes) reactions to inequity during experimental exchange. *Proceedings of the Royal Society of London B, 1560*, 253–258.

Brousseau, G. (1978). Étude locale des processus d'acquisitions scolaires. *Enseignement élémentaire des mathématiques, 18*, 7–21.

Brousseau, G. (1997a). *Theory of didactical situations in mathematics: Didactique des mathématiques, 1970–1990*. N. Balacheff, M. Cooper, R. Sutherland & V. Warfield (Eds. & Trans.). Dordrecht: Kluwer.

Brousseau, G. (1997b). *Theory of didactical situations in mathematics*. Dordrecht: Kluwer.

Brown, T. (2011). *Mathematics education and subjectivity: Cultures and cultural renewal*. London: Springer.

Bruner, J. (1986). *Actual minds, possible worlds*. Cambridge, MA: Harvard University Press.

Bunt, L., Jones, P., & Bedient, J. (1981). *The historical roots of elementary mathematics*. Englewood Cliffs: Prentice-Hall.

Cai, J., Moyer, J. C., & Wang, N. (1997). *Parental roles in students' learning of mathematics: An exploratory study*. Paper presented at the annual meeting of the American Educational Research Association (AERA), Chicago, IL.

Callaghan, T., Rochat, P., Lillard, A., Claux, M. L., Odden, H., Itakura, S., et al. (2005). Synchrony in the onset of mental-state reasoning: Evidence from five cultures. *Psychological Science, 16*(5), 378–384.

Camelo, F., Mancera, G., Romero, J., García, G., & Valero, P. (2010). The importance of the relation between the socio-political context, interdisciplinarity and the learning of the mathematics. In U. Gellert, E. Jablonka, & C. Morgan (Eds.), *Proceedings of the 6th international mathematics education and society conference* (pp. 199–208). Berlin: Freie Universität Berlin.

Canavarro, A. P., Oliveira, H., & Menezes, L. (2012). A framework for mathematics inquiry-based classroom practice: The case of Celia. In *Pre-proceedings of 12th international congress on mathematical education* (pp. 4137–4146). Seoul: ICMI - International Commission on Mathematical Instruction.

Carpenter, T. P., Fennema, E., Peterson, P. L., Chiang, C. P., & Loef, M. (1989). Using knowledge of children's mathematics thinking in classroom teaching: An experimental study. *American Educational Research Journal, 26*(4), 499–531.

Carr, M. (1996). *Motivation in mathematics*. Cresskill: Hampton.

Carr, P., & Kefalas, M. (2009). *Hollowing out the middle: The rural brain drain and what it means for America*. Boston: Beacon.

Carraher, T. N., Carraher, D. W., & Schliemann, A. D. (1985). Mathematics in the streets and in schools. *British Journal of Developmental Psychology, 3*(1), 21–29.

Carter, H. L. (1986). Linking estimation to psychological variables in the early years. In H. L. Schoen & M. J. Zweng (Eds.), *Estimation and mental computation* (pp. 74–81). Reston: NCTM.

Castells, M. (1998). *The information age: Economy, society and culture* (Vol. III, end of millennium (III)). Oxford: Blackwell.

Castells, M. (2004). *The power of identity, the information age: Economy, society and culture* (2nd ed., Vol. II). Cambridge, MA: Blackwell.

Castells, M. (2012). *Redes de indignación y esperanza. Los movimientos sociales en la era de Internet*. Madrid: Alianza.

Castro, A. L. (2011). *Tecnologias Digitais da Informação e Comunicação no Ensino de funções Quadráticas: Contribuições das diferentes representações*. Unpublished Master thesis, Universidade Bandeirante de São Paulo, São Paulo. Available in: http://www.matematicaepraticadocente.net.br/pdf/teses_dissertacoes/dissertacao_AnnaLuisadeCastro_2011_Bette.pdf. Access: 24 Feb 2015

Catsambis, S. (2001). Expanding knowledge of parental involvement in children's secondary education: Connections with high school seniors' academic success. *Social Psychology of Education, 5*(2), 149–177.

Catsambis, S., & Beveridge, A. A. (2001). Does neighbourhood matter? Family, neighborhood, and school influences on eighth grade mathematics achievement. *Sociological Focus, 34*(4), 435–457.

Catterall, J. (2007). Enhancing peer conflict resolution skills through drama: An experimental study. *Research in Drama Education: The Journal of Applied Theatre and Performance, 12*(2), 163–178.

Ceibal. (2010). *Plan Ceibal*. http://www.ceibal.edu.uy/. Accessed 15 July 2013.

Cengiz, N., Kline, K., & Grant, T. J. (2011). Extending students' mathematical thinking during whole-group discussions. *Journal of Mathematics Teacher Education, 14*(5), 355–374.

Chapman, O. (2006). Classroom practices for context of mathematics word problems. *Educational Studies in Mathematics, 62*(2), 211–230.

Chapman, O. (2013). High school mathematics teachers' inquiry-oriented approaches to teaching algebra. *Quadrante, 22*(2), 6–28.

Chapman, O., & Heater, B. (2010). Understanding change through a high school mathematics teacher's journey to inquiry-based teaching. *Journal of Mathematics Teacher Education, 13*(6), 445–458.

Chassapis, D. (1997). The social ideologies of school mathematics applications: A case study of elementary school textbooks. *For the Learning of Mathematics, 17*(3), 24–26.

Chaviaris, P., Stathopoulou, C., & Gana, E. (2011). A socio-political analysis of mathematics teaching in the classroom. *Quaderni di Ricerca in Didattica, 22*(Supplemento 1), 233–237.

Chevallard, Y. (1985). *La transposition didactique*. Grenoble: La Pensée Sauvage.

Chevallard, Y. (1999). L'analyse des pratiques enseignantes en théorie anthropologique du didactique. *Recherches en Didactique des Mathématiques, 19*(2), 221–266.

Chouliaraki, L. (1999). Media discourse and national identity: Death and myth in a news broadcast. In R. Wodak & C. Ludwig (Eds.), *Challenges in a changing world: Issues in critical discourse analysis* (pp. 37–62). Vienna: Passagen.

Chouliaraki, L. (2008). Discourse analysis. In T. Bennett & J. Frow (Eds.), *The SAGE handbook of cultural analysis* (pp. 674–696). London: Sage.

Chouliaraki, L. (2010). Discourse and mediation. In S. Allan (Ed.), *Rethinking communication: Keywords in communication research*. Hampton: Cresskill.

Christenson, S. L., & Sheridan, S. M. (2002). *Schools and families: Creating essential connection for learning*. New York: Guilford.

Christie, F. (2005). *Classroom discourse analysis*. London: Continuum.

Chronaki, A. (2005). Learning about "learning identities" in the school arithmetic practice: The experience of two young minority Gypsy girls in the Greek context of education. *European Journal of Psychology of Education, 20*(1), 61–74.

Chronaki, A. (2008). The teaching experiment: Studying the process of teaching and learning. In B. Svolopoulos (Ed.), *Connections of educational research and practice*. Atrapos: Athens.

Chronaki, A. (2009). An entry to dialogicality in the maths classroom: Encouraging hybrid learning identities. In M. César & K. Kumpulainen (Eds.), *Social interactions in multicultural settings* (pp. 117–143). Rotterdam: Sense.

Chronaki, A. (2011). Troubling essentialist identities: Performative mathematics and the politics of possibility. In M. Kontopodis, C. Wulf, & B. Fichtner (Eds.), *Children, development and education: Cultural, historical and anthropological perspectives* (pp. 207–227). Dordrecht: Springer.

Civil, M. (1999). Parents as resources for mathematical instruction. In M. van Groenestijn & D. Coben (Eds.), *Mathematics as part of lifelong learning (Proceedings of the fifth international conference of adults learning mathematics – A research forum)* (pp. 216–222). London: Goldsmiths College.

Civil, M. (2001). Mathematics for parents: Issues of pedagogy and content. In L. Johansen & T. Wedege (Eds.), *Proceedings of the eighth international conference of adults learning mathematics – A research forum* (pp. 60–67). Roskilde: Centre for Research in Learning Mathematics.

Civil, M., & Bernier, E. (2006). Exploring images of parental participation in mathematics education: Challenges and possibilities. *Mathematical Thinking and Learning, 8*(3), 309–330.

Cline-Cohen, P. (1999). *A calculating people: The spread of numeracy in early America*. New York: Diane.

CNCS (Corporation for National and Community Service). (2013). *National service learning clearinghouse*. http://www.servicelearning.org/what-is-service-learning. Accessed 4 July 2013.

Cobb, P., & Yackel, E. (1998). A constructivist perspective on the culture of the mathematics classroom. In F. Seeger, J. Voigt, & U. Waschescio (Eds.), *The culture of the mathematics classroom* (pp. 158–190). Cambridge: Cambridge University Press.

Cobb, P., Wood, T., & Yackel, E. (1991). A constructivist approach to second grade mathematics. In E. von Glasersfeld (Ed.), *Radical constructivism in mathematics education* (pp. 157–176). Dordrecht: Kluwer.

Cobb, P., Stephan, M., McClain, K., & Gravemeijer, K. (2001). Participating in classroom mathematical practices. *The Journal of the Learning Sciences, 10*(1&2), 113–163.

Cobb, P., McClain, K., Silva Lamberg, T. D., & Dean, C. (2003). Situating teachers' instructional practices in the institutional setting of the school and district. *Educational Researcher, 32*(6), 13–25.

Coben, D., & Weeks, K. (2012). Behind the headlines: Authentic teaching, learning and assessment of competence in medication dosage calculation problem solving in and for nursing. In A. Hector-Mason & D. Coben (Eds.), *Proceedings of the 19th international conference on adults learning mathematics* (pp. 39–55). Hamilton: The National Centre of Literacy & Numeracy for Adults, University of Waikato.

Coben, D., O'Donoghue, J., & FitzSimons, G. E. (Eds.). (2000). *Perspectives on adults learning mathematics: Research and practice*. Dordrecht: Kluwer.

References

Conley, D. (2010). *College and career ready*. San Francisco: Jossey-Bass.
Conquergood, D. (2002). Performance studies: Interventions and radical research. *TDR/The Drama Review, 46*(2), 145–146.
Coon, R. C., Lane, I. M., & Lichtman, R. J. (1974). Sufficiency of reward and allocation behavior: A developmental study. *Human Development, 17*(4), 301–313.
Coray, D., Furinghetti, F., Gispert, H., & Schubring, G. (Eds.). (2003). *One hundred years of L'Enseignement Mathématique: Moments of mathematics education in the twentieth century*. Geneva: L'Enseignement Mathématique.
Corbett, M. (2007). *Learning to leave: The irony of schooling in a coastal community*. Halifax: Fernwood.
Corôa, R. P. (2006). *Saberes construídos pelos professores de matemática em sua prática docente na educação de jovens e adultos*. Unpublished Master thesis, UFPA, Belém.
Cosmides, L. (1989). The logic of social exchange: Has natural selection shaped how humans reason? Studies with the Wason selection task. *Cognition, 31*, 187–276.
Côté, P., Dumoulin, C., & Tremblay, N. (2011). Family-school collaboration: Activities implemented by elementary school teachers involved with high-risk students from underprivileged class. In M. Pieri, A. Pepe, & L. Addimando (Eds.), *Home, school and community: A partnership for a happy life?* (pp. 38–40). Milano: I Libri di Emil.
Crump, T. (1990). *The anthropology of number*. Cambridge: Cambridge University Press.
Cuevas, G., & Driscoll, M. (1993). *Reaching all students with mathematics*. Reston: NCTM.
Cwikla, J. (2014). Can kindergartners do fractions? *Teaching Children Mathematics, 20*(6), 354–364.
D'Ambrosio, U. (1986). *Da realidade à ação – reflexões sobre a Educação Matemática*. São Paulo: Sammus.
Damlamian, A., Rodrigues, J. F., & Sträßer, R. (Eds.). (2013). *Educational interfaces between mathematics and industry: Report on an ICMI-ICIAM-Study*. New York: Springer.
Damon, W. (1975). Early conceptions of positive justice as related to the development of logical operations. *Child Development, 46*, 301–312.
Damon, W. (1977). *The social world of the child*. San Francisco: Jossey-Bass.
Davis, B., & Simmt, E. (2006). Mathematics-for-teaching: An ongoing investigation of the mathematics that teachers (need to) know. *Educational Studies in Mathematics, 61*(3), 293–319.
de Abreu, G. (1993). *The relationship between home and school mathematics in a farming community in rural Brazil*. Unpublished PhD thesis, University of Cambridge, UK.
de Abreu, G., Bishop, A., & Presmeg, N. (2002a). Mathematics learners in transition. In G. de Abreu, A. Bishop, & N. Presmeg (Eds.), *Transitions between contexts of mathematical practices* (pp. 7–22). Dordrecht: Kluwer.
de Abreu, G., Cline, T., & Shamsi, T. (2002b). Exploring ways parents participate in their children's school mathematical learning: Cases studies in multiethnic primary schools. In G. de Abreu, A. Bishop, & N. Presmeg (Eds.), *Transitions between contexts of mathematical practices* (pp. 123–148). Dordrecht: Kluwer.
De La Roche, E. (1993). *Drama, critical thinking and social issues* (ERIC Document Reproduction Service ED379172)
Derman-Sparks, L., & Ramsey, P. G. (2006). *What if all the kids are white: Anti-bias multicultural education with young children*. New York: Teacher College Press.
Desforges, C., & Abouchnar, A. (2003). *The impact of parental involvement, parental support and family education on pupil achievements and adjustment: A literature review*. London: Department for Educational Skills.
Dewey, J. (1979). *Experiência e Educação* (3rd ed.). São Paulo: Morata.
Dewey, J. (2008). *Democracy and education: An educational classic*. Radford: Wilder.
DeYoung, A. J. (1987). The status of American rural education research: An integrated review and commentary. *Review of Educational Research, 57*(2), 123–148.
DeYoung, A. J. (Ed.). (1991). *Rural education: Issues and practice*. New York: Garland.
Diamond, L. J., & Plattner, M. F. (2006). *Electoral systems and democracy*. Baltimore: Johns Hopkins University Press.

Dias, M. C., Tomás, C., & Gama, A. (2011). School achievement in risk situations: The impact of government and non-government programmes in the fight against social exclusion. In M. Pieri, A. Pepe, & L. Addimando (Eds.), *Home, school and community: A partnership for a happy life?* (pp. 35–36). Milano: I Libri di Emil.

Díez-Palomar, J. (2004). *La enseñanza de las matemáticas en la educación de personas adultas: un modelo dialógico*. Unpublished PhD thesis, Universitat de Barcelona.

Díez-Palomar, J. (2011). Being competent in mathematics: Adult numeracy and common sense. In T. Maguire, J. J. Keogh, & J. O'Donoghue (Eds.), *Proceedings of the 18th international conference on adults learning mathematics* (pp. 90–102). Adults Learning Mathematics – A Research Forum.

Díez-Palomar, J. (2013). *Be a teacher in a global society: Challenges and opportunities drawing on a sociological approach to mathematics teacher training*. Paper presented at CIEAEM 65, Torino, Italy.

Díez-Palomar, J., & Kanes, C. (Eds.). (2012). *Family and community in and out of the classroom: Ways to improve mathematics' achievement*. Bellaterra: UAB Press.

Díez-Palomar, J., & Molina, S. (2009). Contribuciones de la educación matemática de las familias a la formación del profesorado. In M. J. González, M. T. González Astudillo, & J. Murillo (Eds.), *Investigación en Educación Matemática XIII* (pp. 211–225). Santander: SEIEM.

Díez-Palomar, J., Gatt, S., & Racionero, S. (2011). Placing immigrant and minority family and community members at the school's centre: The role of community participation. *European Journal of Education, 46*(2), 184–196.

diSessa, A., Hammer, D., Sherin, B., & Kolpakowski, T. (1991). Inventing graphing: Meta-representational expertise in children. *Journal of Mathematical Behavior, 10*, 117–160.

Dixon, S. (2009). *Digital performance: Doubles, cyborgs and multi-identities in digital performance and cyberculture*. https://smartech.gatech.edu/handle/1853/28545. Accessed 15 Jan 2014.

Dowling, P. (1998). *The sociology of mathematics education: Mathematical myths/pedagogic texts*. London: Routledge.

Duatepe, A., & Ubuz, B. (2004, July 4–17). Drama based instruction and geometry. In M. Niss (Ed.), *Proceedings of 10th international congress on mathematical education*, Copenhagen, Denmark.

Duval, R. (1991). Structure du raisonnement déductif et apprentissage de la démonstration. *Educational Studies in Mathematics, 22*(3), 233–261.

Duval, R. (1993). Registres de représentation sémiotique et fonctionnement cognitif de la pensée. *Annales de Didactique et de Sciences Cognitives, 5*, 37–65.

Duval, R. (1995). *Sémiosis et pensée humaine: Registres sémiotiques et apprentissage intellectuels*. Bern: Peter Lang.

Ebbens, S., & Ettekoven, S. (2007). *Actief leren*. Groningen: Wolters Noordhoff.

Eccles, J. S., Midgley, C., Wigfield, A., Buchanan, C. M., Reuman, D., Flanagan, C., et al. (1993). Development during adolescence: The impact of stage environment fit on young adolescents' experiences in schools and families. *American Psychologist, 48*(2), 90–101.

Education, Audiovisual & Culture Executive Agency [EACEA]. (2011). *Mathematics in education in Europe: Common challenges and national policies*. Brussels: EACEA.

Edwards, L., Ferrara, F., & Moore-Russo, D. (Eds.). (2014). *Emerging perspectives on gesture and embodiment in mathematics*. Charlotte: IAP.

Efthimiou, C. J., & Llewellyn, R. A. (2007). Cinema, Fermi problems and general education. *Physics Education, 42*(3), 253–261.

Emilson, A. (2011). Democracy learning in a preschool context. *International Perspectives on Early Childhood Education and Development, 4*, 157–171.

Empson, S. B. (1999). Equal sharing and shared meaning: The development of fraction concepts in a first-grade classroom. *Cognition and Instruction, 17*, 283–342.

Empson, S. B. (2001). Equal sharing and the roots of fraction equivalence. *Teaching Children Mathematics, 7*, 421.

References

Empson, S. B. (2003). Low-performing students and teaching fractions for understanding: An interactional analysis. *Journal for Research in Mathematics Education, 34*, 305–343.

Empson, S. B., & Levi, L. (2011). *Extending children's mathematics: Fractions and decimals*. Portsmouth: Heinemann.

English, L. D. (2006). Mathematical modeling in the primary school. *Educational Studies in Mathematics, 63*(3), 303–323.

English, L., & Sriraman, B. (2010). Problem solving for the 21st century. In B. Sriraman & L. English (Eds.), *Theories of mathematics education: Seeking new frontiers* (pp. 263–290). Berlin: Springer.

Englund, T. (2006). Deliberative communication: A pragmatist proposal. *Journal of Curriculum Studies, 38*(5), 503–520.

Englund, T. (2011). The potential of education for creating mutual trust: Schools as sites for deliberation. *Educational Philosophy and Theory, 43*(3), 236–249.

Entwisle, D. R., & Alexander, K. L. (1996). Family type and children's growth in reading and math over the primary grades. *Journal of Marriage and Family, 58*, 341–355.

Epstein, J. L. (1991). Effects on student achievement of teacher practices of parent involvement. In S. Silvern (Ed.), *Literacy through family, community, and school interaction* (pp. 261–276). Greenwich: JAI.

Epstein, J. L. (2001). *School, family, and community partnerships: Preparing educators and improving schools*. Boulder: Westview.

Epstein, J. L. (2005). *Developing and sustaining research-based programs of school, family, and community partnerships: Summary of five years of NNPS research*. http://www.csos.jhu.edu/p2000/pdf/Research%20Summary.pdf

Erlwanger, S. H. (1973). Benny's conception of rules and answers in IPI mathematics. *Journal of Children's Mathematical Behavior, 1*(2), 7–26.

Ernest, P. (1991). *The philosophy of mathematics education*. London: Falmer.

Ernest, P. (1994). Social constructivism and the psychology of mathematic education. In P. Ernest (Ed.), *Constructing mathematical knowledge: Epistemology and mathematics education* (pp. 62–72). London: RoutledgeFalmer.

Ernest, P. (1997). *Social constructivism as a philosophy of mathematics*. Albany: SUNY Press.

Ernest, P. (2008, November 21–23). What does the new philosophy of mathematics mean for mathematics education. In *Proceedings of the 25th Panhellenic Congress in mathematics education of Hellenic Mathematical Society*, Volos, Greece.

Esteley, C. B., Villarreal, M. E., & Alagia, H. R. (2010). The overgeneralization of linear models among university students' mathematical productions: A long-term study. *Mathematical Thinking and Learning, 12*(1), 86–108.

European Commission. (2004). *Europe needs more scientists. Report by the high level group on increasing human resources for science and technology in Europe*. Brussels: European Commission.

Evans, J. (1999). Building bridges: Reflections on the problem of transfer of learning in mathematics. *Educational Studies in Mathematics, 39*(1–3), 23–44.

Evans, J. (2011). Students' response to images of mathematics in advertising. In T. Maguire, J. J. Keogh, & J. O'Donoghue (Eds.), *Proceedings of the 18th international conference on adults learning mathematics* (pp. 118–128). Adults Learning Mathematics – A Research Forum.

Evans, J., Wedege, T., & Yasukawa, K. (2013). Critical perspectives on adults' mathematics education. In M. A. Clements, A. J. Bishop, C. Keitel, J. Kilpatrick, & F. K. Leung (Eds.), *Third international handbook of mathematics education* (pp. 203–242). New York: Springer.

Eves, H. (1983). *Great moments in mathematics (after 1650)*. Athens: Trohalia.

Fabre, M. (2010). Problématisation des savoirs. In A. Van Zanten (Ed.), *Dictionnaire pédagogique* (pp. 539–541). Paris: Presses Universitaires de France.

Fairclough, N. (1989). *Language and power*. London: Longman.

Fairclough, N. (1995). *Critical discourse analysis: The critical study of language*. New York: Longman.

Fairclough, N. (2003). *Analyzing discourse: Textual analysis for social research.* New York: Routledge.
Fan, X., & Chen, M. (1999). Academic achievement of rural school students: A multi-year comparison with their peers in suburban and urban schools. *Journal of Research in Rural Education, 15*(1), 31–46.
Farley, K. (2002). Digital dance theatre: The marriage of computers, choreography and techno/human reactivity. *Body, Space and Technology, 3*(1), 39–46.
Fehr, H. F., Bunt, L. N. H., & OECE. (1961). *Mathématiques nouvelles.* Paris: OECE.
Fehr, E., Bernhard, H., & Rockenbach, B. (2008). Egalitarianism in young children. *Nature, 454,* 1079–1083.
Feigenson, L., Dehaene, S., & Spelke, E. (2004). Core systems of number. *Trends in Cognitive Sciences, 8,* 307–314.
FitzSimons, G. E. (2000). Section III: Adults, mathematics and work. In D. Coben, J. O'Donoghue, & G. E. Fitzsimons (Eds.), *Perspectives on adults learning mathematics: Research and practice* (pp. 209–227). Dordrecht: Kluwer.
FitzSimons, G. E. (2004). *An overview of adult and lifelong mathematics education.* Keynote presentation at Topic Study Group 6, 10th International Congress on Mathematics Education. http://www.icme10.dk
FitzSimons, G. E. (2005). Numeracy and Australian workplaces: Findings and implications. *Australian Senior Mathematics Journal, 19*(2), 27–40.
FitzSimons, G. E. (2008a). Mathematics and numeracy: Divergence and convergence in education and work. In C. H. Jørgensen & V. Aakrog (Eds.), *Convergence and divergence in education and work* (pp. 197–217). Zurich: Peter Lang.
FitzSimons, G. E. (2008b). A comparison of mathematics, numeracy, and functional mathematics: What do they mean for adult numeracy practitioners? *Adult Learning, 19,* 8–11.
FitzSimons, G. E. (2011). A framework for evaluating quality and equity in post-compulsory mathematics education. In B. Atweh, M. Graven, W. Secada, & P. Valero (Eds.), *Mapping equity and quality in mathematics education* (pp. 105–121). Dordrecht: Springer.
FitzSimons, G. E. (2012). Family math: Everybody learns everywhere. In J. Díez-Palomar & C. Kanes (Eds.), *Family and community in and out of the classroom: Ways to improve mathematics achievement* (pp. 25–46). Bellaterra: Servei de publications de la UAB.
FitzSimons, G. E. (2013). Doing mathematics in the workplace: A brief review of selected literature. *Adults Learning Mathematics: An International Journal, 8*(1), 7–19.
FitzSimons, G. E., & Wedege, T. (2007a). Developing numeracy in the workplace. *Nordic Studies in Mathematics Education, 12*(1), 49–66.
FitzSimons, G. E., & Wedege, T. (2007b). Developing numeracy in the workplace. *Nordic Studies in Mathematics, 12*(1), 49–66.
FitzSimons, G. E., Mlcek, S., Hull, O., & Wright, C. (2005). *Learning numeracy on the job: A case study of chemical handling and spraying.* Final Report. Adelaide: National Centre for Vocational Education Research. http://www.ncver.edu.au/publications/1609.html. Accessed 15 May 2013.
Flavell, J. H. (1988). The development of children's knowledge about the mind: From cognitive connections to mental representations. In J. W. Astington, P. L. Harris, & D. R. Olson (Eds.), *Developing theories of mind* (pp. 244–271). Cambridge: Cambridge University Press.
Flecha, R. (2000). *Sharing words: Theory and practice of dialogic learning.* Lanham: Rowman & Littlefield.
Flecha, A. (2012). Family education improves student's academic performance: Contributions from European research. *Multidisciplinary Journal of Educational Research, 3*(2), 301–321.
Flecha, R., & Soler, M. (2013). Turning difficulties into possibilities: Engaging Roma families and students in school through dialogic learning. *Cambridge Journal of Education, 43*(4), 451–465.
Flecha, A., García, R., Gómez, A., & Latorre, A. (2009). Participation in successful schools: A communicative research study from the INCLUD-ED project. *Cultura y Educación, 21,* 183–196.

Flewitt, R. (2006). Using video to investigate preschool classroom interaction: Education research assumptions and methodological practices. *Visual Communication, 5*(1), 25–50.
Fonseca, M. D. (2005). *Educação de jovens e adultos: especificidades, desafios e contribuições* (2nd ed.). Belo Horizonte: Autêntica.
Formação Brasi, U. C. A. (2009). *Projeto, planejamento das ações, cursos*. Brasilia: Ministério de Educação.
Foucault, M. (1979). On governmentality. *Ideology and Consciousness, 6*(1), 5–22.
Foucault, M. (1980). Truth and power. In C. Gordon (Ed.), *Michel Foucault, power/knowledge: Selected interviews and other writings 1972–1977* (pp. 109–133). New York: Pantheon.
Foucault, M. (1982). The subject and power. *Critical Inquiry, 8*(4), 777–795.
Foucault, M. (1991). The discourse on power. In C. Kraus & S. Lotringer (Eds.), *Michel Foucault: Remarks on Marx. Conversations with Duccio Trombadori (pp. 147–187)*. New York: Semiotexte.
Fox, M. A. (2001). *Pan-organizational summit on the U.S. science and engineering workforce: Meeting summary*. Washington: National Academy of Sciences.
Franke, K. L., Kazemi, E., & Battey, D. (2007). Mathematics teaching and classroom practice. In F. K. Lester (Ed.), *Second handbook of research on mathematics teaching and learning* (pp. 225–356). Charlotte: IAP.
Frankenstein, M. (1989). *Relearning mathematics*. London: Free Association Books.
Fraser, B. J., & Tobin, K. G. (1998). *International handbook of science education*. Dordrecht: Kluwer.
Freire, P. (1981). *Educação como prática de liberdade*. Rio de Janeiro: Paz e Terra.
Freire, P. (1996). *A la sombra de este árbol*. Barcelona: El Roure.
Freire, P. (1998). *Pedagogy of freedom: Ethics, democracy, and civic courage*. Lanham: Rowman & Littlefield.
Freire, P. (2005). *Pedagogia do oprimido* (45th ed.). Rio de Janeiro: Paz e Terra.
Freudenthal, H. (1973). *Mathematics as an educational task*. Dordrecht: Reidel.
Freudenthal, H. (1983). *Didactical phenomenology of mathematical structures*. Dordrecht: Kluwer.
Freudenthal, H. (1991). *Revisiting mathematics education: China lectures*. Dordrecht: Kluwer.
Furtado, E. D. (2008). Políticas públicas de EJA no campo: do direito na forma da lei à realização precária e descontinuidade. In *Anais do XIV Encontro Nacional de Didática e Prática de Ensino*. Porto Alegre: Nexus Soluções e TI.
Gadanidis, G., & Borba, M. (2008). Our lives as performance mathematicians. *For the Learning of Mathematics, 28*(1), 44–51.
Gallese, V., & Lakoff, G. (2005). The brain's concepts: The role of the sensory-motor system in conceptual knowledge. *Cognitive Neuropsychology, 22*(3), 455–479.
Gamon, D., & Bragdon, A. D. (1998). *Building mental muscle: Conditioning exercises for the six intelligence zones*. New York: Barnes & Noble.
García, R. (2011). *Contrato de inclusión dialógica en el colegio La Paz: Una propuesta educativa de éxito para superar la exclusión social*. Unpublished PhD thesis, University of Barcelona, Barcelona.
García, G., Valero, P., Camelo, F., Mancera, G., Romero, J., Peñaloza, G., & Samaca, M. (2010). *Escenarios de aprendizaje de las matemáticas: Un estudio desde la perspectiva de la educación matemática crítica*. Bogotá: Universidad Pedagógica Nacional de Colombia.
Gates, P., & Vistro-Yu, C. (2003). Is mathematics for all? In A. Bishop, M. Clements, C. Keitel, J. Kilpatrick, & F. Leung (Eds.), *Second international handbook of mathematics education* (pp. 31–74). Dordrecht: Kluwer.
Gatt, S., Ojala, M., & Soler, M. (2011). Promoting social inclusion counting with everyone: Learning Communities and INCLUD-ED. *International Studies in Sociology of Education, 21*(1), 33–47.
Gee, J. P. (2003). *What video games have to teach us about learning and literacy*. New York: Palgrave Macmillan.
Geertz, C. (1973). *The interpretation of cultures*. New York: Basic Books.

Gellert, U. (2010). Modalities of local integration of theories in mathematics education. In B. Sriraman & L. English (Eds.), *Theories in mathematics education: Seeking new frontiers* (pp. 537–550). New York: Springer.

Gellert, U. (2012). Pedagogic device: Ein Instrument für die Analyse impliziter Prinzipien mathematischer Unterrichtspraxis. In U. Gellert & M. Sertl (Eds.), *Zur Soziologie des Unterrichts. Arbeiten mit Basil Bernsteins Theorie des pädagogischen Diskurses (pp. 166–190)*. Beltz Juventa: Weinheim.

Gellert, U., & Hümmer, A. M. (2008). Soziale Konstruktion von Leistung im Unterricht. *Zeitschrift für Erziehungswissenschaft, 11*(2), 288–311.

Gellert, U., & Jablonka, E. (2009). "I am not talking about reality": Word problems and the intricacies of producing legitimate text. In L. Verschaffel, B. Greer, W. Van Dooren, & S. Mukhopadhyay (Eds.), *Words and worlds: Modelling verbal descriptions of situations* (pp. 39–53). Rotterdam: Sense.

GeoGebra. (2015). Dynamic mathematics & science for learning and teaching (version 5). http://www.geogebra.org/cms/. Accessed 22 February 2015.

Gerofsky, S. (2006). *Performance space and time*. Symposium discussion paper for digital mathematical performances: A fields institute symposium, University of Western Ontario, London, Canada, June 9–11, 2006. http://www.edu.uwo.ca/mathstory/pdf/GerofskyPaper.pdf.

Gerofsky, S. (2009). Performance mathematics and democracy. *Educational Insights, 13*(1).

Gerofsky, S. (2011). Ancestral genres of mathematical graphs. *For the Learning of Mathematics, 31*(1), 14–19.

Gerofsky, S. (2013). *Learning mathematics through dance*. Proceedings of Bridges 2013, Enschede: Mathematics, Music, Art, Architecture, Culture. http://archive.bridgesmathartorg/2013/bridges2013-337.pdf

Giddens, A. (1991). *Modernity and self-identity: Self and society in the late modern age*. Stanford: Stanford University Press.

Giddens, A. (1993). *New rules of sociological method: A positive critique of interpretive sociologies*. Stanford: Stanford University Press.

Gill, S., & Reynolds, A. J. (1999). Educational expectations and school achievement of urban African American children. *Journal of School Psychology, 37*(4), 403–424.

Giménez, J., Font, V., & Vanegas, Y. M. (2013). Designing professional tasks for didactical analysis as a research process. In C. Margolinas (Ed.), *Proceedings of ICMI Study 22 "Task design in mathematics education"* (pp. 581–590). Oxford: ICMI.

Gispert, H. (2011). Enseignement, mathématiques et modernité au XXème siècle: réformes, acteurs et rhétoriques. *Bulletin de l'APMEP, 494*, 286–296.

Gispert, H., & Schubring, G. (2011). Societal, structural, and conceptual changes in mathematics teaching: Reform processes in France and Germany over the twentieth century and the international dynamics. *Science in Context, 24*(1), 73–106.

Glaser, G., & Strauss, A. L. (1967). *The discovery of grounded theory: Strategies for qualitative research*. Chicago: Aldine.

Goldin-Meadow, S. (2003). *Hearing gesture: How our hands help us think*. Cambridge, MA: The Belknap Press of Harvard University Press.

Goldman, S. (2006). A new angle on families: Connecting the mathematics in daily life with school mathematics. In Z. Bekerman, N. Bubules, & D. Silberman-Keller (Eds.), *Learning in places: The informal education reader* (pp. 55–76). Bern: Peter Lang.

Goldman, S., Pea, R., Blair, K. P., Jimenez, O., Booker, A., Martin, L., & Esmonde, I. (2010). Math engaged problem solving in families. In K. Gómez, L. Lyons, & J. Radinsky (Eds.), *Learning in the disciplines: Proceedings of the 9th International Conference on the Learning Sciences (ICLS 2010)* (pp. 380–388). Chicago: ICLS.

Gomby, D. S., Culross, P. L., & Behrman, R. E. (1999). Home visiting: Recent program evaluations: Analysis and recommendations. *The Future of Children, 9*(1), 4–26.

Gómez, J. (2004). *El amor en la sociedad del riesgo: una tentativa educativa*. Barcelona: El Roure.

References

Gonzales, P., Williams, T., Jocelyn, L., Roey, S., Kastberg, D., & Brenwald, S. (2009). *Highlights from TIMSS 2007: Mathematics and science achievement of U.S. fourth and eighth-grade students in an international context.* NCES 2009–001 Revised US Dept. of Education Report.

González, N., Andrade, R., Civil, M., & Moll, L. (2001). Bridging funds of distributed knowledge: Creating zones of practices in mathematics. *Journal of Education for Students Placed at Risk, 6*(1–2), 115–132.

González, N., Moll, L. C., & Amanti, C. (Eds.). (2005). *Funds of knowledge: Theorizing practices in households, communities, and classrooms.* Mahwah: Lawrence Erlbaum.

Good, T. L., & Grouws, D. A. (1979). The Missouri mathematics effectiveness project: An experimental study in fourth-grade classrooms. *Journal of Educational Psychology, 71*(3), 355–362.

Goold, E., & Devitt, F. (2011). The role of mathematics in engineering practice and in the formation of engineers. In T. Maguire, J. J. Keogh, & J. O'Donoghue (Eds.), *Proceedings of the 18th international conference on adults learning mathematics* (pp. 134–154). Adults Learning Mathematics – A Research Forum.

Goos, M., Kahne, J., & Westheimer, J. (2004). Learning mathematics in a classroom community of inquiry. A pedagogy of collective action and reflection: Preparing teachers for collective school leadership. *Journal for Research in Mathematics Education, 35*(4), 258–292.

Gorgorió, N., Planas, N., & Vilella, X. (2000). The cultural conflict in the mathematics classroom: Overcoming its "invisibility". In A. Ahmed, J. M. Kraemer, & H. Wiliams (Eds.), *Cultural diversity in mathematics (education): CIEAEM 51* (pp. 179–185). Chichester: Horwood.

Goswami, U. (1992). *Analogical reasoning in children.* Hove: Lawrence Erlbaum.

Goswami, U. (2008). *Cognitive development: The learning brain.* New York: Psychology Press.

Greeno, J. (1994). Gibson's affordances. *Psychological Review, 101*(2), 336–342.

Greeno, J. G. (1997). On claims that answer the wrong questions. *Educational Researcher, 26*(1), 5–17.

Grouws, D. A. (Ed.). (1992). *Handbook of research on mathematics teaching and learning.* New York: Macmillan.

Gruenewald, D. (2003). The best of both worlds: A critical pedagogy of place. *Educational Researcher, 32*(4), 3–12.

Gruenewald, D. (2006). Resistance, reinhabitation, and regime change. *Journal of Research in Rural Education, 21*(9), 1–7.

Guedj, D. (2000). *The parrot's theorem.* Athens: Polis.

Gueudet, G., & Trouche, L. (2009). Towards new documentation systems for mathematics teachers? *Educational Studies in Mathematics, 71*(3), 199–218.

Gueudet, G., & Trouche, L. (2010). Ressources vives. *Le travail documentaire des professeurs en mathématiques.* Rennes: PUR & Lyon: INRP.

Gulikers, J. T. M., Bastiaens, T. J., & Martens, R. (2005). The surplus value of an authentic learning environment. *Computers in Human Behavior, 21*(3), 509–521.

Gummerum, M., Hanoch, Y., Keller, M., Parsons, K., & Hummel, A. (2010). Preschoolers' allocations in the dictator game: The role of moral emotions. *Journal of Economic Psychology, 31,* 25–34.

Gur-Ze'ev, I. (1998). Toward a non-repressive critical pedagogy. *Educational Theory, 48*(4), 463–486.

Gutierrez, R. (2013). The sociopolitical turn in mathematics education. *Journal for Research in Mathematics Education, 44*(1), 37–68.

Gutmann, A., & Thompson, D. (1996). *Democracy and disagreement.* Cambridge, MA: Belknap.

Gutstein, E. (2006). *Reading and writing the world with mathematics: Toward a pedagogy for social justice.* New York: Routledge.

Gutstein, E., Lipman, P., Hernandez, P., & de los Reyes, R. (1997). Culturally relevant mathematics teaching in a Mexican American context. *Journal for Research in Mathematics Education, 28*(2), 26–37.

Haarder, B. (2009). Naturfag er almen dannelse. In Undervisningsministeriet (Ed.), *Natur, teknik og sundhed. For alle og for de få. I bredden og i dybden* (pp. 8–13). Copenhagen: Undervisningsministeriet.

Habermas, J. (1984). *The theory of communicative action* (Vol. 1: Reason and the rationalization of society). Boston: Beacon.

Habermas, J. (1987a). *The theory of communicative action* (Vol. 2: Lifeworld and system: A critique of functionalist reason). Boston: Beacon.

Habermas, J. (1987b). *The theory of communicative action (Lifeworld and systems: A critique of functionalist reason* (Vol. 2)). Boston: Beacon.

Hale, C. M., & Tager-Flusberg, H. (2003). The influence of language on theory of mind: A training study. *Developmental Science, 6*, 346–359.

Hall, E. (1966). *The hidden dimension*. New York: Doubleday.

Halle, T. G., Kurtz-Costes, B., & Mahoney, J. L. (1997). Family influences on school achievement in low-income, African-American children. *Journal of Educational Psychology, 89*(3), 527–537.

Halliday, M. A. K. (1975). *Learning how to mean: Explorations in the development of language*. London: Edward Arnold.

Hammond, M. (1999). Issues associated with participation in on line forums: The case of the communicative learner. *Education and Information Technologies, 4*(4), 353–367.

Hammond, M., & Wiriyapinit, M. (2005). Learning through online discussion: A case of triangulation in research. *Australasian Journal of Educational Technology, 21*(3), 283–302.

Harbaugh, W. T., Krause, K., Liday, S. G., & Vesterlund, L. (2003). Trust and reciprocity: Interdisciplinary lessons from experimental research. In E. Ostrom & J. Walker (Eds.), *Trust in children* (pp. 302–322). New York: Russell Sage.

Harding, S. (1998). *Is science multicultural? Colonialisms, feminisms, and epistemologies*. Bloomington: Indiana University Press.

Harindranath, R. (2009). *Audience-citizens: The media, public knowledge and interpretive practice*. New Delhi: Sage.

Harmon, H. L. (2003). Rural education. In J. W. Guthrie (Ed.), *Encyclopedia of education* (2nd ed., pp. 2083–2090). New York: Macmillan.

Harmon, H., Henderson, S., & Royster, W. (2003). A research agenda for improving science and mathematics education in rural schools. *Journal of Research in Rural Education, 18*(1), 52–58.

Harris, M. (1994). *Mathematics in Denman textile courses. Three reports for D College*. National Federation of Women's Institutes. Unpublished manuscript.

Hasan, R. (2001). The ontogenesis of decontextualised language: Some achievements of classification and framing. In A. Morais, I. P. Neves, B. Davies, & H. Daniels (Eds.), *Towards a sociology of pedagogy. The contribution of Basil Bernstein to research* (pp. 185–219). New York: Peter Lang.

Hassi, M. L., Hannula, A., & Saló i Nevado, L. (2010). Basic mathematical skills and empowerment: Challenges and opportunities, Finnish adult education. *Adults Learning Mathematics– An International Journal, 5*(1), 6–22.

Heathcote, D. (1984). Dorothy Heathcote's notes. In L. Johnson & C. O'Neil (Eds.), *Heathcote, Dorothy: Collected writings on education and drama*. London: Hutchinson.

Hedegaard, M., & Chaiklin, S. (2005). *Radical-local teaching and learning: A cultural-historical approach*. Aarhus: Aarhus University Press.

Henderson, A. T., & Berla, N. (Eds.). (1994). *A new generation of evidence: The family is critical to student achievement*. Columbia: National Committee for Citizens in Education.

Henderson, A. T., & Mapp, K. L. (2002). *A new wave of evidence: The impact of school, family, and community connections on student achievement (Annual synthesis)*. Austin: National Center for Family Community Connections with Schools, Southwest Educational Development Laboratory.

Herbel-Eisenmann, B., Lubienski, S. T., & Id-Deen, L. (2006). Reconsidering the study of mathematics instructional practices: The importance of curricular context in understanding local and global teacher change. *Journal of Mathematics Teacher Education, 9*(4), 313–345.

Hersh, R. (1997). *What is mathematics, really?* New York: Oxford University Press.

Hiebert, J. (1999). Relationships between research and the NCTM standards. *Journal for Research in Mathematics Education, 30*(1), 3–19.

Hiebert, J., & Wearne, D. (1986). Procedures over concepts: The acquisition of decimal number knowledge. In J. Hiebert (Ed.), *Conceptual and procedural knowledge: The case of mathematics* (pp. 199–223). Hillsdale: Lawrence Erlbaum.

Higgins, S., Hall, E., Wall, K., Woolner, P., & McCaughey, C. (2005). *The impact of school environments: A literature review.* http://www.ncl.ac.uk/cflat/news/DCReport.pdf.

Hitchcock, G. (1996). Dramatizing the birth and adventures of mathematical concepts: Two dialogues. In R. Calinger (Ed.), *Vita mathematica: Historical research and integration with teaching* (pp. 27–41). Washington, DC: MAA.

Hitt, F. (1998). Difficulties in the articulation of different representations linked to the concept of function. *Journal of Mathematical Behavior, 17*(1), 123–134.

Hitt, F., & González-Martín, A. (2015). Covariation between variables in a modelling process: The ACODESA (collaborative learning, scientific debate and self-reflexion) method. *Educational Studies in Mathematics, 88*(2), 201–219.

Hitt, F., & Kieran, C. (2009). Constructing knowledge via a peer interaction in a CAS environment with tasks designed from a task-technique-theory perspective. *International Journal of Computers for Mathematical Learning, 14*, 121–152.

Hitt, F., Saboya M. et Cortés C. (2015). La pensée arithmético-algébrique dans la transition primaire-secondaire et le rôle des représentations spontanées et institutionnelles. Actes du congrès CIEAEM66, Lyon, France, pp. 252–257. http://math.unipa.it/~grim/CIEAEM%2066_Pproceedings_QRDM_Issue%2024,%20Suppl.1.pdf

Hitt, F., Saboya, M., & Cortés, C. (2013). Structure cognitive de contrôle et compétences mathématiques de l'arithmétique à l'algèbre au secondaire: Les nombres polygonaux. Actes du congrès CIEAEM65, Turin, Italie, pp. 134–146. http://math.unipa.it/~grim/quaderno23_suppl_1.htm

Ho, E. S., & Willms, J. D. (1996). Effects of parental involvement on eight-grade achievement. *Sociology of Education, 69*(2), 126–141.

Hoadley, U. K. (2007). The reproduction of social class inequalities through mathematics pedagogies in South African primary schools. *Journal of Curriculum Studies, 39*, 679–706.

Hogan, T. P., & Brezinski, K. L. (2003). Quantitative estimation: One, two, or three abilities? *Mathematical Thinking and Learning, 5*(4), 259–280.

Holloway, S. (1986). The relationship of mothers' beliefs to children's mathematics achievement: Some effects of sex differences. *Merrill-Palmer Quarterly, 32*(3), 231–250.

Hook, J. (1978). The development of equity and logico-mathematical thinking. *Child Development, 49*, 1035–1044.

Hook, J., & Cook, T. (1979). Equity theory and the cognitive ability of children. *Psychological Bulletin, 86*, 429–445.

Hoops, J. (2011). Developing an ethic of tension: Negotiating service learning and critical pedagogy ethical tensions. *Journal for Civic Commitment, 16*(Jan), 1–15.

Hoover-Dempsey, K. V. (2005). *Research and evaluation of family involvement in education: What lies ahead? Paper presented at the annual meeting of the American Educational Research Association (AERA).* Canada: Montreal.

Hoover-Dempsey, K. V., & Bassler, R. B. (1997). Parents reported involvement in students' homework: Strategies and practices. *The Elementary School Journal, 95*(5), 435–450.

Hoover-Dempsey, K. V.. & Sandler, H. M. (1995). Parent involvement in children's education: Why does it make a difference? *Teachers College Record, 97*(2), 310–331.

Howley, C., & Gunn, E. (2003). Research about mathematics achievement in the rural circumstance. *Journal of Research in Rural Education, 18*(2), 86–95.

Howley, C., & Howley, A. (2010). Poverty and school achievement in rural communities: A social-class interpretation. In K. Schafft & A. Jackson (Eds.), *Rural education for the twenty-first century: Identity, place, and community in a globalizing world* (pp. 34–50). University Park: Pennsylvania State University.

Howley, A., Howley, C., Klein, R., Belcher, J., Tusay, M., Clonch, S., Miyafusa, S., Foley, G., Pendarvis, E., Perko, H., Howley, M., & Jimerson, L. (2010). *Community and place in mathematics education in selected rural schools.* Athens: Appalachian Collaborative Center for Learning, Instruction, and Assessment in Mathematics, Ohio University.

Howley, A., Showalter, D., Howley, M., Howley, C., Klein, R., & Johnson, J. (2011). Challenges for place-based mathematics pedagogy in rural schools and communities in the United States. *Children, Youth and Environments, 21*(1), 101–127.

Hoyles, C., Noss, R., & Kent, P. (2004). On the integration of digital technologies into mathematics classrooms. *International Journal of Computers for Mathematical Learning, 9*, 309–326.

Hoyles, C., Noss, R., Kent, P., & Bakker, A. (2010). *Improving mathematics at work: The need for techno-mathematical literacies.* London: Routledge.

Hudson, J. M., & Bruckman, A. S. (2004). The Bystander effect: A lens for understanding patterns of participation. *The Journal of Learning Sciences, 13*(2), 165–195.

Hunting, R. P., & Davis, G. E. (Eds.). (1991). *Early fraction learning.* New York: Springer.

Iao, L.-S., Leekam, S., Perner, J., & McConachie, H. (2011). Further evidence for nonspecificity of theory of mind in preschoolers: Training and transferability in the understanding of false beliefs and false signs. *Journal of Cognition and Development, 12*, 56–79.

Illeris, K., Katznelson, N., Simonsen, B., & Ulriksen, L. (2002). *Ungdom, identitet og uddannelse.* Frederiksberg: Roskilde Universitetsforlag.

INCLUD-ED. (2012). *Final INCLUD-ED report: Strategies for inclusion and social cohesion in Europe from education* (CIT4-CT-2006-028603). Brussels: European Commission.

INCLUD-ED Consortium. (2009). *Actions for success in schools in Europe.* Brussels: European Commission.

International Fund for Agricultural Development (IFAD). (2010). *Rural poverty report 2011.* Quintily: IFAD.

Jablonka, E. (2010). Contextualised mathematics. In H. Christensen, J. Díez-Palomar, J. Kantner, & C. M. Kinger (Eds.), *Proceedings of the 17th international conference on adults learning mathematics* (p. 3). Adults Learning Mathematics – A Research Forum.

Jablonka, E., & Gellert, U. (2007). Mathematisation – Demathematisation. In U. Gellert & E. Jablonka (Eds.), *Mathematisation and demathematisation: Social, philosophical and educational ramifications* (pp. 1–18). Rotterdam: Sense.

Jablonka, E., & Gellert, U. (2010). Ideological roots and uncontrolled flowering of alternative curriculum conceptions. In U. Gellert, E. Jablonka, & C. Morgan (Eds.), *Proceedings of the sixth international mathematics education and society conference* (pp. 23–41). Berlin: Freie Universität Berlin.

Jablonka, E., & Gellert, U. (2012). Potential, pitfalls, and discriminations: Curriculum conceptions revisited. In O. Skovsmose & B. Greer (Eds.), *Opening the cage: Critique and politics of mathematics education* (pp. 287–308). Rotterdam: Sense.

Jackson, K., & Remillard, J. T. (2005). Rethinking parent involvement: African American mothers construct their roles in the mathematics education of their children. *The School Community Journal, 15*(1), 51–73.

Jacobs, J. E., Lanza, S., Osgood, D. W., Eccles, J. S., & Wigfield, A. (2002). Changes in children's self-competence and values: Gender and domain differences across grades one through twelve. *Child Development, 73*(2), 509–527.

Jahnke, H. N. (1986). Origins of school mathematics in early nineteenth-century Germany. *Journal of Curriculum Studies, 18*(1), 85–94.

Janks, H. (1997). Critical discourse analysis as a research tool. *Discourse: Studies in the Cultural Politics of Education, 18*(3), 329–342.

Janvier, C. (Ed.). (1987). *Problems of representation in the teaching and learning of mathematics.* London: Lawrence Erlbaum.

Jurdak, M. E. (2006). Contrasting perspectives and performance of high school students on problem solving in real world situated, and school contexts. *Educational Studies in Mathematics, 63*(3), 283–301.

Jurdak, M., & Shahin, I. (1999). An ethnographic study of the computational strategies of a group of young street vendors in Beirut. *Educational Studies in Mathematics, 40*(2), 155–172.

Jurdak, M., & Shahin, I. (2001). Problem solving activity in the workplace and the school: The case of constructing solids. *Educational Studies in Mathematics, 47*(3), 297–315.

References

Kahane, R., & Rapoport, T. (1997). *The origins of postmodern youth: Informal youth movements in a comparative perspective.* New York: Walter de Gruyter.

Kahraman, P. B., & Derman, M. T. (2011). The views of primary school and preschool teachers about home visiting: A study in Turkey. In M. Pieri, A. Pepe, & L. Addimando (Eds.), *Home, school and community: A partnership for a happy life?* (pp. 22–24). Milano: I Libri di Emil.

Kalavasis, F. (2012). Modélisation des structures complexes en éducation, en vue de la gouvernance des unités scolaires. *Les Cahiers de Recherche de LAREQUOI, 1,* 46–71. Université de Versailles St Quentin, France.

Kane, P. (2012). Spatial awareness and estimation in recycling and refuse collection. In A. Hector-Mason & D. Coben (Eds.), *Proceedings of the 19th international conference on adults learning mathematics* (pp. 56–68). Hamilton: The National Centre of Literacy & Numeracy for Adults, University of Waikato.

Kant, I. (1781). *Critik der reinen Vernunft.* http://www.deutschestextarchiv.de. Accessed 7 Sept 2013.

Karlberg, M. (2005). Elevating the service in service-learning. *Journal of the Northwestern Communication Association, 34*(Spring), 16–36.

Keeves, J. P., & Darmawan, I. G. N. (2009). Science teaching. In L. J. Saha & A. G. Dworkin (Eds.), *The new international handbook of research on teachers and teaching* (pp. 975–1000). New York: Springer.

Keijzer, R., & Terwel, J. (2003). Learning for mathematical insight: A longitudinal comparative study on modelling. *Learning and Instruction, 13,* 285–304.

Keith, T. Z., Keith, P. B., Troutman, G. C., Bickley, P. G., Trivette, P. S., & Singh, K. (1993). Does parental involvement affect eight-grade student achievement? Structural analysis of national data. *School Psychology Review, 22*(3), 474–496.

Kelly, B. (2011). Learning in the workplace, functional mathematics and issues of transferability. In T. Maguire, J. J. Keogh, & J. O'Donoghue (Eds.), *Proceedings of the 18th international conference on adults learning mathematics* (pp. 37–46). Adults Learning Mathematics – A Research Forum.

Kennedy, N., & Kennedy, D. (2011). Community of philosophical inquiry as a discursive structure, and its role in school curriculum design. *Journal of Philosophy of Education, 45*(2), 265–283.

Kent, P., Bakker, A., Hoyles, C., & Noss, R. (2011). Measurement in the workplace: The case of process improvement in the manufacturing industry. *ZDM – The International Journal on Mathematics Education, 43*(5), 747–758.

Keogh, J. J., Maguire, T., & O'Donoghue, J. (2012). A workplace contextualization of mathematics: Visible, distinguishable and meaningful mathematics in complex contexts. In A. Hector-Mason & D. Coben (Eds.), *Proceedings of the 19th international conference on adults learning mathematics* (pp. 32–38). Hamilton: The National Centre of Literacy & Numeracy for Adults, University of Waikato.

Kilpatrick, J. (1997). *Five lessons from the New Math era.* Paper presented at the conference Reflecting on Sputnik: Linking the past, present, and future of educational reform, National Academy of Sciences of the USA, Washington, DC.

Kilpatrick, J., Swafford, J., & Findell, B. (Eds.). (2001). *Adding in up: Helping children learn mathematics.* Washington, DC: The National Academies Press.

Kilpatrick, J., Martin, G. W., & Schifter, D. (2003). *A research companion to principles and standards for school mathematics.* Reston: NCTM.

Kinzler, K. D., & Spelke, E. S. (2007). Core systems in human cognition. *Progress in Brain Research, 164,* 257–264.

Kirshner, D. (2000). Exercises, probes, puzzles: A crossdisciplinary typology of school mathematics problems. *Journal of Curriculum Theorizing, 16*(2), 9–36.

Kirshner, D. (2002). Untangling teachers' diverse aspirations for student learning: A crossdisciplinary strategy for relating psychological theory to pedagogical practice. *Journal for Research in Mathematics Education, 33*(1), 46–58.

Klein, R. (2007). Educating in place: Mathematics and technology. *Philosophical Studies in Education, 38*, 119–130.

Klein, R. (2008). Forks in the (back) road: Obstacles for place-based strategies. *Rural Mathematics Educator, 7*(1).

Klein, R., & Johnson, J. (2010). On the use of locale in understanding the mathematics achievement gap. In P. Brosnan, D. Erchick, & L. Flevares (Eds.), *Proceedings of the 32nd annual meeting of the North American chapter of the international group for the psychology of mathematics education* (pp. 489–496). Columbus: The Ohio State University.

Klein, R., Hitchcock, J., & Johnson, J. (2013). *Rural perspectives on community and place connections in math instruction: A survey* (Working Paper No. 44). Appalachian Collaborative Center for Learning, Assessment, and Instruction in Mathematics (ACCLAIM). https://sites.google.com/site/acclaimruralmath/Home

Knapp, M. S. (1997). Between systemic reform and the mathematics and science classroom: The dynamics of innovation, implementation and professional learning. *Review of Educational Research, 67*(2), 227–266.

Knijnik, G. (2008). Landless peasants of Southern Brazil and mathematics education: A study of three different language games. In J. F. Matos, K. Yasukawa, & P. Valero (Eds.), *Proceedings of the fifth international mathematics education and society conference* (pp. 312–319). Lisbon: Centro de Investigaçao em Educaçao, Universidade de Lisboa.

Knijnik, G., & Wanderer, F. (2010). Mathematics education and differential inclusion: A study about two Brazilian time–space forms of life. *ZDM – The International Journal on Mathematics Education, 42*(3–4), 349–360.

Knipping, C., & Reid, D. (2013). Have you got the rule underneath? Invisible pedagogic practice and stratification. In A. M. Lindmeier & A. Heinze (Eds.), *Proceedings of the 37th conference of the international group for the psychology of mathematics education* (Vol. 3, pp. 193–200). Kiel: PME.

Knipping, C., Straehler-Pohl, H., & Reid, D. (2011). "I'm going to tell you to save wondering": How enabling becomes disabling in a Canadian mathematics classroom. *Quaderni di Ricerca in Didattica, 22*(Supplemento 1), 171–175.

Knopf, H. T., & Swick, K. J. (2008). Using our understanding of families to strengthen family involvement. *Early Childhood Education Journal, 35*(5), 419–427.

Kooro, M. B., & Lopes, C. E. (2007). *Uma análise das propostas curriculares de matemática para a educação de jovens e adultos.* Paper presented at XI Encontro Nacional de Educação Matemática. UFMG: Belo Horizonte.

Kotarinou, P., Stathopoulou Ch. (2008). Role –playing in Mathematics instruction. In proceedings of the *13th* International congress of Drama/Theatre in Education (pp. 108–118). Ankara, 21–23 November.

Kotarinou, P., Chronaki, A., & Stathopoulou, C. (2010). Debating for 'one measure for the world': Sensitive pendulum or heavy earth? In U. Gellert, E. Jablonka, & C. Morgan (Eds.), *Proceedings of mathematics education and society 6th international conference* (pp. 322–329). Berlin: Freie Universität Berlin.

Krajcik, J. S. (2002). The value and challenges of using learning technologies to support students in learning sciences. *Research in Science Education, 32*(1), 411–414.

Krapp, A. (1999). Interest, motivation and learning: An educational-psychological perspective. *European Journal of Psychology of Education, 14*(1), 23–40.

Kreienkamp, K. (2009). Exemplary project: Math fact memorization in a highly sequenced elementary mathematics curriculum. In R. A. Schmuck (Ed.), *Practical action research: A collection of articles* (2nd ed., p. 143). Thousand Oaks: Corwin.

Kress, G., Jewitt, C., Bourne, J., Franks, A., Hardcastle, J., Jones, K., & Reid, E. (2005). *English in urban classrooms: A multimodal perspective on teaching and learning.* London: RoutledgeFalmer.

Krüger, J., & Michalek, R. (2011). Parents' teachers' cooperation: Mutual expectations and attributions from a parents' point of view. In M. Pieri, A. Pepe, & L. Addimando (Eds.), *Home, school and community: A partnership for a happy life?* (pp. 66–68). Milano: I Libri di Emil.

Krummheuer, G. (2007). Argumentation and participation in the primary mathematics classroom: Two episodes and related theoretical abductions. *Journal of Mathematical Behavior, 26*(1), 60–82.

Kunstler, J. (1993). *The geography of nowhere: The rise and decline of America's man-made landscape*. New York: Simon and Schuster.

Ladson-Billings, G. (1995). Making mathematics meaningful in multicultural contexts. In W. G. Secada, E. Fennema & L. B. Adajian (Eds.), *New directions for equity in mathematics education* (pp. 126–145). Cambridge: Cambridge University Press.

Lagrange, J. B., Artigue, M., Laborde, C., & Trouche, L. (2003). Technology and mathematics education: A multidimensional study of the evolution of research and innovation. In A. J. Bishop, M. A. Clements, C. Keitel, J. Kilpatrick, & F. K. S. Leung (Eds.), *Second international handbook of mathematics education* (pp. 237–269). Dordrecht: Kluwer.

Laisant, C.-A., & Fehr, H. (1899). Préface. *L'Enseignement Mathématique, 1*(1), 1–5.

Lakoff, G., & Núñez, R. (2000). *Where mathematics comes from: How the embodied mind brings mathematics into being*. New York: Basic Books.

Lampert, M. (1990). When the problem is not the question and the solution is not the answer: Mathematical knowing and teaching. *American Educational Research Journal, 27*(1), 29–63.

Larsen, E. (1993). *A survey of the current status of rural education research (1986–1993)* (ERIC Document Reproduction Services No. ED 366 482).

Larsen, G., & Kellogg, J. (1974). A developmental study of the relation between conservation and sharing behavior. *Child Development, 45*, 849–851.

Latour, B. (1987). *Science in action*. Milton Keynes: Open University Press.

Lave, J. (1988). *Cognition in practice: Mind, mathematics and culture in everyday life*. Cambridge: Cambridge University Press.

Lave, J., & Wenger, E. (1991). *Situated learning: Legitimate peripheral participation*. Cambridge: Cambridge University Press.

Lee, J., & McIntyre, W. (2000). Interstate variation in the achievement of rural and nonrural students. *Journal of Research in Rural Education, 16*(3), 168–181.

Lénárt, I. (1996). *Non Euclidean adventures on the Lenart sphere: Activities comparing planar and spherical geometry*. Emeryville: Key Curriculum Press.

Lerman, S. (2000). The social turn in mathematics education research. In J. Boaler (Ed.), *Multiple perspectives on mathematics teaching and learning* (pp. 19–44). Westport: Ablex.

Lerman, S. (2001). Cultural, discursive psychology: A sociocultural approach to studying the teaching and learning of mathematics. *Educational Studies in Mathematics, 46*(1–3), 87–113.

Lerman, S. (2006). Cultural psychology, anthropology and sociology: The developing 'strong' social turn. In J. Maasz & W. Schloeglmann (Eds.), *New mathematics education research and practice* (pp. 171–188). Rotterdam: Sense.

Lerner, R. M. (1984). *On the nature of human plasticity*. New York: Cambridge University Press.

Lesh, R., & Harel, G. (2003). Problem solving, modeling, and local conceptual development. *Mathematical Thinking and Learning, 5*(2), 157–189.

Leung, F. K. (2006). Mathematics education in East Asia and the West: Does culture matter? In F. K. Leung, K. D. Graf, & F. Lopez-Real (Eds.), *Mathematics education in different cultural traditions: A comparative study of East Asia and the West* (pp. 21–46). New York: Springer.

Lim, F. V. (2011). *A systemic functional multimodal discourse analysis approach to pedagogical*. Unpublished PhD thesis, National University of Singapore.

Lim, F. V., O'Halloran, K. L., & Podlasov, A. (2012). Spatial pedagogy: Mapping meanings in the use of classroom space. *Cambridge Journal of Education, 42*(2), 235–251.

Liman, M. A., Salleh, M. J., & Abdullahi, M. (2013). Sociological and mathematics educational values: An intersection of need for effective mathematics instructional contents delivery. *International Journal of Humanities and Social Science, 3*(2), 192–203.

Lindenskov, L., & Wedege, T. (2001a). *Numeracy as an analytical tool in mathematics education and research*. Roskilde: Roskilde University, Centre for Research in Learning Mathematics.

Lindenskov, L., & Wedege, T. (2001b). *Numeracy as an analytical tool in mathematics education and research*. Roskilde: Centre for Research in Learning Mathematics, Roskilde University.

Lipscomb, T. J., Bregman, N. J., & McAllister, H. A. (1983). The effect of words and actions on American childrens prosocial behavior. *Journal of Psychology: Interdisciplinary and Applied, 114*, 193–198.

Lobo da Costa, N. M. (1997). *Funções Seno e Cossenno: Uma sequencia de ensino a partir dos contextos do "Mundo Experimental" e do computador*. Unpublished master thesis, Pontifícia Universidade Católica de São Paulo, São Paulo. Available in: http://www.matematicaepraticadocente.net.br/pdf/teses_dissertacoes/dissertacao_nielce_meneguelo_lobo_da_costa.pdf. Accessed 24 Feb 2015.

Lobo da Costa, N. M. (2010). Reflexões sobre tecnologia e mediação pedagógica na formação do professor de matemática. In W. Beline & N. M. Lobo da Costa (Eds.), *Educação matemática, tecnologia e formação de professores: Algumas reflexões* (pp. 85–116). Campo Mourão: FECILCAM.

Lobo da Costa, N. M., Prado, M. E., & Pietropaolo, R. C. (2013). *Proceedings of CIEAEM 65*. Torino: Italy. *Mathematics teachers continuing education and technology: A necessary practice in a globalized context.*

LoBue, V., Nishida, T., Chiong, C., DeLoache, J. S., & Haidt, J. (2009). When getting something good is bad: Even three-year-olds react to inequity. *Social Development, 20*, 154–170.

Lourenco, O. M. (1993). Toward a Piagetian explanation of the development of prosocial behaviour in children: The force of negational thinking. *British Journal of Developmental Psychology, 11*, 91–106.

Lubienski, S. T. (2000). Problem solving as a means towards mathematics for all: An exploratory look through a class lens. *Journal for Research in Mathematics Education, 31*(4), 454–482.

Lubienski, S. T. (2004). Decoding mathematics instruction: A critical examination of an invisible pedagogy. In J. Muller, B. Davies, & A. M. Morais (Eds.), *Reading Bernstein, researching Bernstein* (pp. 108–122). London: Routledge.

Luminet, J.-P. (2003). *Euclid's bar*. Athens: Livanis.

Lundin, S. (2012). Hating school, loving mathematics: On the ideological function of critique and reform in mathematics education. *Educational Studies in Mathematics, 80*(1), 73–85.

Lusse, M. (2011). How to break through the knowledge paradox in home-school partnership? In M. Pieri, A. Pepe, & L. Addimando (Eds.), *Home, school and community: A partnership for a happy life?* (pp. 68–70). Milano: I Libri di Emil.

Luther, M. (1904). *The life of Luther, written by himself, collected and arranged by M. Michelet translated by William Hazlitt*. London: George Bell & Sons.

Lyotard, J.-F. (1984). *The postmodern condition: A report on knowledge*. Minneapolis: University of Minnesota Press.

Ma, L. (1999). *Knowing and teaching elementary mathematics: Teachers' understanding of fundamental mathematics in China and the United States*. New York: Routledge.

Maaluf, A. (1998). *Les identités meurtrières*. Paris: Grasset & Fasquelle.

Mack, N. K. (1990). Learning fractions with understanding: Building on informal knowledge. *Journal for Research in Mathematics Education, 21*, 16–32.

Mack, N. K. (1998). Building a foundation for understanding the multiplication of fractions. *Teaching Children Mathematics, 5*, 34–38.

Madianou, M. (2005). *Mediating the nation: News, audiences and the politics of identity*. London: UCL Press.

Maia, D. (2007). *Função quadrática: um estudo didático de uma abordagem computacional*. Unpublished master thesis, Pontifícia Universidade Católica de São Paulo, São Paulo.

Mankiewicz, R. (2002). *The story of mathematics*. Athens: Alexandria.

Martin, D. B., Gholson, M. L., & Leonard, J. (2010). Mathematics as gatekeeper: Power and privilege in the production of knowledge. *Journal of Urban Mathematics Education, 3*(2), 12–24.

Marton, F., & Säljö, R. (1984). Approaches to learning. In F. Marton, D. Hounsell, & N. Entwistle (Eds.), *The experience of learning* (pp. 39–58). Edinburgh: Scottish Academic Press.

Maschietto, M., & Trouche, L. (2010). Mathematics learning and tools from theoretical, historical and practical points of view: The productive notion of mathematics laboratories. *ZDM – The International Journal on Mathematics Education, 42*(1), 33–47.

Massey, D. (2005). *For space*. London: Sage.

Matthiessen, C. I. M. (2010). Multisemiosis and context-based register typology: Registeral variation in the complementarity of semiotic systems. In E. Ventola & J. Moya (Eds.), *The world told and the world shown: Multisemiotic issues* (pp. 11–38). Basingstoke: Palgrave Macmillan.

McCoy, M. L., & Scully, P. L. (2002). Deliberative dialogue to expand civic engagement: What kind of talk does democracy need? *National Civic Review, 91*(2), 120–128.

McCrink, K., & Wynn, K. (2007). Ratio abstraction by 6-month-old infants. *Psychological Science, 18*, 740–745.

McCrink, K., Bloom, P., & Santos, L. (2010). Children's and adult's judgments of equitable resource distributions. *Developmental Science, 13*(1), 37–45.

McGregor, J. (2004a). Editorial. *Forum, 46*(1), 2.

McGregor, J. (2004b). Space, power and the classroom. *Forum, 46*(1), 13–18.

McLuhan, M. (2003). *Understanding me*. Toronto: McClelland and Stewart.

McNeill, D. (1992). *Hand and mind: What gestures reveal about thought*. Chicago: University of Chicago Press.

McNeill, D. (2005). *Gesture and thought*. Chicago: University of Chicago Press.

Meagher, M. (2002). *Teaching fractions, new methods, new resources*. Columbus: ERIC Digest.

Mendes, M. (2008). *Introdução do laptop educacional em sala de aula: Indícios de mudanças na organização e gestão de aula*. Unpublished master thesis, Pontifícia Universidade Católica de São Paulo, São Paulo.

Mendes, M., & Almeida, M. E. (2011). Utilização do Laptop em sala de aula. In M. E. Almeida & M. E. Prado (Eds.), *O computador portátil na escola: Mudanças e desafios nos processos de ensino e aprendizagem* (pp. 49–59). São Paulo: Avercamp.

Mendick, H. (2005). Mathematical stories: Why do more boys than girls choose to study mathematics at AS-level in England? *British Journal of Sociology of Education, 26*(2), 236–251.

Menghini, M., Furinghetti, F., Giacardi, L., & Arzarello, F. (Eds.). (2008). *The first century of the international commission of mathematical instruction (1908–2008): Reflecting and shaping the world of mathematics education*. Rome: Istituto della Enciclopedia Italiana.

Meyer, M. R., Delagardelle, M. L., & Middleton, J. A. (1996). Addressing parents' concerns over curriculum reform. *Educational Leadership, 53*(7), 54–57.

Meyrowitz, J. (1997). Shifting worlds of strangers: Medium theory and changes in "them" versus "us". *Sociological Inquiry, 67*(1), 59–71.

Miles, M. B., & Huberman, A. M. (1994). *Qualitative data analysis* (2nd ed.). London: Sage.

Miller, K. (1984). Child as the measure of all things: Measurement procedures and the development of quantitative concepts. In C. Sophian (Ed.), *Origins of cognitive skills: The 18th Carnegie Symposium on Cognition* (pp. 193–228). Hillsdale: Lawrence Erlbaum.

Miller, J. G. (2006). Insight into moral development from cultural psychology. In M. Killen & J. Smetana (Eds.), *Handbook of moral development* (pp. 375–398). Mahwah: Lawrence Erlbaum.

Miller, P., & O'Leary, T. (1987). Accounting and the construction of the governable person. *Accounting, Organizations and Society, 12*(3), 235–265.

Miller, P., & Rose, N. (1990). Governing economic life. *Economy and Society, 19*, 1–31.

Miller, P., & Rose, N. (2008). *Governing the present: Administering economic, social and personal life*. Cambridge: Polity Press.

Mills, K. R. (2012). Some correspondences and disjunctions between school mathematics and the mathematical needs of apprentice toolmakers: A New Zealand perspective. In A. Hector-Mason & D. Coben (Eds.), *Proceedings of the 19th international conference on adults learning mathematics* (pp. 69–83). Hamilton: The National Centre of Literacy & Numeracy for Adults, University of Waikato.

Ministério da Educação. (2007). *Programa de Matemática do Ensino Básico*. Lisbon: DGIDC.

Mix, K., Levine, S., & Huttenlocher, J. (1999). Early fraction calculation ability. *Developmental Psychology, 35*, 164–174.

Moll, L., Amanti, C., Neff, D., & González, N. (1992). Funds of knowledge for teaching: A qualitative approach to developing strategic connections between homes and classrooms. *Theory Into Practice, 31*(2), 132–141.

Montaigne, M. (1854). *Essais*. Paris: Firmin-Didot.

Moore, C. (2009). Fairness in children's resource allocation depends on the recipient. *Psychological Science, 20*(8), 944–948.

Morais, A. M. (2002). Basil Bernstein at the micro level of the classroom. *British Journal of Sociology of Education, 23*(4), 559–569.

Moreau, M.-P., Mendick, H., & Epstein, D. (2010). Constructions of mathematicians in popular culture and learners' narratives: A study of mathematical and non-mathematical subjectivities. *Cambridge Journal of Education, 40*(1), 25–38.

Morency, L.-P., Oviatt, S., Scherer, S., Weibel, N., & Worsley, M. (2013). ICMI 2013 grand challenge workshop on multimodal learning analytics. In *Proceedings of the 15th ACM international conference on multimodal interaction* (pp. 373–378). Sydney: ACM.

Moreno-Armella, L., Hegedus, S., & Kaput, J. (2008). From static to dynamic mathematics: Historical and representational perspectives. *Educational Studies in Mathematics, 68*(2), 99–111.

Moretto, V. P. (2003). *Construtivismo: a produção do conhecimento em aula* (3rd ed.). Rio de Janeiro: DP&A.

Morgan, C. (2009). Questioning the mathematics curriculum: A discursive approach. In L. Black, H. Mendick, & Y. Solomon (Eds.), *Mathematical relationships in education: Identities and participation* (pp. 97–106). New York: Routledge.

Morgan, C. (2012). Studying discourse implies studying equity. In B. Herbel-Eisenmann, J. Choppin, & D. Wagner (Eds.), *Equity in discourse for mathematics education* (pp. 181–192). Dordrecht: Springer.

Moscovici, S. (1988). Notes towards a description of social representations. *Journal of European Social Psychology, 18*(3), 211–250.

Moss, J., & Case, R. (1999). Developing children's understanding of the rational numbers: A new model and an experimental curriculum. *Journal for Research in Mathematics Education, 30*, 122–147.

Moss, J., Beaty, R., McNab, S. L., & Eisenband, J. (2005, September 14). *The potential of geometric sequences to foster young students' ability to generalize in Mathematics*, [PowerPoint file] presented at the Algebraic Reasoning: Developmental, Cognitive and Disciplinary Foundations for Instruction. Washington: The Brookings Institution. Retrieved Aug. 5, 2014, from http://www.brookings.edu/events/2005/09/14-algebraic-reasoning

Mota, R. E., & Tome, T. (2005). Uma nova onda no ar. In A. Barbosa Filho, C. Castro, & T. Tome (Eds.), *Mídias digitais: Convergência tecnológica e inclusão digital* (pp. 51–84). São Paulo: Paulinas.

National Academies. (2007). *Rising above the gathering storm: Energizing and employing America for a brighter economic future*. Washington: National Academies Press.

National Council of Teachers of Mathematics [NCTM]. (2000). *Principles and standards for school mathematics*. Reston: NCTM.

Nattapoj, V. H. (2012). Relationship between classroom authority and epistemological beliefs as espoused by primary school mathematics teachers from the very high and very low socioeconomic regions in Thailand. *Journal of International and Comparative Education, 1*(2), 71–89.

NCTM. (1989). *Curriculum and evaluation standards for school mathematics*. Reston: NCTM.

NCTM. (2000). *Principles and standards for school mathematics*. Reston: NCTM.

Neelands, J. (1984). *Making sense of drama: A guide to classroom practice*. London: Heinemann.

Ni, Y., & Zhou, Y. (2005). Teaching and learning fraction and rational numbers: The Origins and implications of whole number bias. *Educational Psychologist, 40*, 27–52.

Nickson, M. (1994). The culture of the mathematics classroom: An unknown quantity? In S. Lerman (Ed.), *Cultural perspectives on the mathematics classroom* (pp. 7–35). Dordrecht: Kluwer.

Nieto, S. (2004). *Affirming diversity*. New York: Pearson.

Niss, M. (1994). Mathematics in society. In R. W. Scholz, R. Sträßer, & B. Winkelmann (Eds.), *Didactics of mathematics as a scientific discipline* (pp. 367–378). Dordrecht: Kluwer.

Niss, M. (1995). Las matemáticas en la sociedad. *UNO: Revista de Didáctica de las Matemáticas, 2*(6), 45–57.

Niss, M. (1996). Goals of mathematics teaching. In A. J. Bishop, K. Clements, C. Keitel, J. Kilpatrick, & C. Laborde (Eds.), *International handbook of mathematics education* (pp. 11–47). Dordrecht: Kluwer.

Norris, S. (2004). *Analyzing multimodal interaction: A methodological framework*. London: Routledge.

Noss, R. (1994). Structure and ideology in the mathematics curriculum. *For the Learning of Mathematics, 14*(1), 2–10.

Noss, R. (1997). *New cultures, new numeracies. Inaugural professorial lecture*. London: Institute of Education, University of London.

Noss, R., & Hoyles, C. (1996). The visibility of meanings: Modelling the mathematics of banking. *International Journal of Computers for Mathematical Learning, 1*(1), 3–31.

Nunes, T., Schliemann, A. D., & Carraher, D. W. (1993). *Street mathematics and school mathematics*. Cambridge: Cambridge University Press.

O'Halloran, K. L. (2004). Discourse in secondary school mathematics classrooms according to social class and gender. In J. A. Foley (Ed.), *Language, education and discourse: Functional approaches* (pp. 191–225). London: Continuum.

O'Halloran, K. L. (2005). *Mathematical discourse: Language, symbolism and visual images*. London: Continuum.

O'Holloran, K. L. (2010). The semantic hyperspace: Accumulating mathematical knowledge across semiotic resources and modalities. In F. Christie & K. Marton (Eds.), *Disciplinarily: Functional linguistics and sociological perspectives* (pp. 217–236). London: Continuum.

O'Neil, C., & Lambert, A. (1990). *Drama structures: A practical handbook for teachers*. Kingston: Stanley Thornes.

Oblinger, D. (2003). Boomers, gen-xers, and millennials: Understanding the "new students". *EDUCAUSE Review, 38*(4), 36–47.

OECD. (2006). *Evolution of student interest in science and technology studies*. Policy report. Paris: OECD.

Ohanian, S. (1992). *Garbage pizza, patchwork quilts, and math magic: Stories about teachers who love to teach and children who love to learn*. New York: Freeman.

Oliveira, H. (2009a). Understanding the teacher's role in supporting students' generalization when investigating sequences. *Quaderni di Ricerca in Didattica (Matematica), 19*(Supplemento 4), 133–143.

Oliveira, M. K. (2009b). *Cultura e psicologia: questões sobre o desenvolvimento do adulto*. São Paulo: Hucitec.

Oliveira, H., Menezes, L., & Canavarro, A. P. (2012). The use of classroom videos as a context for research on teachers' practice and teacher education. In *Pre-proceedings of 12th international congress on mathematical education* (pp. 4280–4289), Seoul, Korea.

Olson, K. R., & Spelke, E. S. (2008). Foundations of cooperation in young children. *Cognition, 108*, 222–231.

Omniewski, R. (1999). *The effects of an arts infusion approach on the mathematics achievement of second-grade students*. Unpublished PhD thesis, Kent State University, USA.

Ostermeier, C., Prenzel, M., & Duit, R. (2010). Improving science and mathematics instruction: The SINUS project as an example for reform as teacher professional development. *International Journal of Science Education, 32*(3), 303–327.

Pais, A. (2012). A critical approach to equity. In O. Skovsmose & B. Greer (Eds.), *Opening the cage: Critique and politics of mathematics education* (pp. 49–92). Rotterdam: Sense.

Pais, A., & Valero, P. (2012). Researching research: Mathematics education in the political. *Educational Studies in Mathematics, 80*(1), 9–24.

Palm, T. (2006). Word problems as simulations of real-world situations: A proposed framework. *For the Learning of Mathematics, 26*(1), 42–47.

Palm, T. (2008). Impact of authenticity on sense making in word problem solving. *Educational Studies in Mathematics, 67*(1), 37–58.

Papert, S. (1985). *Logo: Computadores e Educação*. São Paulo: Brasiliense.

Parsons, J. E., Adler, T. F., & Kaczala, C. M. (1982). Socialization of achievement attitudes and beliefs: Parental influences. *Child Development, 53*(2), 310–321.

Pasquino, P. (1978). Theatrum politicum. The genealogy of capital: Police and the state of prosperity. *Ideology and Consciousness, 4*, 41–54.

Pearson, W. J. (2008). *Who will do science? Revisited*. Paper presented at the annual meeting of the commission on professionals in science and engineering, Baltimore. www.cpst.org/2008meeting/presentations/pearson-chubin-davis.pdf. Accessed 24 Mar 2014.

Pedauque, R. T. (2006). *Le document à la lumière du numérique*. Caen: C&F.

Pennequin, V., Sorel, O., Nanty, I., & Fontaine, R. (2010). Metacognition and low achievement in mathematics: The effect of training in the use of metacognitive skills to solve mathematical word problems. *Thinking & Reasoning, 16*, 198–220.

Peressini, D. (1998). The portrayal of parents in the school mathematics reform literature: Locating the context for parent involvement. *Journal for Research in Mathematics Education, 29*(5), 555–582.

PhET. (2010). *Interactive simulation project*. http://phet.colorado.edu/. Accessed 15 July 2013.

Piaget, J. (1972). *Judgment and reasoning in the child*. London: Routledge & Kegan Paul. (Original work published 1924)

Piaget, J., & Inhelder, B. (1956). *The child's conception of space*. London: Routledge.

Pickering, A. (2011). H-: Brains, Selves and Spirituality in the History of Cybernetics. *metanexus, 9*(3). Retrieved from metanexus website: http://www.metanexus.net/essay/h-brains-selves-and-spirituality-history-cybernetics

Pierce, J. L. (2009). *A co-construction of space trilogy: Examining how ESL teachers, English learners and classrooms interact*. Unpublished PhD thesis, Indiana University of Pennsylvania.

Pimm, D. (1993). The silence of the body. *For the Learning of Mathematics, 13*(1), 35–38.

Plato (1961). *The collected dialogues of Plato including the letters* (E. Hamilton & H. Cairns, Eds.). New York: Pantheon.

Pólya, G. (1962). *Mathematical discovery: On understanding, learning and teaching problem solving*. New York: Wiley.

Ponte, J. P., Brocado, J., & Oliveira, H. (2003). *Investigações matemáticas na sala de aula*. Belo Horizonte: Autêntica.

Ponza, M. V. (2000). Mathematical dramatisation. In J. Fauvel & J. A. van Maanen (Eds.), *History in mathematics education: The ICMI study* (pp. 335–342). Dordrecht: Kluwer.

Popkewitz, T. S. (2004a). The alchemy of the mathematics curriculum: Inscriptions and the fabrication of the child. *American Educational Research Journal, 41*(1), 3–34.

Popkewitz, T. S. (2004b). School subjects, the politics of knowledge, and the projects of intellectuals in change. In P. Valero & R. Zevenbergen (Eds.), *Researching the socio-political dimensions of mathematics education: Issues of power in theory and methodology* (pp. 251–267). Boston: Kluwer.

Popkewitz, T. S. (2008). *Cosmopolitanism and the age of school reform: Science, education, and making society by making the child*. New York: Routledge.

Popkewitz, T. S. (2009). Curriculum study, curriculum history, and curriculum theory: The reason of reason. *Journal of Curriculum Studies, 41*(3), 301–319.

Porter, T. (1953/1996). *Trust in numbers: The pursuit of objectivity in science and public life*. Princeton: Princeton University Press.

Porter, T. M. (1995). *Trust in numbers: The pursuit of objectivity in science and public life*. Princeton: Princeton University Press.

Pothier, Y., & Sawada, D. (1983). Partitioning: The emergence of rational number ideas in young children. *Journal for Research in Mathematics Education, 14*, 307–317.
Potter, J., Wetherell, M., & Chitty, A. (1991). Quantification rhetoric: Cancer on television. *Discourse and Society, 2*(3), 333–365.
Prado, M. E. (2003). *Educação a distância e formação do professor: Redimensionando concepções de aprendizagem*. Unpublished PhD thesis, Pontifícia Universidade Católica de São Paulo, São Paulo. Available in: http://www.matematicaepraticadocente.net.br/pdf/teses_dissertacoes/tese_Bette_Prado.pdf. Accessed 24 Feb 2015.
Prado, M. E. (2008). Os princípios da informática na educação e o papel do professor: Uma abordagem inclusiva. In D. Raiça (Ed.), *Tecnologias para a educação inclusiva* (pp. 55–66). São Paulo: Avercamp.
Prado, M. E., & Valente. J. A. (2002). A educação a distância possibilitando a formação do professor com base no ciclo da prática pedagógica. In M. C. Moraes (Ed.), *Educação a distância: fundamentos e práticas* (pp. 27–50). Campinas: UNICAMP/NIED.
Prado, M. E., Lobo da Costa, N. M., & Galvão, M. E. (2013). *Proceedings of CIEAEM 65*. Torino: Italy. *Teacher development for the integrated use of technologies in Mathematics teaching*.
Premack, D., & Woodruff, G. (1978). Does the chimpanzee have a "theory of mind"? *Behavioral and Brain Sciences, 4*, 515–526.
PRIMAS. (2013). *Promoting inquiry-based learning in mathematics and science across Europe*. www.primas-project.eu/en/index.do. Accessed 23 Apr 2013.
Prince, J. D. (1904). The code of Hammurabi. *The American Journal of Theology, 8*(3), 601–609.
Proust, C. (2011). Teachers' writings and students' writings: School material in Mesopotamia. In G. Gueudet, B. Pepin, & L. Trouche (Eds.), *From text to 'lived' resources: Mathematical materials and teacher development* (pp. 161–180). Dordrecht: Springer.
Proust, C. (2012). Masters' writings and students' writings: School material in Mesopotamia. In G. Gueudet, B. Pepin, & L. Trouche (Eds.), *From text to 'lived' resources: Mathematics curriculum materials and teacher development* (pp. 161–180). Dordrecht: Springer.
Pruitt, B., & Thomas, P. (2007). *Democratic dialogue: A handbook for practitioners*. Washington, DC: Trydells Tryckeri.
Quine, W. O. (1960). *Word and object*. Cambridge: MIT Press.
Rabardel, P. (1995a). *L'homme et les outils contemporains*. Paris: A. Colin.
Rabardel, P. (1995b). *Les hommes et les technologies, approche cognitive des instruments contemporains*. Paris: Armand Colin.
Radford, L. (2001). The historical origins of algebraic thinking. In R. Sutherland, T. Rojano, A. Bell, & R. Lins (Eds.), *Perspectives on school algebra* (pp. 13–63). Dordrecht: Kluwer.
Radford, L. (2008). The ethics of being and knowing: Towards a cultural theory of learning. In L. Radford, G. Schubring, & F. Seeger (Eds.), *Semiotics in mathematics education: Epistemology, history, classroom, and culture* (pp. 215–234). Rotterdam: Sense.
Radford, L., Edwards, L., & Arzarello, F. (2009). Introduction: Beyond words. *Educational Studies in Mathematics, 70*(2), 91–95.
Ramos, P., Giannella, T. R., & Struchiner, M. (2009). *A pesquisa baseada em design em artigos científicos sobre o uso de ambientes de aprendizagem mediados pelas tecnologias da informação e da comunicação no ensino de ciências: Uma análise preliminar*. Florianópolis: VII Encontro Nacional de Pesquisa em Educação em Ciências.
Rawls, J. A. (1971). *A theory of justice*. Cambridge, MA: Harvard University Press.
Rawls, J. A. (2001). *Justice as fairness: A restatement*. Cambridge, MA: Harvard University Press.
Remillard, J. T., & Jackson, K. (2006). Old math, new math: Parents' experiences with Standards-based reform. *Mathematical Thinking and Learning, 8*(3), 231–259.
Resnick, L. B. (1992). From protoquantities to operators: Building mathematical competence on a foundation of everyday knowledge. In G. Leinhardt, R. Putnam, & R. A. Hattrup (Eds.), *Analysis of arithmetic for mathematics teaching* (pp. 373–429). Hillsdale: Lawrence Erlbaum.
Richardson, J. E. (2007). *Analysing newspapers: An approach from critical discourse analysis*. New York: Palgrave Macmillan.

Rico, L., Gómez, P., & Cañadas, M. C. (2014). Formación inicial en educación matemática de los maestros de Primaria en España 1991–2010. *Revista Española de Pedagogía, 363*, 35–59.

Rochat, P., Dias, M., Liping, G., Broesch, T., Passos-Ferrera, C., Winning, A., & Berg, B. (2009). Fairness in distributive justice by 3- and 5-year-olds across seven cultures. *Journal of Cross-Cultural Psychology, 40*, 416–442.

Rockliffe, F. (2011). The perspectives of parents as students, newly qualified or early career teachers on strategies for developing effective parental partnership in mathematics education. In M. Pieri, A. Pepe, & L. Addimando (Eds.), *Home, school and community: A partnership for a happy life?* (pp. 95–96). Milano: I Libri di Emil.

Roditi, E. (2001). *L'enseignement de la multiplication des déximaux en sixième, étude de pratiques ordinaires*. Unpublished PhD thesis, Université Paris 7, Paris.

Roeser, R. W., Eccles, J. S., & Sameroff, A. J. (2000). School as a context of early adolescents' academic and social-emotional development: A summary of research findings. *The Elementary School Journal, 100*(5), 443–471.

Rogers, R., Malancharuvil-Berkes, E., Mosley, M., Hui, D., & O'Garro Joseph, G. (2005). Critical discourse analysis in education: A review of the literature. *Review of Educational Research, 75*(3), 365–416.

Rose, N. (1988). Calculable minds and manageable individuals. *History of the Human Sciences, 1*, 179–200.

Rose, N. (1991). Governing by numbers: Figuring our democracy. *Accounting Organization and Society, 16*(7), 673–692.

Roth, W.-M., & Tobin, K. (2009). *The world of science education: Handbook of research in North America*. Rotterdam: Sense.

Rouse, J. (1994). Power/knowledge. In G. Gutting (Ed.), *The Cambridge companion to Foucault* (pp. 92–114). Cambridge: Cambridge University Press.

Ruthven, K., Hofmann, R., & Mercer, N. (2011). A dialogic approach to plenary problem synthesis. In B. Ubuz (Ed.), *Proceedings of the 35th conference of the international group for the psychology of mathematics education* (Vol. 4, pp. 81–88). Ankara: ICMI - International Commission on Mathematical Instruction.

Saab, J. (1987). *The effects of creative drama methods on mathematics achievement, attitudes and creativity*. Unpublished PhD thesis, West Virginia University.

Sabena, C., Robutti, O., Ferrara, F., & Arzarello, F. (2012). The development of a semiotic frame to analyse teaching and learning processes: Examples in pre- and post-algebraic contexts. In L. Coulange, J.-P. Drouhard, J.-L. Dorier, & A. Robert (Eds.), *Recherches en Didactique des Mathématiques, Numéro spécial hors-série, Enseignement de l'algèbre élémentaire: bilan et perspectives* (pp. 231–245). Grenoble: La Pensée Sauvage.

Saldanha, R. P. T. (2009). *Indicadores de um currículo flexível no uso de computadores portáteis*. Unpublished master thesis, Pontifícia Universidade Católica de São Paulo, São Paulo.

Sally, D., & Hill, E. (2006). The development of interpersonal strategy: Autism, theory-of-mind, cooperation and fairness. *Journal of Economic Psychology, 27*, 73–97.

Salmon, G. (2004). *E-moderating: The key to teaching and learning online* (2nd ed.). London: Taylor & Francis.

Sawyer, R. K. (2003). *Group creativity: Music, theatre, collaboration*. Mahwah: Lawrence Erlbaum.

Saxe, G. (1981). Body parts as numerals: A developmental analysis of numeration among the Oksapmin in Papua New Guinea. *Child Development, 52*, 306–316.

Schechner, R. (2003). *Performance theory*. London: Routledge.

Schmidt, W. H., McKnight, C. C., & Raizen, S. (Eds.). (1997). *A splintered vision: An investigation of US science and mathematics education*. New York: Springer.

Schoenfeld, A. H. (Ed.). (1983). *Problem solving in the mathematics curriculum: A report, recommendations, and an annotated bibliography*. Washington, DC: MAA.

Schommer, M. (1990). Effects of beliefs about the nature of knowledge on comprehension. *Journal of Educational Psychology, 82*, 498–504.

Schreiner, C. (2006). *Exploring a ROSE garden: Norwegian youth's orientations towards science – Seen as signs of late modern identities*. Unpublished Ph.D. thesis, University of Oslo.

Schreiner, C., & Sjøberg, S. (2007). Science education and young people's identity construction: Two mutually incompatible projects? In D. Corrigan, J. Dillon, & R. Gunstone (Eds.), *The re-emergence of values in science education* (pp. 231–248). Rotterdam: Sense.

Schwarz, B., & Dreyfus, T. (1995). New actions upon old objects: A new ontological perspective of functions. *Educational Studies in Mathematics, 29*(3), 259–291.

Scollon, R., & Scollon, S. W. (2003). *Discourses in place: Language in the material world*. London: Routledge.

Seeger, F., Voigt, J., & Waschescio, U. (Eds.). (1998). *The culture of the mathematics classroom*. Cambridge: Cambridge University Press.

Sen, A. (2009). *The idea of justice*. Cambridge, MA: Harvard University Press.

Serradó, A. (2009). E-forum, a strategy for developing key competences of communication in, with and about mathematics. In L. Gómez Chova, D. Martí Belenguer, & I. Candel Torres (Eds.), *Proceedings of the international conference and new learning technologies (EDULEARN)* (pp. 1–12). Barcelona: IATED.

Serradó, A. (2012). How to question in an on-line forum to promote a democratic mathematical knowledge construction? *International Journal for Mathematics in Education, 4*, 369–374.

Serres, M. (2012). *Petite poucette*. Paris: Le Pommier.

Sfard, A. (2008). *Thinking as communicating*. Cambridge: Cambridge University Press.

Sfard, A., & Prusak, A. (2006). Telling identities: In search of an analytic tool for investigating learning as a culturally shaped activity. *Educational Researcher, 34*(4), 14–22.

Shaver, A. V., & Walls, R. T. (1998). Effect of parent involvement on student reading and mathematics achievement. *Journal of Research and Development in Education, 31*(2), 90–97.

Sheldon, S. B., & Epstein, J. L. (2005). Involvement counts: Family and community partnerships and mathematics achievement. *The Journal of Educational Research, 98*(4), 196–207.

Sheridan, A. (1980). *The will to truth*. London: Tavistock.

Sherman, J. (2009). *Those who work, those who don't: Poverty, morality, and family in rural America*. Minneapolis: University of Minnesota Press.

Sheyholislami, J. (2008). *Identity, discourse and the media: The case of the Kurds*. Unpublished PhD thesis, Carleton University, Ottawa.

Siemens, G. (2012). *Learning analytics: envisioning a research discipline and a domain of practice*. Paper presented at the 2nd International Conference on Learning Analytics & Knowledge, 29 April–2 May, Vancouver, Canada. http://learninganalytics.net/LAK_12_keynote_Siemens.pdf. Accessed 4 Sept 2013.

Sierpinska, A. (1994). *Understanding in mathematics*. London: Falmer.

Sierpinska, A., & Kilpatrick, J. (Eds.). (1998). *Mathematics education as a research domain: A search for identity*. Dordrecht: Kluwer.

Silver, E. (2003). Attention deficit disorder? *Journal for Research in Mathematics Education, 34*(1), 2–3.

Simon, B. S. (2004). High school outreach and family involvement. *Social Psychology of Education, 7*(2), 185–209.

Singer-Freeman, K., & Goswami, U. (2001). Does half a pizza equal half a box of chocolates? Proportional matching in an analogy paradigm. *Cognitive Development, 16*, 811–829.

Sjøberg, S., & Schreiner, C. (2010). *The ROSE project: An overview and key findings*. Oslo: Oslo University.

Skovsmose, O. (1994). *Towards a philosophy of critical mathematics education*. Dordrecht: Kluwer.

Skovsmose, O. (1996). Critical mathematics education. In A. J. Bishop, K. Clements, C. Keitel, J. Kilpatrick, & C. Laborde (Eds.), *International handbook of mathematics education* (pp. 1257–1288). Dordrecht: Kluwer.

Skovsmose, O. (1998a). Linking mathematics education and democracy: Citizenship, mathematics archeology, mathemacy and deliberative interaction. *Zentralblatt für Didaktik der Mathematik, 30*(4), 195–203.

Skovsmose, O. (1998b). Linking mathematics education and democracy: Citizenship, mathematical archaeology, mathemacy and deliberative interaction. *Zentralblatt für Didaktik der Mathematik, 98*(6), 195–203.

Skovsmose, O. (2004a). *Educação matemática critica: a questão da democracia* (2nd ed.). Campinas: Papirus.

Skovsmose, O. (2004b, July 4–11). *Critical mathematics education for the future.* Paper presented at ICME 10, Copenhagen, Denmark.

Skovsmose, O. (2005). *Travelling through education: Uncertainty, mathematics, responsibility.* Rotterdam: Sense.

Skovsmose, O. (2007). Mathematical literacy and globalization. In B. Atweh, A. Calabrese Barton, M. Borba, N. Gough, C. Keitel, C. Vistro-Yu, & R. Vithal (Eds.), *Internationalisation and globalisation in mathematics and science education* (pp. 3–18). Dordrecht: Springer.

Skovsmose, O. (2008). *Desafios da reflexão em educação matemática crítica.* Campinas: Papirus.

Skovsmose, O. (2010). Can facts be fabricated through mathematics? *Philosophy of Mathematics Education Journal, 25*. Online.

Skovsmose, O. (2011). *An invitation to critical mathematics education.* Rotterdam: Sense.

Skovsmose, O., & Valero, P. (2001a). Breaking political neutrality: The critical engagement of mathematics education with democracy. In B. Atweh, H. Forgasz, & B. Nebres (Eds.), *Sociocultural research on mathematics education: An international perspective* (pp. 37–55). Mahwah: Lawrence Erlbaum.

Skovsmose, O., & Valero, P. (2001b). Breaking political neutrality: The critical engagement of mathematics education with democracy. In B. Atweh, H. Forgasz, & B. Nebres (Eds.), *Sociocultural aspects of mathematics education: An international research perspective* (pp. 37–56). Mahwah: Lawrence Erlbaum.

Skovsmose, O., & Valero, P. (2002). Democratic access to powerful mathematical ideas. In L. English (Ed.), *International research in mathematics education* (pp. 383–408). New York: Routledge.

Skovsmose, O., & Yasukawa, K. (2000, March 26–31). Formatting power of mathematics: A case study and questions for mathematics education. In *Proceedings of the 2nd MES conference*, Montechoro, Algarve, Portugal.

Skovsmose, O., Scandiuzzi, P. P., Valero, P., & Alrø, H. (2008). Learning mathematics in a borderland position: Students' foregrounds and intentionality in a Brazilian favela. *Journal of Urban Mathematics Education, 1*(1), 35–59.

Skovsmose, O., Yasukawa, K., & Ravn, O. (2013). Scripting the world with mathematics. In P. Ernest & B. Sriraman (Eds.), *Critical mathematics education: Theory, praxis and reality* (pp. 255–281). Charlotte: IAP.

Sloane, S., Baillargeon, R., & Premack, D. (2012). Do infants have a sense of fairness? *Psychological Science, 23*(2), 196–204.

Smith, T. W., & Colby, S. A. (2007). *Teaching for deep learning.* Washington, DC: Heldref.

Smith, E., & Gorard, S. (2011). Is there a shortage of scientists? A re-analysis of supply for the UK. *British Journal of Educational Studies, 59*(2), 159–177.

Smith, F. M., & Hausafus, C. O. (1998). Relationship of family support and ethnic minority students' achievement in science and mathematics. *Science Education, 82*(1), 111–125.

Smith, T. W., Gordon, B., Colby, S. A., & Wang, J. (2005). *An examination of the relationship between depth of student learning and national board certification status.* Boone: Office for Research on Teaching, Appalachian State University.

Smith, C. E., Blake, P., & Harris, P. L. (2013). I should but I won't: Why young children endorse norms of fair sharing but do not follow them. *PLoS One, 8*(8).

Sobel, D. (2004). *Place-based education.* Great Barrington: Orion Society.

Solomon, I. (1992). *Power and class in modern Greek school (in Greek).* Athens: Alexandria.

Somers, J. (1994). *Drama in the curriculum.* London: Cassell.

Sommerville, J. A., Schmidt, M. F. H., Yun, J., & Burns, M. (2013). The development of fairness expectations and prosocial behavior in the second year of life. *Infancy, 18*, 40–66.

Spelke, E. S. (2000). Core knowledge. *American Psychologist, 55*, 1233–1243.

References 467

Spelke, E. S. (2008). La théorie du 'core knowledge'. *L'Année Psychologique, 108*, 721–756.
Spijkerboer, L., Bootsma, G., & Denijs, W. (2007). *Determinatie en toetsen*. Utrecht: APS.
Spinillo, A., & Bryant, P. (1991). Children's proportional judgments: The importance of 'half'. *Child Development, 62*, 427–440.
Star, S. L. (2010). This is not a boundary object: Reflections on the origin of a concept. *Science Technology Human Values, 35*(5), 601–616.
Starkey, P., & Klein, A. (2000). Fostering parental support for children's mathematical development: An intervention with head start families. *Early Education and Development, 11*(5), 659–680.
Stein, M. K., & Smith, M. S. (1998). Mathematical tasks as a framework for reflection: From research to practice. *Mathematics Teaching in the Middle School, 3*, 268–275.
Stein, M. K., Engle, R. A., Smith, M. S., & Hughes, E. K. (2008). Orchestrating productive mathematical discussions: Helping teachers learn to better incorporate student thinking. *Mathematical Thinking and Learning, 10*(4), 313–340.
Stentoft, D., & Valero, P. (2009). Identities-in-action: Exploring the fragility of discourse and identity in learning mathematics. *Nordic Studies in Mathematics Education, 14*(3), 55–77.
Stentoft, D., & Valero, P. (2010). Fragile learning in mathematics classrooms: Exploring mathematics lessons within a pre-service course. In M. Walshaw (Ed.), *Unpacking pedagogies: New perspectives for mathematics* (pp. 87–107). Charlotte: IAP.
Stinson, D. W. (2004). Mathematics as "gate-keeper"(?): Three theoretical perspectives that aim toward empowering all children with a key to the gate. *The Mathematics Educator, 14*(1), 8–18.
Stinson, D. W., & Bullock, E. C. (2012). Critical postmodern theory in mathematics education research: A praxis of uncertainty. *Educational Studies in Mathematics, 80*(1–2), 41–55.
Stocker, D. (2006). Re-thinking real-world mathematics. *For the Learning of Mathematics, 26*(2), 29–30.
Straehler-Pohl, H., & Gellert, U. (2013). Towards a Bernsteinian language of description for mathematics classroom discourse. *British Journal of Sociology of Education, 34*(3), 313–332.
Straehler-Pohl, H., Gellert, U., Fernandez, S., & Figueiras, L. (2014). School mathematics registers in a context of low academic expectations. *Educational Studies in Mathematics, 85*(2), 175–199.
Streater, A., & Chertkoff, J. (1976). Distribution of rewards in a triad: A developmental test of equity theory. *Child Development, 47*, 800–805.
Sweet, M. A., & Appelbaum, M. I. (2004). Is home visiting as effective strategy? A meta-analytic review of home visiting programs for families with young children. *Child Development, 75*(5), 1435–1456.
Takagishi, H., Kameshima, S., Shug, J., Koizumi, M., & Yamagishi, T. (2010). Theory of mind enhances preference for fairness. *Journal of Experimental Child Psychology, 105*, 130–137.
Tall, D., van Blokland, P., & Kok, D. (1991). *A graphic approach to the calculus*. Warwicks: Rivendell Software.
Taylor, P. C. (1996). Mythmaking and mythbreaking in the mathematics classroom. *Educational Studies in Mathematics, 31*(1–2), 151–173.
Teoh, J. (2012). Drama as a form of critical pedagogy: Empowerment of justice. *The Pedagogy and Theatre of the Oppressed International Journal, 1*(1), 4–26.
Theobald, P. (1997). *Teaching the commons: Place, pride, and the renewal of community*. Boulder: Westview.
Theobald, P. (2009). *Education now: How rethinking America's past can change its future*. Boulder: Paradigm.
Thiollent, M. (1988). *Metodologia da pesquisa-ação*. São Paulo: Cortez.
Thorndike, E. L. (1924). Mental discipline in high school studies. *Journal of Educational Psychology, 15*(1), 1–22.
Thurston, W. P. (1995). On proof and progress in mathematics (reprint). *For the Learning of Mathematics, 15*(1), 29–37.
Tirosh, D. (2000). Enhancing prospective teachers' knowledge of children's conceptions: The case of division of fractions. *Journal for Research in Mathematics Education, 31*, 5–25.

Tobin, K. G. (2006). *Teaching and learning science: A handbook*. Westport: Praeger.
Towers, J. (2010). Learning to teach mathematics through inquiry: A focus on the relationship between describing and enacting inquiry-oriented teaching. *Journal of Mathematics Teacher Education, 13*(3), 243–263.
Tracker. (2015). Video analysis and modeling tool (version 4.87). http://www.cabrillo.edu/~dbrown/tracker/. Accessed 22 February 2015.
Treffers, A. (1978). *Wiskobas Doelgericht*. Utrecht: IOWO.
Treffers, A. (1987). *Three dimensions: A model of goal and theory description in mathematics instruction – The Wiskobas project*. Dordrecht: Reidel.
Troelsen, R. (2005). Unges interesse for naturfag: Hvad ved vi, og hvad kan vi bruge det til? *MONA, 2*, 7–21.
Trouche, L. (2004). Managing the complexity of human/machine interactions in computerized learning environments: Guiding students' command process through instrumental orchestrations. *International Journal of Computers for Mathematical Learning, 9*, 281–307.
Tuckman, B., & Monetti, D. (2010). *Educational psychology*. Belmont: Cengage Learning.
Turner, V. (1986). *From ritual to theatre: The human seriousness of play*. New York: PAJ.
Tyminski, A. M., Land, T. J., Drake, C., Zambak, V. S., & Simpson, A. (2014). Preservice elementary mathematics teachers' emerging ability to write problems to build on children's mathematics. In J.-J. Lo, K. R. Leatham, & L. R. Van Zoest (Eds.), *Research trends in mathematics teacher education* (pp. 193–218). Cham: Springer.
UCA Princípios. (2007). *Orientadores para o uso pedagógico do laptop na educação escolar*. Brasilia: Ministério de Educação.
Ulriksen, L. (2003). Børne- og ungdomskultur og naturfaglige uddannelser. In H. Busch, S. Horst, & R. Troelsen (Eds.), *Inspiration til fremtidens naturfaglige uddannelser. En antologi* (pp. 285–318). Copenhagen: Undervisningsministeriet.
U.S. Census Bureau. (2000). *State & county quickfacts*. http://quickfacts.census.gov
Valente, J. A. (1999). *O computador na sociedade do conhecimento*. Campinas: UNICAM/NIED.
Valente, J. A. (2008). A escola como geradora e gestora do conhecimento: O papel das tecnologias de informação e comunicação. In A. J. H. Guevara & A. M. Rosini (Eds.), *Tecnologias emergentes: Organizações e educação* (pp. 21–40). São Paulo: Cengage Learning.
Valente, J. A. (2011). Um laptop para cada aluno: Promessas e resultados educativos efetivos. In M. E. Almeida & M. E. Prado (Eds.), *O computador portátil na escola: Mudanças e desafios nos processos de ensino e aprendizagem* (pp. 20–33). São Paulo: Avercamp.
Valero, P. (1999). Deliberative mathematics education for social democratization in Latin America. *Zentralblatt für Didaktik der Mathematik, 31*(1), 20–26.
Valero, P. (2004a). Socio-political perspectives on mathematics education. In P. Valero & R. Zevenbergen (Eds.), *Researching the socio-political dimensions of mathematics education* (pp. 5–23). Dordrecht: Kluwer.
Valero, P. (2004b). Postmodernism as an attitude of critique to dominant mathematics education research. In M. Walshaw (Ed.), *Mathematics education within the postmodern* (pp. 35–54). Greenwich: IAP.
Valero, P. (2004c). Socio-political perspectives on mathematics education. In P. Valero & R. Zevenbergen (Eds.), *Researching the socio-political dimensions of mathematics education: Issues of power in theory and methodology* (pp. 5–24). Boston: Kluwer.
Valero, P. (2010). Mathematics education as a network of social practices. In V. Durand-Guerrier, S. Soury-Lavergne, & F. Arzarello (Eds.), *Proceedings of the sixth congress of the European society for research in mathematics education* (pp. 54–80). Lyon: Institut National de Récherche Pédagogique.
Valero, P., & Zevenbergen, R. (Eds.). (2004). *Researching the sociopolitical dimensions of mathematics classroom: Issues of power in theory and methodology (pp. 107–123)*. Dordrecht: Kluwer.
Valero, P., García, G., Camelo, F., Mancera, G., & Romero, J. (2012). Mathematics education and the dignity of being. *Pythagoras, 33*(2), 171–179.

References

Vallin, C. (2004). *Projeto CER: Comunidade Escolar de Estudo, Trabalho e Reflexão.* Unpublished PhD thesis, Pontifícia Universidade Católica de São Paulo, São Paulo.

Van den Heuvel-Panhuizen, M. (2001). Realistic mathematics education in the Netherlands. In J. Anghileri (Ed.), *Principles and practices in arithmetic teaching* (pp. 49–63). Buckingham: Open University Press.

Van den Heuvel-Panhuizen, M. (2005). The role of contexts in assessment problems in mathematics. *For the Learning of Mathematics, 25*(2), 2–10.

Van Dijk, T. A. (1988). *News as discourse.* Hillsdale: Lawrence Erlbaum.

Van Dijk, T. A., & Kintsch, W. (1983). *Strategies of discourse comprehension.* New York: Academic.

Vanegas, Y. M. (2013). Competencias ciudadanas y desarrollo profesional en matemáticas. Unpublished PhD thesis, Universitat de Barcelona.

Vanegas, Y. M., & Giménez, J. (2012). What future mathematics teachers understand as democratical values. *International Journal for Mathematics in Education, 4*, 457–462.

Vanegas, Y., Díez-Palomar, J., Font, V., & Giménez, J. (2013). *Considering extrinsic aspects in the analysis of mathematics teacher-training programs.* Paper presented at the 37th conference of the international group for the Psychology of Mathematics Education (PME), Kiel, Germany

Verschaffel, L. (2002). Taking the modeling perspective seriously at the elementary level: Promises and pitfalls. In A. D. Cockburn & E. Nardi (Eds.), *Proceedings of the 26th PME international conference* (pp. 64–80). Norwich: PME.

Vithal, R. (2003a). Teachers and street children: On becoming a teacher of mathematics. *Journal of Mathematics Teacher Education, 6*(2), 165–183.

Vithal, R. (2003b). *In search of a pedagogy of conflict and dialogue for mathematics education.* Dordrecht: Kluwer.

Vygotsky, L. S. (1978). *Mind in society.* Cambridge, MA: Harvard University Press.

Walkerdine, V. (1988). *The mastery of reason: Cognitive development and the production of rationality.* London: Routledge.

Walkerdine, V. (1990). *The mastery of reason: Cognitive development and the production of rationality.* London: Routledge.

Walshaw, M. (2004). *Mathematics education within the postmodern.* Greenwich: IAP.

Walshaw, M. (2011). Identity as the cornerstone of quality and equitable mathematical experiences. In B. Atweh, M. Graven, W. Secada, & P. Valero (Eds.), *Mapping equity and quality in mathematics education* (pp. 91–102). New York: Springer.

Watson, J. M., & Moritz, J. B. (2003). Fairness of dice: A longitudinal study of students' beliefs and strategies for making judgments. *Journal for Research in Mathematics Education, 34*(4), 270–304.

Watson, J. M., Campbell, K. J., & Collis, K. F. (1999). The structural development of the concept of fraction by young children. *Journal of Structural Learning and Intelligence Systems, 13,* 171–193.

Wedege, T. (2009). The problem field of adults learning mathematics. In G. Griffiths & D. Kaye (Eds.), *Proceedings of 16th adults learning mathematics* (pp. 13–24). London: Adults Learning Mathematics–A Research Forum.

Wedege, T. (2010). People's mathematics in working life. Why is it invisible? *Adults Learning Mathematics – An International Journal, 5*(1), 89–97.

Weinstein, L., & Adam, J. A. (2008). *Guesstimation: Solving the world's problems on the back of a cocktail napkin.* Princeton: Princeton University Press.

Wellman, H. M., & Gelman, S. A. (1998). Knowledge acquisition in foundational domains. In W. Damon (Ed.), *Handbook of child psychology (Cognition, perception, and language* (Vol. 2, pp. 523–573). Hoboken: Wiley.

Wellman, H., & Liu, D. (2004). Scaling of theory of mind tasks. *Child Development, 75,* 523–541.

Wellman, H. M., Hickling, A. K., & Schult, C. A. (1997). Young children's psychological, physical, and biological explanations. In H. M. Wellman & K. Inagaki (Eds.), *The emergence of core domains of thought: Children's reasoning about physical, psychological, and biological phenomena* (pp. 7–25). San Francisco: Jossey-Bass.

Wellman, H., Cross, D., & Watson, J. (2001). Meta-analysis of theory-of-mind development: The truth about false belief. *Child Development, 72*(3), 655–684.

Wells, G. (2004). *Dialogic inquiry: Towards a sociocultural practice and theory of education.* Cambridge: Cambridge University Press.

Wigfield, A., Eccles, J. S., Mac Iver, D., Reuman, D. A., & Midgley, C. (1991). Transitions during early adolescence: Changes in children's domain specific self-perceptions and general self-esteem across the transition to junior high school. *Developmental Psychology, 27*(4), 552–565.

Willis, G. B., Royston, P., & Bercini, D. (1991). The use of verbal report methods in the development and testing of survey questionnaires. *Applied Cognitive Psychology, 5*, 251–267.

Winbourne, P. (2009). Choice: Parents, teachers, children, and ability grouping in mathematics. In L. Black, H. Mendick, & Y. Solomon (Eds.), *Mathematical relationships in education: Identities and participation* (pp. 58–70). New York: Routledge.

Wing, R. E., & Beal, C. R. (2004). Young children's judgments about the relative size of shared portions: The role of material type. *Mathematical Thinking and Learning, 6*, 1–14.

Winter, H. (1994). Modelle als Konstrukte zwischen lebensweltlichen Situationen und arithmetischen Begriffen. *Grundschule, 26*(3), 10–13.

Wood, T., Sullivan, P., Tirosh, D., Krainer, K., & Jaworski, B. (Eds.). (2008). *The international handbook of mathematics teacher education* (Vol. 4). Rotterdam: Sense.

Wu, Z., & Su, Y. (2013). Development of sharing in preschoolers in relation to theory of mind understanding. In M. Knauff, M. Pauen, N. Sebanz, & I. Wachsmuth (Eds.), *Proceedings of the 35th annual conference of the Cognitive Science Society* (pp. 3811–3816). Austin: Cognitive Science Society.

Yackel, E., & Cobb, P. (1996). Sociomathematical norms, argumentation, and autonomy in mathematics. *Journal for Research in Mathematics Education, 27*(4), 458–477.

Yankelovich, D. (1999). *The magic of dialogue: Transforming conflict into cooperation.* New York: Simon & Schuster.

Yassa, N. (1999). High school students' involvement in creative drama: The effects on social interaction. *Research in Drama and Theatre in Education, 4*(1), 37–51.

Yost, P., Siegel, A., & Andrews, J. (1962). Nonverbal probability judgments by young children. *Child Development, 33*, 769–781.

Index

A

Abstract, 45, 48, 52, 59, 60, 69, 74, 94, 104, 134, 136, 157, 196, 202, 204, 206, 242, 264, 265, 268, 283

Abstraction, 104, 202

Access, 7, 8, 20, 36, 54, 68, 70, 80, 87, 91–93, 100, 101, 106, 110, 114, 118, 133, 137, 173–187, 195, 205, 208, 211, 232, 233, 258, 264, 267, 270, 271, 275, 282, 289, 297, 299, 308, 323–339, 346, 352–355, 366, 370, 373, 384, 385, 398, 435

Achievement, 6, 11, 18, 20, 37, 59, 70, 113, 124, 138, 167, 175, 183, 214, 296, 306, 314, 327, 338, 343, 345–347, 354, 357, 388, 393–399, 402

Acquisition, 1, 10, 74, 75, 120, 133, 163, 387

Action-research, 357–360, 415

Activity, 1, 4–10, 17, 23, 24, 34, 41, 47, 50, 52, 54, 60, 63, 70, 72, 74–78, 81, 83, 85, 86, 93, 99, 100, 103, 104, 108, 109, 113, 114, 118, 120, 122, 124, 125, 128, 129, 132, 134, 137, 143–158, 161–163, 169, 170, 179, 192, 195–197, 202, 203, 208, 211, 214–216, 221, 222, 228, 230, 231, 233, 241, 263, 264, 266, 268, 269, 271, 274, 275, 290, 292, 294–296, 302, 306, 308, 323, 326–328, 331, 332, 335, 339, 342–346, 353, 355, 357–361, 367, 368, 373, 379, 380, 384, 385, 388, 389, 394–397, 399–401, 409, 411, 412, 414, 416–418, 421, 422, 425, 428, 432, 434, 435

Adult, 8, 18, 104, 107, 117–128, 131–138, 144, 145, 147, 149–153, 155–157, 174, 177, 192, 195, 202, 207, 211, 263, 272, 394, 399–402

Aesthetic/aesthetics, 9, 196, 204, 208, 211, 243

Affect/affective, 3, 19, 35, 57, 113, 134, 161, 164, 167, 168, 192, 208, 226, 263, 279, 284, 301, 323, 326, 332, 335, 338, 356, 387, 395, 398

Agency, 24, 36, 101, 104, 114, 134, 226, 240, 354, 420

Algebra/algebraic, 3, 5, 35, 46, 48, 59, 60, 63, 202, 311, 312, 367, 388, 413

Algorithm/algorithmic, 4, 148, 203, 204, 372, 374, 375, 432

Allocation, 68, 92, 162, 164, 168, 184

Anthropology/anthropological, 75, 204, 205

Application, 16, 48, 54, 59, 114, 146, 192, 194, 240, 256, 342–345, 352, 353, 358, 360, 367, 385, 386, 389, 411, 417, 422, 433

Applied mathematics, 4, 119

Argumentation, 125, 137, 214, 290, 292, 293, 295, 307

Arithmetic/arithmetics, 3, 24, 46, 48, 84, 89, 90, 102, 103, 146, 148, 378, 413

Artifact, 23–25, 43, 231, 369, 379, 380

Assessment, 39, 43, 81, 109, 113, 180, 230, 232, 233, 237, 354, 415–418, 420, 422, 434

Attainment, 4, 56

Attitude, 162, 163, 193, 206, 214, 226, 292, 293, 295, 301, 310, 334, 335, 352, 360, 395, 416, 417, 419, 423, 432

Authenticity/authentic, 136, 221, 264–266, 269, 275, 283, 284, 296, 301, 302, 358

Authority, 2, 18, 25, 76, 193, 211, 226–229, 231–234, 238, 240, 289, 292, 298, 385, 386, 399, 401, 427
Axiom/axiomatic/axiomatisation, 10, 326–328, 330–332, 432

B

Behavior/behaviour, 6, 51, 67, 79, 80, 127, 129, 162–164, 168, 169, 174, 177, 178, 182, 184, 239, 255, 281, 282, 292, 369, 380, 389, 395, 396
Belief, 48, 49, 51, 57, 61, 156, 162–164, 167, 170, 175, 177, 239, 290, 301, 327, 334, 335, 369, 395, 396, 426, 427
Bolyai, 329–330
Boundary/boundaries, 10, 20–22, 26, 28, 73, 75, 99, 110, 111, 113, 114, 132, 204–208, 211, 213, 242, 243, 249, 260, 265, 368, 409, 425, 427, 428

C

Calculator, 104, 105, 108, 367, 369, 371, 377–380, 387, 388
Career, 16, 17, 19, 27, 43, 59, 131, 420
Cartesian/Descartes, 202, 206, 215
CDA. *See* Critical discourse analysis (CDA)
Certificate, 8, 101, 102, 105
Citizenship/citizen, 10, 122, 143, 146, 267, 279, 284, 285, 292, 293, 301, 323, 324, 336–338, 346, 347, 380
Classification, 70–93, 133, 368, 387
Classroom culture, 7, 67–94, 409, 419
Classroom discourse, 70, 76, 91, 234, 235, 317
Code, 33, 44, 106, 107, 111, 113, 132, 133, 135, 137, 147, 155, 156, 179, 182, 183, 196, 197, 230, 281, 297, 299, 366
Cognition, 27, 175, 177, 184, 214, 215
Cognitive development, 7, 18
Collaboration/collaborative, 9, 38, 40, 50, 51, 54, 56, 58, 59, 70, 167, 169, 170, 209, 211, 231, 234, 241, 243, 293–295, 306, 324, 341, 355, 376, 399, 417, 421, 429
Comenius, 9, 192, 195–197, 378
Commission Internationale pour l'Etude et l'Amélioration de l'Enseignement des Mathématiques (CIEAEM), 1, 6–7, 11, 302, 426, 431–436
Communication, 10, 15, 19, 27, 78, 79, 81, 86–93, 101–105, 108, 110, 112, 143, 155, 195–197, 204, 206, 214, 215, 220, 221, 226–228, 237, 241, 247, 256, 270, 285, 289–293, 296–302, 324, 325, 331, 343, 345, 346, 352, 355, 356, 365–367, 369, 371, 373, 377–380, 386, 400, 401, 420, 426
Community/communities, 4, 7, 33–63, 100, 113, 144, 145, 147–149, 158, 204, 205, 207, 209, 242, 275, 326, 356, 367, 368, 379, 385, 393–395, 397–402, 410, 414, 417–423, 425, 427
Comparative/comparison, 19, 33, 37–43, 56, 70, 71, 107, 110, 162, 175, 178, 238, 240, 256, 275, 307, 310, 346, 378, 413, 414
Competencies/competency, 9, 70, 101, 104, 105, 110–112, 134, 144, 146, 148, 178, 263, 275, 295, 324, 376, 396
Complexity, 10, 37, 104, 110, 111, 144, 145, 150, 158, 197, 221, 234, 235, 263, 283, 290, 295, 306, 344, 367–376, 384, 385, 417, 426, 428, 435
Computer, 5, 8, 10, 26, 101–111, 114, 134, 208, 209, 310, 330, 351–362, 368, 370, 372, 377, 383–385, 387–389, 433
Construction, 1–3, 7, 8, 10, 19, 21–24, 27, 28, 34, 36–38, 42, 51, 68, 69, 82, 94, 111, 121, 133, 144, 146, 151, 156, 169, 170, 175, 194, 195, 198, 221, 222, 226–229, 233–235, 239, 241, 249, 250, 254, 256, 264, 266, 268, 282–284, 290, 291, 294, 295, 302, 306, 308, 311–315, 317, 320, 344, 352, 361, 362, 367, 371, 373, 376, 377, 379, 380, 385, 386, 388, 389, 409, 410, 416, 421, 423, 426–428, 432
Consumer, 112, 359, 409, 410
Contradiction, 9, 22, 24, 53, 107–109, 164, 206, 207, 214, 270–271, 274
Control, 8, 25, 36, 78, 80, 81, 86, 90, 94, 100, 103, 105, 106–109, 132, 176, 177, 215, 232, 249, 285, 292, 353
Cooperative learning, 342, 347
Creativity/creative, 10, 50, 150, 152, 206, 209, 239, 241, 242, 266, 275, 284, 325, 333, 338, 342, 343, 345, 357, 360, 361, 368, 371, 373, 379
Critical discourse analysis (CDA), 250, 253, 254, 280, 282
Critical literacy, 146, 149
Critical mathematics, 9, 60, 143–158, 260, 263, 265, 279, 283, 285, 324–326, 331–338
Critical mathematics education, 9, 60, 143–158, 265, 279, 283, 285, 324–326, 331–336

Critical perspective, 145–149, 323
Critical thinking, 10, 56, 263–275, 279, 284, 285, 324–326, 332, 334, 335, 339, 357, 361, 362, 398, 417
Criticism, 6, 114, 118, 169, 324
Culture/cultural, 8, 15–29, 39, 40, 48, 49, 60, 67–94, 99, 101, 104, 114, 118, 131, 132, 136, 143–158, 164, 193, 197, 204–208, 215, 226, 237, 238, 240, 253, 254, 271, 279–282, 292, 300–302, 306, 328, 345, 354, 355, 362, 395, 396, 409, 410, 417, 419–421, 426, 434
Curriculum/curricula/curricular, 1–6, 9, 11, 18, 22, 24, 29, 35–37, 49, 52, 54, 58–60, 73, 99–104, 108, 111, 113, 114, 119, 128, 130, 132, 134, 137, 143–145, 147–150, 157, 158, 173, 174, 179, 184, 193, 195, 196, 201, 210, 212, 225, 230, 234, 238, 240, 263, 305, 306, 311, 324–326, 345, 352, 354–357, 360, 361, 388, 393–397, 399, 402, 410, 411, 415, 418, 419, 421, 423, 431, 433

D

Decision-making, 16, 101, 104, 110, 112, 169, 243, 263, 264, 275, 292, 300–302, 323, 325, 339, 347, 356, 397
Decontextualized/decontextualised, 87, 88, 90, 104, 107, 137
Deliberate dialogue, 289–302, 341–342, 345, 346
Democracy/democratic/democracies/demcratizing, 5, 6, 8, 10–11, 99–101, 107, 113, 114, 118, 136–138, 145, 146, 161–163, 165, 170, 195, 211, 242, 248, 249, 263, 268, 274, 279, 284, 285, 290–293, 301, 302, 323–339, 346, 393–435
Design, 4–6, 8, 37, 42, 43, 48, 49, 74, 83, 104, 107–109, 121, 128, 134, 137, 144, 145, 147–152, 157, 166, 168, 169, 177, 179, 182, 192, 193, 197, 206, 215, 227, 232, 233, 235, 240, 266, 269–271, 296, 301, 302, 326–332, 343–346, 355, 358, 372, 379, 389, 398, 402, 411, 413, 414, 421, 426, 428
Developmental psychology, 195–197
Dewey, J., 109, 292, 306, 355, 357
Dialectic/dialectics, 82, 215, 366, 367, 371, 374
Dialogue/dialogical/dialogicality, 7, 10, 53, 86, 122, 124, 125, 127, 131, 136, 137, 144, 147, 150, 156, 157, 211, 227, 234, 279, 289–302, 306, 324, 327, 332, 341–347, 370, 371, 377, 385, 398, 401, 421
Didactic contract, 238, 240
Didactics/didactic, 4, 7, 17, 18, 20, 23, 28, 71, 74, 75, 170, 195, 215, 218, 220, 221, 238, 240, 264, 269, 279, 293, 294, 302, 326, 352, 368–371, 376, 379, 380, 386–388, 426, 427, 434, 435
Differential, 20, 52, 70, 100
Digital technology/digital, 9, 26, 162, 201–211, 230, 241–243, 296, 327, 352, 354, 356, 358, 360, 365–369, 371, 379, 380, 386, 388, 434
Dilemma, 162, 166, 169, 260, 334, 335, 419
Discipline, 2, 23, 27, 35, 60, 73, 100, 103, 132, 192, 193, 196, 205, 427, 435
Discourse/discursive, 1, 9, 17, 21, 22, 24, 35, 60, 68, 70, 73–76, 78, 80, 87, 89–94, 100–102, 114, 119, 126, 128, 129, 131–137, 144, 147, 154, 157, 225–235, 241, 248–250, 253, 254, 256, 258, 260, 279–285, 289, 291, 292, 317, 421
Disparity/disparities, 7, 36, 67, 68, 70, 71, 83, 92, 94
Distribution, 2, 7, 41, 50, 52, 54, 61, 91, 99, 100, 105–107, 114, 161, 162, 164–168, 170, 177, 178, 180, 182–184, 226, 227, 254, 266, 271, 273, 281, 282, 295, 300, 312, 356, 369
Diversity, 4, 68, 92, 144, 145, 147, 149, 150, 155–158, 355, 388, 427, 434
Documentational genesis/documentary, 368–369, 374, 379, 380, 385, 386, 389
Domain, 1, 7, 174–175, 177, 193, 195–197, 215, 249, 253, 254, 325, 379
Drama, 10, 26, 247, 259, 323–339, 344–347, 366, 367, 396, 401

E

Early childhood, 144, 177, 178, 195, 411
Ecology/ecological, 6, 294, 346
Economy/economic, 2, 6, 15–17, 20, 22, 24, 26, 33–35, 39, 41, 50, 51, 53, 54, 57–60, 69, 100–101, 103, 121, 124, 125, 127, 170, 184, 196, 247, 249, 250, 252, 259, 260, 282, 297, 300, 301, 311, 324, 328, 355, 356, 366, 367, 394–396, 411
Effectiveness/effective, 18, 34, 74–76, 93, 94, 105, 110, 162, 250, 256, 263, 275, 296, 325, 338, 362, 395, 397, 398, 402

Egalitarian, 36, 51, 56–61, 137, 162, 169, 170, 398, 401
Elementary school, elementary mathematics, elementary, 3, 35, 38, 39, 41, 42, 54, 60, 84, 119, 173, 176, 194, 203, 207, 293, 294, 328, 353, 395, 396, 401
Elite, 2, 3, 56–60, 94
Embodied/embodied cognition/embodiment, 9, 58, 148, 201–204, 210, 211, 214, 215, 218, 220, 221, 227, 237, 241–243
Emotion/emotional, 19, 151, 156, 162, 170, 196, 201, 204, 207, 339, 343, 360, 420
Empirical research, 4, 10, 143, 175, 389, 415
Enculturation, 21, 23, 25, 146, 207, 238, 240, 419
Enhancement, 99, 114
Environment, 1, 9, 25, 35, 37–39, 53, 76, 108, 118, 119, 121, 122, 124, 127–129, 203, 211, 226, 227, 233, 263–265, 269, 270, 275, 283, 293, 294, 297, 302, 307, 310, 324, 325, 328, 329, 335, 342, 345, 346, 353–355, 357, 358, 360, 361, 370, 374, 377, 380, 384, 389, 395, 397, 398, 400, 401, 415, 417, 418, 420, 426–428, 435
Epistemology/epistemological, 4, 5, 11, 131, 143, 145, 193, 221, 239, 327, 334–336, 338, 425, 426, 428
Equality, 36, 41, 51, 54, 56, 89, 90, 99, 126, 145, 161, 162, 164–166, 168, 169, 178, 184, 195, 207, 255, 259, 263, 281, 282, 284, 314, 330
Equation, 54, 89–91, 124, 294, 343, 358, 385, 389
Ethics/ethical, 6, 38, 145, 146, 162, 203, 204, 295, 360, 361, 410, 415, 419–421
Ethnography/ethnographic, 152–153, 227, 230, 237, 326
Ethnomathematics, 23
Euclid/euclidean, 3, 10, 326–333, 335, 338, 367
Everyday, everyday activity, everyday life, 1, 3, 7, 18, 26, 27, 46, 47, 49, 52, 62, 63, 68, 73–77, 80, 85–88, 92–94, 102, 113, 132, 134–137, 147, 148, 158, 162, 163, 175, 203, 207, 256, 264, 265, 283, 379, 384, 410, 414, 419, 422, 427
Evolution/evolutional, 132, 192–194, 196, 218, 297, 299, 305, 366, 426
Experience, 7–9, 11, 21, 22, 26–28, 43, 46, 50, 54, 59–60, 62, 87, 91, 101, 104, 112–114, 117–128, 132, 134, 137, 145, 147, 148, 150, 156, 157, 164, 166, 184, 192–194, 196, 197, 205–207, 213–217, 221, 226–229, 231, 234, 237–240, 254, 264, 267–275, 283, 284, 292–294, 306, 308, 309, 311, 323, 325, 331, 339, 342, 347, 352, 355, 356, 359–361, 366, 368, 371–380, 386–388, 400–402, 409–410, 412, 414–421, 423
Experiment, 3, 9, 10, 134, 136, 152–153, 162, 169, 176, 181, 182, 195, 197, 216–218, 220, 222, 243, 265, 270–273, 297, 302, 326–332, 337, 338, 343, 352, 354, 356, 366–368, 373–380, 383, 386, 389, 420, 432
Explicit pedagogy, 78–80, 87, 91, 92

F
Fairness, 8, 161–170, 176, 178
Fair sharing, 8–9, 173–187
Family/families, 10–11, 19, 21, 23, 27, 35, 39, 41, 42, 47, 50, 55, 63, 69, 87, 125, 136, 144, 147, 178, 196, 197, 203, 208, 209, 241, 242, 252, 267–269, 356, 393–402, 416, 425–429
Fermi problem, 264, 266–267, 269, 275
Field, 1, 2, 4, 15–28, 43, 49, 59, 94, 114, 131, 132, 134, 146, 173, 176, 194, 204, 205, 207, 215, 225, 226, 257, 270, 279, 337, 396, 400, 402, 418, 419, 421, 423, 426, 429
Formation, 1–3, 10, 17, 21, 215, 260, 292, 326, 357, 432
Foucault, M., 24, 71, 249, 260, 280
Fractions/fractional, 8–9, 75, 102, 104, 163, 173–187, 197, 255, 267, 281
Framing, 70, 78–83, 86, 88, 90, 91, 93, 133, 175, 177–178, 250, 421
Froebel, F., 192
Functions/function, 3, 20, 23, 25, 61, 68, 81, 82, 88, 99, 100, 124, 128, 133, 144, 148, 149, 156, 174, 175, 178, 193, 196, 206, 211, 229, 230, 232, 233, 250, 253, 257, 260, 280, 294, 297, 352, 358, 359, 368, 375, 384, 387, 388, 397, 426, 431

G
Game, 9, 23–24, 43, 67–82, 93, 94, 150–157, 162, 164, 167, 170, 177, 180, 196, 208, 218, 221, 222, 237–243, 334, 353, 356, 375, 385, 398, 400, 414
Gauss, 330, 333
Gender, 19, 21, 28, 119, 146, 147, 152, 157, 195, 211, 229, 242, 280
Genre/microgenre, 147–149, 230–233, 256, 258, 281, 282

Index 475

Geometry, 3, 5, 10, 35, 39, 46, 54, 55, 59, 60, 157, 193, 194, 202, 206, 207, 209, 295, 326–333, 335, 338, 361, 365, 367, 385, 431–433
Gesture, 9, 93, 112, 149, 202, 213–222, 237–243, 387
Global, 1–2, 16, 26, 34–36, 51, 56, 58, 59, 61, 145, 146, 197, 226, 263, 292, 319, 324, 352, 353, 356, 386, 425, 428, 435
Governance/governable, 24, 268
Government, 15–17, 25, 100, 120, 193, 248, 249, 254–256, 259, 280, 282, 353, 356, 358, 383, 384, 402
Graph, 39, 51, 69, 107, 165, 167–169, 194, 220, 283, 343, 344, 358, 359, 372, 388, 434

H

Hegel/Hegelian, 240
Hilbert, 328
History/historical, 3, 5, 9, 16, 20–24, 26–28, 68, 103, 111, 133, 134, 136, 146, 148, 149, 156, 157, 191–195, 203, 204, 206, 215, 247, 256, 280–281, 285, 292, 294, 301, 302, 326–330, 338, 356, 362, 365, 367, 385, 386, 418, 426
Horizontal discourse, 92–94, 101, 118–119, 128, 129, 132–136
Humanistic, 2, 23

I

ICT, 5–6, 10, 27
Identity/identities, 1, 7, 8, 21, 25–28, 55, 144, 147–149, 156–158, 178, 334, 427
Ideology, 6, 25, 56, 59, 146, 225, 250, 254, 256, 259, 280, 282
Implementation, 2, 11, 18, 42, 49, 50, 53, 82, 113, 132, 144, 145, 147, 149, 150, 152, 153, 192, 193, 195, 227, 235, 247, 248, 310, 311, 326–331, 347, 352–356, 359, 373, 383–385, 396, 398, 399, 402, 423
Implicit pedagogy, 70, 79, 80, 82, 83, 86, 92–94
Inclusive/inclusion, 7, 68–70, 84, 91–94, 99, 103, 114, 136, 150, 213, 214, 296, 353–356, 420
Individual/individualism/individualization/individualisation, 1–2, 11, 19, 21, 27, 36, 38, 52, 67, 68, 99, 106, 108, 111, 114, 118, 123, 133, 157, 158, 162, 163, 165, 166, 169, 170, 175, 176, 181, 182, 192, 201, 209, 215, 230, 233, 239, 255, 257, 268, 269, 282, 290, 307, 309, 323, 325, 334, 355, 368, 369, 380, 395, 417, 419, 421, 425, 426, 428
Industry, 15, 17, 41, 54, 99–104, 107–109, 112, 131
Inequity, 163, 169, 178
Inquiry-based teaching, 305–308
Inscription, 216, 249
In-service, 38, 174, 302, 347, 398, 420, 423
Institution/institutional, 1, 2, 4, 6, 24, 26, 36–37, 39, 41, 51, 71, 75, 93, 94, 101, 109, 112, 114, 192, 196, 253, 260, 281, 289, 310, 359, 380, 385, 388, 389, 398, 402, 419
Instruction/instructional, 7, 15, 22, 23, 33–63, 69, 78, 79, 106, 114, 133, 136, 174, 176, 177, 192, 197, 201, 228, 233, 238, 297, 302, 308, 326, 384, 397, 401, 402, 409, 420, 422, 427, 428
instrumental/instrument/instrumentalization/instrumentation, 9, 37, 40, 44, 48, 102, 106, 133, 145, 155, 157, 204, 293, 294, 296, 353, 365, 367–369, 371, 373, 379, 384, 386–388
Interaction/interactional/interactivity, 7, 9, 20, 33, 43, 44, 51–53, 67–70, 77–80, 83, 88, 92, 94, 99, 121, 129, 137, 144, 147, 148, 150, 151, 162, 163, 182, 201, 203, 208, 211, 216–218, 221, 222, 226–230, 232, 233, 238–242, 253, 254, 256, 285, 289, 291–293, 306, 307, 309–311, 323, 325, 330, 331, 346, 347, 352, 354, 355, 358, 359, 361, 367, 370, 371, 377, 380, 396, 411, 419, 425–429, 434
Intercultural, 146, 147, 150, 152
interdisciplinarity/interdisciplinary, 1, 5, 60, 193, 326, 361, 426–429
Interest, 3, 17, 38, 70, 108, 118, 135, 143, 162, 174, 192, 208, 213, 232, 238, 248, 267, 279, 295, 315, 329, 345, 357, 365, 383, 411
Internet, 10, 50, 111, 205, 208, 242, 270, 273, 289, 299, 300, 352–354, 358, 361, 384, 385
Interpersonal, 226, 228, 229, 233, 253–254, 337
Interview, 37, 41, 43, 57, 58, 93, 119, 133, 174, 179, 299, 311, 312, 314–315, 327, 328, 332, 338, 360, 369, 377, 379, 395, 400, 401, 414, 415, 417
Invisible, 69, 78–80, 86, 91, 92, 99, 113, 132

J
Justice, 10, 70, 114, 143, 145, 146, 162, 164, 166, 168, 285, 325, 366, 409, 420

K
Kant, Kantian, 373, 374
Kindergarten, 165, 173, 177–179, 184–186, 192, 197

L
Labour-market/labor-market, 68, 69, 92, 247, 248
Language, 3, 8, 23, 24, 58, 69, 82, 84, 87, 88, 113, 121, 133, 137, 145–147, 149, 151, 155, 156, 163, 165, 169, 175, 176, 194–197, 215, 218, 226, 230, 253, 254, 256, 258, 280–282, 302, 312, 326, 327, 329, 330, 352, 353, 372, 385, 419, 432
Learning activities/learning activity, 9, 108, 162, 169, 203, 342–345, 395
Learning opportunities, 67, 68, 70, 83, 92–94, 107, 136, 138, 315
Lecture, 201–203, 210, 232, 327, 347, 359, 422
Liminal space, 9, 201–211, 237–243
Linguistic, 87, 147, 165, 169, 214, 216, 218, 227, 229–233, 250, 253, 254, 256, 280, 291, 302
Literacy, 110–111, 132, 133, 144–146, 148, 149, 156, 196, 417, 434
Logic/logical, 28, 59, 105, 106, 146, 191, 194, 196, 202, 204, 206, 214, 285, 328, 329, 413, 432

M
Manipulation/manipulate, 10, 103, 135, 136, 162, 174, 178, 194, 248, 249, 255, 260, 281
Manipulative, 196, 202, 238, 285
Marginalization/marginalisation/marginalize/marginalise, 8, 70, 80, 144, 148, 157, 410, 421
Mathematical experience, 27, 374, 388
Mathematical knowledge, 1, 3, 6–8, 23, 25, 70, 74, 93, 101, 105, 109, 110, 118, 119, 134, 143, 146, 148–150, 157, 175, 194, 215, 218, 233, 266, 267, 274, 285, 300, 327, 331, 346, 361, 379, 380, 384, 386, 410, 427
Mathematical modeling, 4, 8, 117–129, 264, 266, 275, 384, 389, 409, 422
Mathematical object, 107, 266, 368, 372–374, 376, 380, 388
Mathematical task, 307, 308, 313, 368, 387, 388
Mathematician, 2–4, 7, 23, 209, 211, 243, 329, 336, 365, 372, 410, 413, 416–418, 421, 426
Mathematics classroom, 5, 7, 10, 21, 24, 27, 67–94, 120, 122, 129, 201–204, 206, 207, 213, 216, 238, 239, 242, 265, 284, 289–302, 306, 308, 311–319, 324, 326, 327, 329, 336, 341, 344, 345, 347, 351–362, 383–387, 412, 416, 419, 434
Mathematics for all, 6, 59, 433
Mathematics instruction/mathematical instruction, 7, 15, 33–63, 201, 428
Mathematics learning, 27, 35, 54, 132, 210, 221, 237, 284, 308–310, 315, 319, 393, 394, 402, 419, 421
Mathematization/mathematisation, 91, 132, 134, 135, 409, 422, 433
Meaning/meaningful, 6, 24, 36, 74, 101, 118, 132, 156, 163, 194, 204, 213, 226, 237, 249, 264, 280, 289, 306, 325, 343, 355, 372, 413, 426
Measuring/measurement/measure, 34, 39, 45, 48, 51, 53, 92, 102, 103, 104, 107–109, 174, 177, 178, 194, 217, 219, 248, 256, 257, 259, 266, 271–273, 275, 280, 298, 330, 338, 378
Media, 9, 16, 26, 39, 197, 205, 211, 230, 234, 248–250, 260, 264, 265, 267, 274, 275, 279–284, 308, 311, 319, 353, 354, 356, 366–368, 371, 374, 376, 380, 433
Metaphor/metaphorical, 6, 99, 150, 205, 215, 226, 416
Mixed-methods, 33, 37
Modernization/modernity/modernisation/modern/modernism/modernist, 2, 4, 22, 25, 26, 36, 145, 146, 203, 204, 368
Moral, 2, 8, 54, 104, 112, 161, 162, 170, 176, 291
Motivation/to motivate, 5, 19, 48, 49, 54, 120, 155, 156, 292, 295, 297, 327, 332, 345, 357, 393, 395, 396
Multimedia, 26, 308, 311, 319
Multimodal/multimodality, 9, 214–218, 220, 222, 225–235, 242, 281, 373

N
Narration/narrative, 9, 150, 151, 153, 165, 185, 213–222, 237–243, 331, 356, 360, 415

Index 477

Naturalization/naturalized/naturalised/
 naturalisation/to naturalize/to
 naturalise, 7, 16, 35, 109, 148, 176,
 206, 214, 264, 268, 312, 319, 365, 374
Negotiation/negotiatory/negotiate, 10, 36, 52,
 61, 101, 162, 170, 206, 238, 240, 291,
 293, 294, 297, 298, 300, 301, 325, 327,
 331, 335, 337, 338, 341, 345
Network, 24, 25, 38, 55, 61, 108, 211, 352,
 354, 355, 359, 361, 366, 395, 400, 427
Number symbol, 143, 145, 148, 150, 151,
 154–156
Number system, 194
Numeracy, 100, 110, 111, 118, 119, 128, 131,
 132, 134, 135, 249, 285
Numerical, 16, 86, 194, 202, 248–250,
 253–261, 266, 267, 279–283

O
Obstacle, 7, 8, 49, 52, 60, 197, 301, 387, 398,
 426, 435
Ontology/ontological, 4, 239
Operation, 3–5, 22, 25, 93, 94, 113, 148, 149,
 163, 169, 192, 194, 196, 207, 249, 255,
 266, 318, 357, 361, 368, 385, 386
Operator, 8, 99–102, 104–110, 113, 114,
 410, 416

P
Paradigm, 6, 60, 79, 144, 177, 203, 237,
 367, 425, 426
Paradox, 9, 132, 206, 207, 240, 367, 380, 426
Parents involvement/parent involvement/
 parental, 11, 43, 393, 394, 396,
 397, 402
Participation/participatory, 9–11, 36, 44,
 51, 52, 99, 101, 113, 114, 144, 147,
 150, 156, 164, 169, 170, 201–211,
 213–222, 231, 234, 237, 238, 241–243,
 258, 268, 271, 289–293, 295–302, 310,
 323, 324, 326, 327, 332, 339, 341, 342,
 344–346, 355, 393–402, 409, 415–418,
 420, 421, 435
Pedagogic right, 99–100, 113
Pedagogy/pedagogic, 7, 9, 20–22, 26, 28, 34,
 53, 59, 69, 70, 78–80, 82, 83, 86, 87,
 91–94, 100, 103, 108, 114, 144, 145,
 150, 192, 195, 196, 201–206, 208, 227,
 229, 234, 237, 241, 242, 325, 366, 431
Performance, 9, 18, 48, 69, 102, 103, 133,
 136, 175, 183, 184, 194, 201–211, 233,
 238, 241–243, 260, 324, 328, 329, 332,
 334, 342, 378, 387, 393, 394, 396,
 398, 399, 416
Pestalozzi, 192
Philosophy/philosophical, 2, 60, 84, 145,
 162, 192, 195, 201, 202, 237, 374
Physics, 3, 74, 326, 389
Piaget/Piagetian, 163, 169, 177, 196,
 240, 426
Plato, 35, 60, 191, 202, 215, 385
Play, playful, 2, 7, 9, 23, 37, 50, 57, 72–77,
 80–83, 93, 104, 110, 119, 123, 129,
 133, 136, 143–158, 162, 164, 167, 169,
 181, 182, 184, 193, 196, 201, 203–208,
 213, 238, 241–243, 248, 291, 318, 328,
 329, 331, 338, 373, 384, 409, 410, 415,
 422, 426, 433
Poincaré, 330
Policy/policies, 9, 24, 25, 34, 102, 143,
 144, 147, 247, 248, 254, 259, 260,
 282, 352, 420
Political, politics, 2, 15, 33, 68, 100, 121, 143,
 195, 226, 248, 267, 279, 292, 323, 356,
 410, 433
Politician, 24, 254, 256, 282
Popularization, popularisation, 18, 132
Post-modernity/post-modern, 26
Poststructuralist/poststructuralism, 21, 22, 280
Poverty, 37, 39, 41, 145, 394, 411, 414
Power, 49, 51, 57, 61, 71, 91, 99, 100, 108,
 137, 146, 157, 206, 225, 226, 233, 234,
 239, 240, 247–249, 254, 259, 279, 280,
 282, 285, 291, 318, 323, 352, 409, 410,
 417, 418, 421–423
Praxeology, 75–78, 80–83, 86, 88, 90–93
Pre-schooler/preschooler, 8, 9, 164, 165,
 168–170, 173, 178, 179, 184,
 191–197, 311
Preschool/pre-school, 8, 9, 161–170, 173,
 176–179, 183, 184, 191–197, 294,
 311, 394
Pre-service, 174, 209, 293, 295, 301, 302, 347,
 398, 411, 420
Primary school/primary mathematics/primary,
 2, 4, 5, 18, 33, 37, 41, 42, 71, 91, 153,
 179, 183, 192, 195, 197, 214, 216, 226,
 229, 230, 234, 238, 281, 301, 306, 311,
 319, 338, 385, 411, 432, 434
Privilege, 2, 26, 68, 75, 80, 83, 94, 133, 292,
 396, 423
Probability/probabilistic/probabilities, 4, 5,
 108, 163, 168, 194, 377, 432, 433
Problem solving/problem-solving, 4, 69, 103,
 112, 146, 148–149, 152, 157, 161–170,
 174, 187, 197, 209, 213–222, 237, 240,

242, 265, 268, 269, 284, 301, 308, 325, 359, 376, 416, 419
Process, 1, 18, 35, 68, 99, 119, 134, 144, 161, 174, 191, 213, 226, 239, 249, 264, 284, 291, 306, 323, 341, 352, 367, 384, 393, 411, 425, 433
Product, 1, 56, 100, 102, 103, 105, 106, 109, 147, 179, 216, 226, 250, 360, 386
Professional/profession, 2, 15–17, 20, 25, 56, 61, 108, 110, 117, 119, 128, 131, 132, 196, 197, 205, 210, 291, 293–295, 306, 308, 319, 320, 353, 355, 400, 402, 415, 416, 426
Progressive, 21, 28, 53, 194, 227, 234, 285, 376
Proportion/proportional, 103, 132, 133, 162, 175, 177–178, 255, 267, 294
Prosocial/prosociality, 162, 174, 176–182, 184
Protocol, 43, 44, 180, 230, 231
Psychology/psychological, 7, 19, 174–176, 184, 192, 193, 195–197, 202, 214, 243, 290, 393, 426, 432
Public discourse, 9, 249, 279–285, 290
Pythagoras/Pythagorean, 296, 298, 342, 385

Q
Qualification, 15, 16, 18, 24, 101, 112
Quantification, 248, 280
Questionnaire, 33, 44, 119, 120, 293–295, 327, 360

R
Race, 54, 71–72, 74, 77, 81–83, 93, 94, 146, 147, 157, 211, 280
Random, 33, 44, 100, 108, 121, 165, 167, 181–183, 210
Reality, 1, 16, 35, 68, 99, 118, 134, 145, 162, 184, 192, 202, 214, 226, 237, 249, 263, 280, 292, 311, 325, 343, 351, 373, 384, 398, 417, 433
Real-life problems, 264–266, 283–284
Reasoning, 6, 16, 22, 24, 27, 49, 57, 59, 61, 74, 80, 103, 105, 109, 118, 119, 121, 125, 127, 132, 134, 149, 150, 162–165, 167, 169, 170, 174, 175, 178, 182, 184, 196, 214, 221, 264, 267–269, 272, 274, 285, 291–293, 295, 298, 300, 307–312, 314–317, 319, 325, 327–329, 333–335, 341, 343, 344, 346, 347, 380, 384, 393, 401, 417
Recontextualisation/re-contextualisation, 1, 4, 5, 19, 23, 74–78, 84, 92–94, 144, 234, 281, 360

Reflection, 6, 8, 11, 22, 23, 34, 36, 38, 41, 52, 69, 74, 76, 93, 94, 103, 108, 109, 112, 114, 129, 144, 149, 150, 155, 176, 178, 203, 227, 234–235, 239, 240, 250, 269, 283–296, 300, 301, 305–307, 314, 319, 325–328, 341, 344–346, 351, 352, 355, 356, 359, 361, 378, 379, 386, 410, 415, 417, 419–422
Reflexive/reflexivity/reflexively, 24, 118, 121, 124–129, 145, 360
Reform, 2, 3, 6, 23, 69, 88, 143, 144, 192, 193, 201, 247, 257, 258, 291, 297, 366, 399, 400
Register, 2, 48, 51, 87, 311, 360, 370–372, 374, 375, 379, 387–388
Religion, 157, 204, 205, 366
Representation, 5, 18, 35, 89, 103, 124, 133, 147, 163, 194, 207, 219, 225, 240, 249, 265, 280, 292, 308, 335, 352, 367, 384, 397, 415, 425
Reproduction, 6, 28, 51, 58, 68, 69, 71, 108, 133, 147, 148, 182, 217, 218, 228, 239, 241, 253, 254, 256–258, 260, 342–344, 395
Resistance, 33–35, 110, 267, 398–401
Resources, 8, 9, 15, 16, 23, 34, 36, 38, 45, 49–52, 54, 55, 59, 61, 93, 94, 118, 119, 147, 148, 155, 158, 162, 164, 165, 175–178, 196, 203, 205, 210, 214, 216, 218, 225–227, 229, 230, 232–235, 237–243, 266, 270, 274, 294, 306–308, 310, 311, 325, 334, 335, 352–361, 365, 366, 368–370, 374–376, 379, 380, 385, 386, 388, 389, 396, 398, 402, 411, 416, 423, 435
Revolution/revolutionary, 3, 9, 15, 191, 237–243, 289, 365–367, 396
Rhetoric, 7, 61, 232, 234, 250, 256, 260, 282
Riemann, 329–330
Rousseau, 192, 215
Rural, 33–63, 70, 84, 136, 356, 360

S
School mathematics, 2–6, 9, 17, 19, 21–28, 36, 43, 48, 54, 58, 60, 68, 73–76, 84, 85, 92, 94, 104, 108, 110, 113, 119, 128, 134, 196, 197, 201–204, 206, 209, 222, 275, 279, 283, 285, 296–301, 306, 323, 395, 400, 410, 412, 418, 419, 421, 425, 427
Science, 2, 3, 15, 16, 18–23, 27–29, 34, 35, 38, 39, 54, 74, 84, 112, 129, 133, 203, 209, 214–215, 227, 323, 327, 330, 335, 357, 361, 366, 378, 389, 402, 420, 427, 431

Index

Scientific, 3, 16, 18, 20–22, 108, 113, 146, 176, 192–194, 214, 215, 218, 352, 357, 366, 373, 384, 393
Secondary mathematics, secondary school, secondary education, 2–4, 21, 28, 35, 39, 68, 70, 71, 84, 94, 101, 201–204, 263, 264, 268, 271, 293–301, 326, 345, 385, 411, 431, 432
Selective/selection, 7, 38, 59, 68–70, 74, 78, 81, 92, 94, 100, 101, 128, 165, 166, 168, 169, 230, 249, 259, 261, 283, 307, 309, 311, 315, 359–360, 385, 411, 414
Semantic/semantics, 9, 149, 166, 169, 226, 229–231, 234, 235, 250, 256, 372, 427
Semiotic game, 9, 218, 221, 222, 237–243
Semiotic/semiotics, 9, 214, 216, 218, 221, 222, 226–230, 232–235, 237–243, 367, 370, 372, 379, 387
Skills, 1, 8, 24, 46, 47, 51, 59, 62, 63, 84, 101–106, 108–113, 118, 128, 129, 132–135, 144, 146, 148, 149, 152, 162, 163, 168, 175, 192, 196, 202, 203, 211, 238, 239, 241, 263, 269, 275, 283–285, 291, 297, 298, 324–326, 332, 336, 337, 339, 341, 343, 346, 357, 360, 366, 368, 402, 409–412, 415, 416, 418–423, 427
Social class, 19, 33, 36, 37, 43, 44, 51–54, 56, 57, 60, 133, 268, 352, 394, 395
Social turn, 21
Society, 4–6, 15–17, 19–22, 24–29, 51, 100, 103, 109, 118, 123, 124, 132, 137, 144–149, 162, 176, 178, 192, 193, 196, 204–206, 249, 254, 256, 260, 263, 281, 284, 285, 292, 324, 336, 361, 366, 380, 395, 409, 410, 422, 426–428, 433, 434
Sociocultural/socio-cultural, 19, 21, 68, 70, 71, 99, 101, 131, 132, 155, 253, 254, 279–282, 426
Socio-economic status (SES), 2, 60, 69, 124, 259, 311, 394–396
Sociology/sociological/sociologist, 20, 21, 33, 34, 68, 69, 132–135, 193, 196, 239
Solidarity, 233, 259, 296, 336–338
Space/spatial, 8, 9, 22, 24, 27, 36, 38, 58, 104, 114, 137, 143–158, 184, 194, 196, 197, 201–211, 219, 220, 225–235, 237–243, 249, 272, 273, 279, 292, 312, 344, 354, 355, 368, 401, 413, 414, 422
State, 2, 3, 15, 34–43, 47, 49, 50, 58–61, 72, 79, 81, 88, 93, 117, 120, 132, 135, 137, 144, 162, 163, 168, 193, 202, 205, 206, 225, 227, 247, 263, 268, 285, 295, 355
Statistics/statistical, 4, 5, 44, 59, 104, 108, 110, 114, 194, 297, 300, 377, 378, 414, 432

Stratification, 67–71, 92, 352
Streaming, 69–71, 100, 420
Subjectivation/subjectification, 21, 25
Subjectivity/subjective, 7, 15–29, 132, 144, 164, 166, 167, 169, 280
Sustainability/sustainable, 33–35, 43–45, 48–52, 59, 60, 284
Symbol/sybolic, 8, 103, 110, 133, 135, 145, 147–151, 153–157, 175, 202, 206, 207, 216, 222, 237–239, 241, 284, 312, 316, 374, 434
Systemic/system/systematic, 1, 2, 5–8, 10, 15, 18, 19, 22, 24, 25, 27, 36–37, 68, 69, 76, 83, 92–94, 101, 103, 104, 106, 108–111, 117–129, 133–135, 153, 155, 156, 162, 176, 178, 192–196, 205, 206, 215, 247, 253, 263, 266, 280, 285, 306, 308, 310, 312, 318, 324, 328, 330, 346, 352, 354, 356, 372, 374, 380, 385, 387, 398, 416, 422, 425–429

T

Teacher training/teacher-training, 132, 354, 389, 398, 402, 415, 432
Technique, 5, 10, 33, 74–77, 101, 124, 169, 191–192, 249, 323–339, 366, 367, 419, 427, 431
Technocracy, 5, 6
Technology/technologies, 4–6, 10, 15–17, 22, 23, 26, 28, 41, 58, 74, 76, 104, 107, 108, 110, 112, 113, 144, 170, 194, 205, 241, 249, 260, 281, 292, 301, 351–354, 356–362, 365–380, 383–389, 422
Tension, 5, 17, 36, 55, 56, 58, 99, 107–109, 114, 132, 164, 233, 421, 422
Tertiary, tertiary education, 16, 144
Testing, 18, 19, 36, 37, 39, 43, 45, 52, 55, 56, 58, 60, 61, 71, 72, 74, 82, 84, 100, 103, 104, 106, 110, 136, 165, 167–169, 174, 175, 177–179, 181, 182, 184, 204, 209, 238, 248, 260, 274, 324, 328, 344, 358, 360, 366, 394, 398, 402, 417, 419
Textbook, 2, 24, 25, 37, 49, 69, 84, 86, 91, 147, 201, 230, 232, 233, 384, 386
Text/textual, 79, 80, 103, 112, 205, 216, 230, 250, 253, 254, 256–258, 265, 268, 280–282, 294, 300, 313, 314, 328, 358, 367, 372, 384, 418
Theorem, 82, 298, 328–330, 342, 367, 385, 386, 413
Tool, 21, 22, 28, 44, 57, 71, 101, 108, 113, 132, 134, 136, 146, 147, 157, 192, 205, 214, 220, 263, 266, 273–275, 283, 284, 293, 296, 323–325, 327, 328, 331, 338,

358, 359, 365, 367, 368, 370, 379, 380, 386, 400, 409, 412, 417, 418, 423, 427
Tradition/traditional, 2–3, 5, 9, 10, 20, 27, 28, 36, 49, 52, 53, 55, 68, 73, 92, 110, 135, 192, 201–204, 206, 207, 211, 234, 237, 239, 240, 242, 289, 292, 301, 302, 336, 347, 357, 361, 366, 418, 419, 421, 422
Transferability, 132, 135
Transformation/transformative, 4, 26, 27, 36, 134, 145, 156, 206, 226, 227, 241, 243, 260, 293–295, 305, 306, 345, 366, 367, 369, 374, 379, 380, 387, 399, 412, 418–421, 426
Transmission, 1, 10, 24, 74, 75, 82, 83, 133, 143, 201, 306, 344, 347, 357, 368
Transposition, 1, 23, 29, 74, 75, 371, 427
Trigonometry, 46, 54, 59, 297, 298, 301
Truth, 9, 25, 137, 202, 207, 260, 280, 284, 285, 327, 335, 421

U

University, 2, 16, 18, 19, 34, 38, 40, 46, 50–53, 56, 58–60, 62, 70, 101, 107, 109, 113, 114, 136, 150, 165, 196, 204, 208–209, 211, 354, 355, 358, 374, 395, 402, 411, 412, 415, 419–422, 431, 432, 434
Urban, 34–37, 39, 53, 56, 61, 70, 273, 311, 326, 356, 360, 379, 411, 415, 420

V

Value, 2, 19, 34, 77, 99, 120, 133, 144, 162, 177, 192, 203, 216, 233, 241, 264, 282, 290, 317, 324, 345, 354, 366, 396, 414, 426
Vertical discourse, 93, 101, 119, 132–136, 289
Videotape/video/videotaped, 43, 86, 105, 179, 181, 208–211, 216, 229–231, 233, 235, 308, 311, 312, 327, 356, 377, 389, 414, 415, 417, 418
Vocational, 8, 48, 51, 56, 59, 60, 99–102, 108, 110, 131–132, 242
Voice, 8, 16, 24, 73, 100, 114, 119, 144, 148, 201, 203, 209, 258, 282, 292, 299, 324, 330, 379, 397, 399, 413
Vygotsky. L.S., 68, 215, 221, 238, 386

W

Wittgenstein, 23
Worker/working class/working-class, 8, 57, 99–115, 120, 124, 131, 133, 134, 137, 258, 259, 281, 291, 396
Workforce/work force, 1, 16, 18, 112
Workplace, 99–114, 118, 131–134, 148

Y

Youth, 15–29, 36, 117, 324, 352, 409–414, 418, 420, 421

Author Index

A
Abbott, A., 132
Abreu, G., 60, 61
Adam, J.A., 267
Albarracín, L., 9, 263–275, 279, 283, 284
Aldon, G., 10, 365–380, 385–389
Alexander, K.L., 395
Alrø, H., 19, 291, 297
Andrews, J., 432
Anghileri, J., 307, 308
Appelbaum, M.I., 398
Appelbaum, P., 9, 11, 21, 234, 237–243, 284, 325, 326, 409–423, 426, 427, 429
Apperly, I.A., 175
Apple, M., 144
Appleton, K., 193
Ärlebäck, J.B., 266–268
Armstrong, B.E., 173
Aronowitz, S., 15, 26
Artigue, M., 367, 369, 384, 385, 387–389
Arzarello, F., 214, 216, 218, 221, 373
Ashton, D.N., 15
Astington, J., 163
Atweh, B., 69, 144
Aubert, A., 396
Auslander, P., 205

B
Bailin, S., 325, 334
Bairral, M.A., 291, 297, 358
Bakker, A., 131
Ball, D.L., 393
Balli, S.J., 395
Balzano, E., 398

Barbosa, J.C., 118, 120–124, 137
Bardini, C., 367
Barnett, J.H., 193
Baroody, A.J., 173, 176
Barry, B., 162
Bartolini Bussi, M., 148
Bateson, G., 68
Bauchspies, W., 22
Baudelot, C., 394
Bauman, Z., 35–36, 51
Bazzini, L., 9, 213–222, 237–239
Bebell, D., 353
Berg, P., 35
Berla, N., 393
Berliner, P., 203
Bernier, E., 399
Bernstein, B., 6, 68–70, 73–75, 78, 79, 91–94, 99–101, 107, 110, 113, 114, 119, 128, 132–137, 395
Berry, W., 34
Bezemer, J., 226, 227
Bezuk, N.S., 173
Bieck, M., 173
Biesta, G., 20, 22, 145
Biggs, J., 342
Bishop, A.J., 18, 23, 25, 104, 324
Bissel, J., 226
Black, L., 21, 71–81, 83, 91–94
Blackman, S., 22
Blake, P.R., 178
Block, A., 421
Blömeke, S., 193
Bloom, P., 343
Blum, W., 389
Boaler, J., 19, 201, 417, 418

Bolton, G., 325
Boni, M., 148
Bonotto, C., 113
Borasi, R., 202
Borba, M., 203, 208–209
Bourdieu, P., 133, 395
Bowles, S., 394
Boyd, R., 176, 178
Boyer, C.B., 329–330
Bragdon, A.D., 202
Brosnan, S.F., 175
Brousseau, G., 71, 81–83, 91, 215
Brown, T., 144
Bruckman, A.S., 292
Bruner, J., 214
Bryant, P., 178
Bullock, E.C., 144
Bunt, L., 328

C

Callaghan, T., 165, 168
Camelo, F., 265
Carpenter, T.P., 399
Carr, M., 19
Carr, P., 34, 36, 48, 54, 59
Carraher, T.N., 138
Carter, H.L., 271
Case, R., 173
Castells, M., 36, 289, 410
Catsambis, S., 393, 394
Catterall, J., 325
Cengiz, N., 305–308, 319
Chaiklin, S., 153
Chapman, O., 265, 270, 305, 306, 341
Chassapis, D., 9, 247–261, 279–282, 284
Chen, M., 37
Chertkoff, J., 177
Chevallard, Y., 74–76
Chouliaraki, L., 280, 281
Christenson, S.L., 395
Christie, F., 230
Chronaki, A., 8, 143–158, 197, 326
Civil, M., 395, 396, 399
Cobb, P., 18, 68, 290, 296, 301, 307, 326
Coben, D., 131, 132
Colby, S.A., 342
Conley, D., 60
Cook, T., 177
Coon, R.C., 177
Coray, D., 23
Corbett, M., 34, 36, 54
Côté, P., 395

Crump, T., 148
Cuevas, G., 393
Cwikla, J., 8, 173–186, 197

D

Damlamian, A., 112, 131
Damon, W., 163, 177
Darmawan, I.G.N., 16
Dasmann, R., 35
Davis, B., 207
Davis, G.E., 174, 176
De Fazio, W., 15
Derman, M.T., 398
Derman-Sparks, L., 147
Dewey, J., 109, 292, 306, 355, 357
DeYoung, A.J., 34
Dias, M.C., 393, 395, 398
Díez-Palomar, J., 8, 10, 131–138, 393–402, 416, 426, 429
Dijk, T.A., 250
Dowling, P., 69, 75
Dreyfus, T., 388
Driscoll, M., 393
Duval, R., 372, 387, 388

E

Eccles, J.S., 396
Edwards, L., 221
Efthimiou, C.J., 267
Egéa-Kuehne, D., 22
Emilson, A., 163, 170
Empson, S.B., 174–176
English, L., 4
English, L.D., 266
Englund, T., 292
Entwisle, D.R., 395
Epstein, J.L., 393–397, 400–402
Erlwanger, S.H., 399
Ernest, P., 23, 60, 192, 324
Establet, R., 394
Esteley, C.B., 266
Evans, J., 128, 132, 135
Eves, H., 329

F

Fairclough, N., 250, 253, 280, 281
Fan, X., 37
Farley, K., 205
Fehr, E., 163, 164, 168, 169
Fehr, H., 16

Author Index

Fehr, H.F., 16
Feigenson, L., 174, 175
FitzSimons, G.E., 8, 99–114, 119, 131, 132, 134, 136, 137
Flavell, J.H., 163
Flecha, A., 394, 396
Flecha, R., 135–137, 396, 401
Flewitt, R., 226
Foucault, M., 24, 71, 249, 260, 280
Fox, M.A., 17
Franke, K.L., 305
Frankenstein, M., 145, 146
Fraser, B.J., 18
Freire, P., 118, 136, 145, 351, 352, 356
Freudenthal, H., 134, 135, 265, 283, 435

G

Gamon, D., 202
García, G., 19
García, R., 396, 399
Gates, P., 59
Gatt, S., 396
Gellert, U., 1–11, 71, 74, 75, 87, 135, 221, 240, 435
Gelman, S.A., 174
Gerofsky, S., 9, 201–211, 237, 241–243, 324
Giddens, A., 134, 136, 239
Gill, S., 395
Giménez, J., 10, 289–302, 341
Gintis, H., 394
Giroux, H., 26
Gispert, H., 2, 4
Glaser, G., 183
Goldin-Meadow, S., 215
Goldman, S., 398
Gomby, D.S., 398
González, N., 396
González-Martín, A., 383, 389
Good, T.L., 399
Goos, M., 18
Gorard, S., 17
Gorgorió, N., 9, 263–275, 279, 283, 284, 290
Goswami, U., 174, 178
Greeno, J., 53
Greeno, J.G., 131, 136
Grouws, D.A., 18, 399
Gruenewald, D., 33, 35, 51
Gueudet, G., 368, 385, 386, 389
Gummerum, M., 164, 168
Gunn, E., 37
Gutierrez, R., 21
Gutmann, A., 292, 301
Gutstein, E., 144, 146

H

Haarder, B., 16
Habermas, J., 133–135, 137, 292
Hale, C.M., 163
Halle, T.G., 395
Halliday, M.A.K., 253, 254
Hammond, M., 290, 299
Harding, S., 22
Harel, G., 266
Harmon, H., 34
Harmon, H.L., 34
Hassi, M.L., 118, 119, 128
Hausafus, C.O., 393
Heater, B., 305, 306
Hedegaard, M., 153
Henderson, A.T., 393, 396
Herbel-Eisenmann, B., 202
Hersh, R., 35, 60
Hiebert, J., 173, 400
Hill, E., 163, 164, 168
Hitchcock, G., 328
Hitt, F., 10, 383–389
Ho, E.S., 393, 395
Hoadley, U.K., 69
Holloway, S., 395
Hook, J., 177, 294
Hoops, J., 421
Hoover-Dempsey, K.V., 393, 396, 401
Howley, A., 33, 36, 38, 44, 50, 56
Howley, C., 33, 36–38, 44, 50, 56
Howley, M., 33, 38, 44, 50, 56
Hoyles, C., 110, 135, 388, 389
Huberman, A.M., 37
Hudson, J.M., 292
Hume, J., 173, 176
Hümmer, A.M., 71
Hunting, R.P., 174, 176

I

Iao, L.-S., 175
Illeris, K., 27
Inhelder, B., 163

J

Jablonka, E., 4, 74, 75, 132, 135, 221, 240
Jackson, K., 399, 400
Jacobs, J.E., 396
Janks, H., 280
Janvier, C., 387
Johnson, J., 33, 37
Jurdak, M., 265
Jurdak, M.E., 265

K

Kahane, R., 22, 26
Kahraman, P.B., 398
Kalavasis, F., 11, 425–429
Kane, P., 132
Kanes, C., 393
Karlberg, M., 421
Keeves, J.P., 16
Kefalas, M., 34, 36, 48, 54, 59
Keijzer, R., 176
Keitel, C., 435
Keith, T.Z., 393, 395
Kellogg, J., 177
Kennedy, D., 302
Kennedy, N., 302
Kent, P., 104
Keogh, J.J., 132
Kieran, C., 388
Kilpatrick, J., 16, 136, 400
Kintsch, W., 250
Kinzler, K.D., 174
Kirshner, D., 419
Klein, A., 401
Klein, R., 7, 33–63
Knapp, M.S., 393
Knijnik, G., 23
Knipping, C., 7, 67–94
Kotarinou, P., 10, 323–339, 341
Krajcik, J.S., 358
Krapp, A., 19, 27
Kreienkamp, K., 175
Krüger, J., 396
Krummheuer, G., 290, 296

L

Ladson-Billings, G., 410
Lagrange, J.B., 369
Laisant, C.-A., 16
Lakoff, G., 215
Lampert, M., 399
Larsen, G., 177
Lave, J., 131–132, 135, 296
Lee, J., 37
Lénárt, I., 331
Lerman, S., 21, 68, 74
Lerner, R.M., 177
Lesh, R., 266
Leung, F.K., 19
Levi, L., 174
Lim, F.V., 226–230, 233–235, 240
Liman, M.A., 193
Lipscomb, T.J., 178

Liu, D., 162, 168
Llewellyn, R.A., 267
Lobo da Costa, N.M., 10, 351–362, 385
Lourenco, O.M., 178
Lubienski, S.T., 69, 79, 82
Lundin, S., 6
Lyotard, J.-F., 26

M

Mack, N.K., 173, 174, 186
Madianou, M., 282
Mapp, K.L., 396
Martin, D.B., 70
Marton, F., 342
Maschietto, M., 365
McCoy, M.L., 292, 299
McCrink, K., 163, 168, 177, 178
McIntyre, W., 37
McLuhan, M., 206
McNeill, D., 215
Meagher, M., 173
Mendick, H., 21, 28
Menghini, M., 23
Merzbach, U.C., 329
Meyer, M.R., 399
Meyrowitz, J., 281
Michalek, R., 396
Miles, M.B., 37
Miller, J.G., 39, 162
Miller, K., 174
Miller, P., 249, 260
Mills, K.R., 132
Mix, K., 178
Molina, S., 398, 399, 401
Moll, L., 396
Monetti, D., 193
Moore, C., 164, 168, 178
Morais, A.M., 82
Moreau, M.-P., 18
Morency, L.-P., 367
Moreno-Armella, L., 384
Morgan, C., 18, 71
Moritz, J.B., 163, 168
Moscovici, S., 133
Moss, J., 173, 312
Mota, R.E., 359

N

Ni, Y., 173, 186
Nickson, M., 68
Nieto, S., 147

Author Index

Niss, M., 118, 132, 133
Norris, S., 227
Noss, R., 110, 135, 202
Nunes, T., 265
Núñez, R., 215

O
O'Donoghue, J., 131
O'Dwyer, L.M., 353
O'Halloran, K.L., 226, 227, 230
O'Holloran, K.L., 230
O'Leary, T., 249
Oliveira, H., 10, 305–320, 341
Olson, K.R., 164
Ostermeier, C., 18

P
Pais, A., 6, 23
Paola, D., 15–29
Papert, S., 352–353
Parsons, J.E., 395
Passeron, J.C., 395
Pennequin, V., 173
Peressini, D., 399
Piaget, J, 163, 169, 177, 196, 240, 426
Pólya, G., 4
Ponte, J.P., 357
Ponza, M.V., 328, 329
Popkewitz, T.S., 20, 22, 24, 25
Pothier, Y., 174, 176
Powell, A.B., 291, 297
Prado, M.E., 351–362, 383–384
Premack, D., 163
Prince, J.D., 366
Proust, C., 365
Pruitt, B., 292
Prusak, A., 21

Q
Quine, W.O., 372

R
Rabardel, P., 369, 386, 388
Radford, L., 21, 24, 148, 202
Ramsey, P.G., 147
Rand, D.G., 178
Rapoport, T., 22, 26
Rawls, J.A., 162
Reid, D., 67–94

Remillard, J.T., 399, 400
Resnick, L.B., 256
Reynolds, A.J., 395
Richerson, P.J., 176
Rico, L., 145, 193
Robert, A., 33–63
Robutti, O., 373
Rochat, P., 163, 164, 168
Rockliffe, F., 399
Roeser, R.W., 396
Rogers, R., 253, 254, 280
Rose, N., 249, 260, 282
Roth, W.-M., 18
Ruthven, K., 306, 307

S
Sabena, C., 9, 213–222
Säljö, R., 342
Sally, D., 163, 164, 168
Sandler, H.M., 393, 396, 401
Santos, L., 10, 341–347
Sawada, D., 174, 176
Sawyer, R.K., 203
Schechner, R., 204, 205
Schmidt, W.H., 202
Schoenfeld, A.H., 193
Schommer, M., 193
Schreiner, C., 19, 21, 28
Schubring, G., 2
Schwarz, B., 388
Scollon, R., 226, 227, 230, 234
Scollon, S.W., 226, 227, 230, 234
Scully, P.L., 292, 299
Seeger, F., 68
Serradó, A., 10, 289–302, 341
Sfard, A., 21
Shahin, I., 265
Shaver, A.V., 401
Sheldon, S.B., 396, 397, 400, 402
Sheridan, A., 71
Sheridan, S.M., 395
Sherman, J., 54
Showalter, D., 33
Sierpinska, A., 18, 136
Silver, E., 34
Simmt, E., 207
Simon, B.S., 393
Singer-Freeman, K., 178
Sjøberg, S., 19, 21, 28
Skovsmose, O., 19–21, 61, 122, 145, 146, 249, 285, 291, 297, 302, 323, 324, 357, 361, 410

Sloane, S., 162–164
Smith, E., 17
Smith, F.M., 164, 169, 393
Smith, M., 393
Smith, M.S., 307
Smith, T.W., 342, 344
Sobel, D., 35
Soler, M., 396
Solomon, Y., 225
Somers, J., 325
Sommerville, J.A., 164, 165, 168
Spelke, E.S., 174
Spinillo, A., 178
Sriraman, B., 4
Starkey, P., 401
Stathopoulou, C., 9, 10, 225–235, 237–240, 279–285, 323–339, 341
Stein, M.K., 305–307, 319
Stentoft, D., 21, 22
Stinson, D.W., 70, 144
Straehler-Pohl, H., 7, 67–94
Strauss, A.L., 183
Streater, A., 177
Su, Y., 162, 163, 168
Sung, J., 15
Sweet, M.A., 398

T

Tager-Flusberg, H., 163
Takagishi, H., 162–164, 168
Taylor, P.C., 201
Teoh, J., 325
Terwel, J., 176
Theobald, P., 35, 51
Thiollent, M., 357
Thomas, P., 292
Thompson, D., 292, 301
Thorndike, E.L., 202
Thurston, W.P., 207
Tirosh, D., 173
Tobin, K., 18
Tobin, K.G., 18
Tome, T., 359
Towers, J., 306
Treffers, A., 134, 135
Troelsen, R., 19
Trouche, L., 365, 368, 369, 385, 386, 388, 389
Tuckman, B., 193
Turner, V., 204, 205
Tyminski, A.M., 192

U

Ubuz, B., 338
Ulriksen, L., 27

V

Valente, J.A., 352, 353, 356, 357, 359
Valero, P., 7, 15–29, 68, 144, 291, 292, 297, 298, 300, 302, 324
Vanegas, Y.M., 293–295, 341
Verschaffel, L., 265
Vistro-Yu, C., 59
Vithal, R., 135, 145
Vygotsky, L.S., 68, 215, 221, 238, 386

W

Walkerdine, V., 24, 285
Walls, R.T., 401
Walshaw, M., 21, 144
Wanderer, F., 23, 145
Watson, J.M., 163, 168, 175
Wearne, D., 173
Wedege, T., 101, 111, 118, 128, 132, 134
Weeks, K., 132
Weinstein, L., 267
Wellman, H., 162, 163, 168
Wellman, H.M., 174
Wells, G., 306
Wenger, E., 131–132, 296
Wigfield, A., 396
Willis, G.B., 44
Willms, J.D., 393, 395
Winbourne, P., 18
Wiriyapinit, M., 290
Wood, T., 16
Wu, Z., 162, 163, 168
Wynn, K., 178

Y

Yackel, E., 68, 307
Yankelovich, D., 291
Yassa, N., 325, 339
Yasukawa, K., 323
Yost, P., 163, 169, 178

Z

Zevenbergen, R., 144
Zhou, Y., 173, 186